Symbol	Description	Unit
s	slip	—
S	complex or apparent power	VA
t	time	s
T	average torque	N·m
T_i	instantaneous torque	N·m
v	instantaneous voltage	V
V	electromotive force or electric potential	V
U	velocity	m/s
W	work, energy, heat	J
X	reactance	Ω
Y	admittance	S
Z	impedance	Ω

Greek alphabet

Symbol	Description	Unit
α	firing angle	deg, rad
α	temperature coefficient	1/°C
β	phase angle	deg, rad
γ	conduction angle	deg, rad
δ	torque angle	deg, rad
ε	permittivity	F/m
ε_0	permittivity of free space	F/m
ε_r	relative permittivity	—
η	efficiency	—
Θ	phase angle	deg, rad
λ	flux linkage	Wb
μ	permeability	H/m
μ_0	permeability of free space	H/m
μ_r	relative permeability	—
ρ	resistivity	m
Σ	summation	—
σ	real number	—
τ	time constant	s
ϕ	magnetic flux lines	Wb
ψ	electric flux lines	C/m^2
ω	angular speed	rad/s
Ω	resistance	ohms

Electric Machines

Principles, Applications, and Control Schematics

Second Edition

Dino Zorbas

McGill University

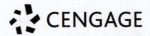

Australia • Brazil • Mexico • Singapore • United Kingdom • United States

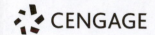

Electric Machines: Principles, Applications, and Control Schematics, Second Edition

Dino Zorbas

Publisher: Timothy Anderson

Senior Developmental Editor: Mona Zeftel

Media Assistant: Ashley Kaupert

Senior Content Project Manager: Kim Kusnerak

Production Director: Sharon Smith

Team Assistant: Sam Roth

Rights Acquisition Director: Audrey Pettengill

Rights Acquisition Specialist, Text and Image: Amber Hosea

Text and Image Researcher: Kristiina Paul

Manufacturing Planner: Doug Wilke

Copyeditor: Betty Pessagno

Proofreader: Pamela Ehn

Indexer: Shelly Gerger-Knechtl

Compositor: MPS Limited

Senior Art Director: Michelle Kunkler

Internal Designer: MPS Limited

Cover Designer: Rose Alcorn

Cover Image: ©Rui Frias/iStockphoto

For product information and technology assistance, contact us at **Cengage Customer & Sales Support, 1-800-354-9706.**
For permission to use material from this text or product, submit all requests online at **www.cengage.com/permissions**. Further permissions questions can be emailed to **permissionrequest@cengage.com**.

Library of Congress Control Number: 2013954200

ISBN-13: 978-1-133-62851-4

ISBN-10: 1-133-62851-6

Cengage
20 Channel Center Street
Boston, MA 02210
USA

Cengage is a leading provider of customized learning solutions with office locations around the globe, including Singapore, the United Kingdom, Australia, Mexico, Brazil, and Japan. Locate your local office at: **international.cengage.com/region**.

Cengage products are represented in Canada by Nelson Education, Ltd.

To learn more about Cengage platforms and services, register or access your online learning solution, or purchase materials for your course, visit **www.cengage.com**.

Printed in Mexico
Print Number: 04   Print Year: 2021

To all those who believe that the preparation of tomorrow's professionals must include organized practical training.

Fortune favors the prepared mind.
Louis Pasteur

Don't talk to me of your Archimedes lever . . . .
Give me the right word and the right
accent and I will move the world.
Joseph Conrad

CONTENTS

Preface   xi

CHAPTER 1   Basic Electromagnetic Concepts   1

1.0   Introduction   3

1.1   Electric Circuit Concepts   3
- 1.1.1   Electric Potential   3
- 1.1.2   Complex Numbers   7
- 1.1.3   Electric Field Parameters   10
- 1.1.4   Basic Elements   11
- 1.1.5   Impedance   32
- 1.1.6   Principles of Electric Circuits   35
- 1.1.7   Power of a Single-Phase Source   41
- 1.1.8   Power Factor   47
- 1.1.9   Effects of the Power Factor   49
- 1.1.10   Power-Factor Improvement   51
- 1.1.11   Harmonics   57
- 1.1.12   Theorems   61

1.2   Magnetic Circuit Concepts   64
- 1.2.1   Magnetic Flux   64
- 1.2.2   Magnetic Flux Density   69
- 1.2.3   Magnetomotive Force   70
- 1.2.4   Magnetic Field Intensity   71
- 1.2.5   The *B-H* Curve and Magnetic Domains   75
- 1.2.6   Permeability   78
- 1.2.7   Reluctance   82
- 1.2.8   The Concept of Generated Voltage   85
- 1.2.9   Energy Content of Magnetic Materials   93
- 1.2.10   Principles of Magnetic Circuits   95
- 1.2.11   Magnetic Losses   103
- 1.2.12   Equivalent Circuit of a Coil   109
- 1.2.13   Mathematical Relationships of Self- and Mutual Inductances   111
- 1.2.14   Polarity and Equivalent Circuits of Magnetically Coupled Coils   119
- 1.2.15   Force, Energy, and Torque   123

1.3   Force, Energy, and Torque   125
- 1.3.1   Force Exerted by an Electromagnet   127
- 1.3.2   Energy Stored in the Magnetic Field of Coils   131

1.3.3   Torque Developed by Rotating Transducers   133

1.3.4   Applications of Electromagnetism   134

1.4   Summary   135

1.5   Review Questions   143

1.6   Problems   143

CHAPTER 2   Transformers   151

2.0   Introduction   153

2.1   Single-Phase Transformers   154

2.1.1   Principle of Operation   155

2.1.2   Ideal Transformer Relationships   157

2.1.3   Derivation of the Equivalent Circuit   160

2.1.4   Waveform of Excitation Current   169

2.1.5   Components of Primary Current and Corresponding Fluxes   171

2.1.6   Transformer Characteristics   173

2.1.7   Per-Unit Values   178

2.2   Three-Phase, Two-Winding Transformers   188

2.2.1   Introduction   188

2.2.2   Review of Three-Phase Systems   189

2.2.3   One-Line Diagram   193

2.2.4   Types of Three-Phase Transformers   194

2.2.5   Harmonics of the Exciting Current   210

2.3   Autotransformers   211

2.4   Parallel Operation of Transformers   219

2.4.1   Transformers Must Have Equal Leakage Impedances   220

2.4.2   Transformers Must Have Equal Turns Ratios   220

2.4.3   Equal Phase Shift Between the Voltages of Paralleled Units   221

2.4.4   Same Phase Rotation   222

2.5   Instrument Transformers and Wiring Diagrams   225

2.5.1   Instrument Transformers   225

2.5.2   Wiring Diagrams   230

2.6   Transformer's Nameplate Data   236

2.7   Conclusion   238

2.8   Summary   239

2.9   Review Questions   243

2.10   Problems   244

CHAPTER 3   Three-Phase Induction Machines   249

3.0   Introduction   251

3.1   Three-Phase Induction Motors   251

3.1.1   Stator and Rotor   251

3.1.2   Principles of Operation   255

3.1.3   Rotating Magnetic Field   256
3.1.4   Slip   261
3.1.5   Equivalent Circuit   264
3.1.6   Torque and Power Relationships   269

3.2   Industrial Considerations   277
3.2.1   Classification of Induction Motors   277
3.2.2   Mechanical Loads   279
3.2.3   Mechanical Load Changes and Their Effects on a Motor's Parameters   281
3.2.4   Voltage, Efficiency, and Power-Factor Considerations   282

3.3   Measurement of Equivalent-Circuit Parameters   290

3.4   Asynchronous Generators   298

3.5   Controls   302
3.5.1   Reduction of the Motor's High Starting Current and Torque   302
3.5.2   Variable Frequency Drives   307
3.5.3   Soft Start   317
3.5.4   Plugging   319

3.6   Conclusion   319

3.7   Tables   320

3.8   Review Questions   324

3.9   Problems   325

CHAPTER 4   Single-Phase Motors   331

4.0   Introduction   332

4.1   Revolving Fields   333
4.1.1   Rotor Stationary   334
4.1.2   Rotor Turning   334

4.2   Equivalent Circuit   334

4.3   Torque Developed   335

4.4   Methods of Starting   337
4.4.1   Auxiliary Winding, Permanently Connected   337
4.4.2   Auxiliary Winding with Capacitor Start   339
4.4.3   Auxiliary Winding, Removed After Starting   340
4.4.4   Auxiliary Winding, with a Starting and a Running Capacitor   341

4.5   Magnetic Fields at Starting   341

4.6   Types of $1\text{-}\phi$ Motors   345
4.6.1   Split-phase   345
4.6.2   Shaded-Pole Motors   346
4.6.3   Revolving Field   346
4.6.4   Stepper Motors   347
4.6.5   Control   348
4.6.6   Series AC/DC Motors   349

4.7   Conclusion   350
4.8   Problems   351

CHAPTER 5   Synchronous Machines   353

5.0   Three-Phase Synchronous Machines   354
5.1   Three-Phase Cylindrical Rotor Machines: Motors   355
   5.1.1   Rotor   359
   5.1.2   Rotating Fields   359
   5.1.3   Principle of Operation   361
   5.1.4   Starting   362
   5.1.5   Equivalent Circuits   363
   5.1.6   Field Current   366
   5.1.7   Phasor Diagrams   367
   5.1.8   Power and Torque Developed   374
   5.1.9   Effects of Field Current on the Characteristics of the Motor   378
5.2   Three-Phase Cylindrical Rotor Machines: Generators   389
   5.2.1   Equivalent-Circuit and Phasor Diagrams   389
   5.2.2   Regulator of Alternators   392
   5.2.3   Characteristics of Alternators   393
   5.2.4   Measurement of Parameters   397
5.3   Salient-Pole Synchronous Machines   405
   5.3.1   Introduction   405
   5.3.2   Phasor Diagrams   407
   5.3.3   Power Developed   409
   5.3.4   Torque Angle for Maximum Power   411
   5.3.5   Stiffness of Synchronous Machines   412
5.4   Conclusion   416
5.5   Review of Important Mathematical Relationships   418
5.6   Manufacturer's Data   420
5.7   Review Questions   422
5.8   Problems   422

CHAPTER 6   DC Machines   425

6.0   Introduction   426
6.1   Steady-State Analysis   427
   6.1.1   General   427
   6.1.2   Principles of Operation   429
   6.1.3   Power Considerations   433
   6.1.4   Voltage and Torque Relationships as Functions of Mutual Inductance   436
   6.1.5   Magnetic System, Flux Distribution, and Armature Reaction   438
   6.1.6   Commutation   442
   6.1.7   Equivalent Circuits and External Machine Characteristics   444

6.1.8   Starting   462
6.1.9   Open-Circuit Characteristics and DC Generators   464

6.2   Modern Methods of Speed Control   477
6.2.1   Rectifiers   477
6.2.2   Single-Phase, Full-Wave Controlled Rectifiers   478
6.2.3   Three-Phase, Full-Wave Controlled Rectifiers   481

6.3   Conclusion   489
6.4   Review Questions   493
6.5   Problems   493

CHAPTER 7   Control Schematics   497
7.0   Introduction   498
7.1   Basic Devices and Symbols   499
7.1.1   Electromagnetic Relays   499
7.1.2   Thermal Overload Relays   503
7.1.3   Electrical Contacts   504
7.1.4   Indicating Lights   506
7.1.5   Start-Stop Pushbuttons   506
7.1.6   Industrial Timers   507
7.1.7   Temperature Sensors   509

7.2   The Concept of Protection   512
7.2.1   General   512
7.2.2   Protective Devices   514

7.3   Actual Control Schematics   519
7.3.1   Control Schematics of Three-Phase Induction Motors   519
7.3.2   Synchronous Motors   524

7.4   Conclusion   530
7.5   Review Questions   533
7.6   Problems   533

**CHAPTER 8   Electrical Safety and Reduction in
Energy Consumption   537**
8.1   Electrical Safety   539
8.1.1   Introduction   539
8.1.2   Basic Protective Devices   540
8.1.3   Grounding Systems   542
8.1.4   The Utility's Ground   542
8.1.5   Residence's Ground   543
8.1.6   Equipment's Enclosure Ground   543
8.1.7   Bonding   543
8.1.8   Ground Resistance   544
8.1.9   Short Circuits   544

8.1.10   Floating Neutral   544
8.1.11   Touching a Live Conductor   545
8.1.12   Open Neutral   546
8.1.13   Arc Flash   546
8.1.14   Energy Stored   546
8.1.15   Energy Stored in Inductive Circuits   547
8.1.16   Energy Stored in Capacitive Circuits   547
8.1.17   Stray Voltages   548
8.1.18   Mishandling Medical Equipment   549
8.1.19   Cathodic Corrosion   549
8.1.20   Lightning Strikes   551
8.1.21   Conclusion   551

8.2   Reduction in Energy Consumption   555
8.2.1   Unbalanced Voltages   556
8.2.2   Unbalanced Line Currents   557
8.2.3   Lighting   557
8.2.4   Synchronous Motors   557
8.2.5   Reducing the Power Demand   558
8.2.6   Power-Factor Improvement   558
8.2.7   High-Efficiency Motors   558
8.2.8   Variable-Speed Drives   558
8.2.9   Reduce Harmonics   559
8.2.10   Voltage to DC Motors   559
8.2.11   Recuperating Rejected Heat   559
8.2.12   Heat Losses   559
8.2.13   Heat Pumps   560
8.2.14   New Technologies   560
8.2.15   Remarks   560

APPENDIXES   561

A   Three-Phase Systems   563
B   Per-Unit System   587
C   Laplace Transforms   597
D   Solid-State Devices   601
E   Basic Economic Considerations   603
F   Photovoltaics   607
G   Tables   611
H   Bibliography   615

Answers to Problems   617

Index   621

The purpose of this revision to *Electric Machines: Principles, Applications, and Control Schematics* is threefold:

1. To update the technical data and applications of the electrical equipment discussed (transformers, motors, controls, and energy reduction per unit output) since the first printing.

2. To both expand upon and further simplify the basic concepts of electrical engineering. For example, the basic law of nature (action is equal and opposite to reaction) is used in conjunction with Ohm's law in magnetic circuits to explain the principle of induction and the operation of transformers and electric machines. This revision expands on important concepts such as harmonics, induction currents, energy conservation, electrical safety, and others. These are summarized in a tabular form for easy understanding, review, and industrial usage. Where possible, all concepts are illustrated graphically.

3. To provide detailed additional information in the web section on programming logic controllers (PLC), electrical safety, and energy conservation.

In summary, this book aims to describe and simplify the basic concepts of the design and operation of electric machines, to ease the transition from classroom work to that of the practicing engineer, and to give an efficient start to those who will find themselves in engineering departments of industrial or commercial enterprises.

The following list is a brief summary of the chapter contents, together with a description of corresponding web chapters.

Chapter 1 Basic Electromagnetic Concepts

The basic elements (resistors, inductors, and capacitors) are each illustrated with more than 10 highlights with emphasis placed on their practical aspects. The various forms of power (complex, apparent, real, and reactive) are introduced including the concepts of actual and apparent power factor. Examples are given with actual utility penalties for low-power factor operation. The generation and effects of harmonics are also discussed. For easy understanding, KVL is illustrated by using exercises from electronic circuits. The section on magnetism discusses the concepts and applications of induction currents, magnetic equivalent circuits, and the force produced by electromagnets.

To ease the student's introduction to electric machines, summary tables are included on electric and magnetic circuit parameters, power, and energy. These points are included in order to facilitate understanding of their analogous magnetic circuit concepts. One complements the other.

Finally, we use a simple circuit (Problem 1-28) as a real-world example. If it could be realized, it would produce all humanity's energy requirements.

Chapter 1W on the web includes additional solved and unsolved problems on the theory of circuits and forces that are produced by rotating transducers.

Chapter 2 Transformers

Transformers change the voltage from one level to another, constituting a bottle-neck effect on a power distribution system. One has to be aware of the continuous improvement of their characteristics. Transformers depend on the selection of the downstream protective devices. Their applications require knowledge of the per-unit system that is further detailed in the Appendixes.

A detailed description is included on standard 1-ϕ, 3-phase transformers, parallel operation, and special winding connections.

Chapter 2W on the web includes wiring diagrams and analog meters found in substation apparatus.

Chapter 3 Three-Phase Induction Motors

Three-phase induction motors are the working horse of industry and have evolved not so much in their design but on the electronic control of their speed. As such, they replace the older design with variable speed controls resulting in higher efficiency and a better match of their torque-speed characteristics to that of the load. More specifically, the following are included:

- derivation of the equivalent circuit,
- methods of starting,
- torque-speed characteristics,
- matching of the load characteristics to that of the motor, and
- detailed analysis of the asynchronous generator that transform wind power to electrical power.

Chapter 3W on the web includes the transient response and special applications of three-phase induction motors.

Chapter 4 Single-Phase Motors

This chapter discusses the development of two-rotating magnetic fields from 1-ϕ voltage supply and how eliminating one of them reduces the noise associated with their interaction. The various types of 1-ϕ machines, phasor diagrams, and the corresponding torque-speed characteristics are also examined. The 1-ϕ phase type of motors are mainly used for control and ventilation applications.

Chapter 4W on the web explains the development of magnetic fields produced by a two-phase voltage supply and the speed control of two-phase motors.

Chapter 5 Synchronous Machines

Three-phase synchronous machines are suitable for loads that require constant speed. All power generation is through three-phase synchronous generators; their constant operating speed ensures constant voltage frequency. A detailed analysis is included on cylindrical and salient rotor machines, on the derivation of their equivalent circuits, on the VEE curves, and on their phasor diagrams. The latter are simplified using a clearly identified unique sequential approach.

Chapter 5W on the web describes the transient waveforms of stator and rotor currents and the resulting current when a three-phase fault occurs. The latter constitutes the basics of selecting circuit breakers in a power distribution system.

Chapter 6 DC Machines

DC machines have widespread applications because their torque-speed characteristics can be adjusted easily to match those of the driven load. Their operating efficiency and flexibility improved with the advent of variable speed controls.

Chapter 6W on the web covers the transient response of dc machines providing an additional insight on their operation.

Chapter 7 Control Schematics

The classical relay-contact diagrams are described along with their usage in the control of 1-ϕ, MV induction motors, and synchronous motors. Understanding of an industrial process requires the design and selection of controls. This chapter is a prerequisite to understanding the drawings that depict the operation of simple and/or complicated industrial controls.

Chapter 7W on the web discusses the basic controls of 3-ϕ induction motors and those of dc machines. Logic circuits and programmable logic controllers (PLC) are introduced.

Chapter 8 Electrical Safety and Reduction in Energy Consumption

The electrical safety section covers the safety of personnel in various situations: at home, within industries, in swimming pools, in plumbing systems, at farms, and so on. As per recent statistics, there are 10,000 electrical injuries per year in the United State alone.

The energy conservation section outlines many areas of energy conservation in commercial buildings and industrial plants.

Chapter 8W presents a detailed analysis with many solved and end-of-chapter problems on the issues of electrical safety and energy conservation.

Appendices

The appendices include prerequisite concepts that are required for a better understanding of the subject matter under consideration (three-phase systems, per unit system of values, economic aspects, and protective device designation).

Chapter AP-W on the web is an extension of the book's appendices. It includes additional examples on 3-ϕ systems, per unit values, photo-voltaics, and a detailed description of the basic blocks of electronic speed control (i.e., diodes, bipolar transistors, insulated gate bipolar transistors and thyristors).

MindTap Online Course and Reader

In addition to the print version, this textbook is also available online through MindTap, a personalized learning program. Students who purchase the MindTap version will have access to the book's MindTap Reader and will be able to complete homework and assessment material online, through their desktop, laptop, or iPad. If your class is using a Learning Management System (such as Blackboard, Moodle, or Angel) for tracking course content, assignments, and grading, you can seamlessly access the MindTap suite of content and assessments for this course.

In MindTap, instructors can:

- Personalize the Learning Path to match the course syllabus by rearranging content, hiding sections, or appending original material to the textbook content
- Connect a Learning Management System portal to the online course and Reader
- Customize online assessments and assignments
- Track student progress and comprehension with the Progress app
- Promote student engagement through interactivity and exercises

Additionally, students can listen to the text through ReadSpeaker, take notes and highlight content for easy reference, and check their understanding of the material.

Acknowledgments

The author wishes to acknowledge the many helpful suggestions offered by

- Todd Batzel, *Pennsylvania State University*
- Patrick Chapman, *University of Illinois at Urbana-Champaign*
- Christopher Doss, *North Carolina A&T State University*
- Mehrdad Ehsani, *Texas A&M University*
- Ali Emadi, *Illinois Institute of Technology*
- Herman Hill, *Ohio University*
- Rick Hoadley, *Milwaukee School of Engineering*
- Walid Hubbi, *New Jersey Institute of Technology*
- Mahesh Krishnamurthy, *Illinois Institute of Technology*
- Ahmad Nafisi, *Cal Poly San Luis Obispo*
- Alvernon Walker, *North Carolina A&T State University*

The author is indebted to all his teachers, students, and authors (see bibliography listing at the end of the book) who prepared the road for the realization of this undertaking. The constructive comments, guidance, and encouragement provided by the editorial team of Swati Meherishi and Mona Zeftel are greatly appreciated. I express my sincere thanks to Mrs. Colette Julien for her great patience in typing the manuscript.

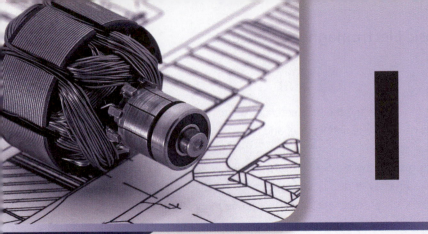

Basic Electromagnetic Concepts

1.0   Introduction

1.1   Electric Circuit Concepts

1.2   Magnetic Circuit Concepts

1.3   Force, Energy, and Torque

1.4   Summary

1.5   Review Questions

1.6   Problems

What You
Will Learn in
This Chapter

A. Theoretical Aspects

1 The concept of voltage and the parameters of electric fields.

2 The concepts of resistance, inductance, and capacitance.

3 The resonance condition, the time constant, and the principles of electric circuits.

4 The various forms of power, power factor, and power-actor correction.

5 The harmonics in a power distribution system.

6 The magnetic flux lines, magnetic field density, and magnetic field intensity.

7 The permeability and the principles of magnetic circuits.

8 The equivalent circuit of coils, their self and mutual inductances.

9 The magnetic energy stored in a coil, the force of an electromagnet, and the torque produced by a rotating coil.

10 The general concepts of power and energy.

B. Practical Highlights—Interface

1 The failures of the Spaceship Apollo 13 (1970) and the Russian *Salyut I* (1971).

2 The frequently ineffective current drawn from the utility, the corresponding utility penalty, and how to remove it (for the same reasons that it is not very effective when trying to lose weight while walking in the park and why in the 1940s the German antitank rockets could not destroy the Russian T-34 tanks).

3 The advantages and disadvantages of using capacitors to improve the power factor.

4 The alternate definition of the time constant supplements the accurate but incomplete truths of the equivalent concepts that are found in publications of chemistry, physics, and even electrical engineering.

5 The concept of the induction principle and its associate advantages/disadvantages are simplified and illuminated by just using the concept of "Action is equal and opposite to reaction."

6 The principle of motors, generators, and relays operation is analogous to that of "boy attracts girl and vice versa."

7 The analog meters of the utilities cannot measure all the power consumed by nonlinear loads such as computers, variable speed drives, lighting fixtures, and other digital equipment.

8 The resistance and inductance of electric circuits is analogous to the friction and mass/polar moment of inertia of mechanical moving/rotating bodies.

9 The degaussing of the British fleet, the potentially catastrophic effects of the induction current, and the broader overview of power and energy.

10 Magnetic levitation and high-speed trains.

C. Additional Students' Aid on the Web

Examples and Problems on Electric Circuits

Example of Force Developed by a Rotating Electromagnet

1.0 Introduction

General Overview

This chapter discusses the basic parameters and principles of electric and magnetic circuits and the development of force and torque. Electric circuit parameters are normally covered in other basic courses, but they are briefly reviewed here in order to facilitate the understanding of magnetic circuit parameters.

Each electric circuit concept is analogous to a corresponding magnetic circuit concept. As a result, mastering the theory of electric circuits will help you grasp the more abstract concepts of magnetism. In order to understand electric machines, you first need a very thorough knowledge of both magnetism and electric circuits.

1.1 Electric Circuit Concepts

This section covers the concepts of electric potential, electric field intensity, resistance, capacitance, inductance, impedance, principles of electric circuits, the characteristics of the power drawn from a single-phase voltage source, the apparent and actual power factors, theorems, and harmonics.

1.1.1 Electric Potential

Electric potential, electromotive force (emf), electrical pressure and voltage are synonymous terms. In the SI system of units, they are expressed or measured in volts. When the voltage at a power outlet is 120 volts, the electric potential, electric voltage, and the electric pressure at the outlet is of the same magnitude:

$$1 \text{ volt} = \frac{1 \text{ Joule}}{\text{Coulomb}}$$

$$10^3 \text{ V} = 1 \text{ kV}$$

$$10^{-3} \text{ V} = 1 \text{ mV}$$

Voltage is the intensity of an electrical energy source just as temperature is the intensity of a heat source. Voltage and temperature are destructive only when they are associated with high-energy sources. For example, the temperature of a lighted match is about 95°C. When you place your finger on the match, it hurts, but it does not kill you. In contrast, somebody immersed in a home's hot water tank (60°C) would be seriously injured. Similarly, the threshold of static sensation, which is about 4000 V, does not injure because it is not accompanied by sufficient electrical charges. In contrast, the 120 V across a home's receptacle may be fatal to anyone who comes in contact with it. Voltage has a meaning when and only when it is measured with respect to the reference point. The reference terminal is often

the neutral or the ground wire. These two wires as per electrical code requirements are connected together in the premises' electrical entrance. Stating that the
voltage at a particular point is 120 V implies what this voltage is with respect to
ground.

Designation

Refer to Fig. 1-1. The voltage at terminal B is at 10 V with respect to that of terminal A. Mathematically,

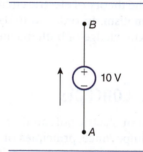

FIG. I-I Voltage designation.

$$V_{BA} = 10\,V$$

The voltage, however, at A is at $-10\,V$ with respect to point B.

$$V_{AB} = -10\,V$$

Instead of the polarities $+$ and $-$, you may use an arrow to indicate relative polarities, as shown in Fig. 1-1.

All main and secondary power distribution networks are characterized by a voltage source. The voltage source may be of constant amplitude (a dc voltage source)
or of alternating waveform (an ac voltage source). The various waveforms—whether
they represent voltage, current, or power—can be described in terms of their instantaneous, effective, or average values.

Instantaneous Value

The instantaneous value of a waveform is given by a general equation that
describes the waveform as a function of time. For example, the instantaneous
value of the voltage waveform that is available in a residential single-phase power
outlet is

$$v = V_m \sin \omega t \qquad\qquad (1.1)$$

where v is the instantaneous value of the voltage, V_m is its maximum value,
and ω is its angular frequency of oscillation.

The angular frequency of oscillation is given by

$$\omega = 2\pi f = \frac{2\pi}{T} \text{ rad/s} \tag{1.2}$$

where f and T are, respectively, the frequency and the period of oscillation of the voltage waveform. A sinusoidal voltage waveform is shown in Fig. 1-2.

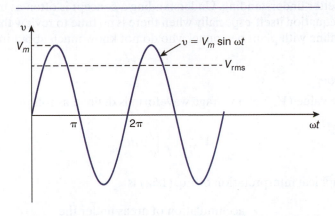

FIG. 1-2 A sinusoidal voltage waveform.

The standard frequency of oscillation of the current and voltage waveforms of single-phase and three-phase power sources in North America is usually 60 Hz. In some other parts of the world it is 50 Hz.

The instantaneous values of the various functions are in general represented by the lower-case letters with or without the time-function notation. Thus, v (or $v(t)$) and i (or $i(t)$) represent the instantaneous values of voltage and current.

Effective Value

The effective or root mean square (rms) of a voltage waveform is given by

$$V_{\text{rms}}^2 = \frac{1}{T} \int_0^T v^2 \, dt \tag{1.3}$$

where v is the instantaneous value of the voltage waveform.

Root-mean-square values have been introduced because the heating effect of an ideal sinusoidal voltage waveform in a resistor is equal to that produced by a dc source that has the same voltage as the rms value of the given ac waveform. In addition, the usage of rms values simplifies the final form of voltage, current, and power formulas. In designating rms values, the capital letter of the parameter is usually used without any subscript. Thus, V and I correspond, respectively, to the rms values of the voltage and current. The voltage rating of all ac apparatuses is given by the equivalent rms values. Thus, when a voltage is indicated as being 208 V or 480 V, it is understood that the magnitude is an rms value. The single-phase voltages delivered to a typical North American residence are 120 V (line-to-neutral) and 240 V (line-to-line).

The graphical interpretation of Eq. (1.3) is

$$V_{rms}^2 = \frac{\text{accumulation of areas under the voltage squared-time diagram over a complete cycle}}{\text{duration of one cycle}} \qquad (1.4)$$

The graphical interpretation of Eq. (1.3), or of any other equation, helps you develop a better understanding. Understanding a concept is often as important as solving the equation itself, especially when there is no time to review the equations or when dealing with plant personnel who do not know much about integration.

Average Value

The average value (V_{av}) of a voltage waveform is defined as follows:

$$V_{av} = \frac{1}{T} \int_0^T v \, dt \qquad (1.5a)$$

The graphical interpretation of Eq. (1.5a) is

$$V_{av} = \frac{\text{accumulation of areas under the } v\text{-}t \text{ diagram over a complete cycle}}{\text{duration of one cycle}} \qquad (1.5b)$$

The average value of a symmetrical sinusoidal waveform is, as seen by inspection, equal to zero. The average value is also referred to as a dc value.

EXAMPLE 1-1   Determine the rms value of the sinusoidal voltage waveform shown in Fig. 1-2.

SOLUTION

The instantaneous value of the voltage is

$$v = V_m \sin \omega t$$

Substituting into Eq. (1.3), we get

$$V_{rms}^2 = \frac{1}{2\pi} \int_0^{2\pi} (V_m \sin \omega t)^2 \, d\omega t$$

$$= \frac{V_m^2}{2\pi} \int_0^{2\pi} \left(\frac{1 - \cos 2\omega t}{2} \right) d\omega t$$

$$= \frac{V_m^2}{4\pi} \left[\omega t \Big|_0^{2\pi} - \frac{1}{2} \sin 2\omega t \Big|_0^{2\pi} \right]$$

$$= \frac{V_m^2}{2}$$

from which

$$V_{\text{rms}} = \frac{V_m}{\sqrt{2}}$$

The rms value of the voltage waveform is shown in Fig. 1-2.

Find the rms and average values of the waveform shown in Fig. 1-3. Verify the answer graphically.

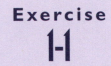

Exercise

1-1

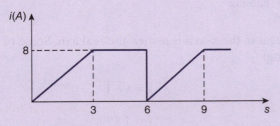

FIG. 1-3

Answer 6.53 A rms, 6A

1.1.2 Complex Numbers

The designation and significance of the letter J go back several centuries. At that time, mathematicians, when finding the roots of equations with negative square roots, thought that their physical interpretation was <u>impossible</u> or <u>imaginary</u>. Further studies by many mathematicians, and in particular Demoivre and Euler, demonstrated that a phasor can be represented by its polar form or rectangular form incorporating the factor j. For example, a phasor can be written as follows

$$\underbrace{re^{j\theta} = r < \theta}_{\text{Polar form}} = \underbrace{r(\cos\theta + j\sin\theta)}_{\text{Rectangular form}} \qquad (1.6)$$

The factor r is the magnitude of the phasor. This equation can be derived by using the Maclaurin series or the exponential representation of the trigonometric functions.

Figure 1-4 illustrates the polar and rectangular form of phasors.

About j

In engineering, the coefficient j represents the component of a phasor in the y-axis, which is referred to as the imaginary axis. It is not imaginary, however. It is

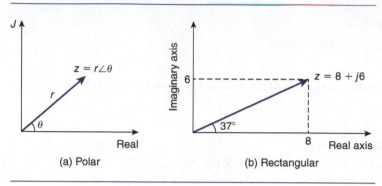

FIG. 1-4 Phasors.

so designated because the x-axis is named the real axis. Some of the highlights of j are the following:

I)
$$j = \sqrt{-1} \tag{1.7}$$

II)
$$j = 1\underline{/90} \tag{1.8}$$

III) If
$$A + jB = C + jD \tag{1.9}$$

Then

$$A = C, \qquad B = D$$

IV) It changes differential equations of linear systems to algebraic form. It is part of the Laplace operator(s)

$$s = j\omega + \alpha \tag{1.10}$$

where α approaches zero and ω is the angular speed of the phasor in radians/second (voltage, current, etc.).

When a linear differential equation (DE) is written in terms of s (see Appendices), then you can use <u>algebra</u> to solve it. In general, a time-domain function has its equivalent Laplace domain representation and vice versa. These representations or transformations are available in Laplace transform tables.

Besides that, representation of a function by its equivalent Laplace equivalent reveals (as will be demonstrated in the follow-up sections) the value of the function at time equal to zero and at time equal to infinite.

Properties of Complex Numbers

When two phasors Z_1 and Z_2 are as follows

$$Z_1 = R_1 + jX_1 = r_1\underline{/\Theta_1}\ \Omega \tag{1.11}$$

$$Z_2 = R_2 + j X_2 = r_2 \underline{/\Theta_2} \ \Omega \qquad\qquad (1.12)$$

Then from the properties of complex numbers,

$$Z_1 + Z_2 = R_1 + R_2 + j (X_1 + X_2) \ \Omega \qquad\qquad (1.13)$$

$$Z_1 Z_2 = r_1 r_2 \underline{/\Theta_1 + \Theta_2} \ \Omega \qquad\qquad (1.14)$$

$$\frac{Z_1}{Z_2} = \frac{r_1}{r_2} \underline{/\Theta_1 - \Theta_2} \ \Omega \qquad\qquad (1.15)$$

The addition, multiplication, and division of complex numbers are used in many aspects of Electrical Engineering and, as such, will be used throughout this book.

EXAMPLE 1-2

When $Z_1 = 3 + j 4 \ \Omega$ and $Z_2 = 10 \underline{/-37} \ \Omega$, determine $Z_1 + Z_2$, $Z_1 Z_2$ and $\dfrac{Z_1}{Z_2}$

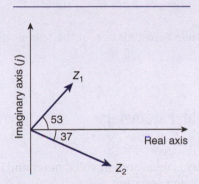

FIG. I-5 Complex phasors.

SOLUTION

Refer to Fig. 1-5.

$$Z_1 = 3 + j4 = 5 \underline{/53} \ \Omega, \quad Z_2 = 8 - j6 \ \Omega$$

and

$$Z_1 + Z_2 = 3 + 8 + j(4-6) = 11 - j2 = 11.18 \underline{/-10.3} \ \Omega$$

$$Z_1 Z_2 = 5 \underline{/53} \ (10 \underline{/-37}) = 50 \underline{/16} \ \Omega$$

$$\frac{Z_1}{Z_2} = \frac{5 \underline{/53}}{10 \underline{/-37}} = 0.5 \underline{/90} = j0.5 \ \Omega$$

Exercise
1-2

Given

$$Z_1 = 4 - j\,3 \ \Omega, \quad Z_2 = 10 \underline{/53} \ \Omega$$

Find

$$Z_1 + Z_2, \quad Z_1 Z_2, \quad \frac{Z_1}{Z_2}$$

Answer $11.18 \underline{/26.6}, 50 \underline{/16}, -j0.5$

Exercise
1-3

Prove Euler's identity

$$e^{j\Theta} = \cos\Theta + j\sin\Theta$$

(*Hint*: Use the exponential representation of the trigonometric functions.)

1.1.3 Electric Field Parameters

Electric Field Intensity*

The electric field intensity is measured in volts/meter and is related to the electric potential (v) by the following equation:

$$v = \int E\,d\ell \tag{1.16}$$

The minimum electric field intensity for successful radio and TV operation is about:

- Radio, AM: 0.5 m V/m
- Radio, FM: 0.5 m V/m
- Digital TV: 15 m V/m, frequency dependent
- Analog TV: 60 m V/m, frequency dependent

In the vicinity of radio or television stations, for example, the electric field intensity is relatively large. In such places, an incandescent lamp attached to a wire may—depending on the lamp's location and the length of the wire loop—produce a glow.

---

*In some publications the symbol E is used to designate energy. In this text, energy is designated by the letter W.

where $d\ell$ is the differential of length. The electric field intensity at 10 cm in front of the computer's screen or on the ground level below high-voltage transmission lines is about 30 V/m.

Electric field intensity is also present between two points on Earth (ground) through which a leakage current flows. The corresponding voltage is referred to as a "stray voltage." Such voltages, as explained in Chapter 8 on electrical safety, can cause injuries and malfunction of sensitive equipment.

The electric field intensity being a force or vector is reversed in direction when the associated voltage changes from positive to negative. This reversing force, depending on its magnitude, duration, and frequency, may be hazardous to humans. At 60 Hertz, it was incorrectly claimed that the threshold of its maximum safety level was 30 V/m. Presently, the actual safety limit is unknown. The electric field intensity, as per Maxwell's equation, is related to the magnetic field intensity (H) as follows:

$$\frac{E}{H} = 377 \qquad\qquad (1.17)$$

This equation is used to measure E. (See page 73.)

Electric Field Flux Density

Electric flux is defined as a quantitative evaluation of electric field lines. It is represented by the number of electric field lines passing through a surface.

Electric flux density (D) is equal to the electric flux divided by the area (A) perpendicular to it. That is,

$$D = \frac{\Psi}{A} \qquad\qquad (1.18)$$

The unit of electric flux density is coulomb/m^2.

The electric flux density of any given material is related to the electric field intensity and the permittivity of the material by

$$\varepsilon = \frac{dD}{dE} \qquad\qquad (1.19)$$

where ε is the permittivity of the material. Thus, when the flux density–field intensity curve of a material is given, its slope at the operating flux density is equal to the relative permittivity of the material (Fig. 1-6).

1.1.4 Basic Elements

The basic building blocks of Electrical Engineering are:

Resistance, Inductance, and Capacitance

An understanding of their characteristics will simplify many concepts and applications in power distribution, magnetism, and electronics. Their analogy to mechanical systems will drive the point home and will also facilitate the understanding of some theoretical and practical aspects of heat transfer and hydraulics.

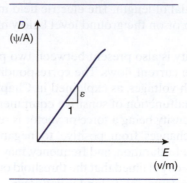

FIG. I-6 Electric flux density versus flux intensity characteristic.

Resistance

I Physical Significance

In a dc circuit, the opposition to the flow of current is called resistance. It is equivalent to the friction in a moving body. The higher a material's temperature, the higher the agitation and collision of its atoms and thus the higher its opposition (resistance) to the passage of electrons (current).

2 Symbols and Designation

The symbol of a fixed and variable resistor is shown in Fig. 1-7(a).

3 Unit

The resistance is measured in ohms (Ω).

$$10^{-3} \, \Omega = 1 \, m\Omega$$

$$10^3 \, \Omega = 1 \, k\Omega$$

The minimum resistance in a 240/120 V, 100 A home is 3 Ω, while that of a human body is in the range of hundreds to thousands of ohms.

In all power distribution drawings, however, the ohmic values of resistances (motors, transformers, cables, etc.) are replaced by their equivalent "per unit values." The latter reveal by inspection many characteristics of the device under consideration. See Appendixes.

4 Design

The resistance (R) of a conductor whose cross-sectional area is constant is given by

$$R = p \frac{\ell}{A} \tag{1.20}$$

where p is the resistivity of the conductor's material expressed in $\Omega \times m$, ℓ is its length expressed in m, and A is the conductor's cross-sectional area perpendicular to the flow of the current, expressed in m^2.

By inspecting Eq. (1.20), it is clear that conductors of a larger cross-sectional area have lower resistance than conductors of a smaller cross-sectional area. Thus, in order to minimize the resistance of a winding (wire forming one or more loops, often referred to as a coil or inductor), a wire of as large a cross-sectional area as can be physically accommodated must be used.

When caught outdoors during a thunderstorm, running to the nearest depression in the ground increases the body's resistance and thus lessens the destructive effects of any lightning.

The resistivity of any material depends on its atomic structure and on its temperature. At 20°C, the resistivity of aluminum is 1.64 times greater than that of copper.

5 Effects of Temperature

The resistance of a wire as a function of temperature is given by

$$R = R_o \left[1 + \alpha_{20} \left(T - 20\right)\right] \tag{1.21}$$

where R_o is the resistance of the wire at 20°C, R is its resistance at T°C, and α_{20} is the temperature coefficient at 20°C. The temperature coefficient for copper is 0.00393/°C; for tungsten, it is 0.0045/°C. Then from Eq. (1.21), the temperatures at which copper and tungsten become superconductive (i.e., have resistance equal to zero) are −234.45°C and −202.2°C, respectively. Recently, a new class of ceramic material has been developed that becomes superconductive at much higher temperatures.

Did You Know?

The phenomenon of superconductivity was first observed in 1911 by a Dutch physicist Heike Kamerlingh Onnes in 1911. He was the first scientist to liquefy helium in 1908, which set the stage for his best-known work, the discovery of superconductivity three years later. He noted that the resistance of a frozen mercury rod abruptly dropped to zero when cooled to the boiling point of helium (4.2 Kelvin). Superconductivity is further discussed in Section 1.3.4.

6 Voltage across a Resistor

From Ohm's law, the voltage (V) across a resistor is given by

$$V = IR \tag{1.22}$$

where I and R are, respectively, the current and the circuit's resistance.

7 Phasor

The voltage and the current in a resistor, as can be concluded from Eq. (1.22), are in phase. They are sketched in Fig. 1-7(b).

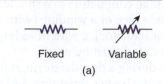

Fixed   Variable

(a)

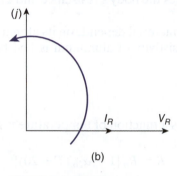

(b)

FIG. I-7 Resistors.
(a) Fixed and variable resistor.
(b) Voltage–current phasors.

8 Power

The power (P) consumed by a resistor is

$$P = VI = \frac{V^2}{R} = I^2R \tag{1.23}$$

The power, or the rate of heat loss, in a winding depends on its resistance, which in turn depends on its operating temperature. When speaking of the losses of an apparatus, the temperature to which these losses correspond must be specified.

9 Energy

By definition, the energy (W) is related to the power (P) by

$$W = P\,(\text{time}) = I^2R\,(\text{time})$$

It is measured in kWh.
(See Tables 1-6 and 1-7 at the end of the chapter for a summary of important highlights of power, energy, and their corresponding costs.)

10 Combination of Resistors

Given the resistors R_1, R_2, and R_n, their equivalent (R_e) series and parallel combinations are as follows.

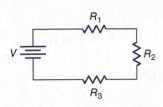

FIG. I-8 Resistors in series.

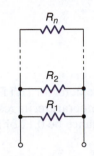

FIG. I-9 Resistors in parallel.

(a) *Series combination* (Fig. 1-8)

$$R_e = R_1 + R_2 + \cdots R_3 \tag{1.24}$$

(b) *Parallel combination* (Fig. 1-9)

$$\frac{1}{R_e} = \frac{1}{R_1} + \frac{1}{R_2} + \cdots \frac{1}{R_n} \tag{1.25}$$

The equivalent resistance of two parallel resistors R_1 and R_2 is

$$R_e = \frac{R_1 R_2}{R_1 + R_2} \tag{1.26}$$

EXAMPLE **I-3**

a. Given three copper resistors, each of 10 Ω, find their equivalent resistance when they are connected in parallel.

b. Find the percentage change in the equivalent resistance when their temperature increases by 20°C.

SOLUTION

a. $\dfrac{1}{R_e} = \dfrac{1}{R_1} + \dfrac{1}{R_1} + \dfrac{1}{R_1} = \dfrac{1}{10} + \dfrac{1}{10} + \dfrac{1}{10}$

and

$$R_e = 3.33 \ \Omega$$

b. $R_e = 3.33(1 + 0.00393)(20)$
   $= 3.60 \ \Omega$

and the percentage change is:

$$= \frac{3.60 - 3.33}{3.33}(100) = \underline{7.86\%}$$

Exercise 1-4

A platinum resistance temperature detector (RTD) is used to monitor a motor's winding temperature. At room temperature, its resistance is 0.10 ohm. At what temperature increase will its resistance be 0.12 ohm? Given, $\alpha = 0.00375/°C$.

Answer 53.33°C

Self-Inductance

The inductance is associated with a magnetic field, and since the latter is due to flow of current, it can be said that conductors that carry current always have an inductance or simply a self-inductance.

When the flux lines of one coil link an adjacent coil and vice versa, then there is a mutual coupling between the coils and/or a mutual inductance. (See Section 1.2.13.)

I Physical Significance

The inductance (L) of a coil or of a winding is analogous to the mass (M) of a moving object or to the polar moment of inertia (J) of a rotating body.

The electrical momentum (LI) corresponds to the mechanical momentum (MU). Mathematically,

$$LI \equiv MU \tag{1.27}$$

and

$$LI \equiv J\omega \tag{1.28}$$

where ω is the angular rotation of the rotating body.

The flow of current through a circuit of high inductance cannot be interrupted instantaneously, just as a large moving object cannot be stopped. Interrupting the flow of current in a highly inductive circuit is as destructive as attempting to stop a moving truck.

2 Unit

The inductance is measured in the Henry (H).

$$10^{-3}\, H = 1\, mH$$

$$10^{-6}\, H = 1\, \mu H$$

The inductance of a transformer could be several mH, while that of a relay could be in the range of Henries.

3 Design

The inductance of a coil depends on its physical characteristics (K_1), on the permeability of its material (μ), and on the number of turns squared. Mathematically,

$$L = K_1\, \mu N^2 \tag{1.29}$$

For constant permeability,

$$L = KN^2 \tag{1.30}$$

where K is a constant.

Since almost all loads have winding turns (motors, transformers, etc.), it can be said that the loads in industrial plants are inductive.

4 Voltage

The voltage (v) across an inductor is given by

$$v = L\frac{di}{dt} \tag{1.31}$$

or

$$v = N\frac{d\phi}{dt} \tag{1.32}$$

where i is the current as a function of time, N is the number of winding turns, and ϕ is the magnetic flux. When the slope or the time rate of change of the current or that of the flux are high, the voltage across the inductor will be high.

5 Impedance

The impedance (Z_L) or the opposition of an ideal inductor to the flow of current is

$$Z_L = j2\,\pi\, fL \tag{1.33}$$

$$= j\omega L \tag{1.34}$$

$$= jX \tag{1.35}$$

where f and ω are, respectively, the supply voltage's natural frequency of oscillation and the angular frequency of oscillation. X is called the reactance of the coil and is measured in ohms. The impedance is a phasor, while the reactance is a scalar quantity.

The impedance of an inductor can also be represented by the Laplace notation as follows

$$Z_L = sL \tag{1.36}$$

where s is the Laplace operator.

6 Phasor Diagram

From Ohm's law, the current (I) in an inductor is given by

$$I_L = \frac{V_L}{Z} \tag{1.37}$$

From above,

$$I_L = \frac{V_L}{jX} \tag{1.38}$$

$$= \frac{V_L}{X} \angle{-90} \tag{1.39}$$

That is, the current is lagging the voltage by 90 degrees. This is indicated in Fig. 1-10(b).

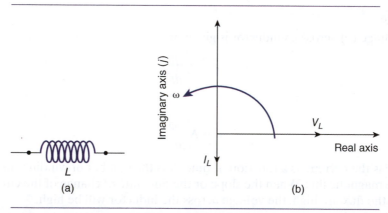

FIG. I-I0 (a) Inductor. (b) Voltage-current phasor.

Physical Justification

An inductor is a characteristic of a magnetic field that is due to current flow. Current flow follows, as per Ohm's law, the application of voltage. That is, the voltage across an inductor proceeds (leads) the current or the current through an inductor lags the voltage.

7 Power

As will be developed in a later section, the average power (P) is given by

$$P = VI \cos \theta \tag{1.40}$$

Since the phase angle (θ) between the voltage and the current is 90 degrees, the inductors consume no power.

8 Energy Stored

The expression for the energy (W) stored in a coil is derived as follows:

$$dW = P \, dt \tag{1.41}$$

$$W = \int v \iota \, dt \tag{1.41a}$$

$$= \int L \frac{di}{dt} \iota \, dt \tag{1.41b}$$

$$= \int_0^I L \iota \, di \tag{1.41c}$$

$$= \frac{1}{2} LI^2 \text{ watts-seconds} \tag{1.41d}$$

The energy in a coil or that in any magnetic field contrary to that of capacitors is released on current interruption. This property of the inductive loads may damage electronic circuits and poses a safety hazard for personnel. It is for that reason that in electronic circuits a discharge resistor should be connected in parallel with the inductors.

For example, pulling the cable of an operating toaster from its power supply receptacle, causes a spark. This could be attributed to the momentary discharge of the energy stored in the inductance of the circuit across the receptacle.

9 Starting Current

When a coil or a motor at rest is connected to a voltage supply, its starting currents are zero. In general, the inductor is open circuit at $t = 0$.

For a coil,

$$Z_L = j\omega L \tag{1.42a}$$

$$= sL \text{ ohms} \tag{1.42b}$$

At $t \to 0$, $s \to \infty$ and the coil's impedance is infinite.

To prevent electrical disturbances, such as lightning strokes, from reaching a sensitive electronic device, a coil of high inductance is used upstream of it. (In Europe, an inductance is inserted in series with the supply power lines, while in some parts of North America, a transformer of turns ratio equal to unity is used.)

Furthermore, medical devices that are used to apply a high voltage to a patient with a defibrillating heart usually include an inductance so that when they are switched ON, the current is zero and desired levels are reached with further adjustments.

10 Time Constant

The current in every dc or ac circuit has a dc component that is of the form

$$I_t = A_o \, e^{-at} \tag{1.43}$$

where A_o and a are the circuit's constants. $\frac{1}{a}$ is termed the circuit's time constant (*tc*). That is,

$$tc = \frac{1}{a} \tag{1.44}$$

This definition of *tc* makes the exponent equal to minus 1. In an *R-L* circuit,

$$tc = \frac{L}{R} \tag{1.45}$$

There are other designations that describe the decay of exponential functions.

Chemists and physicists speak of the half-life expectancy of the variable under consideration, while often in applied science, the time constant is defined as the value of t, at which the function's value is 0.37 of its original value.

In contrast to these alternatives, use the definition given previously, because in practice, it is important to know how long it will take for the exponential decaying function to become zero. The corresponding time interval is about four time constants (Fig. 1-11).

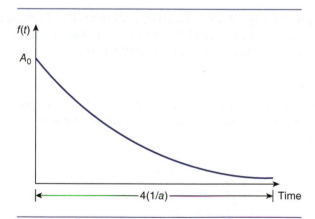

FIG. I-II Exponential decaying function

When a standard ac motor is switched on to its nominal voltage supply, its starting current (See Chapter 3, Tables 3-1 and 3-2.) is about six times its rated value and its duration is four time constants. That is, its rate of heat loss (I^2R) is 36 times its rated value during the starting period. The time constant of the human eye is about 1 m-sec. When a utility transfers a city's distribution system from one generating station to the other, the corresponding circuit breakers open and close in less than 4 m-sec, and thus a human's eyes do not notice the change in illumination levels.

Similarly, when a room's temperature should be decreased or increased to a desired value through the air supply, it will take four time constants to reach the desired temperature. Since the inductance is equivalent to mass, the system's time constant is equal to

$$\frac{\text{room's equivalent air volume}}{\text{rate of supply air volume}}$$

A better method is to develop and solve the corresponding differential equation.

II Resonance Condition

Refer to Fig. 1-12. At resonance, the circuit's current is maximum. This implies that the impedance (Z) is minimum.

$$Z = R + j\omega L + \frac{1}{j\omega C} \tag{1.46}$$

where $\frac{1}{j\omega C}$ is the impedance of the capacitor.

For the current to be <u>maximum</u>,

$$j\omega L + \frac{1}{j\omega C} = 0 \tag{1.47}$$

From which,

$$\omega^2 = \frac{1}{LC} \tag{1.48}$$

or

$$f_o = \frac{1}{2\pi \sqrt{LC}} \tag{1.49}$$

where f_o designates the natural frequency of oscillation.

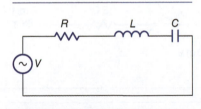

FIG. 1-12 An R-L-C circuit.

In general, resonance is a condition of <u>maximum</u> current or voltage or love or hate or patriotism, and so on. Every circuit or building or individual has its own unique resonant frequency or sensitivity.

A general description of the various types of resonance is given in Chapter 2.

12 Combination of Inductors

The series and parallel equivalent of individual inductors is, as shown in Fig. 1-13, identical to those of resistors.

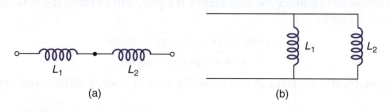

(a)   (b)

FIG. 1-13 Ideal Inductors. (a) Series connection. (b) Parallel connection.

EXAMPLE **1-4**

Refer to Fig. 1-14.

a. When the switch S is closed for a long time, determine the energy stored in the inductance.

b. When the switch becomes open, find how long it will take for the energy to be dissipated.

SOLUTION

a. $I = \dfrac{V}{R} = \dfrac{10}{5} = 2A$

$$W = \frac{1}{2} LI^2 = \frac{1}{2} (2)^2 (2 \times 10^{-3}) = 4 \text{ m watts-seconds}$$

b. The energy will take about four time constants to decay to zero.

$$4 \frac{L}{R} = 4 \frac{L}{\infty} = 0.0 \text{ second}$$

(When the switch is open, its resistance is infinite).

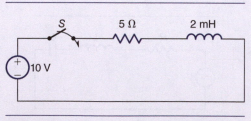

FIG. 1-14

Exercise

1-5

In the following cases, determine the time constant and the time it will take for the radiation to reach negligible levels.

a. The half-life expectancy of the radiation emitted from the Chernobyl nuclear accident has a half-life expectancy of 5000 years.

b. The deadly radon gas that can be found in the basements of some homes has a half-life expectancy of 3.7 days.

Answer (a) 28,854 years, **(b)** 21.35 days

Capacitance

I What Is a Capacitance?

Capacitance is associated with the force between two objects, one of which has free positive charges and the other free negative charges. This force is directed from the positive to the negative charges. When moving along a room, the capacitance changes with respect to the encountered objects because the object's free electron charges are different in polarity (positive or negative) with respect to the moving individual. As such, there is always a capacitance between an overhead transmission line and the ground, between an individual and a radio, between the room's lighting fixtures and a student in the classroom. In other words, there is almost always a capacitance between two objects. This physical phenomenon is used in measurements and analysis of transmission line problems, in the noncontact voltage detecting "stick" of the electricians, etc.

Figure 1-15(a) shows two capacitor plates that are connected to a dc voltage supply of V volts. The upper plate is positively charged, and the lower plate negatively charged.

The vector ψ between the plates represents the electric flux or the electric lines of force. The electric flux emerges from the positively charged plate and terminates in the negatively charged plate. According to Gauss's law, the electric flux is equal to the number of free charges in one plate.

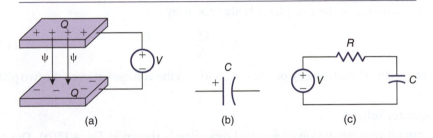

(a)   (b)   (c)

FIG. 1-15 Capacitor. (a) Charged capacitor plates. (b) Ideal capacitor. (c) Capacitive circuit.

2 Unit

The unit of a capacitor is the Farad (F)

$$10^{-3}\ F = 1\ mF$$

$$10^{-6}\ F = 1\ \mu F \quad \text{and} \quad 10^{-12}\ F = 1\ \mu\mu F$$

The capacitance between a ceiling's lighting fixture in a classroom and a student is about 50 $\mu\mu F$, while that of industrial capacitors is in the range of mF.

3 Design

In an ideal case, the capacitance of two parallel plates is given by

$$C = \varepsilon \frac{A}{d} \tag{1.50}$$

where

C is the capacitance of the two plates (F)

ε is the permittivity of the material between the capacitor plates (F/m)

A is the cross-sectional area of one plate (m^2)

d is the perpendicular distance between the plates (m).

The permittivity (ε) of a material is equal to the permittivity of free space (ε_0) times the relative permittivity (ε_r) of the material. That is,

$$\varepsilon = \varepsilon_r\,\varepsilon_0 \tag{1.51}$$

The relative permittivity of paper is about 80 times larger than that of air. In the tuning of some radios, the parallel capacitor plates are moved in relation to each other, and thus their effective area changes, resulting in a new capacitance. The new capacitance, according to the theory of resonant circuits, produces a new resonant frequency, and thus the radio can pick up a new station. A capacitor is sometimes referred to as a condenser.

4 Capacitor-Charge Relationship

The capacitance of the two plates is also given by

$$C = \frac{Q}{v_c} \tag{1.52}$$

where Q is the electric charge of one plate and v_c is the voltage between the two plates.

5 Capacitor Voltage

The circuit representation of an ideal capacitor is shown in Fig. 1-15(b). The current (i_c) to the capacitor is, by definition, equal to the time rate of change of the electric charge. That is,

$$i_c = \frac{dQ}{dt} \qquad (1.53)$$

From Eq. (1.52) and (1.53) we get

$$v_c = \frac{1}{C} \int i_c \, dt \qquad (1.54)$$

Equation (1.54) is a general expression for the voltage across a capacitor. When the current is constant, the voltage across the capacitor increases linearly with time.

6 Impedance

The impedance (Z_c) or the opposition to the current presented by a capacitor is

$$Z_c = \frac{1}{j\omega c} \text{ ohms} \qquad (1.55a)$$

$$= j \frac{1}{j^2 \omega c} \qquad (1.55b)$$

$$= -j \left(\frac{1}{\omega c} \right) \qquad (1.55c)$$

$$= -j X_c \text{ ohms} \qquad (1.55d)$$

where ω is the angular frequency (rad/s) of the voltage source and X_c is the reactance of the capacitor measured in ohms. The impedance of a capacitor can also be represented in terms of the Laplace operator

$$Z_c = \frac{1}{sC} \text{ ohms} \qquad (1.56)$$

7 Phasor

From Ohm's law, for a pure capacitor

$$I_c = \frac{V_c}{Z_c} \qquad (1.57a)$$

$$= \frac{V_c}{1/j\omega C} \qquad (1.57b)$$

$$= j\omega CV \qquad (1.57c)$$

$$= \omega CV \, \underline{/90} \qquad (1.57d)$$

That is, the current in a capacitor leads the voltage by 90 degrees (see Fig. 1-16).

Physical Justification.
The voltage across a capacitor is due to electronic charges on its plates. These charges were carried to the plates through some kind of current flow. That is, the

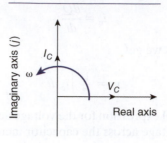

FIG. I-16 Capacitor's
voltage-current phasors.

current to a capacitor precedes the development of voltage across it or the voltage lags behind the current.

8 Power

The average power consumed by a capacitor is

$$P = VI \cos \theta \tag{1.58}$$

where θ is the phase angle between the voltage and the current. This phase angle is 90 degrees and thus,

$$P = 0$$

That is, an ideal capacitor does not consume any power.

9 Energy

The incremental energy (dW) gained by a capacitor in an infinitesimal amount of time dt is

$$dW = v_c i_c \, dt \tag{1.59}$$

From the above and Eq. (1.53), we obtain

$$dW = v_c \, dQ \tag{1.60}$$

Substituting for dQ, its equivalent from Eq. (1.52), we get

$$dW = C v_c \, dv_c \tag{1.61}$$

Taking integrals and assuming that the voltage changes from zero up to V volts, we obtain the energy stored within the electric field of a capacitor:

$$W = C \int_o^v v_c \, dv_c = \frac{1}{2} CV^2 \tag{1.62}$$

where W is the energy stored in the electric field and V is the voltage across the capacitor.

When the current in a capacitor is interrupted, the capacitor remains charged. Depending on the magnitude of its voltage, it may be hazardous to personnel. When the switch is closed, the supply voltage may aid the capacitor voltage. As a result, the voltage across the switch will be, instantaneously, twice that of the supply voltage. This again may be hazardous to personnel and may also carbonize the contacts of the switch, causing it to malfunction.

Carbonization is probably the most common cause of failure of equipment that is called upon to start and stop frequently: air conditioners, clothes washers, and the like. An international company that operates public clothes washing machines reported an annual failure rate of about 12%. The major cause, as determined by consultants, was the carbonization of the timer's contact that resulted from the circuit capacitors.

To eliminate these problems the capacitors should be connected in parallel to properly sized resistors so that the stored energy may be safely dissipated as heat. For these reasons, the electrical code requires that a capacitor in a power distribution network that is rated at less than 750 V should be equipped with a parallel-connected resistor so that the stored energy may be dissipated in less than 60 seconds.

10 Starting Current

Equation (1.56) is important in evaluating the effect of a capacitor on a distribution system. Theoretically, s approaches infinity at $t = 0$. That is, the impedance of a capacitor is zero, and thus from Ohm's law the current is infinite.

In practice, ideal capacitors do not exist. Manufacturers report that at $t = 0$ (starting current) could be up to 300 times rated. In order to eliminate this adverse effect of the capacitors, their power supply cable should be rolled up in two to three turns so that an inductance is introduced into the circuit.

11 Time Constant

The time constant of an RC circuit is

$$tc = CR \text{ seconds} \qquad (1.63)$$

(This mathematical relationship is derived in the section on Laplace transforms in the Appendices.)

12 Displacement Current

In an attempt to explain the existence of magnet fields in empty space, James Maxwell postulated that there must be a capacitor current, which he called displacement current, (I_d). This current is not like the conduction current (rate of electrons' flow), but it is given by the rate of change of electric flux lines. That is,

$$I_d = \frac{d}{dt}(\Psi) \qquad (1.64)$$

This current does not flow through the air between the capacitor plates. Maxwell justified it by reasoning that the time rate of change of the electric flux lines stretch or modify the atoms' electron configuration—a form of current—and thus there must also be a changing magnetic field within the air space.

13 Resonance Condition

The capacitor is used to adjust the resonant frequency (f_o) of an electronic circuit. The resonant frequency given by Eq. (1.49) is rewritten here as

$$f_o = \frac{1}{2\pi \sqrt{LC}} \tag{1.65}$$

where L and C are, respectively, the circuit's inductance and capacitance. In selecting a radio station or a TV channel, it is usually done by adjusting the capacitance between two spherical disks, one of which moves with respect to the other.

14 Combination of Capacitors

The equivalent of a series-connected capacitor is similar to that of parallel-connected resistors. Similarly, parallel-connected capacitors are equivalent to series-connected resistors.

15 Types of Capacitors

There are the following three types of capacitors: electrostatic, electrolytic and supercapacitors.

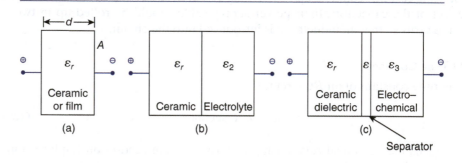

FIG. 1-17 Capacitors. (a) Electrostatic (dry separator). (b) Electrolytic (moist separator). (c) Supercapacitor organic electrolyte.

- **Electrostatic**

  The electrostatic or classical capacitors have a capacitance in the range of $\mu\mu F$ to several μF. Their capacitance, depends, as previously described, on the permittivity of the insulating material between the capacitor plates, the separation distance between the plates, and the area of one of the plates. It is used in radio and in tuning circuits. The energy storage capability of electrostatic capacitors is very small.

- **Electrolytic**

  Electrolytic capacitors have a capacitance that is much larger than that of the classical capacitors. As the name implies, there is an electrolytic chemical between the capacitor plates that are marked positive (+) and negative (−). When the applied dc voltage has opposite polarity to that of the plates, the capacitor will be instantly damaged. The derivation of the capacitance between the plates is derived by using the Laplace equation, which describes the electric fields. Alternatively, the capacitance of an electrolytic capacitor is equal to the parallel combination of two capacitors, each of whose capacitance is given by Eq. (1-50), modified according to the dimensions and permittivity of each layer.

- **Supercapacitor**

  The Supercapacitor has two layers of carbon-based organic electrolytes that are separated by a nonconducting thin plate whose thickness is in the range of nanometers. The supercapacitor is the latest improvement in storing energy in an electric field, and because its capacitance could be up to several thousand Faradays, in some applications, it can replace the standard batteries. Fig. 1-19 shows the development of a supercapacitor in China. Table 1-1 compares some characteristics of the supercapacitors with those of standard Li-ion batteries.

Mathematical Considerations

A supercapacitor can be represented by the resistance R_1 (Fig. 1-18) in series with a capacitor that in turn is connected in parallel with the resistor R_2.

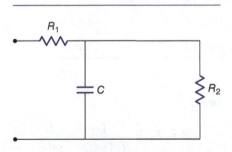

FIG. 1-18 Approximate equivalent circuit of a supercapacitor.

The input impedance Z to the circuit is //

$$Z = R_1 + R_2 \,// \frac{1}{j\omega C} = R_1 + \frac{R_2}{1 + j\omega CR_2}$$

$$= R_1 + \frac{R_2\,(1 - j\omega CR_2)}{1 + (\omega CR_2)^2}$$

$$= R_1 + \frac{R_2}{1 + (\omega CR_2)^2} - j\,\frac{\omega CR_2^2}{1 + (\omega CR_2)^2}$$

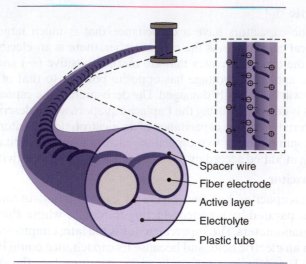

FIG. 1-19 Supercapacitors. *Based on NextBigFuture .com*

- Spacer wire
- Fiber electrode
- Active layer
- Electrolyte
- Plastic tube

No.	Characteristic	Battery	Supercapacitors
	TABLE 1-1 Relative characteristics of Li-ion batteries and supercapacitors		
1	Voltage range	Low	Lower
2	Energy stored per unit volume	Low	Lower
3	Charging time	High	Higher
4	Discharging time	Short	Shorter

For $(\omega CR_2)^2 > 1$, we obtain

$$Z = R_1 + \frac{1}{R_2\,(\omega C)^2} - \frac{j}{\omega C}$$

The effective circuit resistance (R_e) and capacitance (C_e) are, respectively,

$$R_e = R_1 + \frac{1}{(\omega C)^2 R_2}$$

$$C_{ef} = C$$

The circuit's time constant (tc) on capacitor charging is

$$(tc)_{ch} = C\left(R_1 + \frac{1}{(\omega C)^2 R_2}\right)$$

The charging time is several seconds long.

On opening the circuit, the capacitor C is discharged through R_2 and the discharge time constant is

$$(tc)_{\text{dis}} = CR_2$$

When the capacitor is switched OFF, the discharge time is very small, and therefore the electrodes may overheat and produce sparks.

Did You Know?

Scientists Build Incredible Supercapacitor Using Ink from Ordinary Pens

Researchers from Peking University in Beijing, China built the device by coating two long, thin, carbon fibers with the ink, then wrapping them in a flexible plastic casing, filled with electrolyte. The pen ink was used in the supercapacitor after the same team discovered that it contains carbon nanoparticles—perfect for storing charge. When applied to the carbon electrode, it provides an enormous surface area for holding charge: 27 m^2 per gram of ink.

EXAMPLE 1-5

a. The voltage between the plates of a 2 $\mu\mu F$ capacitor is 10 V. Estimate the number of electric flux lines between the plates.

b. An electronic circuit's capacitor and its discharge resistor are, respectively, 15 mF and 2 KΩ. Determine the time it will take for the capacitor to be discharged.

SOLUTION

a. $Q = CV$

$$= 2 \times 10^{-12} \, (10) = 20 \times 10^{-12} \text{ Coulombs}$$

(1 Coulomb is equal to 6.25×10^{18} electrons)

Thus,

$$Q = 20 \times 10^{-12} \, (6.25 \times 10^{18}) = 12.5 \times 10^7 \text{ electrons (flux lines)}$$

b. The discharge time $= 4 \, RC$

$$= 4(2 \times 10^3)(15 \times 10^{-3}) = \underline{120 \text{ seconds}}$$

Exercise

1-6

Refer to Fig. 1-20.

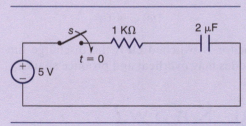

FIG. 1-20

a. How many flux lines are across the capacitor at the instant of closing the switch and after steady state has been established?

b. On opening the switch, how long will it take for the capacitor to discharge?

c. Determine the energy stored in the capacitor.

Answer (a) 62.5×10^{12} lines   (b) ∞ Sec   (c) $25 \, \mu$ w-Sec

Exercise

1-7

Why should oscilloscope measuring probes when used to measure unknown voltage signals incorporate an in series-connected capacitor?

Should the value of the capacitor be large or small? Briefly explain.

1.1.5 Impedance

Physically, the impedance of a circuit represents the opposition of the circuit to the flow of current. The current may be of any waveform, but in this text, only ac waveforms are considered. The impedance of various apparatuses and machines varies widely. For example, the magnitude of the impedance of a TV set, as seen from the 120 V, 60 Hz voltage source, is several ohms, while the impedance of a large induction machine, as seen from two of its supply wires, is in many cases a fraction of an ohm.

In general, impedance (Z) has a real and a positive or negative imaginary component. That is,

$$Z = R \pm jX \text{ ohms} \qquad (1.66)$$

where R and X are, respectively, the resistance and reactance of the circuit.

The positive reactance ($X = \omega L$) is called inductive, while the negative reactance is called capacitive ($X = 1/\omega C$). The angular frequency of oscillation (ω) is expressed in rad/s, the inductance (L) in henries, and the capacitance C in farads.

The inverse of the impedance is called admittance (Y), measured in A/V. The admittance has a real part and a negative or positive imaginary part.

That is,

$$Y = G \mp jB = \frac{1}{Z} \, A/V \qquad\qquad (1.67)$$

The real part (G) of the admittance is called conductance, while its imaginary part (B) is called susceptance. The positive and negative signs in the equation represent capacitive and inductive susceptance, respectively. The real part of Y is not necessarily equal to the reciprocal of the real part of Z.

For example, if an impedance is equal to $3 + j4$ ohms, then its conductance is, as shown below, $^{3}/_{25}\, A/V$.

$$Y = \frac{1}{Z} = \frac{1}{3 + j4} = \frac{3 - j4}{(3 + j4)(3 - j4)}$$

$$= \frac{3 - j4}{(3^2 + 4^2)} = \frac{3}{25} - j\frac{4}{25} \, A/V$$

The $R\text{-}L$ circuit shown in Fig. 1-21(a) is connected to a 120 V, 60 Hz voltage supply; determine the circuit:

EXAMPLE 1-6

a. Impedance.
b. Current.

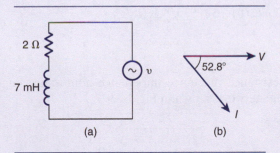

(a)   (b)

FIG. 1-21

SOLUTION

a. The impedance is

$$Z = R + jX_L = R + j2\pi fL = 2 + j2\pi(60)(7 \times 10^{-3}) = 2 + j2.64$$

$$= 3.31\,\underline{/52.8^\circ}\,\Omega$$

b.   The current is

$$I = \frac{V}{Z} = \frac{120\ \underline{/0}}{2 + j2.64}$$

$$= 36.24\ \underline{/-52.8°}\ A$$

The voltage and current phasors are shown in Fig. 1-21(b).

Exercise 1-8

The *R-C* circuit shown in Fig. 1-22 is connected to a 120 V, 60 Hz supply. When the phase angle of the current is 36.9° leading, determine the circuit:

a.   Capacitance.

b.   Impedance.

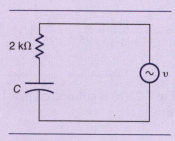

FIG. 1-22

Answer (a) 1.77 μF; (b) 2.5 $\underline{/-36.9°}$ KΩ

Exercise 1-9

Determine the resistance, reactance, impedance, admittance, conductance, and susceptance of the network shown in Fig. 1-23.

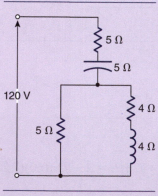

FIG. 1-23

Answer $Z = (7.68 - j3.97)\ \Omega$; $Y = (0.10 + j0.05)$ A/V

1.1.6 Principles of Electric Circuits

This section explains Ohm's and Kirchhoff's laws and their application in solving problems of electric circuits.

Ohm's Law

According to the generalized Ohm's* law, the current that will flow in an electric circuit is given by the ratio of the circuit potential divided by the circuit impedance. Referring to Fig. 1-24, we have

$$i = \frac{v}{Z} \tag{1.68}$$

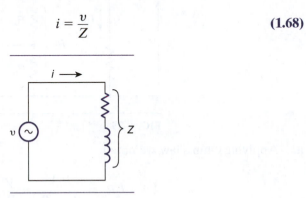

FIG. 1-24

where i, v, and Z are, respectively, the current, voltage, and impedance of the circuit in phasor form.

The impedance is a characteristic of the circuit and in electric machines—depending on the type of machine and the operating condition—may be a variable. The current through an electric circuit driven by a zero-impedance voltage source can be increased to infinity when the circuit impedance is shorted.

In the following example, Ohm's law is used to derive general expressions for the voltage across two series resistors as a function of the supply voltage (the voltage-divider concept) and the current through two parallel resistors as a function of the supply current (the current-divider concept). The derived mathematical equations simplify the solution of many problems in electric and magnetic circuits.

Referring to the circuit of Fig. 1-25, prove the following:

EXAMPLE **1-7**

a. $V_x = V \dfrac{R_b}{R_2}$

b. $I_1 = I \dfrac{R_2}{R_1 + R_2}, \quad I_2 = I \dfrac{R_1}{R_1 + R_2}$

---

* Ohm's experiments and conclusions dealt with dc circuits only.

The expression in (a) is the voltage-divider concept; the expressions in (b) are current-divider concepts.

SOLUTION

The currents through the various resistors are identified as shown in Fig. 1-25.

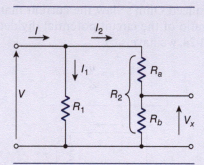

FIG. 1-25

a. Applying Ohm's law, we obtain

$$V_x = I_2 R_b = \frac{R_b V}{R_a + R_b} = \frac{V}{R_2} R_b$$

b. Again from Ohm's law, we get the following relationships:

$$V = I(R_1 \parallel R_2) = I \frac{R_1 R_2}{R_1 + R_2} \qquad \textbf{(I)}$$

Also,

$$V = I_1 R_1 \qquad \textbf{(II)}$$

and

$$V = I_2 R_2 \qquad \textbf{(III)}$$

From (I), (II), and (III), we obtain

$$I_1 = I \frac{R_2}{R_1 + R_2}$$

and

$$I_2 = I \frac{R_1}{R_1 + R_2}$$

Find the current through the capacitor of Fig. 1-26.

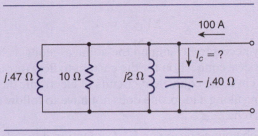

FIG. 1-26

Answer $1.54 \underline{/142°}$ kA

NOTE: Figure 1-26 represents the per-phase equivalent circuit of a 20 MW plant to the flow of the fifth harmonic current at a particular load condition. The 100 A current is the fifth harmonic current produced by the solid state control devices of ac and dc machines. The 0.47 Ω reactance represents the reactance of the utility (source), the 10//j2 Ω impedance represents the load of the plant, and the 0.40 Ω reactance represents the plant's *incorrectly* sized capacitor banks, all at the 4.16 kV level and 300 Hz.

Kirchhoff's Laws

Voltage Law

According to Kirchhoff's voltage law (KVL), the algebraic sum of the electric voltages along a closed loop is equal to zero. That is,

$$\Sigma V_{\text{Loop}} = 0 \qquad\qquad (1.69)$$

In applying KVL, the following procedure is recommended:

a. Draw or assign loop currents (a loop current is a current that flows in a closed path) in whatever direction is preferable. That is, the clockwise (CW) or counterclockwise (CCW) current direction is arbitrary.

b. Assign positive (+) and negative (−) signs to the passive elements. The terminals where the current is entering should be marked positive, and the terminals where the current is leaving, negative. In other words, the terminal of the passive element where the current emerges is at a relatively lower potential than the terminal where the current enters. A good analogy is that current, like free-running water, flows from a higher elevation (potential) to a lower one.

c. Identify the relative polarity of the sources (active elements). This is normally given. An arrow, placed between the terminals of a voltage source, indicates the direction of rising potential, and the tip of the arrow points toward the terminal of positive polarity.

d.   Starting at any convenient point, proceed along a loop summing up all the voltages encountered. According to KVL, these voltages must be equal to zero. That is,

$$\Sigma V_{\text{Loop}} = 0$$

It is absolutely necessary to sum up the voltages along a closed loop* even if there is no physical connection for the current to flow through. This condition is encountered in some equivalent circuits of electronic devices.

In proceeding along a loop, one does not have to follow the same direction as the loop current.

e.   The sign of each voltage term in a KVL equation depends on the polarity of the element first seen as one proceeds along the loop. In going from the positive to the negative terminal of an element, the voltage across this element may be taken as negative. In contrast, in going from a negative to a positive terminal of an element, the voltage across this element must be taken as positive. However, the opposite sign convention could be selected.

f.   When the correct solution of the loop equations yields a negative current, this means only that the assumed current direction is wrong. The problem does not have to be redone.

> Getting out of bed in the morning, doing the day's work, then returning to the same bed in the evening does not change a person's potential energy, thus KVL is satisfied.

Similarly, a pump's power requirement to recirculate water from say a 100 m underground reservoir to surface and vice versa is, because of KVL, almost negligible. (Electric potential is analogous to water elevation—hydrostatic head—in hydraulics).

EXAMPLE **1-8**   Write the loop equation for the circuit shown in Fig. 1-27(a).

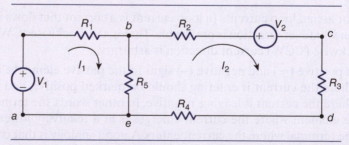

FIG. 1-27

---

* A closed loop, in this context, is simply the path that is followed in writing KVL; it may have nothing to do with the closed loop around which a loop current flows.

SOLUTION

Currents I_1 and I_2 flow in loop 1 and loop 2, respectively. Clockwise direction is arbitrarily chosen for all loop currents. The sign, or the relative polarity across each element, is designated as shown.

    Loop 1 $(a - b - e - a)$

Applying KVL, we obtain

$$V_1 - I_1 R_1 - I_1 R_5 + I_2 R_5 = 0$$

The voltage across R_5 is due to the currents I_1 and I_2. These currents pass in opposite directions through the same resistance and will result in voltage drops of opposite polarity.

    Loop 2 $(b - c - d - e - b)$

Applying KVL again, we get

$$I_2 R_2 + V_2 + I_2 R_3 + I_2 R_4 + I_2 R_5 - I_1 R_5 = 0$$

Current law

According to Kirchhoff's current law (KCL), the algebraic sum of the currents into a junction is equal to zero. That is,

$$\Sigma I_{\text{junction}} = 0 \tag{1.70}$$

In other words, the incoming currents to a junction must be equal to the outgoing currents.

    In applying KCL, the following procedure is recommended:

a. Identify and designate all the points on the circuit that are at different potentials. These points are referred to as the *nodes* of the circuit. For this reason, Eq. **(1.70)** is often referred to as the nodal equation.

b. Select a reference node—that is, a point of zero potential.

c. Apply KCL to each node. It is very important to note that *the node under consideration has the highest potential* in the circuit, and thus the currents leave the node unless they are injected into it by a current source.

d. The currents that leave the node must be designated with a sign opposite to those approaching the node.

Determine a general expression for the voltage at node B in Fig. 1-28.

EXAMPLE **1-9**

SOLUTION   *Node A*

There are three nodes—A, B, and C. Select node C as reference. Applying KCL in node B, we obtain

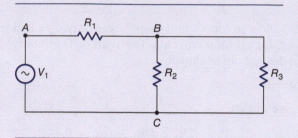

FIG. 1-28

$$\frac{V_B}{R_2} + \frac{V_B - V_A}{R_1} + \frac{V_B}{R_3} = 0$$

$$V_A = V_1$$

From above,

$$V_B\left(\frac{1}{R_1} + \frac{1}{R_2} + \frac{1}{R_3}\right) = \frac{V_1}{R_1}$$

or

$$V_B = \frac{V_1}{R_1}\left(\frac{R_1 R_2 R_3}{R_2 R_3 + R_1 R_3 + R_1 R_2}\right)$$

Ohm's and Kirchhoff's laws are the outstanding principles of electrical engineering. The design of all circuits must always satisfy these laws.

The high starting current in three-phase induction machines, the high short-circuit currents of ac machines, the sparking that accompanies the interruption of current through various coils, and the occasional fires that result from electric circuits are all caused by reactions of nature that attempt to satisfy these basic principles.

Exercise
1-11

The current through the resistor of the circuit in Fig. 1-29 is

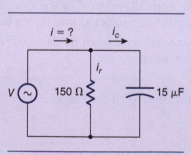

FIG. 1-29

$$i_r = 2 \sin\left(600t + \frac{\pi}{6}\right) \text{A}$$

a. Determine the rms value of the total current.
b. Determine the power factor of the circuit.
c. Draw the phasor diagram.

Answer (a) 2.38 A; (b) 0.60 leading

a. Prove KVL.
b. Refer to Fig. 1-30. Determine the voltage V_{GS}.

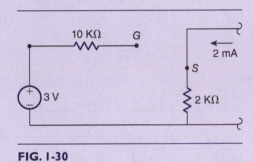

FIG. I-30

Answer −1 V

1.1.7 Power of a Single-Phase Source

In this section, the various forms of power and their units will be discussed, namely, the instantaneous power, the average power, the reactive power, the complex power, and the apparent power. The emphasis will be placed on the average power because of its importance in all power distribution systems. It is for this reason that its governing equation will be derived by using three different methods.

I *Instantaneous Power*

Consider the circuit of Fig. 1-31(a). The input voltage waveform is shown in Fig. 1-31(b). The instantaneous power ($p(t)$) drawn by the load, in terms of the instantaneous values of the voltage and current, is given by

$$p(t) = vi \qquad\qquad (1.71)$$

where

$$v = V_m \sin \omega t \qquad\qquad (1.72)$$

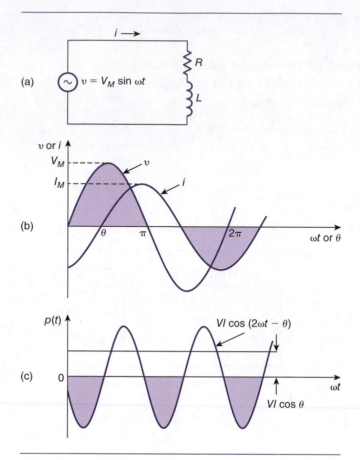

FIG. 1-31 (a) *R-L* circuit. (b) Instantaneous supply voltage and current. (c) Instantaneous power.

and

$$i = \frac{v}{Z} \tag{1.73}$$

$$= \frac{V_m}{|Z|} \sin (\omega t - \theta) \tag{1.73a}$$

$$= I_m \sin (\omega t - \theta) \tag{1.73b}$$

where θ is the phase angle of the inductive impedance Z.

Substituting the instantaneous values of the voltage (Eq. (1.72)) and current (Eq. (1.73b)) into Eq. (1.71), we obtain

$$p(t) = V_m I_m \sin \omega t \sin (\omega t - \theta) \tag{1.74}$$

Replacing the maximum values by their equivalent rms values, and using the identity

$$\sin \alpha \cdot \sin \beta = -\frac{1}{2} [\cos(\alpha + \beta) - \cos(\alpha - \beta)] \tag{1.75}$$

we obtain

$$p(t) = VI \cos \theta - VI \cos (2\omega t - \theta) \tag{1.76}$$

Thus, the instantaneous power drawn from a single-phase network has a constant component ($VI \cos \theta$) and a pulsating component ($VI \cos (2\omega t - \theta)$) that oscillates at a frequency twice that of the supply voltage.

The variation of instantaneous power as a function of time is shown in Fig. 1-31(c).

The constant component is independent of time and is referred to as the average power. It provides the power to motors and heaters while the pulsating power does no work. It produces noise, and is analogous to the foam of a glass of beer that goes back and forth between a bartender and the customer. The latter does not drink it but must pay for it. Similarly, pulsating power although it does no work has to be paid for because of the cost of the cables and equipment associated with its generation and transportation.

2 Average Power

The average value (P) of the power drawn by an inductive load can also be derived as follows:

$$P = \frac{1}{T} \int_0^T p(t) dt \tag{1.77}$$

$$= \frac{1}{T} \int_0^T (V_m \sin \omega t) [I_m \sin (\omega t - \theta)] \, d\omega t \tag{1.77a}$$

$$= \frac{V_m I_m}{4\pi} \int_0^{2\pi} [\cos \theta - \cos (2\omega t - \theta)] \, d\omega t \tag{1.77b}$$

The second term in the integrand, being a sinusoidal function, has an average value equal to zero. Thus,

$$P = \left(\frac{V_m I_m}{4\pi} \cos \theta \right) \omega t \Big|_0^{2\pi} = \frac{V_m I_m}{2} \cos \theta \tag{1.77c}$$

or

$$P = VI \cos \theta \tag{1.78}$$

$\cos \theta$ is known as the power factor.

As seen from the above, the average power consumed by a load is equal to its power factor times the product of the rms value of its voltage and current.

A third method to derive the equation of the average power is to use the following definition of the dot product

$$P = V \cdot I \tag{1.79}$$

The dot signifies the projection of the current phasor on the axis of voltage. Thus,

$$P = VI \cos \theta \tag{1.80}$$

The average power is measured in watts and is used to designate the nominal power or the rating of motors and heaters. Average power is often referred to as the real or active or name-plate or output power.

$$1000 \text{ W} = 1 \text{ kW}$$

$$1000 \text{ kW} = 1 \text{ MW}$$

The average power of an adult is 60–100 W, while the requirements of a city with four million inhabitants is about 3000 to 6000 MW (see Table 1-5).

The utilities do not measure the average power consumed by a home. They do, however, measure the power consumed by industries. The power charges are, depending on the utility, 5 to 40% of the total electrical costs.

The utilities measure the average power consumed over 15-minute or 60-minute time intervals and select for billing purposes the monthly highest of these measurements. Their charges are in the range of $2 \to $15/kW/month. Rotary meters cannot measure all the power consumed by nonlinear loads (variable speed drives, computers, etc.) See Problem 1-14.

3 Reactive Power

The reactive power, designated by the letter Q, is measured in Volt $\times$ Amps Reactive:

$$Q = \text{VAR} \tag{1.81}$$

The reactive power is used to designate the nominal or rated reactive power of inductors (Inductive kVAR) and capacitors (capacitive kVAR):

$$1000 \text{ VARs} = 1 \text{ kVAR}$$

$$1000 \text{ kVAR} = 1 \text{ MVAR}$$

The reactive power of a load whose voltage-current phase angle is θ degrees is

$$Q = VI \sin \theta \tag{1.82}$$

The inductive loads draw reactive power from the utilities, while the capacitor sends reactive power to them. Another explanation is to say that when a capacitor is connected in parallel to an inductive load, it reduces the current supplied by the utility.

4 Complex Power

The complex power is designated by the bolted upper-case letter S. It is defined as follows:

$$S = VI^* \tag{1.83}$$

$$S = P + jQ \tag{1.84}$$

I^* is the conjugate of the current.

$$\text{When } I = 10 \underline{/-30°} \text{ A, then } I^* = 10 \underline{/30°} \text{ A.} \tag{1.85}$$

This definition is used to justify the fact that inductive loads draw reactive power (positive) from the utility, while the capacitive loads give reactive power to the utility (negative).

The magnitude of the complex power is called *apparent* power (S). That is,

$$S = |S| \tag{1.86}$$

$$= VI \tag{1.87}$$

$$= \sqrt{P^2 + Q^2} \tag{1.88}$$

The apparent power is measured in volt-amps (VA) and is used to designate the nominal or the rated power of generators and transformers.

EXAMPLE 1-10

Prove that the electrical distribution system of a factory can be represented by the parallel combination of a resistor R and an inductive reactance X as follows:

$$R = \frac{V^2}{P}, \quad X = \frac{V^2}{P \tan \theta}.$$

where V is the voltage supplied
P is average power consumed
θ is the phase angle between the voltage and the current.

SOLUTION

Refer to Fig. 1-32.

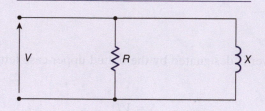

FIG. 1-32 Parallel R-X representation of a factory's inductive load.

$$P = VI \cos \theta$$

$$= I_R^2 R$$

$$= \left(\frac{V}{R}\right)^2 R$$

$$= \frac{V^2}{R}$$

and

$$R = \frac{V^2}{P}$$

$$I_x = \frac{V}{X}$$

$$Q = VI \sin \theta$$

$$= VI_x \sin 90°$$

$$= V \left|\frac{V}{X}\right|$$

$$X = \frac{V^2}{Q} = \frac{V^2}{P \tan \theta}$$

The derived relationships and the load's circuit representation are of extreme practical importance when evaluating the possible resonant conditions that may arise due to power-factor correction capacitors.

1.1.8 Power Factor

The power factor is the cosine of the angle between the voltage and current or the cosine of the impedance's angle (Fig. 1-33).

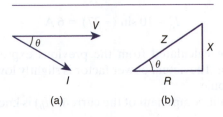

FIG. I-33 Power-factor angle.
(a) Voltage–current phasors.
(b) Inductive impedance diagram.

When the current is lagging the voltage (inductive loads), the power factor is said to be lagging, and when the current is leading the voltage (capacitive loads), the power factor is said to be leading. The range of the power factor is between 40 and 100%.

A low or a poor power factor is a characteristic of digital communication transmitters, small motors, partially loaded large motors, welders, and the like.

A 100% power factor is, of course, associated with electric heaters.

Mathematically, the power factor could be given by any of the following relations:

$$\cos\theta = \frac{R}{|Z|} \qquad (1.91)$$

where R and Z are, respectively, the load's resistance and the magnitude of the impedance.

Also,

$$\cos\theta = \frac{P}{S} \qquad (1.92)$$

where P and S are, respectively, the average and apparent power of the load.

The previous relationship can be also written as follows:

$$\cos\theta = \frac{VI_P}{\sqrt{(VI_P)^2 + (VI_r)^2}} \qquad (1.93)$$

where V is the load's voltage and I_P and I_r are, respectively, the magnitudes of real and reactive components of the load's current (I).

$$I = I_P + I_r \qquad (1.94)$$

For example,

when $$I = 10 \underline{/-37}\,\text{A}$$

then

$$I_P = 10 \cos 37 = 8 \text{ A}$$

and

$$I_r = 10 \sin(-37) = 6 \text{ A}$$

The power factor calculated from the previous expressions is termed the apparent power factor. The actual power factor is slightly lower than that because of the system's harmonics.

When the harmonic component of the current (I_h) is known, the actual power factor is given by

$$\cos \theta = \frac{VI_P}{\sqrt{(VI_P)^2 + (VI_r)^2 + (VI_h)^2}} \tag{1.95}$$

EXAMPLE **1-11**

The apparent power factor of a 208 V, 5 kW load is 0.75 lagging, and its harmonic current is 15% to that of the fundamental. Calculate the actual power factor.

SOLUTION

$$I = \frac{5000}{(208)(0.75)} = 32.05 \underline{/-41.4} \text{ A}$$

$$= 24.01 - j21.20$$

and

$$I_h = 0.15 \,(32.05)$$

$$= 4.81 \text{ A}$$

Substituting in Eq. (1.95), we obtain

$$\cos \theta = \frac{208(24.01)}{\sqrt{(208(24.01))^2 + (208(21.20))^2 + (208(4.81))^2}}$$

$$= 0.74 \text{ lagging}$$

In this case, there is no substantial difference between the actual and apparent power factor.

A 480 V, one-phase load draws 20 kW at apparent and actual power factors of 0.92 and 0.91, respectively.

a. Determine the actual line current and its harmonic content.
b. What will rotary and digital power meters indicate?

Answer (a) 14.9%, (b) 20 kW, 20.22 kW

Exercise

1-13

1.1.9 Effects of the Power Factor

1 From Eq. (1.80), the current is given by

$$I = \frac{P}{V \cos \theta} \tag{1.96}$$

Say,

$$\frac{P}{V} = 100$$

Then

for $\theta = 60°$, $\cos 60° = 0.5$, $I = 200$ A

for $\theta = 0$, $\cos 0° = 1$,   and $I = 100$ A

That is, the power factor controls the current to a load. The lower the power factor, the higher the current. The technical and economical disadvantages of the low power factor (higher current) are the following:

1. Higher generator ($I^2 R$)
2. Larger generator
    (The generators are purchased or rated in volt x amps.)
3. Higher cable losses ($I^2 R$)
4. Larger diameter of cable
    (The larger the diameter of a cable, the more current it can carry.)
5. Larger transformer losses ($I^2 R$)
6. Larger size of transformer
    (The transformers are purchased or rated in volt x amps.)
7. The lower the voltage across the load (V_L),

$$V_L = V_{source} - IZ \tag{1.97}$$

where Z is the impedance of the source.

The lower the load voltage, the lower the torque (T). In Chapter 3, we explain why the torque is proportional to voltage square $T \alpha V^2$. Also, the lower the voltage, the lower the light output of the lighting fixtures. It is for the above reasons that the utilities penalize customers who operate at low power factors.

In general, it can be said that the power factor is the effectiveness of the current, just as in a mechanical system, the cosine of the angle between the force and the direction of displacement is a measure of the effectiveness of the force. (T-34 tanks were built with an inclination, and as a result, the standard antitank rockets could not destroy them.)

Similarly, a person expends less calories by walking on level ground than by walking uphill because the phase angle between the force (weight) and the direction (horizontal) is 90° degrees. Energy is equal to the force times its displacement times the cosine of the angle between the force and its displacement.

Calculation of Penalty

The utilities use one of the following two techniques to penalize consumers for poor power factor.

I) They charge for kVA registered regardless of phase angle θ. Since the kVA depends on θ, there is always an additional charge for θ larger than zero degrees.

II) They charge for kVA registered in conjunction with θ above a given magnitude (usually 26°) as follows:

They measure the month's apparent (S) and average power (P). Then based on their minimum power-factor requirements ($\cos \theta_m$), they compare the following:

$$S (\cos \theta_m) \text{ and } P$$

For billing purposes, they select the largest of the two. The kW penalty is the difference of the two. For example, assume that a utility's minimum power factor is 90%, while a plant operates at 0.80 Pf and draws 100 kVA and 80 kW. Then the penalty is calculated as follows:

$$S \cos \theta_m = 100 (0.9) = 90 \text{ kW}$$

The customer is charged an extra

$$90 - 80 = 10 \text{ kW}$$

The term $S \cos \theta_m$ is often referred to as the corrected power in kW. Its cost depends on the monthly charges/kW.

The minimum power-factor requirement is a commercial term chosen by the utility.

1.1.10 Power-Factor Improvement

Since the current through a capacitor leads its voltage and that through an inductive load lags its corresponding voltage, the parallel hook-up of a capacitor to such a load will reduce the current drawn from the utility.

Refer to Fig. 1-34.

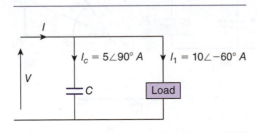

FIG. 1-34 Usage of a capacitor to reduce the line current.

The current supplied by the source is given by

$$I = I_c + I_1$$

$$= 5 \underline{/90} + 10 \underline{/-60}$$

$$= 6.20 \underline{/-36.2} \text{ A}$$

That is, by connecting the capacitor in parallel to the load, the current supplied by the utility was reduced from 10 A to 6.20 A and its phase angle was also reduced from 60 degrees to 36.2 degrees. The power factor of the load, however, does not change. A similar conclusion can be drawn by adding up the current phasors graphically.

Consider Fig. 1-35.

The powers of an inductive load are represented graphically by the triangle $A_1A_2A_4$, which is referred to as the load's power triangle. The voltage-current phasor of a capacitor and its corresponding power representation A_5A_6 ($S = Q$ and $P = 0$) are shown in Fig. 1-35(b).

In order to increase the power factor from $\cos \theta_2$ to $\cos \theta_1$, the reactive power represented by the length A_3A_4 must be removed by connecting a capacitor in parallel to the load.

From the triangle,

$$A_3A_4 = A_2A_4 - A_2A_3$$

or

$$= P \tan \theta_2 - P \tan \theta_1$$

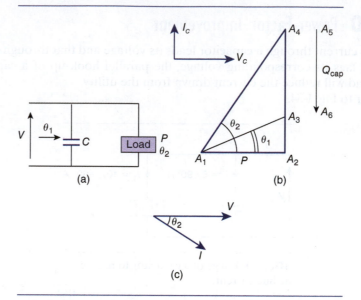

FIG. 1-35 Power-factor correction. (a) Circuit. (b) Power representation. (c) Voltage-current waveforms.

When P is kW and A_3A_4 is in capacitive kVAR (Q), then

$$Q = P (\tan \theta_2 - \tan \theta_1) \tag{1.98}$$

Equation (1.98) can be used to select the capacitor requirements to change θ_2 to θ_1.

EXAMPLE **I-12** Refer to circuit Fig. 1-36. Determine the circuit's
a. Impedance.
b. Inductance
c. Current.
d. Power factor.
e. Average power supplied.
f. Complex, apparent, and reactive power.

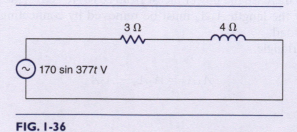

FIG. 1-36

SOLUTION

a. The impedance is

$$Z = R + jX$$

$$= 3 + j4 = 5 \underline{/53.1}\ \Omega$$

b. The inductive reactance is

$$X = 4\ \Omega$$

$$\omega L = 4\ \Omega$$

$$L = \frac{4}{377} = 10.61\ \text{mH}$$

c. The current is

$$I = \frac{V}{Z}$$

$$= \frac{170\sqrt{2}}{3 + j4} = 24.04 \underline{/-53.1}\ \text{A}$$

d. The power factor is

$$\cos\theta = \cos 53.1$$

$$= 0.60\ \text{lagging}$$

e. The average power is

$$P = VI\cos\theta = \frac{170}{\sqrt{2}}\,(24.04)\cos 53.1 = 1734.0\ \text{watts}$$

or

$$P = I^2 R = (24.04)^2 (3) = 1734.0\ \text{watts}$$

f. The complex power is

$$S = VI^* = \frac{170}{\sqrt{2}}\,24.04 \underline{/53.1} = 2890.0 \underline{/53.1}\ \text{VA}$$

and the apparent power is

$$S = |VI|$$

$$= 2890.0\ \text{VA}$$

The reactive power is

$$Q = VI \sin \theta$$

$$= \frac{170}{\sqrt{2}} (24.04) \sin 53.1$$

$$= 2312.00 \text{ VARs}$$

or

$$Q = I^2 X$$

$$= (24.04)^2 (4)$$

$$= \underline{2312.00 \text{ VARs}}$$

EXAMPLE 1-13 A 480 V, 40 kW load operates at a *Pf* of 80%. Determine:

a. The utility's penalty per year when the charges for the power factor are less than 100% are $11/kW/month.

b. The size and cost of the capacitor bank improve the power factor to 100% when the purchasing cost of the capacitor is $40/kVAR.

SOLUTION

a. $S = \dfrac{P}{\cos \theta} = \dfrac{40}{0.8} = 50 \text{ kVA}$

Penalty $= 50 - 40 = 10(1) = 10 \text{ kW}$

Its cost (C)

$C = 10(11)(12) = \underline{\$1320/\text{year}}$

b. The size of the capacitor bank is

$Q = P (\tan \theta_2 - \tan \theta_1)$

$= 40 (\tan 36.9 - \tan 0) = 30 \text{ kVAR}$

Its cost (C) is

$C = 30(40) = \underline{\$1200.00}$

That is, the payback period is less than one year.

EXAMPLE 1-14 A 230 V, 60 Hz voltage source supplies rated power* to three loads whose characteristics are shown in Fig. 1-37(a). Assuming that each load's efficiency is 90%, determine:

---

* According to NEMA (National Electrical Manufacturers Association) section MG 1-10.40, the name plate power is the "horsepower output."

a.   The overall power factor of the system.

b.   The magnitude of the current drawn from the voltage source.

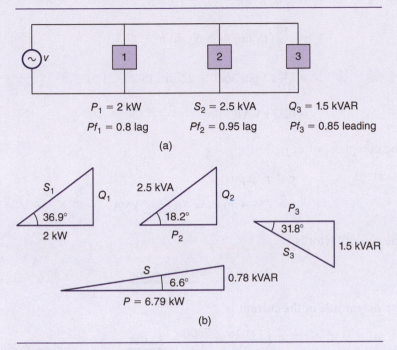

$P_1 = 2$ kW $S_2 = 2.5$ kVA $Q_3 = 1.5$ kVAR

$Pf_1 = 0.8$ lag $Pf_2 = 0.95$ lag $Pf_3 = 0.85$ leading

(a)

(b)

FIG. 1-37

SOLUTION

a.   The individual power triangles are constructed from the given data. The input overall power triangle can be obtained graphically from the individual power triangles (see Fig. 1-37(b)) or mathematically as follows. The input real power is

$$P = \frac{1}{\eta}(P_1 + P_2 + P_3)$$

where η is the efficiency (output power/input power) of each load. Substituting the equivalent equations for P_2 and P_3, we obtain

$$P = \frac{1}{\eta}\left(P_1 + S_2 \cos\theta_2 + \frac{Q_3}{\tan\theta_3}\right)$$

$$= \frac{1}{0.9}\left[2 + 2.5(.95) + \frac{1.5}{\tan 31.8°}\right]$$

$$= 7.55 \text{ kW}$$

The input reactive power is

$$Q = \frac{1}{\eta}(Q_1 + Q_2 - Q_3)$$

$$= \frac{1}{0.9}(P_1 \tan\theta_1 + S_2 \sin\theta_2 - Q_3)$$

$$= \frac{1}{0.9}(2 \tan 36.9° + 2.5 \sin 18.2° - 1.5)$$

$$= 0.87 \text{ kVAR}$$

The complex power is

$$S = P + jQ$$

$$= 7.55 + j0.87 = 7.6 \underline{/6.6°} \text{ kVA}$$

The power factor is

$$\cos(6.6°) = \underline{0.99 \text{ lagging}}$$

b. The magnitude of the current is

$$I = \frac{|S|}{V} = \frac{7600}{230} = \underline{33.0 \text{ A}}$$

Exercise 1-14

Refer to Fig. 1-38. Determine the circuit's

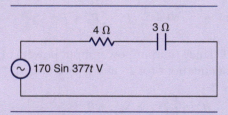

FIG. 1-38

a. Impedance.
b. Capacitance.
c. Current.
d. Power factor.

e. Average power and reactive power.
f. Apparent power.

Answer (a) $5\,\underline{/-36.9}\ \Omega$ (b) 884.17 μF (c) $24.04\,\underline{/36.9}\ A$ (d) 0.8 leading
(e) 2312.0 watts Q = 1734.0 VAR capacitive.

An impedance of $Z = 12 + j12$ ohms is connected to a 240 V, 60 Hz supply.
Determine:

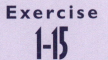

Exercise

1-15

a. The voltage and current as trigonometric functions of time, taking the voltage
   as reference.
b. The instantaneous power as a function of time.
c. The maximum value of the instantaneous power.
d. The average power.
e. The volt-amperes and the VARs drawn from the 240 V supply.

Answer (a) 339.41 sin 377t V, 20 sin (377t − 45°) A
      (b) 2.4 − 3.39 cos (754t − 45°) kW
      (c) 5.79 kW
      (d) 2.4 kW
      (e) 3.39 kVA, 2.4 kVAR

1.1.11 Harmonics

Harmonics

Harmonics are sinusoidal waveforms whose frequency is a fraction or a multiple
of the 60 Hz fundamental frequency of the generating stations (in some other
parts of the world the fundamental frequency is 50 Hz).

The third harmonic (180 Hz) is of particular importance because of resonance concerns and because of its overheating effects.

The adverse effects of the harmonics are as follows:

- Increasing the device's power losses, and thus contributing to its overheating.
- Modifying the voltage waveforms throughout the distribution network.
- Overloading the circuits.
- Can set up harmful resonant conditions.
- Inducing voltages on adjacent metallic objects that may cause the malfunc-
  tion of the 5 V microprocessors.

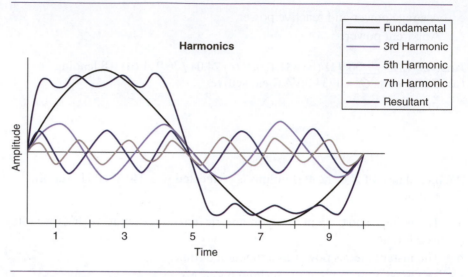

FIG. 1-39 Harmonics.

One method that can mitigate these effects is to filter them out or, where possible, prevent their development. A rather inexpensive method is to supply sensitive equipment through an isolating transformer. Harmonics could be of a transient or steady state.

Harmonics are an inherent part of power distribution systems and can be either transient or permanent.

Transient Harmonics

Any time a switch is switched ON or OFF (control of lighting, of electric heaters, etc.), the device's current is momentarily changed, and as a result, the voltages and currents throughout the distribution system are suddenly altered in order to satisfy the basic principles of electricity. After all, the voltage in a coil depends on the slope of the current waveform $\left(v = L\,\dfrac{di}{dt} \right)$ at the instant under consideration.

The supply transformer windings voltage is also altered, and the otherwise sinusoidal waveforms are momentarily distorted and thus full of harmonics. Their development is shown diagrammatically in the Ladder diagram representation (Fig. 1-40).

These phenomena are of a transient nature, and their duration, depending on the circuit's time constant, lasts several m-sec. They are very important in understanding the operation of motors, transformers, and the like.

Steady-State Harmonics

At nominal operating conditions and in order to minimize the energy usage, the variable speed drives, high-efficiency lighting fixtures, computers, and so on, draw

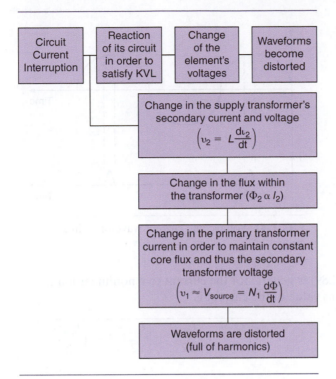

FIG. 1-40 Ladder diagram representation of the generation of transient harmonics.

from their voltage supplies nonsinusoidal current waveforms. That is, at steady state, these waveforms contain many harmonics of a wide range of magnitudes and frequencies. In the follow-up discussion, the harmonics associated with the operation of computers, variable speed drives, and lighting fixtures are very briefly described.

1 Computerized Equipment

A typical voltage and current waveform of a computer is as shown in Fig. 1-41.

These devices incorporate an electronic switch that is timed to let current through only for a small part of the voltage cycle which minimizes the energy consumed and prevents overheating.

These current pulses (ι) can be represented, as per Fourier Series analysis, by the following equation:

$$\iota = \underset{1}{I_m} \cos \omega t + \underset{2}{I_m} \cos 2\omega t + \underset{3}{I_m} \cos 3\omega t + \ldots \underset{n}{I_m} \cos n\omega t \qquad \textbf{(1.89)}$$

where I_{m1} is the amplitude of the fundamental, I_{m2} and $I_{m3} \ldots I_{mn}$ are the maximum values of the second, third, nth harmonic.

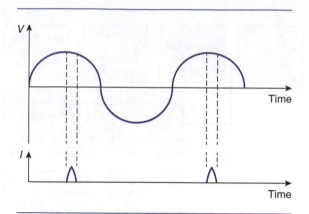

FIG. 1-41 Voltage and current waveforms in a computer.

Equation (1.89) is general for the current to a nonlinear load. The total rms value of the current is

$$I = \sqrt{\left(\frac{I_{m1}}{\sqrt{2}}\right)^2 + \left(\frac{I_{m2}}{\sqrt{2}}\right)^2 + \left(\frac{I_{m3}}{\sqrt{2}}\right)^2 + \cdots \left(\frac{I_{mn}}{\sqrt{2}}\right)^2} \tag{1.90}$$

2 Speed Control of Motors

In contrast to these unique current pulses, the motor's controls incorporate circuits that develop variable width current pulses—and thus produce harmonics—that are used to vary the speed of motors.

3 Lighting Circuits

Another source of harmonics are the lighting fixtures and electric machines in Table 1-2. The electric machines are designed to operate for reasons of efficiency at the top of the linear part of their B-H curve. Any increase in the nominal voltage would result in current harmonics.

Voltage and/or current waveforms that contain harmonics can be accurately measured by the so-called true rms meters. Their design is based on Eq. (1.3). In

TABLE 1-2 Lighting fixtures harmonics and their frequency range

Lighting Fixture	Approximate Total Harmonic Current in Percent of the Fundamental	Frequency of Harmonics
With electromagnetic ballasts	19%	< 1000
With electronic ballasts	49%	> 20,000
Plug-in compact fluorescent lamps	120%	> 20,000

contrast, the low-priced meters that are sensing the average but indicating the rms, depending on the waveform, are not accurate.

It is given that the current in a circuit is

EXAMPLE 1-15

$$\iota = 10 \sin \omega t + 5 \sin 3 \omega t + 3 \sin 5 \omega t$$

Determine:

a. The rms value of the current.
b. The total harmonic current in percent of the fundamental.

SOLUTION

a. $I = \sqrt{\left(\dfrac{10}{\sqrt{2}}\right)^2 + \left(\dfrac{5}{\sqrt{2}}\right)^2 + \left(\dfrac{3}{\sqrt{2}}\right)^2}$

   $= 8.19\ \text{A}$

b. $I_n = \sqrt{\left(\dfrac{5}{\sqrt{2}}\right)^2 + \left(\dfrac{3}{\sqrt{2}}\right)^2} = 4.12\ \text{A}$

   In percent of the fundamental,

   $$\frac{4.12}{10/\sqrt{2}}\,(100) = \underline{58.31\%}$$

1.1.12 Theorems

Every branch of engineering, depending on the type of problems, developed for their solution some shortcuts referred to as theorems. They are not principles, but techniques to minimize the time it takes to solve a problem. Here, only the superposition and Thevénin's theorems will be explained.

Superposition Theorems

Given a circuit with several voltage and/or current sources, discard all the sources except one and then find its effects on a given element. Repeat the procedure as many times as there are sources. Finally, add the results algebraically. (One encounters current sources in electronics because some transistors deliver almost a constant current regardless of the load resistor.)

Thevénin's Theorem

Thevénin's theorem states that any network or circuit of the electrical distribution system of a home, factory, and the like, can be represented by a resistance (or impedance) identified as R_{th} and a voltage source (V_{th}) in series with the resistance (Fig. 1-42).

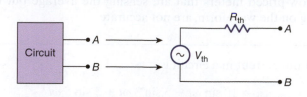

FIG. 1-42 Thévenin's equivalent.

To Find V_{th}

Open up the terminals under consideration and calculate the voltage across them. In other words, V_{th} is what a voltmeter will indicate across the open circuit terminals under consideration. ($R_{voltmeter}$ is very high.)

To Find R_{th}

Discard all independent sources and calculate the resistance looking from the terminals under consideration.

The Thévenin's equivalent resistance can be also calculated as follows:

$$R_{th} = \frac{V_{oc}}{I_{sc}}$$

where V_{oc} is the open circuit voltage across the terminals under consideration and I_{sc} is the short circuit current through the terminals under consideration.

Note: By far one of the simplest techniques to solve problems in photovoltaics is to apply Thévenin's theorem (see Web documentation).

According to where Thévenin's theorem is to be applied, the selected terminals may be identified as A-B.

EXAMPLE **1-16**   Refer to Fig. 1-43(a). Determine the current through the 5 Ω resistor by using Thévenin's theorem.

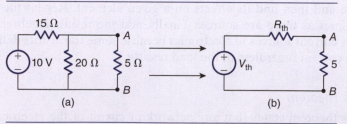

FIG. 1-43 (a) Circuit.   **(b)** Equivalent Thévenin's.

SOLUTION

The terminals across the 5 Ω resistor are identified as *A-B*. The circuit is redrawn just to simplify the presentation.

$$I_{AB} = \frac{V_{th}}{R_{th} + 5}$$

To Find R_{th}

Looking from the terminals *A-B* toward the source,

$$R_{th} = 20 /\!/ 15 = 8.57 \ \Omega$$

To Find V_{th}

The open-circuit voltage across *A-B* is

$$V_{AB\,(open)} = 10 - I\,(15)$$

$$= 10 - \frac{10}{15 + 20}\,(15) = 5.71 \ V$$

or

$$V_{AB} = I\,(20)$$

$$= \frac{10}{15 + 20}\,(20) = 5.71 \ V$$

Then

$$I = \frac{V}{R_{total}} = \frac{5.71}{8.57 + 5} = \underline{0.42 \ A}$$

Refer to Fig. 1-44. Determine the current through the 10 Ω resistor by using Thévenin's theorem.

Exercise

1-16

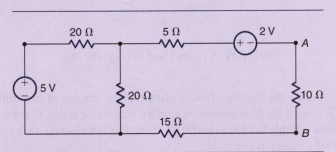

FIG. 1-44

Answer 12.5 m A.

1.2 Magnetic Circuit Concepts

This section discusses the fundamental concepts and principles of magnetic circuits. Knowledge of these elementary concepts will provide the basis for understanding the design and operation of electric machines.

1.2.1 Magnetic Flux

Magnetic flux is a characteristic of magnetic fields, just as electric flux is a characteristic of electric fields. Magnetic flux may be thought of as representing the lines of force between a north magnetic pole and a south magnetic pole. Magnetic flux is represented by the Greek letter ϕ (phi), and in the SI system of units it is measured in, or has the unit of, webers (Wb).

Refer to Fig. 1-45. The magnetic flux emerges from the north magnetic pole, goes through the south pole, and returns again to the north pole. In other words, the magnetic flux or the magnetic lines of force are continuous, and thus they form a "*closed loop*." This flux always exists between the poles of a magnet and will arrange itself to conform to the shape of any magnet.

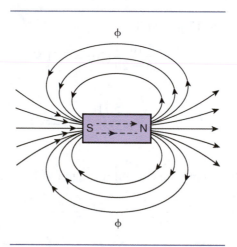

FIG. 1-45 A magnet and its magnetic flux.

Fig. 1-46 shows the tracing of magnetic fields by means of iron fillings.

Fig. 1-47 shows an illustration of earth's magnetic field. The magnetic field intensity of the earth's field is about 0.31 A/m (0.0038 Gauss) at the equator and 0.7 A/m (0.009 Gauss) at the poles.

The purpose of these approximate values is to provide a basis for comparing the strength of magnetic fields.

The flux produced by a coil wound around a magnetic material depends on the properties of the magnetic material, the number of turns in the winding, and

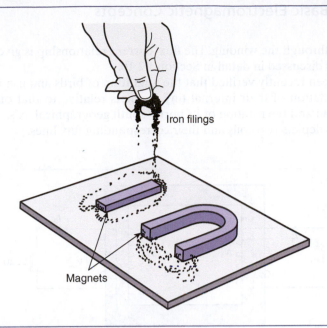

FIG. I-46 Tracing the magnetic fields by means of iron filings. *Based on AvStop, Aviation Maintenance Technician Handbook, http://avstop.com/ac/Aviation_Maintenance_ Technician_Handbook_General/images/fig10-16.jpg*

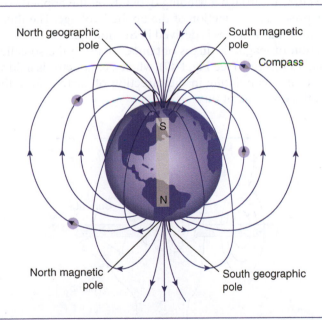

FIG. I-47 Illustration of Earth's magnetic field. The geomagnetic field of the Earth is very similar to that of a large bar magnet placed at the center of the Earth, with its south end oriented toward the north magnetic pole. *Based on EPA, Geomagnetic Field, http://www.epa.gov/esd/cmb/GeophysicsWebsite/pages/reference/properties/Magnetic_ Susceptibility/Geomagnetic_Field.htm*

the current through the winding. The flux–current relationship is given by Ohm's law, which is discussed in detail in Section 1.2.10.

It has been recently verified that the migration of birds and fish is controlled by the orientation of their internal tiny magnets relative to that of the Earth's magnetic field and the rotation of its north–south geographical axis.

Fig. 1-48 depicts two coils and their corresponding flux lines.

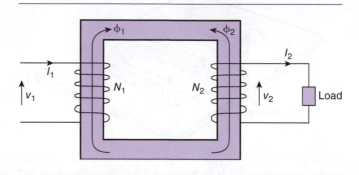

FIG. 1-48 Interaction of magnetic fields.

Since the current through a winding depends on the applied voltage, the flux can also be expressed as a function of the applied voltage. The flux-voltage relationship for sinusoidal voltages is derived in Section 1.2.8.

The direction of magnetic flux is obtained by using the so-called right-hand rule. According to this rule (see Fig. 1-49), when a conductor is held with the right hand and the thumb is pointing in the direction of current flow, then the other fingers curl in the direction of flux.

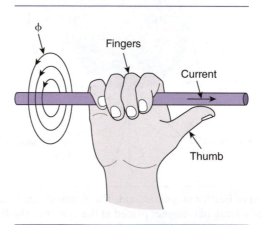

FIG. 1-49 Illustration of the right-hand rule.

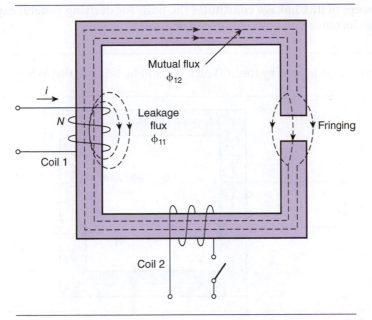

FIG. I-50 An illustration of the concept of mutual flux, leakage flux, and fringing.

Consider coil 1, shown in Fig. 1-50. The flux generated by the coil's current has two components. Component ϕ_{12} links both coils; component ϕ_{11} links only coil 1. ϕ_{12} and ϕ_{11} are called the mutual and leakage flux, respectively. Thus, the total flux (ϕ_1) generated by coil 1 is

$$\phi_1 = \phi_{11} + \phi_{12} \tag{1.99}$$

The spreading out of the mutual flux around the air gap is called fringing. This is further explained in Section 1.2.10.

The product of the winding turns and the flux that links them is called the flux linkage. Flux linkage is generally represented by the Greek letter λ (lambda). Thus,

$$\lambda_{11} = N\phi_{11} \tag{1.100}$$

$$\lambda_{12} = N\phi_{12} \tag{1.101}$$

where λ_{11} and λ_{12} are the leakage and mutual flux linkages of winding 1, respectively.

The total flux linkage (λ_1) of coil 1 is then

$$\lambda_1 = \lambda_{11} + \lambda_{12} \tag{1.102}$$

The concept of flux linkage constitutes the basis for deriving general expressions for the inductance and voltage induced in a coil.

EXAMPLE **1-17** The total flux produced by the 100-turn coil in Fig. 1-51 is 0.005 Wb.

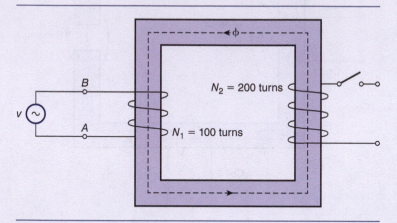

FIG. 1-51

Assuming that the leakage flux is 4% of the total flux, determine:

a. The flux linkage of each winding.
b. The relative polarity of the voltage source.

SOLUTION

a. Designating with λ_1 and λ_2 the flux linkage of coil 1 and coil 2, we have

$$\lambda_1 = N_1 (\phi_{11} + \phi_{12})$$

$$\lambda_2 = N_2 \phi_{12}$$

where ϕ_{11} is the flux that links only coil 1, and ϕ_{12} is the flux that links both coils. Substituting the given values, we obtain

$$\lambda_1 = 100(0.005) = 0.5 \text{ Wb}$$

and

$$\lambda_2 = 200(0.005)\left(1 - \frac{4}{100}\right) = 0.96 \text{ Wb}$$

b. The relative polarity of the voltage supply controls the direction of the current flow, which in turn controls the direction of the flux flow. Using the right-hand rule, we see that the current must flow from terminal A to terminal B in a counterclockwise direction. Thus, the potential of point A is higher than the potential of point B.

1.2.2 Magnetic Flux Density

Magnetic flux density is defined as the ratio of the magnetic flux divided by the area perpendicular to the flux. In mathematical symbols,

$$B = \frac{\phi}{A} \qquad (1.103)$$

where B is the flux density.

The unit of flux density in the SI system of units is the tesla,* abbreviated by the letter T.

Considering incremental changes in the area, from Eq. (1.103) we get

$$\phi = \int B \, dA \qquad (1.104)$$

The flux density of a material is a measure of its magnetization, which reveals the magnetic status of the material.

The nominal magnetization level of an apparatus's magnetic material depends on the design of its magnetic circuit. For example, the flux density of the coils used for communication purposes is maintained at about 0.001 T, while the flux density of power transformers, under normal operating conditions, is about 0.9 T.

The flux density in an air gap can be easily measured by using a Gauss meter. The design of such a meter may be based on the "Hall effect."

According to this principle, the voltage induced in a probe—usually made from a thin semiconductor material—that carries current (i) perpendicular to the magnetic field density (B) is given by

$$v = KiB \qquad (1.105)$$

where K is a constant of proportionality that depends on the physical dimensions of the probe.

When the probe of a Gauss meter that carries a constant current is placed perpendicular to a magnetic field, the induced voltage will be indicated on the scale of the meter. The manufacturer of the meter gives the constants that relate the voltage to the magnetic flux and to the flux density.

---

* 1 tesla = 1 weber/m^2

  = 10^4 gauss

1 gauss = 1 maxwell/cm^2

  = 1 line/cm^2

1 weber = 10^8 lines

1.2.3. Magnetomotive Force

The magnetomotive force (mmf), or the magnetic potential of a coil, is given by the product of the winding's current (i) times its number of turns (N). That is,

$$\text{mmf} = Ni \qquad (1.106)$$

The mmf is also represented by the following symbols:

$$\text{mmf} = \mathcal{F} = U$$

The unit of the mmf in the SI system of units is ampere-turns (A).

When a coil is connected to an electric voltage, as shown in Fig. 1-52, a current will flow, which in turn will produce an mmf.

The mmf, like the current, has an instantaneous, average, and effective value. When the mmf is of sinusoidal waveform, it can be represented by a phasor having the same phase angle as that of the current that produces it.

The mmf may be thought of as the driving force of transformers and electric machines, just as the emf constitutes the driving force in incandescent lamps and electric heaters. The polarity of the mmf is as critical to magnetic circuits as the polarity of the emf is to electric circuits.

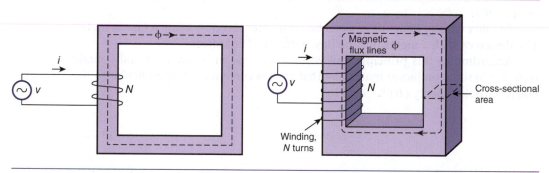

FIG. 1-52 Physical representation of a magnetic circuit. (Leakage flux is neglected.)

EXAMPLE **1-18** When a sinusoidal voltage source of 60 Hz and 120 V is connected to a 50-turn coil, a current of 15 A rms circulates. Assuming a linear magnetic circuit, determine the maximum value and the instantaneous value of the mmf.

SOLUTION

The maximum value of the current is

$$I_m = \sqrt{2}\, I_{\text{rms}}$$
$$= \sqrt{2}\,(15) = 21.21 \text{ A}$$

By definition, therefore, the maximum value of the coil's mmf is

$$\mathscr{F}_{max} = NI_{max} = 50(21.21) = \underline{1060.66 \text{ A}}$$

Since the current is a sinusoidal function, the mmf will also be a sinusoidal function. Thus,

$$\mathscr{F} = \mathscr{F}_m \sin \omega t$$
$$= \underline{1060.66 \sin 377\, t\text{A}}$$

1.2.4 Magnetic Field Intensity

The magnetic field intensity (H) for a given magnetic circuit is given by the ratio of the mmf divided by the length of the mean magnetic path. Thus,

$$\text{magnetic field intensity} = \frac{\text{ampere-turns}}{\text{length of mean magnetic path}}$$

Using mathematical symbols,

$$H = \frac{Ni}{l} \tag{1.107}$$

Considering incremental length changes, we have

$$Ni = \int H\, dl \tag{1.108}$$

The unit of magnetic field intensity is

$$\frac{\text{ampere-turns}}{\text{meter}} \left(\frac{\text{A}}{\text{m}}\right)$$

$$1 \text{ A/m} = 4\pi \times 10^{-3} \text{ oersteds (Oe)}$$

Magnetic field intensity is often referred to as magnetizing force.
   Magnetic field intensity is related to flux density by

$$\mu = \frac{dB}{dH} \tag{1.109}$$

where μ represents the permeability of the material. This parameter is discussed extensively in Section 1.2.6.
   Magnetic field intensity is used in the analysis of magnetic circuits. It is of unique importance in finding the voltage induced in a plane that is located at a given distance from a current-carrying conductor. The general expression for magnetic field intensity can be derived as follows. Consider Fig. 1-53. The conductor carries current of amperes away from the reader into the plane of the paper.

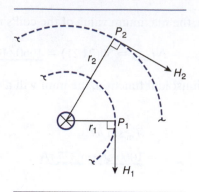

FIG. I-53 The concept of field intensity.

The conductor's current produces a magnetic flux whose direction can be easily obtained from the right-hand rule. The mean magnetic paths of the flux at points P_1 and P_2 are given, respectively, by

$$I_{m1} = 2\pi r_1 \qquad (1.110)$$

and

$$I_{m2} = 2\pi r_2 \qquad (1.111)$$

From Eq. (1.75), the magnetic field intensity at the points P_1 and P_2 is

$$H_1 = \frac{i}{2\pi r_1} \qquad (1.112)$$

$$H_2 = \frac{i}{2\pi r_2} \qquad (1.113)$$

The direction of the vector H_1, as shown in the diagram, is perpendicular to the radius r_1 at the point P_1.

In general, the magnetic field intensity at a point P that is at a radius r from the center of the current-carrying conductor is

$$H = \frac{i}{2\pi r} \qquad (1.114)$$

The voltage induced in a plane is due only to the perpendicular component of the H on the plane under consideration.

Relationship between Electric and Magnetic Field Intensities

Starting from Maxwell's equations, Poiynting showed that in free space the electric field intensity (E) is related to the magnetic field intensity (H) by the following equation.

$$E = nH \qquad (1.115)$$

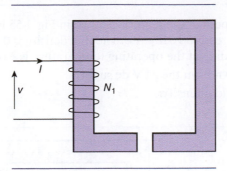

FIG. I-54 Electromagnet.

where n is the impedance of space given by

$$n = \sqrt{\frac{\varepsilon_0}{\mu_0}} \qquad (1.116)$$

Substituting for the values of permittivity and permeability of free space, we obtain

$$n = 377 \text{ ohms} \qquad (1.116a)$$

From above,

$$E = 377H \text{ V/m} \qquad (1.117)$$

In terms of flux density, we obtain

$$E = 377 \frac{B}{\mu} \qquad (1.118)$$

Measuring B in mG, we obtain

$$E = \frac{377}{4\pi \times 10^{-7}} (10^{-3})(10^{-4})B \qquad (1.119)$$

and

$$E = 30 \, B \text{ V/m} \qquad (1.120)$$

where B is in mG. That is,

$$\underline{1 \text{ mG} = 30 \text{ V/m}} \qquad (1.121)$$

Since it is relatively easy and inexpensive to measure B, calibrate such meters to indicate the electric field intensity.

EXAMPLE **1-19** The mean magnetic path of the material shown in Fig. 1-55 is 0.3 m, and its effective cross-sectional area is 5×10^{-4} m². The flux density is 0.007 T, and the dc resistance of the winding at the operating temperature is 2 ohms. Determine:

a. The current drawn from the 24 V dc supply.

b. The magnetic field intensity.

c. The flux.

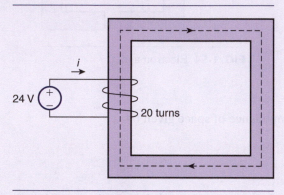

FIG. 1-55

SOLUTION

a. $i = \dfrac{V}{R} = \dfrac{24}{2} = \underline{12\ A}$

b. $H = \dfrac{Ni}{l} = \dfrac{20(12)}{0.3} = \underline{800\ A/m}$

c. $\phi = AB = 5 \times 10^{-4}\ (0.008) = \underline{4 \times 10^{-6}\ Wb}$

Exercise

1-17

The flux through the toroidal magnetic material shown in Fig. 1-56 is 6 μWb, and the magnetic field intensity is 12 A/m. The length of the magnetic path is 0.5 m, and its effective cross-sectional area is 1.5×10^{-4} m². Determine:

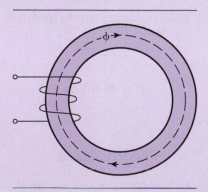

FIG. 1-56

a.   The coil's mmf.
b.   The flux density.

Answer (a) 6.0 A; (b) 0.04 T

1.2.5 The *B-H* Curve and Magnetic Domains

The intrinsic properties of a magnetic material are, to a certain extent, revealed by
the material's *B-H* curve. Typical *B-H* curves for cast iron and cast steel are shown
in Fig. 1-57(a). The flux density (*B*) is the ordinate (*y*-axis) of the curves, and the
field intensity (*H*) is their abscissa (*x*-axis).

Since the flux density is proportional to the voltage (*v*), and the field intensity
is proportional to the current (*i*), then the *B-H* curve of a machine can be represented, as shown in Fig. 1-57(b), by its equivalent *v-i* characteristic curve. The *v-i*
characteristic of a machine is referred to as its "open-circuit" or "magnetization"
characteristic (MC).

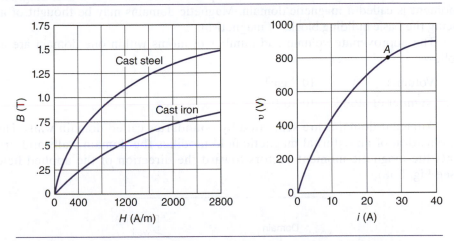

FIG. I-57 Characteristics of magnetic materials: (a) *B-H* curve. (b) Open-circuit
characteristic.

Under normal operating conditions, a machine operates at a flux level that
is very close to point *A*. Point *A* is referred to as the knee* of the magnetization
characteristic. When the parameters of electric machines are measured at a flux
density lower than that of point *A*, the resulting values are referred to as *unsaturated*. When they are measured at a flux density that is higher than that of point *A*,
they are referred to as *saturated* values.

---

* The knee point voltage may be defined as the voltage at which a 10% increase in voltage will result
in a 50% increase in current.

The intrinsic properties of magnetic materials, and thus their particular *B-H* characteristics, depend on the orientation of the materials' magnetic moments.

Magnetic moments are caused by the spinning and orbiting of an atom's electrons. Rotating electrons (electric charges) constitute a flow of current, and thus produce flux and a corresponding magnetic moment whose direction can be determined by using the right-hand rule. The electrons' configuration depends on the temperature and on the chemical composition of the material under consideration.

Magnetocaloric effect refers to a material's temperature increase when it is magnetized and to a temperature decrease when it is demagnetized. This is due to the material's unique orientation of its magnetic domains. Temperature changes up to 5°C have been reported. As such, it replaces and/or supplements some of the standard compressor refrigerant systems. Its advantages are low operating costs and maintenance costs.

At a very high temperature known as the Curie point, magnetic moments cease to exist, and thus the magnetic material loses its magnetic properties. The Curie point for iron is about 775°C.

The strength of magnetic moments is also attenuated by time and ambient conditions. The degradation of a material's magnetic properties as a function of time and temperature is known as the disaccommodation factor.

The smallest region in which a group of atoms has a common magnetic moment is called a magnetic domain. Magnetic domains may be thought of as being the basic building blocks of magnetism.

The approximate volume and number of atoms within one domain are as follows:

Volume: $10^{-8}\,\text{cm}^3$
Number of atoms: 10^9 to 10^{12}

Adjacent domains are separated by boundaries called domain walls. The application of an external magnetic field modifies the domain walls and orients the magnetic moment vectors toward the direction of the applied field (see Fig. 1-58).

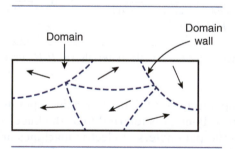

FIG. 1-58 Orientation of magnetic domains.

Figure 1-59(a) is an elementary representation of a magnetic material's random magnetization. For simplicity, only part of the magnetic material is considered. The orientation of the individual magnetic moments is such that no external

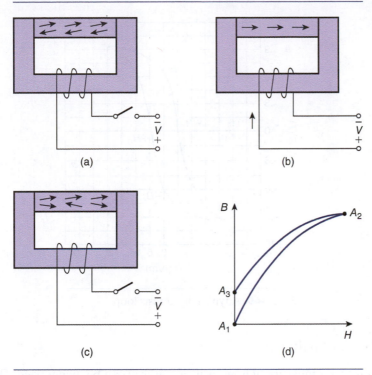

FIG. 1-59 Orientation of magnetic domains and the *B-H* curve:
(a) Opposite orientation. (b) Parallel orientation toward the induced
field. (c) Partial orientation. (d) Corresponding *B-H* curve.

field is produced. This is indicated by point A_1 in the flux density–field intensity
characteristic of the material shown in Fig. 1-59(d).

In Fig. 1-59(b), the magnetic domains are completely oriented in the direction
of the applied field, and the magnetization of the material is indicated by point
A_2 in Fig. 1-59(d).

When the external field is removed, the magnetic domains remain partially
oriented, as shown in Fig. 1-44(c). As a result, the material is said to have residual
magnetism. This corresponds to point A_3 in the *B-H* curve of the material.

Hysteresis Loop

When a material is subjected to a cyclic variation of a magnetizing force, its *B-H*
characteristic at steady state forms a closed loop, commonly referred to as the
hysteresis loop. A typical hysteresis loop for cast iron is shown in Fig. 1-60.

The magnetizing force (H_c) that corresponds to zero flux density is referred
to as the coercive force.

The product of the residual flux density (B_r) times the coercive force (H_c) is used
to designate the quality of permanent magnet material. The higher the $B_r H_c$ product
of a material, the better the material is suited for permanent-magnet applications.

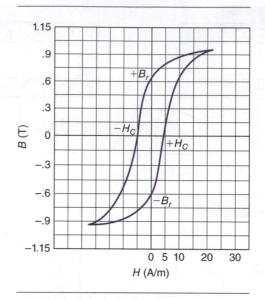

FIG. 1-60 A typical hysteresis loop.

1.2.6 Permeability

Permeability is a measure of how easily a material can be magnetized. The higher the permeability of a material, the easier it is to magnetize.

Magnetizing a material means aligning all microscopic magnetic forces or moments within the material in the direction of the externally applied field. The alignment of the tiny and random inner magnetic moments toward a particular direction results in the magnetization of the material or the development of a stronger external magnetic field.

In other words, permeability is a measure of how easily the magnetic domains within a material can be aligned in the direction of the magnetizing force. Permeability may also be thought of as indicating how much flux density will be produced within a material by the application of a given magnetomotive force. The higher the permeability of a material, the greater the amount of flux density produced in a given value of field intensity.

Permeability is an intrinsic property of all materials and is designated by the Greek letter (μ) (mu). The unit of permeability in the SI system of units is the henry/meter (H/m).

The actual permeability (μ) of a magnetic material is given by the following equation:

$$\mu = \mu_0 \mu_r \tag{1.122}$$

where μ_r is the relative permeability of the material and μ_0 is the permeability of free space.

$$\mu_0 = 4\pi \times 10^{-7} \text{ H/m} \tag{1.123}$$

TABLE 1-2(a) Relative permeability of some materials at low flux densities

Relative Permeability	Material			
	Air	Powdered Iron	Typical Transformer Iron	Permalloy
(μ_r)	1	100	3000	100,000

As shown in Table 1-2(a), the relative permeability of magnetic materials varies over a wide range.

The incremental permeability of magnetic materials is given by the slope of density versus intensity characteristic curve. That is,

$$\text{permeability} = \frac{\text{change in magnetic flux density}}{\text{corresponding change in magnetic field intensity}}$$

or

$$\mu = \frac{dB}{dH} \tag{1.124}$$

Permeability is not a constant parameter but depends on the flux density or on the applied mmf that is used to energize the magnetic circuit, as shown in Fig. 1-61. Thus, when speaking of the relative permeability of a material, one has to specify the flux level at which it corresponds to.

The permeability of a material can also be expressed as a function of a coil's inductance. From the induction principle, we have

$$v = -N\frac{d\phi}{dt} = L\frac{di}{dt} \tag{1.125}$$

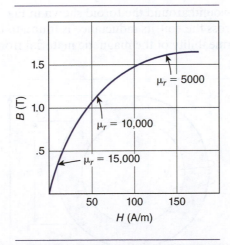

FIG. 1-61 Typical *B-H* characteristic.

From the above, we obtain

$$N\phi = Li \tag{1.126}$$

where L and i are, respectively, the self-inductance and the instantaneous value of the coil's current. Solving for the flux, we get

$$\phi = \frac{Li}{N} \tag{1.127}$$

For operation along the linear section of the B-H curve,

$$\mu = \frac{B}{H} \tag{1.128}$$

From the above, we obtain

$$\mu = \frac{Li}{NA}\left(\frac{l}{Ni}\right) \tag{1.129}$$

or

$$\mu = \left(\frac{l}{N^2A}\right)L \tag{1.130}$$

or

$$\mu = KL \tag{1.131}$$

The constant of proportionality (K) can be calculated from the number of winding turns and the geometry of the magnetic core under consideration. By knowing the constant of proportionality and measuring the inductance of a given coil, the permeability of the magnetic material can be easily established.

EXAMPLE **1-20**   A coil of 50 turns is wound around the toroid shown in Fig. 1-62. When a certain voltage is applied across the coil, its inductance is found to be 200 μH. Determine the relative permeability of the magnetic material from which the toroid is made.

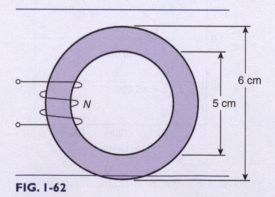

FIG. 1-62

SOLUTION

The cross-sectional area perpendicular to the flux is

$$A = \frac{\pi}{4}(0.5 \times 10^{-2})^2 = 0.196 \times 10^{-4}\ \text{m}^2$$

The mean magnetic path is

$$l = 2\pi r_m = 2\pi(0.0275) = 0.17\ \text{m}$$

From Eq. (1.130), the permeability is

$$\mu = \frac{Ll}{N^2 A} = \frac{(200 \times 10^{-6})(0.17)}{(50)^2(0.196 \times 10^{-4})} = 0.70 \times 10^{-3}\ \text{mH/m}$$

Thus, the relative permeability is

$$\mu_r = \frac{\mu}{\mu_0} = \frac{0.70 \times 10^{-3}}{4\pi \times 10^{-7}} = \underline{560.23}$$

REMARKS The example would be more illuminating if the test voltage were specified. After all, the value of the permeability is meaningful if and only if the corresponding magnetization level is known.

The coil shown in Fig. 1-63 has, at the operating temperature, a resistance of 4 ohms and draws 12 amperes from a 60 Hz, 120 V supply. The material's mean magnetic path is 0.75 m, and its cross-sectional area is uniform at $80 \times 10^{-4}\ \text{m}^2$.

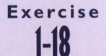

Exercise

1-18

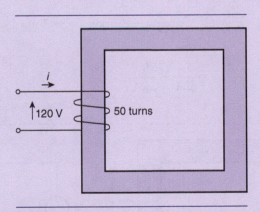

FIG. I-63

Determine:

a. The relative permeability of the material.
b. The flux and its direction.
c. The power factor of the coil.

Answer (a) 725.5; (b) 5.83 mWb cw; (c) 0.4 lagging.

1.2.7 Reluctance

The opposition to the flux in a given magnetic circuit is termed reluctance ($\mathcal{R}$). As in the case of the resistance of electric circuits, it is given by the ratio of the applied magnetic potential divided by the flux through the magnetic path under consideration. Mathematically,

$$\mathcal{R} = \frac{NI}{\phi}\left(\frac{\text{A}}{\text{Wb}}\right) \tag{1.132}$$

The previous expression, as will be explained in Section 1.2.10, constitutes Ohm's law in magnetic circuits. The reluctance of a magnetic path, being analogous to the resistance of an electric circuit, is also given by the following relationship:

$$\mathcal{R} = \frac{\ell}{\mu A} \tag{1.133}$$

where ℓ is the length of the mean magnetic path, μ is the permeability of the material, and A is the material's cross-sectional area perpendicular to the flux.

Reluctance, like electric resistance, varies with the impurity and temperature of the materials. Unsaturated magnetic materials have a relatively low reluctance because their permeability is thousands of times greater than that of the air.

The equivalent series or parallel reluctances of a magnetic circuit are calculated in the same way as the equivalent series or parallel resistances of an electric circuit. For example, the total reluctance of the magnetic circuit shown in Fig. 1-64(b) is the sum of the reluctances of the magnetic material and of the air gap. That is,

$$\mathcal{R}_{\text{total}} = \mathcal{R}_{\substack{\text{magnetic} \\ \text{material}}} + \mathcal{R}_{\text{air gap}} = \mathcal{R}_m + \mathcal{R}_g \tag{1.134}$$

$$= \frac{l_m}{\mu A} + \frac{l_g}{\mu_0 A} \tag{1.135}$$

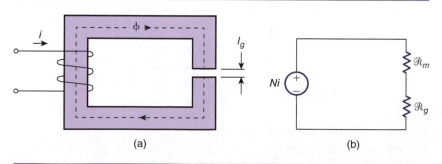

(a)   (b)

FIG. 1-64 Electromagnet. (a) Physical representation. (b) Equivalent magnetic circuit.

Thus,

$$\mathcal{R}_{total} = \frac{1}{\mu_0 A}\left(\frac{l_m}{\mu_r} + l_g\right) \tag{1.136}$$

As seen from Eq. 1.136, the relative effect of the air gap on the magnetic circuit depends on how the length of the air gap compares to the ratio l_m/μ_r of the magnetic material.

The closed-loop property of magnetic flux and the concept of reluctance will be demonstrated by the following practical illustrations.

A Laboratory Case

In a laboratory, the arrangement shown in Fig. 1-65(a) was used to measure the magnetic properties of E-type laminations as a function of temperature. Although the applied voltage to the coil was gradually increased to maximum, and the temperature of laminations was varied over a wide range, no reading could be obtained in the measuring bridge.

The mistake was that the setup used did not provide a path of low resistance to allow for the existence of flux. Owing to its large reluctance, air presents very high opposition to the flow of magnetic flux.

The problem was solved by rearranging the laminations as shown in Fig. 1-65(b). In this case, magnetic materials were placed in the air path of the flux, and so the magnetic resistance was reduced a thousandfold. As a result, the quantity of flux through the coil increased. The measuring bridge then began to give readings that varied according to the ambient temperature and the applied voltage.

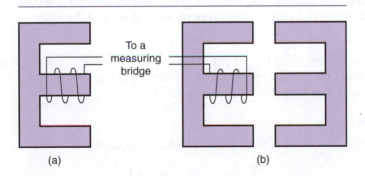

(a)   (b)

FIG. 1-65 Illustrating the closed-loop property of magnetic flux: (a) Incorrect setup for measuring properties of magnetic materials. (b) Correct setup.

Demagnetizing the British Fleet

Refer to Fig. 1-65(c). During the first phase of the Second World War, the German air force dropped magnetic mines in the navigated rivers and coastal waters of England. When a ship moved above a magnetic mine, the flux of its residual magnetism—a property of steel structure, closed its loop by penetrating through the mine. The movement of the ship was accompanied by a change of the flux

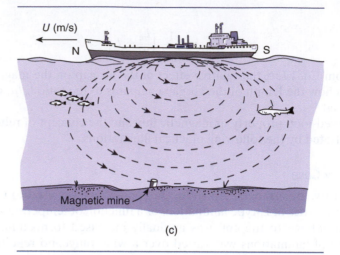

(c)

FIG. 1-65 (c) The induction principle and the triggering of a magnetic mine.

through the magnetic mine, and thus a voltage was induced that triggered an explosion. In all, 37 allied ships were destroyed.

The problem was solved by demagnetizing, or *degaussing*, the British Fleet. That is, flux of opposite direction and equal in strength to that produced by the ship's residual magnetism was passed through the ship's structure. As a result, the ship could not effectively emit any flux, and thus the magnetic mines remained untriggered. The neutralizing flux was generated by passing a specific amount of dc current through loops of wire properly wound around the ship.

EXAMPLE **1-21**

Determine the length of the air gap (Fig. 1-64(a)) that will make its reluctance equal to that of the magnetic material. Assume that the cross-sectional areas are the same and that $l_m = 0.60$ m and $\mu_r = 1200$.

SOLUTION

Rewriting the expression for the circuit reluctance, we have

$$\mathcal{R} = \frac{1}{\mu_0 A}\left(\frac{l_m}{\mu_r} + l_g\right)$$

In the above equation, the reluctance of the air gap will be equal to the reluctance of the magnetic material, provided that

$$\frac{l_m}{\mu_r} = l_g$$

Thus,

$$l_g = \frac{0.60}{1200} = \underline{0.5\ \text{mm}}$$

Understanding the "degaussing of the British Fleet" requires knowledge of four basic concepts or principles of magnetic circuits. Give the descriptive mathematical relationships and the equivalent equations from the electric circuits.

Exercise

1-19

1.2.8 The Concept of Generated Voltage

A voltage can be induced in a conductor either by a time-varying magnetic field or as a result of a conductor's motion in a stationary magnetic field. The first type is referred to as *transformer voltage*, the second as *speed voltage*.

The instantaneous value of each of these voltages will be designated, for the purpose of minimizing the symbol usage, by the same letter (v).

Transformer Voltage

Refer to Fig. 1-66. When a conductor is placed in a time-varying magnetic field, a voltage is induced in it. This voltage is commonly known as the transformer voltage. This voltage, first observed by Faraday, is given by

$$v = -N\frac{d\phi}{dt} \tag{1.137}$$

where v is the instantaneous value of the voltage induced in the conductor, N is the number of the conductor's turns, and $d\phi/dt$ is the time-rate of change of the magnetic flux. As per Lenz's law, the minus sign in the equation indicates that the voltage induced is of a polarity that will produce a current whose flux opposes the flux that induced the voltage. The polarity of the induced voltage is further explained in the section on magnetically coupled circuits (see Section 1.2.14).

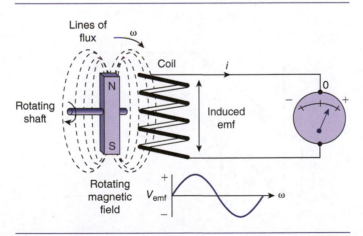

FIG. 1-66 Rotating magnetic field and induced emf. *Based on Basic Electronics Tutorials, http://www.electronics-tutorials.ws/electromagnetism/mag25.gif*

Equation (1.137) can also be written as

$$v = -\frac{d\lambda}{dt} \tag{1.138}$$

where λ is the conductor's flux linkage.

Equations (1.137) and (1.138) both describe what is often referred to as the *induction principle*.

EXAMPLE **1-22** a. Derive the expression for the voltage induced per unit length in the telephone lines shown in Fig. 1-67(a). The single-conductor power line runs parallel to the telephone lines and carries a current given by

$$i = I_m \sin \omega t \ \text{A}$$

b. Evaluate the induced voltage, given that

$$I_m = 1000 \ \text{A}$$

$$\omega = 377 \ \text{rad/s}$$

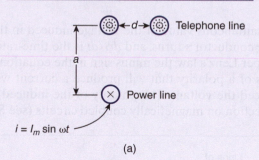

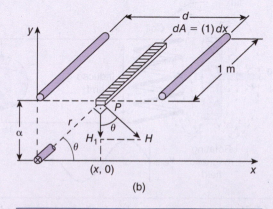

FIG. 1-67

$$\mu_0 = 4\pi \times 10^{-7} \text{ H/m}$$

$$\alpha = 3 \text{ m}$$

$$d = 0.5 \text{ m}$$

SOLUTION

a. A voltage has meaning if it is specified between two points. In this example, the requirement is to find the voltage induced per unit length in the telephone lines. Rewriting the induction principle equation, we have

$$v = -N\frac{d\phi}{dt} \tag{I}$$

Thus the essential part of the problem is how to find the flux through the area between the telephone lines.

By definition,

$$\phi = \int B \, (dA) \tag{II}$$

where dA is an increment of area perpendicular to the flux. (The induced voltage depends on the component of the magnetic field perpendicular to the area between the telephone wires.)

For linear variation of flux density, we have

$$B = \mu H \tag{III}$$

Also,

$$H = \frac{Ni}{l_m} \tag{IV}$$

The number of turns is equal to unity, and the mean magnetic path at point P, shown in Fig. 1-67(b), is

$$l_m = 2\pi r \tag{V}$$

Thus,

$$H = \frac{i}{2\pi r}$$

Its component (H_1), perpendicular to the area under consideration, is

$$H_1 = H \cos\theta = \left(\frac{i}{2\pi r}\right)\cos\theta \tag{VI}$$

From the geometry of Fig. 1-67(b), we obtain

$$\cos\theta = \frac{x}{r} = \frac{x}{\sqrt{\alpha^2 + x^2}} \tag{VII}$$

From Eq. (VI) and (VII), we get

$$H_1 = \frac{i}{2\pi}\left(\frac{x}{\alpha^2 + x^2}\right) \tag{VIII}$$

Then Eq. (III) and (VIII) give

$$B = \mu\frac{i}{2\pi}\left(\frac{x}{\alpha^2 + x^2}\right) \tag{IX}$$

Considering the shaded strip, we have

$$dA = (1)\, dx \tag{X}$$

The corresponding incremental flux through this element of area is

$$d\phi = d(BA) = B\, dx \tag{XI}$$

Substituting Eq. (IX) into Eq. (XI), we obtain

$$d\phi = \mu\left[\frac{i}{2\pi}\left(\frac{x}{\alpha^2 + x^2}\right)dx\right] \tag{XII}$$

Taking the integrals, we find:

$$\phi = \mu\frac{i}{2\pi}\int_0^d \frac{x\, dx}{\alpha^2 + x^2}$$

The variation of x or the limits of integration are from $x = 0$ to $x = d$. After integrating, we obtain

$$\phi = \mu\frac{i}{2\pi}\left(\frac{1}{2}\right)\ln\left(\alpha^2 + x^2\right)\Big|_0^d$$

$$\phi = \frac{\mu}{4\pi}i\ln\left(\frac{\alpha^2 + d^2}{\alpha^2}\right)$$

Substituting for the current, we get

$$\phi = \frac{\mu}{4\pi}\ln\left(\frac{\alpha^2 + d^2}{\alpha^2}\right)I_m\sin\omega t$$

Taking the derivative of the flux with respect to time, and substituting the result into Eq. (I), we obtain

$$v = -\frac{\mu}{4\pi}\ln\left(\frac{\alpha^2 + d^2}{\alpha^2}\right)I_m\,\omega\cos\omega t \text{ V/m}$$

b. Substituting the given values, we have

$$v = -\frac{4\pi \times 10^{-7}}{4\pi}\ln\left(\frac{9 + 0.25}{9}\right)1000(377)\cos 377t$$

$$= \underline{-1.03\cos 377t \text{ mV/m}}$$

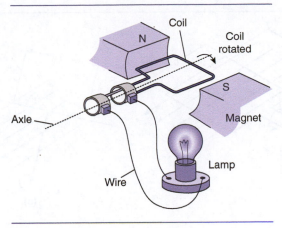

FIG. 1-68 Generation of voltage.

Speed Voltage

The voltage induced in a conductor that moves through a stationary magnetic field (Fig. 1-68) is referred to as the motional or speed voltage.

Speed voltage (v) is given by the following equation:

$$v = \int (U \times B)\, dl \qquad (1.139)$$

where $U \times B$ is the cross product of the velocity and flux density vectors and dl, is a length element along the conductor.

When $U, B,$ and l are perpendicular to each other, then Eq. (1.139) is written as

$$v = BUl \qquad (1.140)$$

where $B, U,$ and l are the magnitudes of the flux density, velocity, and length of the conductor, respectively.

EXAMPLE **1-23**

Figure 1-69(a) shows one coil turn of an ac generator that rotates at ω rad/s through a constant magnetic field of B tesla. Determine the voltage generated at the terminals of the generator if its coil has N turns connected in series.

SOLUTION

Figure 1-69(b) shows the position of the conductor after it is rotated θ radians with respect to its position at $t = 0$. It should be noted that in the two sides of the conductor whose individual length is α meters, a voltage is induced, while in the other two sides of the conductor whose individual length is b meters, no voltage is induced. The area swept by each b-side of the conductor is parallel to the direction of the magnetic flux density; thus there is, in effect, no voltage induced in these two sides of the conductor.

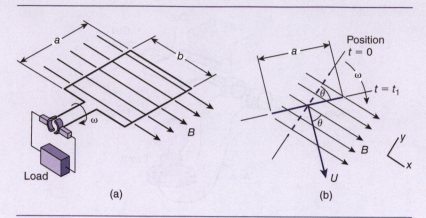

(a)   (b)

FIG. 1-69

The magnitude of the tangential component of the linear velocity is

$$U = \frac{b}{2}\omega \text{ m/s}$$

In vector form,

$$\vec{U} = \frac{b}{2}\omega(\cos\theta\,\tilde{\alpha}_x - \sin\theta\,\tilde{\alpha}_y)$$

where $\tilde{\alpha}_x$ and $\tilde{\alpha}_y$ are the unit vectors in the x and y direction, respectively.

The flux density vector is

$$\vec{B} = B\tilde{\alpha}_x$$

The cross product of the velocity and flux density is

$$\vec{U} \times \vec{B} = \begin{vmatrix} \tilde{\alpha}_x & \tilde{\alpha}_y & \tilde{\alpha}_z \\ \omega\frac{b}{2}\cos\theta & -\omega\frac{b}{2}\sin\theta & 0 \\ B & 0 & 0 \end{vmatrix}$$

$$= \frac{b}{2}\omega\tilde{\alpha}_z \begin{vmatrix} \cos\theta & -\sin\theta \\ B & 0 \end{vmatrix}$$

$$= \frac{b}{2}\omega B \sin\theta\,\tilde{\alpha}_z$$

Substituting the above into Eq. (1.139), we obtain

$$v = \frac{b}{2}\omega B \sin\theta \int dl$$

The effective length of the conductor is $2a$ m. Thus,

$$v = abB\omega \sin\theta$$

Since $\theta = \omega t$ and the generator has N turns, the instantaneous value of the voltage generated is

$$v = \underline{abBN\omega} \sin \omega t \; \text{V}$$

For the ac generator of Example 1-23, determine the power furnished by the prime mover if the generator's efficiency is 90% and the resistance of the load at operating temperature is R ohms.

Answer $\dfrac{0.56}{R}(\alpha bBN\omega)^2$

Voltage–Flux Relationship

When the coil of a magnetic circuit is connected to an ac voltage source, a current circulates. As a result, flux is produced. The equation that describes the flux generated as a function of the coil's voltage is derived as follows.

Refer to the magnetic circuit in Fig. 1-70. If losses are neglected, the voltage induced is equal to the applied voltage (v_1). Assuming sinusoidal waveforms, we have

$$v_1 = V_m \sin \omega t \tag{1.141}$$

Equating the supply voltage and the voltage induced, we obtain

$$V_m \sin \omega t = -N\frac{d\phi}{dt} \tag{1.142}$$

from which

$$d\phi = -\frac{1}{N} V_m \sin \omega t \; dt \tag{1.143}$$

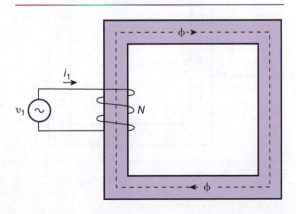

FIG. I-70 Magnetic circuit.

Integrating, we get

$$\phi = \frac{V_m}{\omega N} \cos \omega t \qquad \textbf{(1.144)}$$

The flux, then, is a sinusoidal function whose maximum value is given by

$$\phi_m = \frac{V_m}{\omega N} \qquad \textbf{(1.145)}$$

From the above,

$$V_m = 2\pi f N \phi_m \qquad \textbf{(1.146)}$$

or

$$\sqrt{2}V = 2\pi f N \phi_m \qquad \textbf{(1.147)}$$

where V is the rms value of the applied voltage. From the last equation, we obtain

$$V = 4.44 N f \phi_m \qquad \textbf{(1.148)}$$

Equation (1.148) relates magnetic flux to applied electric volts and is of fundamental importance to the understanding of transformers and electric machines. This equation is often used to calculate the flux within a toroidal magnetic core.

Exercise

1-21

When the 40-turn coil of Fig. 1-71 is connected to the 120 V, 60 Hz voltage supply, the voltages induced in the other coils are as shown. By neglecting leakage flux and winding losses, determine:

a. The number of turns of each coil.
b. The rms value of the mutual flux.

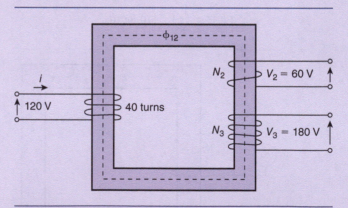

FIG. 1-71

Answer (a) 20 and 60 turns; (b) 7.96 mWb

1.2.9 Energy Content of Magnetic Materials

The energy content of a magnetic material depends on its level of magnetization, its physical size, and its intrinsic characteristics.

Refer to Fig. 1-72. The incremental change of its electrical input energy (dw) is

$$dw = vi \, dt \qquad (1.149)$$

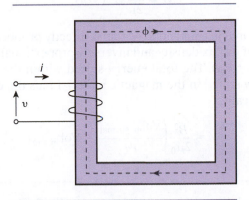

FIG. I-72

where v is the voltage induced on its coil and i is the instantaneous value of the coil's current.

The magnitude of the voltage induced is

$$v = N\frac{d\phi}{dt} = NA\frac{dB}{dt} \qquad (1.150)$$

and

$$H = \frac{Ni}{l} \qquad (1.151)$$

or

$$i = H\frac{l}{N} \qquad (1.152)$$

From Eqs. (1.149), (1.150), and (1.152) we obtain

$$dw = (Al)H \, dB \qquad (1.153)$$

Recognizing that the product Al has the unit of volume, and assuming that the magnetic material is energized from zero up to B tesla, the energy stored per given volume of magnetic material is

$$\frac{W}{\text{volume}} = \int_0^B H \, dB \qquad (1.154)$$

Assuming an operation along the linear part of the magnetization characteristic, we obtain

$$\frac{W}{\text{volume}} = \frac{1}{\mu} \int_0^B B \, dB = \frac{1}{2\mu} B^2 \qquad (1.155)$$

The energy stored in a magnetic material is

$$W = \frac{1}{2\mu} B^2 \, (\text{Vol}) \qquad (1.156)$$

Thus, the energy stored in a magnetic system is directly proportional to its volume and to the square of its flux density and inversely proportional to its permeability.

Consider Fig. 1-64(a). The total energy stored within the magnetic circuit is equal to the energy stored in the magnetic material plus the energy stored in its air gap. Mathematically,

$$W = \frac{B^2}{2\mu_0} \left(\frac{\text{Vol}_{\text{magn. mat.}}}{\mu_r} + \text{Vol}_{\text{air gap}} \right) \qquad (1.157)$$

EXAMPLE 1-24

The relative permeability of the material shown in Fig. 1-64(a) is 1200. The mean length of the magnetic material is 0.75 m, and that of the air gap is 0.75 cm. The coil has 60 turns and draws 8 A from the 120 V ac supply. The volume of the magnetic material is 6×10^{-3} m³, and the volume of the air gap is 60×10^{-6} m³. Determine the energy stored:

a. In the magnetic material.
b. In the air gap.

SOLUTION

The magnetic flux density is calculated as follows:

$$B = \frac{\phi}{A}$$

$$= \frac{Ni}{\mathcal{R}_T A}$$

$$= \frac{Ni}{A \left(\dfrac{l_m}{\mu_m A_m} + \dfrac{l_g}{\mu_g A_g} \right)}$$

Disregarding fringing, we obtain

$$B = \frac{Ni\mu_0}{\dfrac{l_m}{\mu_r} + l_g}$$

Substituting the given values, we get

$$B = \frac{60(8)(4\pi \times 10^{-7})}{\dfrac{0.75}{1200} + 0.0075}$$

$$= 74.24 \times 10^{-3}\,\text{T}$$

a. The energy stored in the magnetic material is

$$W_m = \frac{B^2}{2\mu}\,\text{Vol}_{(m)} = \frac{(74.24 \times 10^{-3})^2}{2 \times 1200(4\pi \times 10^{-7})}\,(6 \times 10^{-3})$$

$$= \underline{10.96\,\text{mJ}}$$

b. The energy stored in the air gap is

$$W_g = \frac{(74.24 \times 10^{-3})^2}{2(4\pi \times 10^{-7})}\,(60 \times 10^{-6})$$

$$= \underline{131.6\,\text{mJ}}$$

Comparing the results, we see that the energy stored in the air gap is about 12 times greater than the energy stored in the magnetic material.

1.2.10 Principles of Magnetic Circuits

The basic principles of magnetic circuits are Ohm's law and Kirchhoff's laws. These laws are analogous to those of the electric circuits that are identified by the same name.

Ohm's Law

The flux of magnetic circuits is analogous to the current of electric circuits. The flux within a simple magnetic circuit is given by the ratio of the circuit's magnetic potential divided by the reluctance of the magnetic path. Mathematically,

$$\phi = \frac{Ni}{\mathscr{R}} \tag{1.158}$$

This equation is Ohm's law for magnetic circuits. In a complex electromagnetic system, the numerator of the equation should include the algebraic sum of all mmf's in the particular loop under consideration, and the denominator should include the sum of the loop's reluctances.

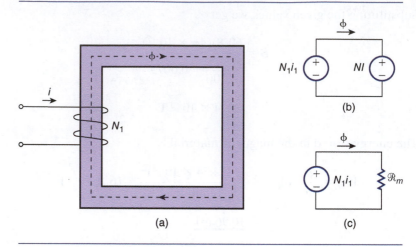

FIG. 1-73 Magnetic circuit. (a) Elementary physical representation. (b) Equivalent circuit in terms of mmf's. (c) Equivalent circuit in terms of reluctance.

An ideal magnetic circuit (Fig. 1-73(a)) can be represented in terms of either the circuit mmf (Fig. 1-73(b)) or the circuit reluctance (Fig. 1-73(c)). The applied mmf is N_1i_1, and that absorbed by the magnetic material is NI. The latter is, as per Ohm's law, equal to ϕR_m, where ϕ and R_m are, respectively, the flux and the reluctance of the magnetic circuit.

The magnetic circuit, unlike the electric circuit, cannot be easily shorted because, as the flux increases above the knee of the magnetization curve, the permeability of the magnetic material decreases, and thus the reluctance of the magnetic path increases. Hence, there is a limit to the quantity of flux that can exist in a magnetic circuit.

The intrinsic properties of any given magnetic material are such as to limit the quantity of flux within its structure.

Exercise

1-22

The magnetic circuit shown in Fig. 1-73(a) has a mean magnetic path of 0.5 m, a relative permeability of 2000, and a uniform cross-sectional area equal to $5 \times 10^{-4}\,\text{m}^2$. The coil has 100 turns and, when connected to a 120 V, 60 Hz voltage source, draws 5 A. Determine:

a. The inductance of the coil.
b. The maximum flux through the magnetic material.

Answer (a) 25.13 mH; (b) 1.78 mWb

Kirchhoff's Laws

Voltage Law

The KVL of magnetic circuits is similar to that of electric circuits. That is, the sum of the mmf's along a closed loop is equal to zero. Mathematically,

$$\Sigma \, (Ni)_{\text{loop}} = 0 \qquad\qquad \textbf{(1.159)}$$

This equation is analogous to Eq. (1.69), according to which the sum of the emf's along a closed loop in an electric circuit is equal to zero.

Current Law

According to Kirchhoff's current law, the sum of the magnetic flux in a junction is equal to zero. Mathematically,

$$\Sigma = \phi_{\text{junction}} = 0 \qquad\qquad \textbf{(1.160)}$$

That is, the sum of the flux entering a magnetic junction is equal to the sum of the flux leaving the junction.

　　For magnetic circuit calculations, take into consideration the leakage flux, the stacking factor, and the fringing of the flux through the circuit's air gaps.

Stacking Factor

The effective cross-sectional area of a magnetic core is less than the actual cross-sectional area because of the insulation on the surface of the laminations. The reduction in the actual area is accounted for by using what is called a stacking factor. The value of the stacking factor ranges between 0.85 and 0.95, depending on the thickness of the laminations and on the amount of their surface insulation.

Fringing

The spreading out of the flux around the edges of an air gap (Fig. 1-50) is called fringing, or diverging of the flux. In calculating the flux density through a short air gap, fringing is accounted for by increasing the given cross-sectional area of the air gap as shown below.

$$\text{Effective cross-sectional area} = (m + l_g)(n + l_g) \qquad\qquad \textbf{(1.161)}$$

where m and n are the dimensions of the magnetic core perpendicular to the flux lines and l_g is the length of the air gap.

Leakage Flux

The leakage flux is relatively small, owing to its large reluctance path. Thus, it is normally neglected in magnetic circuit calculations. In the analysis of transformers and electric machines, however, the effects of the leakage flux are included in their equivalent electric circuits.

EXAMPLE **1-25**   In the circuit shown in Fig. 1-74(a), l_g and l_m represent, respectively, the length of the air gap and the total length of the mean magnetic path. Determine:

a.   An expression for the total reluctance.

b.   The length of the air gap that will absorb one-half of the applied mmf.

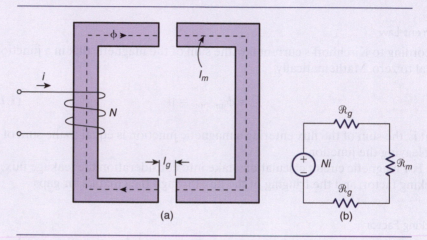

FIG. 1-74

SOLUTION

The equivalent circuit is shown in Fig. 1-74(b).

a.   The total reluctance is the sum of the reluctance of the magnetic material and that of the air gap. Thus,

$$\mathscr{R}_T = \mathscr{R}_m + \mathscr{R}_{\text{air gap}} = \frac{l_m}{\mu A} + \frac{2l_g}{\mu_0 A}$$

$$= \frac{1}{\mu_0 A}\left(\frac{l_m}{\mu_r} + 2l_g\right)$$

where l_g is the length of each air gap.

b.   From the equivalent circuit, the reluctances are in series. For one-half of the input mmf to be absorbed by the air gap, the reluctances of the magnetic material should be equal to that of the air gap. Thus,

$$\frac{l_m}{\mu} = \frac{2l_g}{\mu_0}$$

From which

$$l_g = \frac{l_m}{2\mu_r}$$

For the schematic shown in Fig. 1-75(a), draw the equivalent circuit in terms of: EXAMPLE 1-26

a.   mmf's

b.   Reluctances.

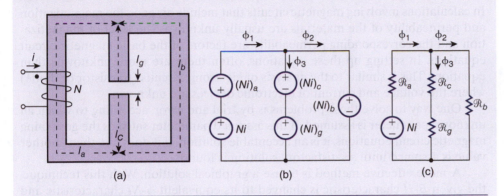

FIG. 1-75

SOLUTION

a.   The equivalent circuit, in terms of mmf's, is shown in Fig. 1-75(b). The mmf's in the air gap and in the other magnetic paths are identified with the same subscripts as the mean magnetic lengths.

Applying KVL, we have

$$Ni = (Ni)_a + (Ni)_c + (Ni)_g$$

or

$$Ni = (Ni)_a + (Ni)_b$$

It should be noted that, as in electric circuits, magnetic circuits and magnetic potentials across parallel branches of reluctances are equal.

b.   The equivalent circuit with the various reluctances is shown in Fig. 1-75(c). Applying Ohm's law for magnetic circuits, we obtain

$$\phi_1 = \frac{Ni}{\mathcal{R}_a + \mathcal{R}_b /\!/ (\mathcal{R}_c + \mathcal{R}_g)}$$

Applying the current-divider concept, we find

$$\phi_3 = \phi_1 \left(\frac{\mathcal{R}_b}{\mathcal{R}_b + \mathcal{R}_c + \mathcal{R}_g} \right)$$

and

$$\phi_2 = \phi_1 \left(\frac{\mathcal{R}_c + \mathcal{R}_g}{\mathcal{R}_b + \mathcal{R}_c + \mathcal{R}_g} \right)$$

Graphical Analysis of Electromagnetic Circuits

In calculations involving magnetic circuits that include air gaps, the magnetization and permeability of the materials are usually unknown. The level of magnetization and the corresponding permeability are factors of the basic magnetic circuit equations. In setting up these equations, often there are more unknowns than equations. This is similar to the analysis of electronic circuits (transistor or diode) where the voltage and current of electronic devices are unknown.

One way to solve these problems is by trial and error, according to which an unknown parameter is assumed. If this assumed parameter satisfies the governing magnetic circuit equations, it is an acceptable solution. If it does not, then another value is assumed until a satisfactory solution is found.

A more effective method is to use a graphical solution. With this technique, the given *B-H* characteristic is changed to its equivalent ϕ-*Ni* characteristic and the flux of the magnetic circuit is found as a *function* of the magnetomotive forces within a closed loop.

The resulting equation is drawn on the same coordinate system as the ϕ-*Ni* characteristic. The point of intersection of the two characteristics gives the actual flux and the magnetic potential drop within the magnetic material.

For a demonstration, consider the simple magnetic circuit of Fig. 1-76(a). Its *B-H* characteristic is shown in Fig. 1-76(b). The lengths of the magnetic material (l_m) and of the air gap (l_g) as well as their effective areas $(A_m$ and $A_g)$ are known. In order to find the flux through the magnetic material, proceed as follows:

Steps

1. *Draw the ϕ-$(Ni)_m$ characteristic.*

    Change the coordinates of the *B-H* characteristic to its equivalent ϕ-$(Ni)_m$ characteristic ($\phi = BA$ and $(Ni)_m = H_m \times l_m$).

2. *Write the governing equations* (Fig. 1-76(c)).

$$(Ni)_s = (Ni)_m + (Ni)_g \tag{1.162}$$

$$(Ni)_g = \phi \mathcal{R}_g \tag{1.163}$$

where $(Ni)_s$, $(Ni)_m$, and $(Ni)_g$ represent the magnetic potentials of source, magnetic material, and air gap, respectively.

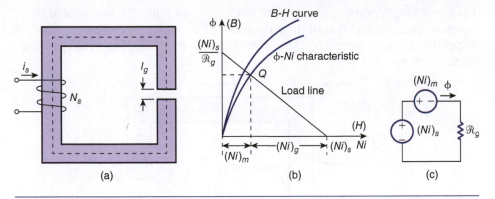

FIG. 1-76 Graphical solution of magnetic circuits: (a) Simple magnetic circuit. (b) *B-H* and ϕ-*Ni* characteristic. (c) Equivalent circuit.

3. *Draw the load-line equation.*
    From Eqs. (1.162) and (1.163) we obtain

    $$(Ni)_s = (Ni)_m + \phi\mathcal{R}_g \qquad\qquad (1.164)$$

    from which

    $$\phi = \frac{(Ni)_m}{\mathcal{R}_g} + \frac{(Ni)_s}{\mathcal{R}_g} \qquad\qquad (1.165)$$

    Equation (1.165) is of the slope (m)-intercept (b) form $(y = mx + b)$. Its intercepts are

    $$y\text{-intercept} \quad \phi = \frac{(Ni)_s}{\mathcal{R}_g}, \qquad x\text{-intercept} = (Ni)_s$$

    Since the intercepts are known, the equation is drawn as shown in Fig. 1-76(b). Equation (1.165) is known as the load-line equation. Its intersection with the ϕ-*Ni* characteristic gives the operating point, which is designated by the letter *Q*. The operating point gives the actual flux, and the mmf drop within the magnetic material and through the air gap, as shown in Fig. 1-76(b).

The coordinates of the *B-H* characteristic of the magnetic material shown in Fig. 1-77 are as follows:

Exercise

1-23

B (T)	0.10	0.45	0.78	0.96	1.08
H (A/m)	50	150	350	800	1440

The mean length of the magnetic path is 0.75 mm, and the effective area of the magnetic material and that of the air gap are 1.1×10^{-3} m^2 and 1.2×10^{-3} m^2, respectively. For an air-gap length of 0.60 mm, determine the flux density in the magnetic material.

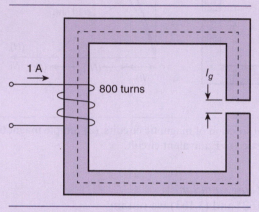

FIG. I-77

Answer 0.90 T

Exercise 1-24

The *B-H* characteristic of the magnetic material shown in Fig. 1-78 is the same as that of Exercise 1-23. Determine the flux through the coil, given that

$$l_{bad} = 0.64 \text{ m}, \qquad l_{bd} = 0.2 \text{ m}, \qquad l_{bcd} = 0.4 \text{ m}$$

$$l_g = 1.0 \text{ mm}, \qquad \text{stacking factor} = 0.95$$

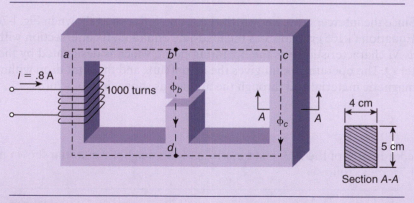

FIG. I-78

Answer 1.94 mWb

1.2.11 Magnetic Losses

Magnetic losses can be separated into eddy-current losses and hysteresis losses. The total magnetic losses (P_m) of a coil are given by

$$P_m = P_e + P_h \qquad (1.166)$$

where P_e and P_h represent, respectively, eddy-current and hysteresis losses. Eddy currents are produced within the magnetic material because of the alternating nature of the flux, which in turn is produced by the winding's ac current. When the magnetic material has infinite resistance (zero conductivity), however, there will not be any eddy-current losses.

Hysteresis represents the friction among the walls of the magnetic domains. The magnetic domains are constantly being reoriented because of the alternating nature of the applied field.

Magnetic losses are also referred to as excitation losses or open-circuit losses. They are discussed in more detail in Section 2.1.3.

Eddy-Current Losses

The voltage applied to a coil wound around a toroidal magnetic material will be accompanied, as can be predicted from the induction principle, by a flux. Mathematically,

$$v = -N\frac{d\phi}{dt} \qquad (1.167)$$

For operations along the linear section of the magnetization characteristic, the flux, according to Ohm's law for magnetic circuits, is proportional to the current. That is, for a linear change in the flux there will be a linear change in the current. Using mathematical symbols:

$$\phi \propto i \qquad (1.168)$$

According to Lenz's law, the flux within the magnetic material will induce opposing voltages that produce circular currents (I_e, eddy currents). As shown in Fig. 1-79, these induced currents flow around the flux and produce eddy-current loss. The eddy-current loss is proportional to eddy current squared. That is,

$$P_e \propto I_e^2 \qquad (1.169)$$

In terms of the applied voltage, the last expression becomes

$$P_e = K_1 V^2 \qquad (1.170)$$

where V is the effective value of the applied voltage and K_1 is a constant of proportionality. This constant is directly proportional to the volume of the material and inversely proportional to its resistivity.

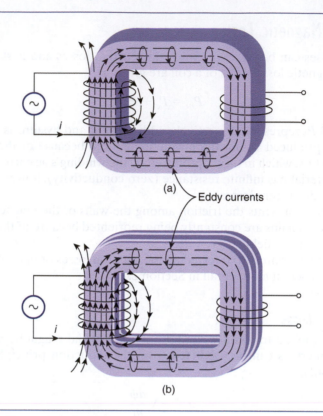

(a)

Eddy currents

(b)

FIG. 1-79 An illustration of eddy currents.
(a) Nonlaminated core. (b) Laminated core (notice the relative reduction in the eddy currents).

Substituting for the voltage its equivalent expression from Eq. (1.148), we obtain

$$P_e = K_2 \, (f\phi)^2 \tag{1.171}$$

The constant of proportionality K_2 is directly proportional to K_1 and to the square of the number of turns of the coil.

Eddy-current losses can be reduced by increasing the resistance to the flow of eddy currents and/or by decreasing the magnitude of the flux.

Resistance to the flow of eddy currents (R_e) can be increased by decreasing the thickness of the laminations and by increasing their surface insulation.

For a given coil voltage, the flux can be reduced by increasing the number of winding turns.

Hysteresis Losses

Hysteresis losses are heat losses due to the friction between magnetic domains. Each time the input voltage changes polarity, the magnetic field within the core of the coil also changes polarity. Thus the flux—or the lines of force—change direction. The consequent continuous alignment and realignment of the magnetic domains in the direction of the alternating flux produce friction, which manifests itself as heat within the material.

Hysteresis loss (P_h) is related to the flux of the core and to the frequency (f) of the voltage source by

$$P_h \propto f\phi^h \qquad (1.172)$$

where h is a constant characteristic of the material. Its usual range is from 1.4 to 2.5.

From Eqs. (1.148) and (1.172), we obtain

$$P_h \propto f\left(\frac{V}{f}\right)^h \qquad (1.173)$$

From the above,

$$P_h = K_3 \frac{V^h}{f^{h-1}} \qquad (1.174)$$

The constant of proportionality K_3 is directly proportional to the volume of the magnetic material under consideration.

Etymologically, hysteresis is a Greek word that means lagging. In fact, owing to hysteresis, the induced emf is not at 90° to the flux but lags it by (90° + θ_h), where θ_h is due to the hysteresis phenomenon (see Fig. 1-80). This deviation in the phase angle is used to advantage in the production of torque in so-called hysteresis motors.

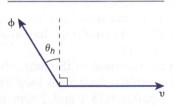

FIG. 1-80 Effect of hysteresis on the phase angle between flux and emf.

The hysteresis losses of a magnetic material can also be evaluated from the area of its hysteresis loop as follows (see Fig. 1-81):

$$\text{area of hysteresis loop} = \int H \, dB \qquad (1.175)$$

Substituting the values

$$H = \frac{Ni}{l} \quad \text{and} \quad dB = \frac{d\phi}{A} = -\frac{v}{AN} \, dt$$

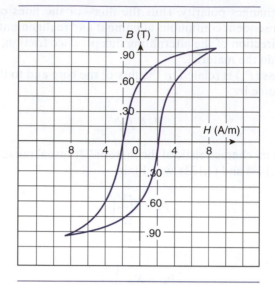

FIG 1-81 Hysteresis loop.

we obtain

$$\text{area of hysteresis loop} = -\frac{1}{Al}\int vi\,dt \qquad \textbf{(1.176)}$$

As seen from the above, the area of a hysteresis loop is equal to the energy dissipated in the material per cycle per unit volume due to the hysteresis phenomenon.

An alternative method of evaluating the hysteresis losses of a material subjected to a cyclic variation of its magnetizing force is to measure the area of the hysteresis loop in proper units and to multiply that area by the period of the magnetizing force expressed in seconds.

For example, if the cyclic variation of the magnetizing force is 60 times per second and the area of the resulting hysteresis loop has 25 squares, the perpendicular sides of which are equal to 0.15 T and 2 A/m, respectively, the hysteresis losses are

$$25(0.15)(2)(60) = 450 \text{ W/m}^3$$

EXAMPLE **1-27**

a. Figure 1-82 shows a section of an underground, three-phase, three-single-conductor, 4.16 kV power distribution network. The contractor mistakenly installed *steel* instead of *plastic* elbows as adaptors. As a result, the cables were damaged during the plant's operation, owing to induction currents. Explain the underlying principles and the nature of magnetic losses.

b. Prove that no induction currents are induced in steel conduits through which one three-conductor cable is installed.

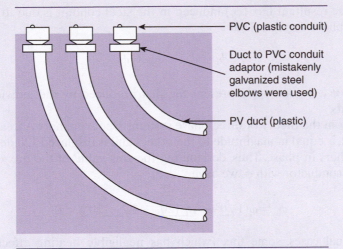

PVC (plastic conduit)

Duct to PVC conduit
adaptor (mistakenly
galvanized steel
elbows were used)

PV duct (plastic)

FIG. I-82 Three-phase, single-conductor, power distribution.

*As per industrial practices the PVC conduit is used for exposed cable
installations (such as going from the secondary transformer terminals to the
underground cable distribution), while the PV duct is used only in concealed
underground cable distribution. Both the conduit and the duct are made from
plastic material, but the conduit is much more rugged than the duct.*

SOLUTION

a. Owing to Ohm's law in magnetic circuits ($\phi = Ni/\mathcal{R}$), the conductor's current
    produced flux, which according to the "right-hand rule," encircled the area
    outside the conductor's current. As a result, flux was induced in the galva-
    nized steel elbows.

    The sinusoidal variation of the steel elbow's flux produced eddy currents
    and continuous realignment of the material's magnetic domains. According
    to Lenz's law, eddy currents represent the natural reaction of the material to
    oppose the induced flux. The realignment of the magnetic domains is a natu-
    ral tendency of the basic building blocks of magnetism to follow the direction
    of the external field.

    Eddy currents produced eddy-current losses ($P_e \propto V^2$) and the alter-
    nating realignment of the magnetic domains produced hysteresis losses
    ($P_h \propto f\phi^{1.6}$). This resulted in overheating of the steel elbows, which led to
    the melting of the conductor's insulation, generation of sparks, and electri-
    cal shorting of the multiconductor distribution. Thus, owing to the induction
    principle, the cables were damaged and the plant's production was interrupt-
    ed for the time necessary to identify the problem, remove the steel elbows,
    and purchase and install new cables.

b. Induction currents are not induced in a steel conduit through which a three-
    conductor cable (phases A, B, and C) passes because the net induced flux is
    theoretically equal to zero. The proof of this is as follows:

The resultant flux (ϕ_r) induced in the steel conduit is due to the flux of three-phase currents. That is,

$$\phi_r = \phi_A + \phi_B + \phi_C$$

where ϕ_A, ϕ_B, and ϕ_C represent the flux produced by the individual phase currents.

From the theory of three-phase systems (see Appendix A), each current, or flux, is equal in magnitude to the others and is displaced 120 degrees from the others in phase. Thus, designating the rms value of the flux of the one-phase conductor with ϕ, we have

$$\phi_r = \phi(1\ \underline{/0^\circ} + 1\ \underline{/120^\circ} + 1\ \underline{/-120^\circ}) = 0$$

In actual cases, some flux (which has negligible heating effects) will be induced, owing to imperfect symmetry in the phase conductor's spacing relative to the steel conduit.

This industrial problem occurred in a factory in the Southwestern United States in August 1981. Damage was estimated at several hundred thousand dollars. It is astonishing that such a faulty installation took place, considering that (according to internal company memos) the consultants who designed the electrical distribution system, as well as the plant's electrical engineers, were very much aware of the induction problems associated with single-conductor distribution. What is more, the construction manager who supervised the installation was an experienced mechanical engineer.

EXAMPLE **1-28** A firm that manufactures transformers requires magnetic materials that have a relative permeability equal to 100. A salesman of magnetic materials claims and demonstrates in the lab that his company's product has a permeability equal to 100. After testing the proposed material, however, the manufacturing firm's engineer rejects it. Give two possible technical reasons for his decision.

SOLUTION

The permeability of a material has meaning only when the flux density at which it is measured is specified. As shown in Fig. 1-61, each magnetic material has a wide range of permeabilities. The engineer may have rejected the materials because, at the *level of flux density* he or she was interested in, the permeability was not 100.

A second reason may have been that the *losses* of the proposed materials were too high. Additional reasons for the engineer's rejection of the proposed material may have been the unacceptable change in its permeability with *time* (disaccommodation factor) at high *temperatures*. The *cost* of the material is also of paramount importance.

This example demonstrates the various interrelated factors that tend to be considered when a decision is to be made in selecting and/or designing a product. The phrasing of the specifications, or the description of the problem, is also important. In the above example, the manufacturer's requirements were not explicitly stated; an experienced salesperson should have known the implied restrictions when dealing with the properties and characteristics of magnetic materials.

1.2.12 Equivalent Circuit of a Coil

A coil and its equivalent electric circuit are shown in Figs. 1-83(a) and (b), respectively. The impedance $R_1 + jX_1$ is referred to as the coil's leakage impedance, and the impedance $R_m + jX_m$ is the coil's magnetizing impedance. Leakage impedance is usually a small fraction of magnetizing impedance.

The resistance R_1 represents the dc resistance of the winding. As can be seen from Eq. (1.20), the winding resistance depends on the resistivity of the material, the length of the conductor, and its cross-sectional area. Resistivity depends on the conductor's chemical composition and temperature. The cross-sectional area of a conductor plays an important role in controlling the losses. There are, however, physical and economic restraints that determine how large a conductor's diameter can be.

The dc resistance per unit length of any gauge of wire is readily available in published tables.

The leakage reactance X_1 is due, as the name implies, to the coil's leakage flux.

The resistance R_m represents the equivalent electrical resistance of the magnetic material and depends on the material's eddy-current and hysteresis losses.

The reactance X_m is referred to as the magnetizing reactance of the coil. This reactance is due to the flux within the magnetic material, which is directly proportional to the applied voltage, as can be seen from Eq. (1.148).

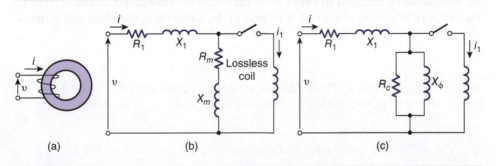

(a)   (b)   (c)

FIG. 1-83 Electromagnet. (a) Elementary representation. (b) Its equivalent electrical circuit. (c) Equivalent circuit of (b).

The power consumed by R_1 constitutes the *electrical losses* of the coil; the power consumed by R_m represents the *magnetic losses* of the iron core. Electrical losses are often referred to as copper losses, and magnetic losses as core, or iron, losses.

The coil's power consumption manifests itself as heat, which, if not promptly removed, may cause insulation breakdown and fire.

Winding losses are discussed further in Section 2.1.3.

The magnetizing impedance Z_m can also be represented by a resistance (R_c) connected in parallel to a reactance (X_ϕ), as shown in Fig. 1-83(c).

Equating the two equivalent impedances, we obtain

$$R_m + jX_m = \frac{jR_cX_\phi}{R_c + jX_\phi} \tag{1.177}$$

or

$$R_m + jX_m = \frac{jR_cX_\phi(R_c - jX_\phi)}{R_c^2 + X_\phi^2} \tag{1.178}$$

The equivalence of series and parallel representation of a coil's impedance (Eqs. 1.179 and 1.180) can be derived by using the common parameters (P, V, and I) to the input of a coil (see Example 2-2).

$$R_c = \frac{R_m^2 + X_m^2}{R_m} \tag{1.179}$$

and

$$X_\phi = \frac{R_m^2 + X_m^2}{X_m} \tag{1.180}$$

The current through R_c provides the magnetic losses, while the current through X_ϕ magnetizes the magnetic material. The series representation of the magnetizing impedance of a coil is simpler to obtain than its parallel representation. The latter, however, reveals more about the magnetization process and for this reason is more common.

The lossless coil of Fig. 1-83(b) represents the capability of the winding to be magnetically coupled to other coils that may be connected to the system of Fig. 1-83(a). When no other coil is wound in the same core, the ideal coil is open-circuited and thus $i_1 = 0$.

Exercise 1-25

When connected to a 120 V, 60 Hz power supply, a coil draws 4.8 A and consumes 100 W of power. If the dc resistance of the coil is 2 ohms, determine:

a. The magnetic losses.
b. The equivalent series electrical resistance of the magnetic material.

Answer (a) 53.92 W; (b) 2.34 Ω

1.2.13 Mathematical Relationships of Self- and Mutual Inductances

The self-inductance of a single coil is given by the following relationships:

$$\text{self-inductance} = \frac{\text{flux linkage of the coil}}{\text{current through the coil}}$$

$$= \frac{(\text{number of turns})(\text{quantity of flux})}{\text{current through the coil}} \qquad \textbf{(1.181)}$$

Using mathematical symbols, we obtain

$$L = \frac{\lambda}{i} = \frac{N\phi}{i} \qquad \textbf{(1.182)}$$

For an electromagnetic system with two coils, such as the one shown in Fig. 1-84(a), the flux linkage of each coil is defined as follows:

$\lambda_{11} = N_1\phi_{11} = $ flux linkage of coil 1 produced by its own current i_1

$\lambda_{12} = N_1\phi_{12} = $ flux linkage of coil 1 produced by the current i_2 of coil 2

$\lambda_{22} = N_2\phi_{22} = $ flux linkage of coil 2 produced by its own current i_2

$\lambda_{21} = N_2\phi_{21} = $ flux linkage of coil 2 produced by the current i_1 of coil 1

The total flux linkages of coil 1 (λ_1) and coil 2 (λ_2) are given by

$$\lambda_1 = N_1\phi_1 = N_1(\phi_{11} + \phi_{12}) \qquad \textbf{(1.183)}$$

$$\lambda_2 = N_2\phi_2 = N_2(\phi_{22} + \phi_{21}) \qquad \textbf{(1.184)}$$

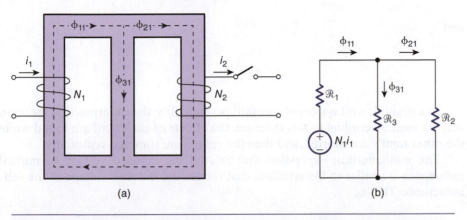

(a)   (b)

FIG. 1-84 Magnetic system with negligible leakage flux. (a) Magnetic coupled coils ($i_2 = 0$). (b) Its equivalent magnetic circuit.

The self- and mutual inductances are defined as follows:

$$L_{11} = \frac{\lambda_{11}}{i_1} = \frac{N_1\phi_{11}}{i_1} = \text{self-inductance of coil 1}$$

$$L_{12} = \frac{\lambda_{12}}{i_2} = \frac{N_1\phi_{12}}{i_2} = \text{mutual inductance of coil 1 due to current in coil 2}$$

$$L_{22} = \frac{\lambda_{22}}{i_2} = \frac{N_2\phi_{22}}{i_2} = \text{self-inductance of coil 2}$$

$$L_{21} = \frac{\lambda_{21}}{i_1} = \frac{N_2\phi_{21}}{i_1} = \text{mutual inductance of coil 2 due to current in coil 1}$$

Physically, the mutual inductance (L_{12}) of coil 1, with respect to coil 2, is the flux linkage that links coil 1 due to the current in coil 2.

In a two-subscript parameter, the first subscript identifies the affected coil, while the second identifies the coil that produced this effect.

From the outlined mathematical relationships, it is evident that, for the determination of self- and mutual inductances of an electromagnet, the equivalent magnetic circuit must be drawn, and then basic principles must be used to obtain ϕ_{11}, ϕ_{12}, and ϕ_{21}.

Applying Ohm's law in Fig. 1-84(b), for example, we obtain

$$\phi_{11} = \frac{N_1 i_1}{\mathcal{R}_1 + \mathcal{R}_2 /\!/ \mathcal{R}_3}$$

Using the current-divider technique,

$$\phi_{21} = \phi_{11} \frac{\mathcal{R}_3}{\mathcal{R}_2 + \mathcal{R}_3}$$

and

$$\phi_{31} = \phi_{11} \frac{\mathcal{R}_2}{\mathcal{R}_2 + \mathcal{R}_3}$$

In a multiple-coil system of constant permeability, the superposition theorem may be used. According to this theorem, the effects of each mmf are found while the other mmf's are shorted, and then the results are summed vectorially.

The mathematical expression that related the induced voltage to mutual inductance is similar to the equation that relates the induced voltage to the self-inductance. That is,

$$v_1 = \frac{d}{dt}(L_{12}i_2) \tag{1.185}$$

For constant mutual inductance, we have

$$v_1 = L_{12} \frac{d}{dt} i_2 \qquad \textbf{(1.186)}$$

where v_1 is the voltage induced due to mutual inductance L_{12}.

Coefficient of Coupling

The coefficient of coupling is a factor that indicates the degree of magnetic coupling between two coils. It is always less than 1. The coefficient of coupling K is given by

$$K = \frac{L_{12}}{\sqrt{L_{11}L_{22}}} \qquad \textbf{(1.187)}$$

In general, the higher the coefficient of coupling, the lower the leakage flux of the coils.

Using the data shown in Fig. 1-85, determine:

a. The flux linkage of coil 1.
b. The inductance, as seen from the terminals of coil 1.
c. The mutual inductance of coil 2, with respect to coil 1.

EXAMPLE 1-29

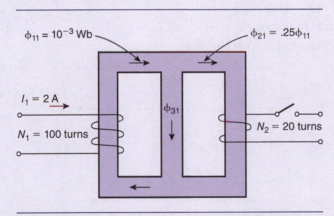

$\phi_{11} = 10^{-3}$ Wb $\phi_{21} = .25\phi_{11}$

$I_1 = 2$ A

ϕ_{31}

$N_1 = 100$ turns $N_2 = 20$ turns

FIG. I-85

SOLUTION

a. The flux linkage for coil 1 is

$$\lambda_{11} = N_1\phi_{11} = 100(10^{-3}) = \underline{0.1 \text{ weber-turn}}$$

b.   The inductance of coil 1 is

$$L_{11} = \frac{\lambda_{11}}{I_1} = \frac{0.1}{2} = \underline{50 \text{ mH}}$$

c.   The mutual inductance of coil 2 with respect to coil 1 is

$$L_{21} = N_2 \frac{\phi_{21}}{I_1} = \frac{20(0.25 \times 10^{-3})}{2} = \underline{2.5 \text{ mH}}$$

EXAMPLE **1-30**   Laboratory measurements at 60 hertz for the two coils shown in Fig. 1-86 gave the following results.

Test	Coil Condition		Voltage in Volts	Current in Amperes, rms
1	Coil 1	connected to a voltage source	80	1.5
	Coil 2	open-circuited	30	0
2	Coil 2	connected to a voltage source	60	1
	Coil 1	open-circuited	20	0

Neglecting the resistance of the coils, determine:

a.   The self-inductance of each coil.
b.   The mutual inductances.
c.   The coefficient of mutual coupling.

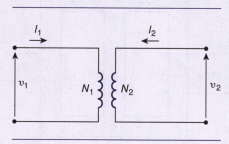

FIG. 1-86

SOLUTION

a.   The self-inductance of coil 1 can be found from the following basic relationship:

$$v_1 = L_{11} \frac{di_i}{dt}$$

Using rms values and considering magnitudes only,

$$V_1 = L_{11}\omega I_1$$

Thus,

$$L_{11} = \frac{80}{2\pi(60)(1.5)} = \underline{0.14\ \text{H}}$$

Similarly,

$$L_{22} = \frac{60}{2\pi 60(1)} = \underline{0.16\ \text{H}}$$

Alternatively, from Eq. (1.182), using rms values, we obtain

$$L = \frac{N\phi}{I}$$

Since the current is given in rms, the flux must also be expressed in rms. Then from the above and Eq. (1.148), the self-inductance of coil 1 is

$$L_{11} = \frac{N}{I}\left(\frac{\phi_m}{\sqrt{2}}\right) = \frac{N}{I\sqrt{2}}\left(\frac{V}{4.44Nf}\right)$$

Thus,

$$L_{11} = \frac{80}{1.5\ \sqrt{2}\ (4.44 \times 60)} = \underline{0.14\ \text{H}}$$

b. The mutual inductance of coil 1 with respect to coil 2 is found as follows:

$$v_1 = L_{12}\frac{d}{dt}i_2$$

Using rms values,

$$V_1 = L_{12}\omega I_2$$

From the above,

$$L_{12} = \frac{V_1}{\omega I_2} = \frac{20}{377(1)} = \underline{53.05\ \text{mH}}$$

Similarly,

$$L_{21} = \frac{V_2}{\omega I_1} = \frac{30}{377(1.5)} = \underline{53.05\ \text{mH}}$$

c. The coefficient of mutual coupling is

$$K = \frac{L_{12}}{\sqrt{L_{11}L_{22}}} = \frac{53.05 \times 10^{-3}}{\sqrt{0.14 \times (0.16)}} = \underline{0.354}$$

EXAMPLE **1-31** For coil 1 of the electromagnetic system shown in Fig. 1-87(a), neglect fringing, leakage flux, and the mmf within the magnetic material, then:

1. Determine:
   a. The flux linkage.
   b. The self-inductance.
   c. The mutual inductance.
2. If the current in coil 2 is reversed, repeat 1.

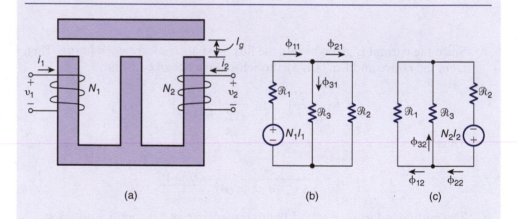

(a)   (b)   (c)

FIG. I-87

SOLUTION

1. a. *Flux linkage.* By definition, the flux linkage of coil 1 is

$$\lambda_1 = \lambda_{11} + \lambda_{12}$$
$$= N_1\phi_{11} + N_1\phi_{12} \tag{I}$$

The term $N_1\phi_{11}$ represents the flux linkage of coil 1 due to its own current. To find $N_1\phi_{11}$, discard the current in coil 2 and draw the equivalent circuit, as shown in Fig. 1-87(b). Thus,

$$\lambda_{11} = N_1\phi_{11} = N_1\left(\frac{N_1I_1}{\mathcal{R}}\right) \tag{II}$$

The reluctance (as seen by the source of mmf $N_1 I_1$) is

$$\mathscr{R} = \mathscr{R}_1 + \mathscr{R}_2 \, // \, \mathscr{R}_3$$

Since the air gaps have the same lengths, the reluctances $\mathscr{R}_1$, $\mathscr{R}_2$, and $\mathscr{R}_3$ are equal. Thus,

$$\mathscr{R} = \mathscr{R}_1 + \frac{\mathscr{R}_1}{2} = \frac{3}{2}\mathscr{R}_1 = \frac{3}{2}\left(\frac{l_g}{\mu_0 A}\right) \qquad \textbf{(III)}$$

From Eq. (II) and (III), we obtain:

$$N_1 \phi_{11} = N_1^2 I_1 \frac{2\mu_0 A}{3 l_g} \qquad \textbf{(IV)}$$

The term $N_1 \phi_{12}$ of relationship (I) represents the **flux linkage** of coil 1 due to the current in coil 2. To find the effects of the second magnetic voltage source, apply the superposition principle and draw the equivalent circuit as shown in Fig. 1-87(c).

From the equivalent circuit,

$$N_1 \phi_{12} = N_1 \phi_{22}\left(\frac{\mathscr{R}_3}{\mathscr{R}_1 + \mathscr{R}_3}\right) = N_1 \frac{\phi_{22}}{2}$$

$$N_1 \phi_{12} = \left(\frac{N_1}{2}\right)\frac{N_2 I_2}{\mathscr{R}_2 + \mathscr{R}_1 \, // \, \mathscr{R}_3} = \frac{N_1 N_2 I_2}{2\left(\frac{3}{2}\mathscr{R}_1\right)}$$

$$= N_1 N_2 I_2 \frac{\mu_0 A}{3 l_g} \qquad \textbf{(V)}$$

The total flux linkage of coil 1 is obtained by adding Eq. (IV) and (V). Thus,

$$\lambda_1 = \frac{\mu_0 A}{3 l_g}\left(2 N_1^2 I_1 + N_1 N_2 I_2\right) \qquad \textbf{(VI)}$$

b) *Self-inductance.* By definition, the self-inductance of coil 1 is

$$L_{11} = \frac{\lambda_{11}}{I_1} = \frac{N_1 \phi_{11}}{I_1}$$

From the last relationship and from Eq. (IV) we obtain

$$L_{11} = N_1^2 \frac{2}{3}\frac{\mu_0 A}{l_g} \qquad \textbf{(VII)}$$

c.   *Mutual Inductance.* The mutual inductance of coil 1, with respect to coil 2, is

$$L_{12} = \frac{\lambda_{12}}{i_2} = \frac{N_1 \phi_{12}}{i_2}$$

Using rms values, from Eq. (V) and the last relationship, we obtain

$$L_{12} = N_1 N_2 \frac{\mu_0 A}{3 l_g} \qquad \text{(VIII)}$$

2.   *Reversing current I_2*

a.   If the current in coil 2 changes direction, its flux will oppose the flux produced by the current in coil 1. Thus,

$$\lambda_1 = \frac{\mu_0 A}{3 l_g} \left(2 N_1^2 I_1 - N_1 N_2 I_2 \right) \qquad \text{(IX)}$$

b. and c.   The self- and mutual inductances are scalar quantities, independent of current direction.

Exercise

1-26

Figure 1-88 shows an electromagnet and its equivalent magnetic circuit. Prove that

$$\phi_{12} = \frac{N_2 I_2}{\mathcal{R}_2 + \mathcal{R}_1 \, /\!/ \, \mathcal{R}_3} \left(\frac{\mathcal{R}_3}{\mathcal{R}_1 + \mathcal{R}_3} \right)$$

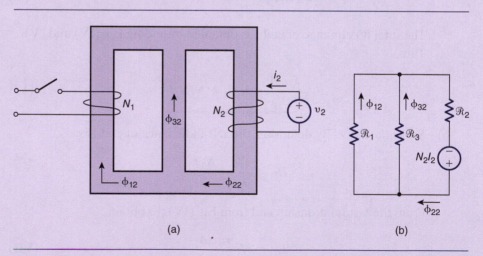

(a)

(b)

FIG. 1-88

Figure 1-89(b) shows the idealized *B-H* characteristic of the magnetic material shown in Fig. 1-89(a). For a flux density of 1.4 *T*, determine:

a.   The relative permeability of the magnetic material.

b.   The inductance of the coil.

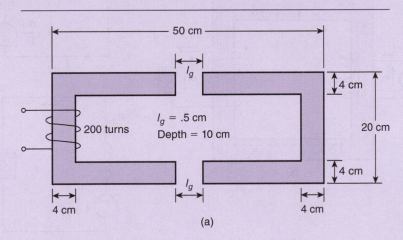

(a)

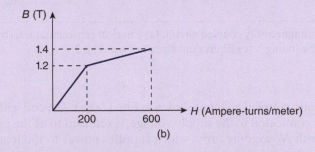

(b)

FIG. 1-89

Answer (a) 397.9;   (b) 15.36 mH

1.2.14 Polarity and Equivalent Circuits of Magnetically Coupled Coils

Understanding magnetically coupled windings is very important because it provides the basis for understanding the operation of transformers and electric machines. As will be explained in the chapters that follow, these devices have two or more magnetically coupled windings. The primary concern here is the polarity of the induced voltages and the equivalent circuit of each winding.

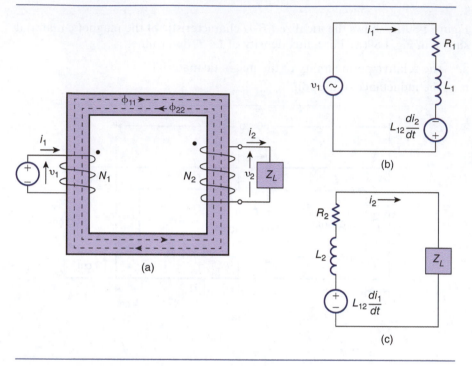

FIG. I-90 A magnetically coupled circuit. (a) Physical representation. (b) Equivalent circuit of winding 1. (c) Equivalent circuit of winding 2.

Polarity

Two magnetically coupled coils are shown in Fig. 1-90(a). The coil with N_1 winding turns, which is connected to the supply voltage, is referred to as the primary winding. The coil with N_2 winding turns, which supplies power to the load impedance Z_L, is the secondary winding.

The polarity of the voltage induced in a winding is indicated by a dot or by the positive $(+)$ and negative $(-)$ signs. A dotted terminal is at higher potential than an undotted terminal.

If the polarity of the induced voltage is known, the direction of current flow and that of the flux lines can easily be established.*

The significance of the dotted terminals is that the fluxes produced by the currents in the two coils are in opposition when the current in one coil enters the dotted terminal while the current in the other coil leaves the dotted terminal.

In Fig. 1-90(a), notice that, although the upper terminals of both windings are positive and the polarity of the voltage v_1, v_2 is the same, the currents flow in opposite directions. The current in the primary winding enters the dotted terminal because of

---

*When energy is drawn from a voltage source, the current comes out of its positive terminal. If a source is absorbing energy, current enters its positive terminal first.

the polarity of the input voltage, while the current in the secondary winding flows out of the dotted terminal because of the polarity of the induced voltage (v_2).

Equivalent Circuit

Assuming negligible magnetic losses, the equivalent circuits of coil 1 and coil 2 are shown in Fig. 1-90(b) and (c), respectively. The resistance and self-inductance of each coil are as shown.

From basic definitions, the voltages induced in the primary winding due to the flux of the secondary, and vice versa, are given by the following equations:

$$V_{12} = \frac{d}{dt}(L_{12}i_2)$$

$$= \frac{d}{dt}(N_1\phi_{12}) \tag{1.188}$$

and

$$V_{21} = \frac{d}{dt}(L_{21}i_1)$$

$$= \frac{d}{dt}(N_2\phi_{21}) \tag{1.189}$$

Assuming equal mutual inductances, we have

$$V_{12} = L_{12}\frac{d}{dt}i_2 \tag{1.190}$$

and

$$V_{21} = L_{12}\frac{d}{dt}i_1 \tag{1.191}$$

Using rms values, we may write the above equations in their phasor form as follows:

$$V_{12} = j\omega L_{12}I_2 \tag{1.192}$$

$$V_{21} = j\omega L_{12}I_1 \tag{1.193}$$

where ω is the angular frequency of the supply voltage.

As shown in the equivalent circuits, the polarities of these voltages are opposite to the polarities of the voltage drops across the leakage impedances because of Lenz's law. The equivalent circuits of magnetically coupled circuits whose core losses are not negligible are discussed in Chapter 2, Section 2.1.3.

EXAMPLE **1-32**   Figure 1-91(a) shows two magnetically coupled coils (coil 1 and coil 2). When the coils are connected in series cumulatively (same sense):

a.   Mark their polarity and show the direction of the flux.

b.   Prove that the total equivalent inductance is given by the following relationship:

$$L_{\text{total}} = L_{11} + L_{22} + 2L_{12}$$

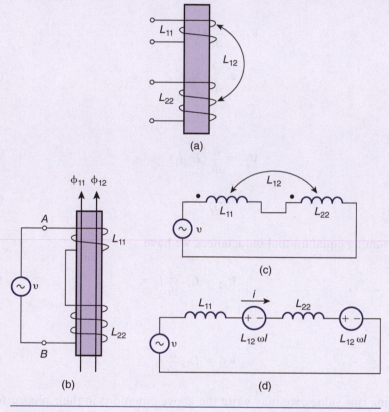

FIG. 1-91

SOLUTION

a.   Cumulative connection of the coils means that each coil's flux is in the same direction as the other's. The coils are said to be of the same sense and are represented as shown in Fig. 1-91(b).

   Since the flux of coil 2 aids, or is in the same direction as, the flux of coil 1, the schematic with the dotted terminals will be as shown in Fig. 1-91(c). The corresponding equivalent circuit is shown in Fig. 1-91(d).

b. Neglecting resistances from the equivalent circuit of Fig. 1-91(d), at steady state, we have

$$V = Ij\omega(L_{11} + L_{22} + 2L_{12})$$

or

$$V = I(j\omega L_T)$$

Thus,

$$L_T = L_{11} + L_{22} + 2L_{12}$$

Exercise

1-28

Refer to Fig. 1-91(a). When the coupled coils are connected in series differentially (opposite sense):

a. Mark their polarity and show the direction of flux.
b. Prove that their total equivalent inductance is given by the following equation:

$$L_{total} = L_{11} + L_{22} - 2L_{12}$$

1.2.15 Force, Energy, and Torque

In this section, general expressions are derived for the following: the force exerted on a current-carrying conductor when it is placed in a magnetic field; the force exerted by an electromagnet; the energy stored in independently excited coils; and the torque developed by rotating transducers.

Force Exerted on a Current-Carrying Conductor When It Is Placed in a Magnetic Field

Ampere's experiments on the forces between current-carrying loops of thin wire led to the formulation of Eq. (1.194), which gives the force exerted on a current-carrying conductor when it is placed in a magnetic field.

$$\vec{f} = \vec{il} \times \vec{B} \tag{1.194}$$

where f is the force, in newtons, exerted on a conductor whose length is l meters and which carries a current of i amperes in a magnetic field whose flux density is B tesla. The factors il and B are vectors.

In the rectangular coordinate system, these vectors may have components in the x, y, and z directions. That is,

$$\vec{il} = (il)_x \tilde{a}_x + (il)_y \tilde{a}_y + (il)_z \tilde{a}_z \tag{1.195}$$

and

$$\vec{B} = B_x \tilde{a}_x + B_y \tilde{a}_y + B_z \tilde{a}_z \tag{1.196}$$

where $\tilde{a}_x, \tilde{a}_y,$ and $\tilde{a}_z$ are, respectively, the unit vectors in the x, y, and z directions; $(il)_x, (il)_y,$ and $(il)_z$ are, respectively, the magnitudes of the current-length product of the conductor under consideration in the x, y, and z directions; and $B_x, B_y,$ and B_z are the magnitude of the components of the flux density in the $x, y,$ and z directions, respectively.

Equation (1.194), being the cross product of two vectors, can be evaluated as follows:

$$\vec{f} = \begin{vmatrix} \tilde{a}_x & \tilde{a}_y & \tilde{a}_z \\ (il)_x & (il)_y & (il)_z \\ B_x & B_y & B_z \end{vmatrix} \tag{1.197}$$

From the above,

$$\vec{f} = \tilde{a}_x \begin{vmatrix} (il)_y & (il)_z \\ B_y & B_z \end{vmatrix} - \tilde{a}_y \begin{vmatrix} (il)_x & (il)_z \\ B_x & B_z \end{vmatrix} + \tilde{a}_z \begin{vmatrix} (il)_x & (il)_y \\ B_x & B_y \end{vmatrix} \tag{1.198}$$

$$= \tilde{a}_x (B_z (il)_y - B_y (il)_z) - \tilde{a}_y (B_z (il)_x - B_x (il)_z) + \tilde{a}_z (B_y (il)_x - B_x (il)_y) \tag{1.199}$$

Using the properties of the cross product of two vectors, Eq. (1.194) can also be written as follows:

$$f = il \, B \sin \theta_{il,B} \tag{1.200}$$

where

$$il = \text{the magnitude of the current-length vector}$$
$$B = \text{the magnitude of the flux density vector}$$
$$\theta_{il,B} = \text{the smaller of the two angles between the vectors } (il) \text{ and } B$$

When the flux density is perpendicular to the conductor's length, then from Eq. (1.200), we obtain

$$f = ilB \tag{1.201}$$

1.3 Force, Energy, and Torque

Equation (1.194) gives the magnitude and the direction of the force on a current-carrying conductor when it is placed in a magnetic field. The force is perpendicular to the current-carrying conductor and acts in the direction of the weaker magnetic field. This is illustrated in Fig. 1-92.

Refer to Fig. 1-92(a). The conductor's current flows into the page and perpendicular to the magnetic flux density. The direction of the conductor's field is determined by using the right-hand rule. To the right of the conductor, the field of its current opposes that of the existing magnetic field; as a result, the effective field in that region is weaker than the field to the left of the conductor. The force, as shown, always acts toward the region of the weaker field. Similar reasoning leads to the determination of the direction of the force when the current emerges out of the page (see Fig. 1-92(b)).

The torque (T) developed on a rotating conductor is equal to the cross product of the force and the perpendicular distance (r) from the conductor's axis of rotation to the point of application of the force. Mathematically,

$$T = \vec{r} \times \vec{f} \tag{1.202}$$

From Eq. (1.194) and the above, we obtain

$$T = \vec{r} \times (\vec{il}) \times \vec{B} \tag{1.203}$$

Equation (1.203) gives the torque (motor action) produced by current-carrying conductors that rotate in a magnetic field. The opposite is also true; that is, when an external torque rotates a closed-loop conductor in a constant magnetic field, a voltage is induced in the conductor and thus a current is circulated (generator action). Thus, Eq. (1.203) is the basis of motor and/or generator design. In a generator, the input is a torque and the output is a voltage; in a motor, the input is a voltage and the output is a torque. In a generating system, the larger the torque on the generator's shaft, the larger is its output current. Thus, when an electric

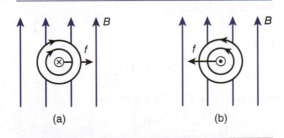

(a)   (b)

FIG. 1-92 The direction of the force on a current-carrying conductor when placed in a magnetic field. (a) Current into page. (b) Current out of page. Force acts toward reduced magnetic field.

load, such as a student's desk lamp, is switched on, the generator's output current increases in order to meet the new requirements; as a result, the generator's input torque must increase accordingly.

In a motoring system, the higher the conductor's current (derived from an external source), the higher its output torque. When the shaft's mechanical load increases (increase in the requirements of torque), the current drawn by the motor's conductors must increase accordingly.

The motor's and generator's governing equations are also derived in Chapter 6 (Eqs. (6.13 and (6.22)).

EXAMPLE **1-33**   Find the force exerted on the 6.5 m long, 4-segment conductor shown in Fig. 1-93.

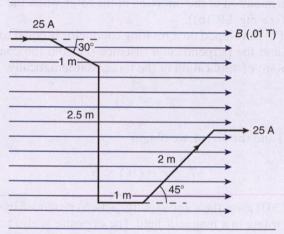

FIG. 1-93

SOLUTION

From the given diagram, the current-length vector is

$$\vec{il} = 25(\cos 30°\tilde{a}_x - \sin 30°\tilde{a}_y) - 25(2.5)\tilde{a}_y + 25a_x$$
$$+ 50(\cos 45°\tilde{a}_x + \sin 45°\tilde{a}_y)$$
$$= 82\tilde{a}_x - 39.65\tilde{a}_y$$

The flux density factor is

$$\vec{B} = 0.01\tilde{a}_x$$

Substituting the above into Eq. (1.197), we obtain

$$\vec{f} = \begin{vmatrix} \tilde{a}_x & \tilde{a}_y & \tilde{a}_z \\ 82 & -39.65 & 0 \\ 0.01 & 0 & 0 \end{vmatrix} = \tilde{a}_z \begin{vmatrix} 82 & -39.65 \\ 0.01 & 0 \end{vmatrix}$$

$$= 0.396\tilde{a}_z \text{ N}$$

Figure 1-94 shows a 1.0 ohm conductor moving along the conducting rails at a constant speed of 70 m/s. Assuming that the flux density is constant, determine the force on the moving conductor and the current through the resistor.

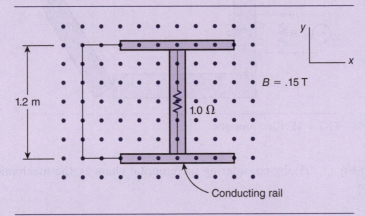

FIG. 1-94

Answer $-2.27\tilde{a}_x$ N, 12.6 A

1.3.1 Force Exerted by an Electromagnet

When a metallic body is placed in the air gap of an electromagnet, the electromagnet exerts a force on the body and attempts to move it along the air gap in such a way as to *maximize* the coil's inductance. When this condition is satisfied, the electromagnet cannot be disturbed any further. Alternatively, the north magnetic pole attracts the south pole.

Referring to Fig. 1-95, from the principle of energy conservation we have

$$\begin{bmatrix} \text{energy supplied} \\ \text{by the source} \\ (W_s) \end{bmatrix} = \begin{bmatrix} \text{energy stored} \\ \text{in the field} \\ (W_f) \end{bmatrix} + \begin{bmatrix} \text{heat dissipated} \\ (I^2R)(\text{time}) \end{bmatrix} + \begin{bmatrix} \text{mechanical} \\ \text{work done} \\ (W_m) \end{bmatrix} \quad \textbf{(1.204)}$$

Using mathematical symbols, we obtain

$$W_s = W_f + I^2R(t) + W_m \quad \textbf{(1.205)}$$

The resistance r represents the losses in the magnetic material and in the winding of the coil. For simplicity, these losses are lumped together and are represented by the resistance R, which is connected in parallel across the voltage source. In this section, however, the losses will be assumed to be negligible.

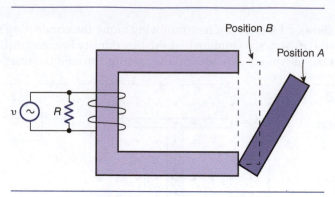

FIG. 1-95 Electromagnet.

From Eq. (1.205), by considering incremental changes, the mechanical work is given by

$$dW_m = dW_s - dW_f \qquad (1.206)$$

For linear displacements, the force (f) developed is given by the partial derivative of the mechanical work with respect to the displacement in whose direction the force is acting. Mathematically,

$$\vec{f} = \frac{\partial}{\partial x} W_m \qquad (1.207)$$

Equation (1.207) can be used to find the force developed by a linear transducer, or the weight an electromagnet can lift.

DC Excitation

When the coil of an electromagnet is connected to a dc voltage source, the current drawn is, for all practical purposes, constant. Thus from Eq. (1.206), we obtain the following

$$\vec{f} = -\frac{\partial}{\partial x} W_f \qquad (1.208)$$

The electromagnetic force, when it is applied to a moving body, produces a linear acceleration that is related to the mass M of the body by Newton's equation:

$$\Sigma F_n = Ma_n \qquad (1.209)$$

where ΣF_n represents the vectorial sum of all forces acting in the n direction, and a_n is the body's acceleration in the same direction.

Knowledge of the acceleration and the velocity of a body subjected to electromagnetic forces is a prerequisite for optimum design of weight-lifting electromagnets.

EXAMPLE 1-34

The electromagnet shown in Fig. 1-96(a) is used to control the position of the cylindrical plunger. The coil has N turns and carries a constant current of i amperes. Neglecting fringing, the mmf within the material and air gaps l_{g2}, determine general expressions for the following as a function of the length of the air gap l_{g1}:

a. The self-inductance of the coil.
b. The energy stored in the magnetic field.
c. The electromagnetic force acting on the plunger.

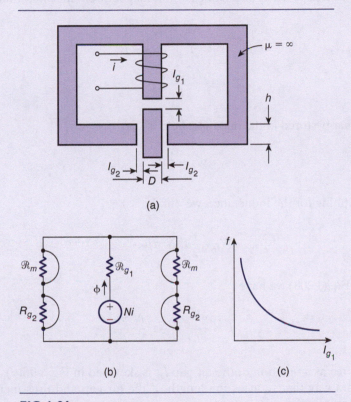

(a)

(b)

(c)

FIG. 1-96

SOLUTION

a. The equivalent magnetic circuit of the electromagnet is shown in Fig. 1-96(b). The reluctance of the magnetic material and that of the air gaps l_{g1} are shorted due to the given assumption. By definition, the reluctance $\mathscr{R}_{g1}$ is

$$\mathscr{R}_{g1} = \frac{l_{g1}}{\mu_0 A}$$

The area A perpendicular to flux is given by

$$A = \pi r^2$$

where r is the radius of the cylindrical plunger. From the above,

$$\mathscr{R}_{g1} = \frac{l_{g1}}{\pi \mu_0 r^2}$$

Then from Eq. (1.182) and Ohm's law in magnetic circuits, we obtain

$$L = \frac{N\phi}{i} = \frac{N}{i}\left(\frac{Ni}{\mathscr{R}}\right)$$

$$= \frac{N^2 \pi \mu_0 r^2}{l_{g1}}$$

b. The energy stored in the magnetic field is

$$W_f = \frac{1}{2}Li^2$$

Substituting for the inductance, we obtain

$$W_f = \frac{1}{2}i^2 N^2 \pi \mu_0 \frac{r^2}{l_{g1}}$$

c. From Eq. (1.208) we have

$$\vec{f} = \frac{\partial W_f}{\partial l_{g1}} = \frac{1}{2}i^2 \frac{N^2 \pi \mu_0 r^2}{l_{g1}^2}$$

The force as a function of the air gap l_{g1} is sketched in Fig. 1-96(c). The force acts in a way that reduces the length of the air gap and thus increases the inductance of the coil.

REMARKS

1. Determination of the mmf within the air gaps l_{g2} requires an estimate of the area through which the flux perpendicularly enters the moving cylindrical plunger. A reasonable approximation is obtained by assuming that this area is equal to 60% of the outside area of a cylindrical surface whose height and diameter are, respectively, h and D meters (see Fig. 1-96(a)).

2. The stated assumptions simplify the solution. They yield, an unrealistic answer, however, for when $l_{g1} = 0$, the force is equal to infinity.

1.3.2 Energy Stored in the Magnetic Field of Coils

The general expressions for the energy stored in the magnetic field of single and double excited coils are derived as follows.

Single Coil

The general expression for the energy stored in a single coil is derived in Section 1.1.4 (see Eq. (1.41d)). Here, the same equation will be derived using the concept of flux linkage.

Refer to Fig. 1-97(a). After the switch is closed, the instantaneous power $p(t)$ supplied by the electrical source to the coil is

$$p(t) = vi \qquad \text{(1.210)}$$

and

$$p(t) = i\,\frac{d\lambda_{11}}{dt} \qquad \text{(1.211)}$$

Neglecting losses, the *energy* supplied to the field of the coil (W_f) is equal to the energy drawn from the source (W_s). Considering incremental magnitudes, we have

$$dW_f = dW_s \qquad \text{(1.212)}$$

$$dW_f = p(t)\,dt = \left(i\,\frac{d\lambda_{11}}{dt}\right)dt \qquad \text{(1.213)}$$

$$= i\,d\lambda_{11} \qquad \text{(1.214)}$$

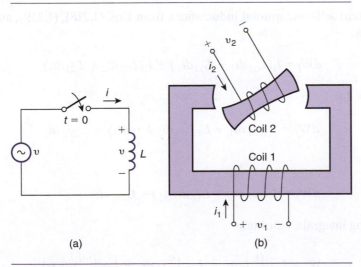

(a)   (b)

FIG. I-97 Electromechanical systems. (a) Single coil. (b) Two coils.

For a linear system, while the current increases from zero to i amperes, the energy stored in the field of the coil is

$$W_f = \int_0^i i\, d\lambda_{11} = L_{11} \int_0^i i\, di = \frac{L_{11}\, i^2}{2} \tag{1.215}$$

where L_{11} is the self-inductance of the coil.

Two-Coil System

Similarly, for the two-coil electromagnetic system shown in Fig. 1-97(b), the energy supplied by the electrical source is equal to the energy stored in the field. Considering incremental magnitudes and neglecting losses, we have

$$dW_f = dW_s \tag{1.216}$$

Thus,

$$dW_f = v_1 i_1 dt + v_2 i_2 dt \tag{1.217}$$

and

$$dW_f = i_1\, d\lambda_1 + i_2\, d\lambda_2 \tag{1.218}$$

The flux linkages of coil 1 and 2 are, by definition,

$$\lambda_1 = \lambda_{11} + \lambda_{12} = L_{11} i_1 + L_{12} i_2 \tag{1.219}$$

$$\lambda_2 = \lambda_{22} + \lambda_{21} = L_{22} i_2 + L_{21} i_1 \tag{1.220}$$

For constant self- and mutual inductances, from Eqs. (1.218), (1.219), and (1.220) we obtain

$$dW_f = i_1(L_{11} di_1 + L_{12} di_2) + i_2(L_{22} di_2 + L_{21} di_1) \tag{1.221}$$

For $L_{12} = L_{21}$,

$$dW_f = L_{11}\, i_1\, di_1 + L_{12}\, (i_1\, di_2 + i_2\, di_1) + L_{22} i_2\, di_2 \tag{1.222}$$

and

$$dW_f = L_{11}\, i_1\, di_1 + L_{12}\, d(i_1 i_2) + L_{22} i_2\, di_2 \tag{1.223}$$

Taking integrals, we obtain

$$W_f = L_{11} \int_0^{i_1} i_1\, di_1 + L_{22} \int_0^{i_2} i_2\, di_2 + L_{12} \int_0^{i_1 i_2} d(i_1 i_2) \tag{1.224}$$

Equation (1.224) yields

$$W_f = \frac{1}{2} L_{11} i_1^2 + \frac{1}{2} L_{22} i_2^2 + L_{12} i_1 i_2 \qquad (1.225)$$

When the coil inductances change while the currents remain constant, then from Eq. (1.225), the incremental change in the energy stored in the field of the two coils will be

$$dW_f = \frac{i_1^2}{2} dL_{11} + \frac{i_2^2}{2} dL_{22} + i_1 i_2 \, dL_{12} \qquad (1.226)$$

1.3.3 Torque Developed by Rotating Transducers

The instantaneous value of the torque developed by a body that rotates in a magnetic field is

$$T_i = \frac{\partial}{\partial \beta} (W_f) \qquad (1.227)$$

where T_i is the instantaneous value of the torque, W_f is the energy stored in the field, and β is the angular displacement of the rotating body with respect to a reference axis. Field energy is a function of this angle.

From basic definitions, the average value of the torque (T) is

$$T = \frac{1}{2\pi} \int_0^{2\pi} T_i \, dt \qquad (1.228)$$

In order to find the average value of the torque produced on a rotating body, first determine the energy stored in the magnetic system as a function of the body's angular position, then find its derivative and, finally, use Eq. (1.228).

Figure 1-98 shows an elementary representation of a reluctance motor. The coil is connected to a 60 Hz voltage source and draws a sinusoidal current of 20 A rms at a power factor of 0.6 lagging. The approximate self-inductance of the coil is $L = 0.1 + 0.01 \cos 2\beta$ H. Determine an expression for:

Exercise

1-30

a. The instantaneous value of the torque.
b. The speed at which the motor will develop average torque.
c. The average torque.

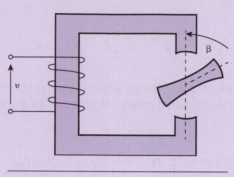

FIG. I-98

Answer (c) 1.92 N·m

1.3.4 Applications of Electromagnetism

Throughout this chapter the basic characteristics of electromagnets are described. The design of all motors and generators is, of course, based on these fundamentals. In this section, we briefly discuss electromagnetic levitation and robotics.

Magnetic Levitation

Magnetic levitation (maglev) is used in the design and operation of high-speed trains, toys, and so on.

Principle of Operation

The principle of operation of magnetic levitation is based on the property of superconductors to repel the impinging magnetic flux lines and on the natural tendency of the same polarity (*N-N*) magnetic poles to repel each other.

A typical *B-H* characteristic of a superconducting material is shown in Fig. 1-99(a). Below a certain magnetic field intensity (H_c), the magnetic flux density (B) is zero.

For this condition, the material's permeability $\left(\mu = \dfrac{dB}{dH}\right)$ is zero, and the resistance (reluctance) to the passage of magnetic flux lines is infinite.

As described previously, the reluctance $\mathcal{R}$ is given by

$$\mathcal{R} = \frac{l}{\mu A}$$

where *l* and *A* are, respectively, the length of the magnetic path and the area perpendicular to the flow of flux lines.

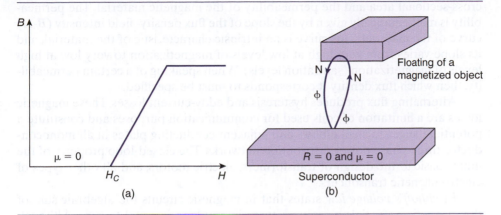

FIG. 1-99 Principles of operation of maglev. (a) *B-H* characteristic of superconductor. (b) Floating of a magnetized object above a superconductor.

From above, the magnetic flux lines cannot penetrate through the superconductor material and thus bounce back to complete their closed-loop property (Fig. 1-99(b)). The polarity of the flux lines that exit the magnetic object and the superconducting are both, by the laws of magnetism, of north pole nature and as such repel each other. The magnetized object, depending on its weight, floats above the superconductor.

Use of Electromagnets in Robots

Most robots, whether they are tiny toys or big industrial machines, use electromagnets. Electromagnets create a magnetic force whenever you power them with an electric current. This simple principle has led to the invention of items critical to robot design.

1.4 Summary

Each magnetic circuit parameter is analogous to a corresponding electric circuit's parameter. For example, the inductance, mmf, and reluctance of magnetic circuits correspond to the capacitance, emf, and resistance of electric circuits. The equivalent magnetic circuit of a particular system provides an in-depth view of its operation and limitations, while its equivalent circuit supplements and often simplifies the conclusions derived from the magnetic circuit.

The three fundamental laws of electric and magnetic circuits, on which the operation of electric machines is based, are Ohm's law, Kirchhoff's voltage law (KVL), and Kirchhoff's current law (KCL). According to *Ohm's law*, in the absence of saturation, the flux in magnetic circuits is proportional to the mmf and inversely proportional to the reluctance of the circuit. The reluctance, in turn, is proportional to the length of the magnetic path and inversely proportional to the cross-sectional area and the permeability of the magnetic material. The permeability is mathematically given by the slope of the flux density–field intensity (*B-H*) curve of the material. This curve is an intrinsic characteristic of the material, and its slope varies from very high at low levels of magnetization to very low at high levels of magnetization (saturation levels). When speaking of a certain permeability, then which flux density it corresponds to must be specified.

Alternating flux produces hysteresis and eddy-current losses. These magnetic losses are a limitation of coils used for communication purposes and constitute a potential danger to steel elbows and adjacent conducting plates in all monoconductor three-phase power distribution networks. The closed-loop property of the flux is basic to the design of transformers, electric motors, and all other types of electromagnetic transducers.

Kirchhoff's voltage law states that in magnetic circuits the algebraic sum of the magnetic potentials or ampere-turns in any loop is equal to zero. Although the magnetic potential absorbed within the magnetic material is relatively small, it plays an important role in transformers and electric machines. *Kirchhoff's current law* states that in magnetic circuits the algebraic sum of the flux in a junction must be equal to zero.

The *inductance* of a coil is analogous to the mass of an object or to the moment of inertia of a rotating body. As such, it represents the natural tendency of a coil to oppose those external forces that try to change its status. For this reason, precautions must be taken (use of properly sized switches, fuses, and circuit breakers) when the current in an inductive load is to be interrupted.

The *energy stored* in the magnetic field of a coil is proportional to the coil's inductance. The *partial derivative* of the energy stored, with respect to an associated linear or angular displacement, gives the instantaneous value of the force, or torque, developed. The force, or torque, acts in a direction that increases the system's inductance. Thus, the concept of inductance is of fundamental importance to the understanding of all force-, or torque-, developing mechanisms, such as electric machines and other electromechanical transducers.

Alternatively, the electromagnetic force can be derived from Ampere's experimental observations. According to these experiments, the force produced on one current-carrying conductor by another is equal to the cross product of the current–length vector of one conductor and the flux density produced by the other. In other words, there is a force produced on a current-carrying conductor when that conductor is properly placed in a magnetic field.

A voltage is generated in a given coil either when an alternating flux cuts through the stationary coil (transformer voltage) or when the coil is properly rotated through a stationary magnetic field (speed voltage).

The Principle of Induction

Overview

Permanent magnets and their corresponding flux lines are due to the spinning of their atoms' electrons, that is, to the flow of current. Similarly, the current in a wire or in a coil or in an apparatus is always accompanied by closed-loop magnetic flux lines that when passing through a metallic object, disturb it Because action is equal and opposite to reaction, the metallic object generates its own flux lines (θ_ε) in order to oppose the disturbance. The subscript ε designates, in general, "eddy-current" parameters.

These opposing flux lines are produced, as per Ohm's law in magnetic circuits, by a circular current (I_ε) within the metallic object. The flow of this current, of course, depending on the encountered resistance (R_e), produces power (P_ε) or rate of heat loss ($P_\varepsilon = I_\varepsilon^2 R_e$). This over time may overheat the metallic object, leading to potentially serious problems.

The changing of the flux's direction within the material—due to the frequency of the voltage source—aligns and realigns the building blocks of magnetism (magnetic domains) to the direction of the flux lines (θ_ε). The accompanied friction losses are called hysteresis losses (P_h). Then the total of heat loss (P_T) within the material depends on the circulating current and on the realignment of the magnetic domains. That is,

$$P_T = P_\varepsilon + P_h$$

These power losses are of concern to designers and operators of electrical equipment (motors, transformers, communication coils, etc.). As such, research is continually being done to determine how to minimize them.

Furthermore, the rate of heat generated within the material in the vicinity of metallic objects through which magnetic flux lines pass is of paramount importance to monoconductor power distribution because it can lead to enormous damage.

In such cases, use the following two approaches to prevent such a problem.

- The most effective way is to use a three-conductor cable for power distribution instead of 1×3 monoconductors.

  The net flux exterior of a three-conductor cable is negligible because the current's phasors and the associate flux in a three-phase system are 120 degrees to each other, and thus the net exterior flux is theoretically zero.

  For economical reasons, however, when the current is, say, more than 300 A—monoconductors are used whose total external flux (due to spacing) is not zero.

- The second way is to verify that no metallic objects are in the vicinity of monoconductor power distribution, to oversize the junction boxes, and to ventilate where possible the route along the passage of the cables.

  When in doubt, use an infrared-temperature sensor—thermal imaging—to measure the temperature in all suspected metal objects while the cable carries full-load current.

Mathematical Considerations

The events associated with the induction currents are understood better when, as described below, they are accompanied with their corresponding mathematical relationships, that is, the laws of nature.

1. ***Source of current***

   The current (I) in a circuit is the result of Ohm's law in electric circuits and is given by

   $$I = \frac{V}{Z}$$

   where V and Z are, respectively, the source voltage and the impedance to the current path.

2. ***Generation of flux lines***

   Rewriting Ohm's law in magnetic circuits, we have

   $$\phi = \frac{NI}{\mathscr{R}}$$

   or

   $$\phi \propto I$$

   where N and $\mathscr{R}$ are, respectively, the number of winding turns and the reluctance of the closed path through which the flux lines go through.

3. ***Closed-loop property of magnetic flux lines***

   The currents' flux lines form a closed loop as per Maxwell's equation:

   $$\oint BdA = 0$$

   where B and dA are, respectively, the magnetic flux density and an element of area.

4. ***Action is equal and opposite to reaction***

   The flux lines produced disturb whatever is in their path. From mechanics,

   $$\text{Action} = \text{opposite to reaction}$$

5. ***Disturbing flux (ϕ) and its opposition (ϕ_ε)***

   When the disturbing flux goes through a metallic object, the object produces its own flux lines (ϕ_ε) in order to satisfy the previous principle. These flux lines are designated, as previously stated, with the subscript ε to signify what is commonly known as eddy-current flux.

6. ***Ohm's law in magnetic circuits***

   The generation of these flux lines are due to the materials internally circulated by eddy currents (I_ε).

   $$\phi_\varepsilon \propto I_\varepsilon$$

7. *Eddy-current losses*

The rate of heat losses within the material (P_ε) is

$$P_\varepsilon = I_\varepsilon^2 R_\varepsilon$$

where R_ε is the resistance of the material to the flow of eddy currents.

8. *Hysteresis losses*

The hysteresis power losses (P_h) are due to the frictional losses associated with the alignment and realignments of the materials' magnetic domains.

9. *Total losses (P_T)*

Then the total losses associated with the principle of induction are

$$P_T = P_\varepsilon + P_h$$

The development of the formulas for the eddy-current and hysteresis losses are given in Section 1.2.11.

Figure 1-102 shows diagrammatically the concepts associated with the induction principle.

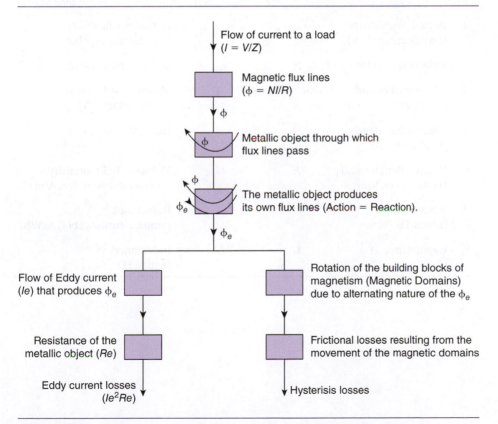

FIG. 1-102 Block diagram representation of the induction principle and its effects.

Induction Effects on Electronic Circuits

When a coil, a relay, a contactor, and such. are switched off, the current decays to zero within some μ seconds. The current's corresponding flux can induce ground voltages to the distribution system that may cause malfunctioning of the 5 V microprocessors of the control systems. This can lead to injuries and equipment damage. To prevent such damages a snubber circuit (*R-C* in series) should be connected in parallel to the coils.

Tables 1-3 and 1-4 summarize the various magnetic and electric circuit parameters, their units, and their interrelationships.

TABLE 1-3	Summary of the electric and magnetic field parameters and their units			
	Electric		**Magnetic**	
No	Description	Symbol	Symbol	Description
1	Electric flux (coulombs)	ψ	ϕ	Magnetic flux (weber, Wb = 10^8 lines = 10^8 maxwells)
2	Actual permittivity (farads/meter, F/m)	$\varepsilon = \varepsilon_0 \varepsilon_r$	$\mu = \mu_0 \mu_r$	Actual permeability (henries/meter, H/m)
3	Relative permittivity	ε_r	μ_r	Relative permeability
4	Electric volts, emf (volts, V)	emf = V	mmf = $U = \mathscr{F}$ = NI	Magnetic volts, mmf (ampere-turns, A)
5	Electric flux density (coulombs/m^2)	D	B	(tesla, Wb/m^2, T)
6	Electric field intensity (volts/meter, V/m)	E	H	Magnetic field intensity (ampere-turns/meter, A/m)
7	Resistance (ohms, Ω)	R	$\mathscr{R}$	Reluctance (ampere-turns/weber, A/Wb)
8	Capacitance (farads, F)	C	L	Inductance (henry, H)

TABLE 1-4   Basic concepts in mathematical form

No.	Electric	Magnetic	Remarks
1	$\varepsilon = \dfrac{dD}{dE}$	$\mu = \dfrac{dB}{dH}$	Physical constants
2	$D = \dfrac{\psi}{A}$	$B = \dfrac{\phi}{A}$	Flux densities
3	$E = \dfrac{dV}{dl}$	$H = \dfrac{NI}{l}$	Field intensities
4	$R = \rho \dfrac{l}{A}$	$\mathcal{R} = \dfrac{l}{\mu A}$	Opposition to flow
5	$I = \dfrac{V}{R}$	$\phi = \dfrac{NI}{\mathcal{R}}$	Ohm's law
6	$\Sigma V_{\text{loop}} = 0$	$\Sigma (NI)_{\text{loop}} = 0$	KVL
7	$\Sigma I_{\text{junction}} = 0$	$\Sigma \phi_{\text{junction}} = 0$	KCL
8		$V = 4.44 N f \phi_{\text{max}}$	Relates the electric and magnetic field parameters

TABLE 1-5   Some highlights about power*

	Per-unit cost[†] of purchasing an electrical power source				
		P —	100 W		Manual power capacity of an average human adult
Large diesel generator	1	— O —	(100–400) W/m²		Power available from sun during sunlight hours (Chicago)
Large hydropower station	8	— W —	25 W/m²		Lighting requirements of an office building
Wind power system	10	— E			
Atomic reactor	10	— R —	80 W/m²		Requirements of average home for heating, cooling, cooking, lighting, etc. (Chicago)
Uninterruptible power supply	8				
Photovoltaic cells	15		To produce[‡] a ton per year		Cement: 40 W Paper: 200 W Aluminum: 500 W

*The given power requirements and costs are approximate.
[†]The per-unit cost in 2010 was about $1000/kW.
[‡]Usual rating of plants.

TABLE 1-6   Some highlights about energy*

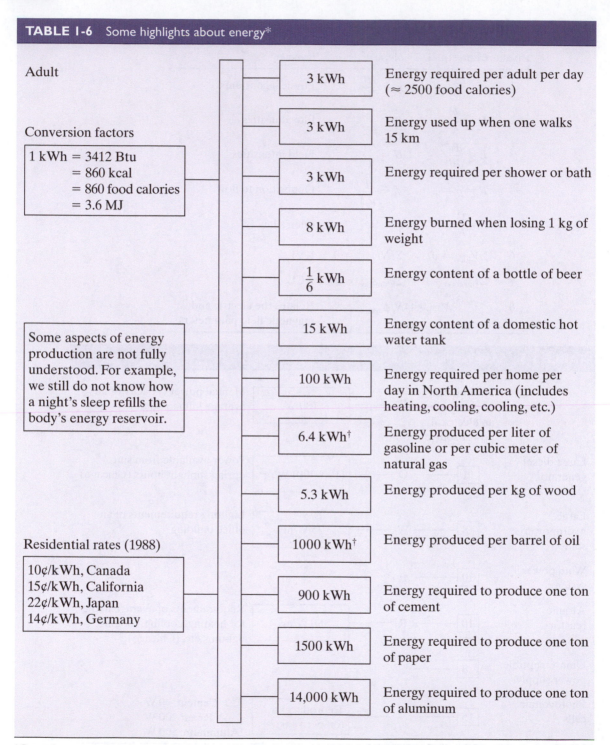

Adult

Conversion factors

1 kWh = 3412 Btu
     = 860 kcal
     = 860 food calories
     = 3.6 MJ

Some aspects of energy production are not fully understood. For example, we still do not know how a night's sleep refills the body's energy reservoir.

Residential rates (1988)

10¢/kWh, Canada
15¢/kWh, California
22¢/kWh, Japan
14¢/kWh, Germany

3 kWh	Energy required per adult per day ($\approx$ 2500 food calories)
3 kWh	Energy used up when one walks 15 km
3 kWh	Energy required per shower or bath
8 kWh	Energy burned when losing 1 kg of weight
$\frac{1}{6}$ kWh	Energy content of a bottle of beer
15 kWh	Energy content of a domestic hot water tank
100 kWh	Energy required per home per day in North America (includes heating, cooling, cooling, etc.)
6.4 kWh†	Energy produced per liter of gasoline or per cubic meter of natural gas
5.3 kWh	Energy produced per kg of wood
1000 kWh†	Energy produced per barrel of oil
900 kWh	Energy required to produce one ton of cement
1500 kWh	Energy required to produce one ton of paper
14,000 kWh	Energy required to produce one ton of aluminum

*Cost and statements about energy requirements are approximate.
†Assumed efficiency of combustion is 62%.

1.5 Review Questions

1. "A formula or an equation represents—in mathematical symbols—a physical or a natural law." Explain this statement and give two examples to demonstrate its significance.

2. Why were the effective or rms values of voltages and current introduced?

3. What are the two essential differences between electric and magnetic flux?

4. What is the difference between emf and mmf?

5. In a laboratory, how would you measure the magnetic flux that crosses a given air gap of a magnetic circuit? How could you measure the mmf within a toroidal electromagnet?

6. Why do the resistance and the inductance of a material or a circuit change as a func-tion of temperature? What is the Curie temperature of a material?

7. How would you short-circuit an electric and a magnetic circuit?

8. How would you explain inductance or permeability to a mechanical engineering student?

9. Define permeability and give five factors that should be considered when magnetic materials are to be purchased.

10. How would you increase the permeability of a material?

11. Is the energy stored within the components of a magnetic circuit inversely proportional to their permeabilities?

12. Differentiate between transformer and speed voltage.

1.6 Problems

1-1 a. One of the most expensive mishaps (2.5×10^8) in the American space exploration program occurred in 1970 because of an error in the basic design of Apollo 13. The dc supply to the resistive heat element of the liquid oxygen tank was 48 V while the resistor was rated 24 V, 5 kW,* with a maximum power capacity of 300%.* As the Apollo astronauts were approaching the gravitational field of the Moon, they switched on the 48 V supply, and the oxygen tank was blown away. This is how the famous "drama in the skies" started. Determine by what percentage the maximum power capability of the heating element was exceeded when the 48 V dc supply was switched on.

b. One of the worst fatal mishaps in the Russian space exploration program took place during the flight of Salyut I in 1971. The astronauts' cabin was equipped with a valve, which was to open automatically on landing, thereby equalizing the cabin and the ambient pressures. After several weeks of record-breaking events in space, the astronauts were returning home. The valve opened accidentally on reentry, and the breathing atmosphere started to escape. The three astronauts attempted to close the valve but failed because, according to official statements, closing the valve manually required two minutes, while the cabin's atmosphere was emptied in one minute. If the valve had been

---

* Assumed

electrically operated, what should the time constant (L/R) of its solenoid have been for a safety factor of 200%?

1-2 For the voltage waveform shown in Fig. P1-2, determine:

   a. Its rms value.

   b. Its average value.

   c. The average power dissipated when this voltage is applied across a 4 ohm resistor.

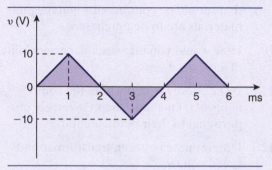

FIG. P1-2

1-3 **a.** For the circuit shown in Fig. P1-3, determine *graphically* the voltage across the 5 ohm resistor. Calculations with complex numbers must not be used. [*Hint*: In an *RL* or *RC* circuit, across a voltage *V*, the current is given by the intersection of two semicircles whose diameters are V/R and V/X.]

   b. Check your answer *mathematically*.

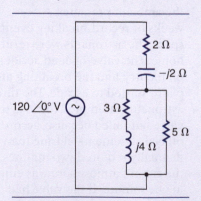

FIG. P1-3

1-4 For the circuit shown in Fig. P1-4, determine the currents I_1, I_2, and I_3 by using loop equations (KVL).

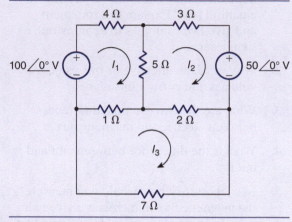

FIG. P1-4

1-5 For the circuit in Fig. P1-5, determine the voltage V_{AB} by using nodal equations (KCL). Check your answer by using loop equations (KVL).

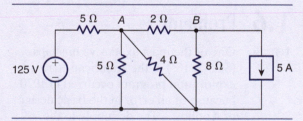

FIG. P1-5

1-6 Refer to Fig. P1-6. Determine the current through R_1 by using:

   a. KVL.

   b. KCL.

   c. The superposition theorem.

   d. Thévenin's theorem.

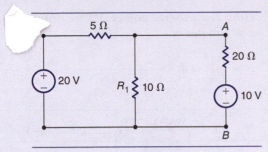

FIG. PI-6

1-7 Refer to Fig. P1-7.

 a. The switch is closed for a long time. On opening it, determine the circuit's time constant and the energy dissipated across the switch.

 b. A discharge resistor of 10 ohms is connected in parallel to the inductor. When the switch becomes open, determine the energy stored in the coil and the circuit's time constant.

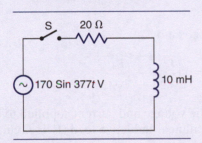

FIG. PI-7

1-8 Refer to Fig P1-8. Determine the circuit's time constant when the switch is closed and when it is opened.

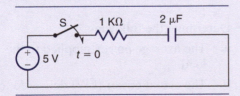

FIG. PI-8

1-9 The power characteristics of a factory's three departments A, B, and C (see Fig. P1-9) are as shown in the following table.

Calculate and complete the table's missing information.

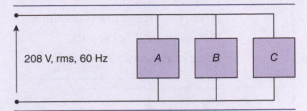

FIG. PI-9

Department	Complex Power (kVA)	Average Power (kW)	Impedance (Ohms)	Current (A)
A	$12\underline{/37}$			
B			$3.59 + j1.91$	
C				$80\underline{/90}$
Total				

1-10 The power distribution characteristics of a factory's two departments A and B, in Fig. P1-10, are as shown in the following table. Each of the loads consists of resistors and inductors in series. Calculate and complete the table's missing information.

Department	Complex Power (kVA)	Average Power (kW)	Inductance (H)	Current (A)
A				$50\underline{/-18°}$
B	$20\underline{/25.84°}$			
Total				

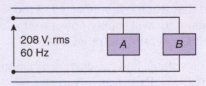

FIG. PI-10

1-11 For the circuit shown in Fig. P1-11, determine:

 a. The size of the capacitor bank so that when it is connected in parallel to the loads the circuit will operate at a Pf of 0.95 lagging.

b. The annual extra charges of the utility and the cost of the capacitor bank. They are given as:

> Utility charges $12/kW/month
>
> Minimum utility Pf: 95%
>
> Cost of capacitor $30/kVAR

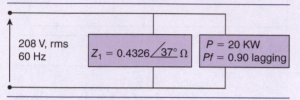

FIG. P1-11

1-12 A commercial building has a continuous power demand of 50 kW at a power factor of 85%. Determine:

a. The annual penalty if the utility's minimum power factor is 0.95% and the cost of power is $10/kW/month.

b. The cost of the capacitor, which when connected in parallel to the load will improve the power factor to 100%. (Assume the cost of the capacitor to be $30.0 kVAR.)

1-13 A single-phase 100 kVA generator supplies power to an 80 kW load at 0.8 power lagging. An additional load of 15 kW and 90% power factor lagging is to be added to the generator load.

Determine the size of the capacitor bank so that when it is connected to the load the generator will supply power to the loads while operating at rated kVA.

1-14 The voltage and the current of a nonlinear load (Fig. P1-14), are given by:

$$v = V_m \sin \omega t, \qquad i = I_m = \sin n\,\omega t$$

Prove that such loads draw no average power. In other words, the electromechanical utility meter cannot register all the power absorbed by nonlinear loads.

Hint: The average power (P) measured by a wattmeter is

$$P = \frac{1}{T} \int^{T} vi \, dt$$

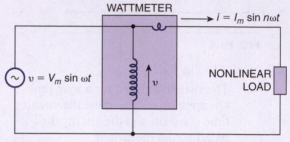

FIG. P1-14

1-15 Show that the rms value of the current

$$I = I_{m1} \sin\left(\omega t - \frac{\pi}{6}\right) + I_{m3} \sin\left(3\omega t - \frac{\pi}{12}\right)$$

is given by

$$I = \sqrt{\left(\frac{I_{m1}}{\sqrt{2}}\right)^2 + \left(\frac{I_{m3}}{\sqrt{2}}\right)^2}$$

1-16 The voltage and current supplied to a nonlinear inductive load (R-L circuit) are as follows:

$$v = 170 \sin 377t \text{ V}$$

$$i = 10 \sin\left(377t - \frac{\pi}{6}\right) + 3 \sin\left(1131t - \frac{\pi}{12}\right) \text{ A}$$

Determine the following:

a. The average power supplied to the load.

b. The actual power (kW) delivered to the load.

c. The percentage of the total power due to harmonic current.

d. The apparent power delivered to the load.

1-17 The line current to a 5 kW, one-phase, 208 V, 0.9 *Pf* motor contains 10% of third harmonic current. Determine the apparent and actual:

   a. Line currents.

   b. Motor power.

1-18 Refer to the schematic of the metallic reed switch shown in Fig. P1-18. Such switches are inserted in very small air gaps to detect weak magnetic signals. When the applied field is cw, as shown, the switch becomes closed.

   a. When the direction of the flux lines becomes ccw, will the switch remain closed or will it open?

      Briefly describe the reasons, using simple terminology.

   b. When the flux lines are parallel to the plates, will the switch close?

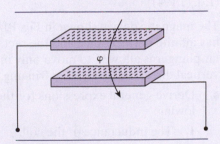

FIG. P1-18

1-19 The telephone line shown in Fig. P1-19 runs parallel to a balanced three-phase power line.

   a. Show that the induced voltage per meter of the telephone line, due to current in phase "a" only, is given by

$$v = I_m\,\omega\,\frac{\mu_0}{4\pi}\,\ln\left[\frac{d_1^2 + (d_2 + d_3)^2}{d_1^2 + d_2^2}\right]\cos\omega t \text{ V/m}$$

   b. Determine the rms value of the net voltage induced in the telephone line per unit length, assuming

$$i_a = 1000 \sin 377t \text{ A}$$

$$d = 10 \text{ m}, \quad d_1 = 5 \text{ m}, \quad d_2 = 15 \text{ m}, \quad d_3 = 1 \text{ m}$$

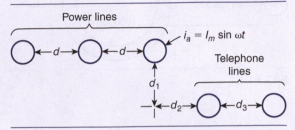

FIG. P-19

1-20 The conductor shown in Fig. P1-20 moves at a velocity of *U* m/s in a uniform magnetic field. The magnitude of the magnetic flux density is *B* tesla, and its direction is as shown. Determine the voltage induced in the conductor.

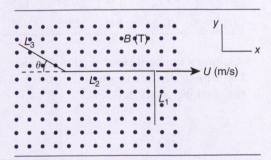

FIG P1-20

1-21 The toroidal coil shown in Fig. P1-21 has a circular cross-sectional area and an inductance of 0.5 H. Neglecting leakage flux and losses, determine:

   a. The relative permeability of the magnetic material.

   b. The magnetic flux.

   c. The magnetic flux density.

d. The magnetic flux intensity.

e. The magnetic potential drop within the magnetic material.

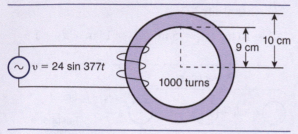

FIG. P1-21

1-22 In the magnetic circuit of Fig. P1-22,

$$l_{bad} = 0.5 \text{ m}, \quad l_{bd} = 0.15 \text{ m},$$
$$l_{bcd} = 0.60 \text{ m}, \quad l_g = 0.8 \text{ mm}$$

The coordinates of the *B-H* characteristic of the magnetic material are as follows.

B(T)	0.12	0.54	0.94	1.15
H(A/m)	100	300	700	1600

Assume a uniform cross-sectional area. Determine the flux through the path *bcd* if the stacking factor is 0.95.

1-23 Determine the force between the two coplanar conductors shown in Fig. P1-23.

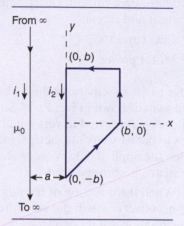

FIG. P1-23

1-24 Derive the general expression for the speed voltage (Eq. 1.140)) generated in a moving conductor by using the principle of energy conservation and Ampere's law (Eq. (1.194)).

1-25 The magnetic device shown in Fig. P1-25 has infinite permeability, and the rectangular plunger is allowed to move only in the vertical direction. Neglecting fringing,

a. Derive general expressions for the following:

1. The inductance of the coil.

2. The energy stored in the magnetic field.

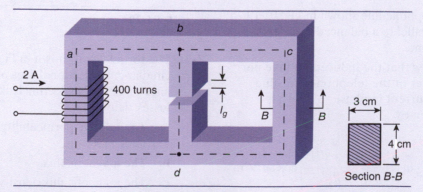

FIG. P1-22

3. The force acting on the rectangular plunger.

4. The flux density through the center leg of the electromagnet.

b. Evaluate the derived expressions using the following data:

$l_{g1} = 10$ mm, $l_{g2} = 2$ mm

$D = 50$ mm, $W = 60$ mm, $h = 40$ mm

$N = 1000$ turns, $I = 2$ A

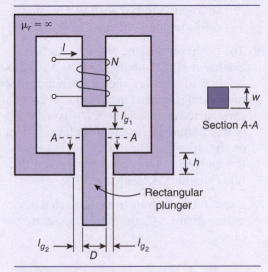

FIG. P1-25

1-26 The coil shown in Fig. P1-26 is energized from a dc voltage source. The mmf-ϕ characteristic of the magnetic material is given by

$$\mathcal{F} = 6.5 \times 10^4 \phi \text{ ampere-turns}$$

For an air gap of 10 mm, determine:

a. The inductance of the coil.

b. The energy stored in the magnetic field.

c. The mechanical work done, if the air gap is slowly reduced to 1 mm. Sketch the variation of the force as a function of the air-gap length.

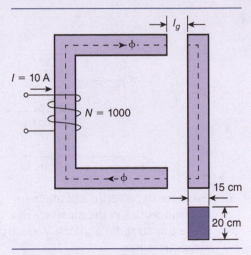

FIG. P1-26

1-27 Two coils, one mounted on the stator and the other on the rotor of a two-winding machine, have the following self- and mutual inductances:

$$L_s = L_1$$
$$L_r = L_2$$
$$L_{sr} = L_0 + L_{12} \cos 2\beta$$

where β is the angle between the axis of the two coils. The coils are connected in series, and their current is $i = I_m (\omega t - \theta)$. Determine:

a. The instantaneous torque.

b. The average torque.

1-28 Assume that the supply current to the R-L circuit is as shown in Fig. P1-28(a).

a. Show that circuit's efficiency is infinite when

$$R = \frac{3}{4} \left(\frac{L}{T} \right)$$

b. For what reasons can such a circuit not be realized?

(*Hint*: It has nothing to do with the second law of thermodynamics.)

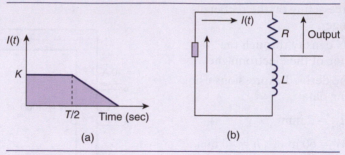

FIG. P1-28

1-29 a. Estimate the electric and magnetic field intensities in the air space of a home due to radio and/or TV electromagnetic waves.

(*Hint*: See page 64.)

b. Compare the strength of Earth's magnetic field intensity to that of a radio wave in a residential area.

c. The utilities' large power transformers have a magnetic flux density of 0.9 *T*. Estimate the transformers' magnetic field intensity.

d. Briefly explain the reasons that superconductors can be said to be electrically short-circuited ($R = 0$) and magnetically are open-circuited ($R = \infty$).

1-30 The first project of the physicist Andrei Zakharov (1942)—father of Russia's hydrogen bomb—was to design a magnetic probe for locating shrapnel in wounded horses. He constructed an assembly of *H*-shaped sheets of transformer iron with the indicating meter on the cross-bar. The device was not sensitive enough and was never put into production (Memoirs. A Zakharov, Page 42).

Explain briefly how to make such a device more sensitive. Consider the two extreme cases.

2

Transformers

2.0   Introduction

2.1   Single-Phase Transformers

2.2   Three-Phase, Two-Winding Transformers

2.3   Autotransformers

2.4   Parallel Operation of Transformers

2.5   Instrument Transformers and Wiring Diagrams

2.6   Transformer's Nameplate Data

2.7   Conclusion

2.8   Summary

2.9   Review Questions

2.10   Problems

What You
Will Learn in
This Chapter

A **Theoretical Aspects**
1 Principle of operation of transformers
2 1-θ Transformers
3 Three-phase, two-winding transformers
4 Special transformers (1-θ to 3-θ and 2-θ to 3-θ)
5 Three-phase, three-winding transformers
6 Autotransformers
7 Parallel operation of transformers
8 Optical and hard-wired instrument transformers
9 Hook-up of three-phase digital meters
10 Ferroresonance
11 K-rated transformers

B **Practical Highlights—Interface**
1 Application of transformers
2 Measurement of the transformers' equivalent circuit parameters
3 Transformers' starting current and harmonics
4 How to draw phasor diagrams
5 Efficiency of transformers
6 Rating and one-line diagram representation (Almost all industrial transformers have a nominal rating and an increased capacity of 1.33 × nominal by just adding cooling fans at about a 10% increase in the cost of the transformer.)
7 How to solve problems involving transformers, their upstream and downstream cables, and various types of loads
8 How to solve problems of Δ-Y transformers that supply L-N, L-L, and three-phase loads
9 Why a hard-wired CT will be damaged and may cause fires when it is open-circuited while connected to an operating distribution system. Contrary to it, a PT will be destroyed when its secondary winding is shorted.
10 How the optical instrument transformers operate and how you can hook up a digital multimeter
11 Ferroresonance and preventing transformer overheating due to system harmonics
12 Governing equations
13 Manufacturers' published losses, efficiency, and impedance of transformers

C **Additional Student Aid on the Web**
1 Wiring diagrams for analog meters
2 Analog meters

2.0 Introduction

This chapter covers the principles of operation, equivalent circuits, losses, and applications of the various types of transformers.

Transformers, as the name implies, transform or change, from one level to another, the current and voltage that are applied to their input windings. An increase, or step-up, in the voltage across one winding is accompanied by an equal decrease, or step-down, of the current in the same winding.

Depending on the distance between the generating station and the user of electricity, the voltage—through a setup transformer—is increased, so the transmission line current is decreased and the line's energy loss and voltage drop are decreased.

Almost all power distribution within factories, homes, and elsewhere is for economical reasons of sinusoidal voltages. (It simplifies the design and operation of motors and generators and the step-up or step-down of a given voltage).

Single-phase transformers are covered in the greatest detail in this chapter because every other type of transformer is either a slightly modified single-phase transformer or a combination of single-phase transformers.

Two-winding, three-phase transformers are also discussed because they are of primary importance to industry.

Autotransformers and *transformers in parallel operation* have limited applications, but they are extensively analyzed because serious field problems may result from misunderstanding their operation.

Instrument transformers are an essential part of any distribution network. Although covered in texts on electrical measurements, they are briefly discussed here in order to point out some of the common problems that may arise during their installation and operation, and to emphasize the effects of magnetic saturation. The recently developed optical transformers are also described.

Analysis of transformers requires knowledge of *per-unit values* and of *three-phase networks*, which are common to all types of ac machines. Both of these topics are briefly discussed in this chapter and are extensively covered in the Appendices of this revision of the book.

The end of this chapter includes some practical highlights of transformer applications, such as nameplate data (technical data inscribed on a metal plate fastened on an easily seen part of the transformer) and typical manufacturers' test results.

The photos of Fig. 2-1 show a large outdoor substation transformer and other transfers with small power ratings.

Fig. 2-2 is a photo of one of the 1st transformers built in North America.

The property of induction was discovered in the 1830s, but it wasn't until 1886 that William Stanley, working for Westinghouse, built the first refined, commercially used transformer. His work was built on some rudimentary designs by the Ganz Company in Hungary (ZBD Transformer 1878), and Lucien Gaulard and John Dixon Gibbs in England.

FIG. 2-1 Various types of transformers. *DSBfoto/Shutterstock.com and Courtesy of mikroElektronika*

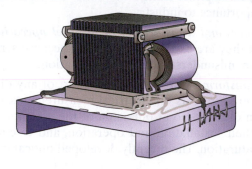

Stanley's first transformer, which was used in the
electrification of Great Barrington, Massachusetts in 1886.

FIG. 2-2 Transformer of 1886. *Based on Stanley Transformer, http://edisontechcenter.org/ LocalSites/StanleyTransformer1.jpg*

2.1 Single-Phase Transformers

As shown in Fig. 2-3, single-phase transformers usually have one input and one output winding, often referred to as the transformer's primary and secondary windings. These windings are not electrically connected but are magnetically coupled. The primary winding draws energy from a voltage source, whereas the secondary winding delivers energy to a load.

Transformers are not power amplifiers. For all practical purposes, the apparent input power ($|S|$) to a transformer's primary winding is equal to the apparent power delivered to the load by its secondary winding. In other words, the volt-amperes of the primary winding ($V_1 I_1$) are approximately equal to the

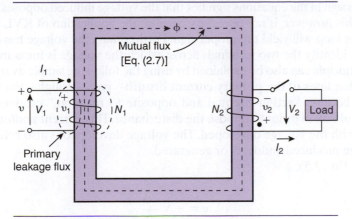

FIG. 2-3 Single-phase transformer.

volt-amperes of the secondary winding ($V_1 I_2$). Mathematically,

$$|S| = V_1 I_1 \approx V_2 I_2 \qquad (2.1)$$

In actual cases—because of transformer losses—the volt-amperes delivered to a load by a transformer are slightly less than the volt-amperes drawn from the voltage source.

2.1.1 Principle of Operation

The operation of single-phase transformers (and all other transformers) is based on the principle of induction. According to this principle, a voltage is induced in a winding when the winding's flux linkages (λ) change as a function of time. The instantaneous value of the flux linkages is defined as

$$\lambda = N\phi \qquad (2.2)$$

where N and ϕ are, respectively, the coil's number of turns and the instantaneous value of the flux per turn. Alternatively,

$$\lambda = Li \qquad (2.3)$$

where L and i are, respectively, the coil's inductance and instantaneous current.

According to Faraday's law, the instantaneous value of the induced voltage (v) is given by

$$v = -\frac{d\lambda}{dt} \qquad (2.4)$$

or

$$v = -\frac{d}{dt}(N\phi) \qquad (2.5)$$

The minus sign in the equations signifies that the voltage induced opposes the supply voltage. This, however, is not necessary because the application of KVL in the corresponding loop will yield the proper sign. Furthermore, the voltage has a meaning when you identify the two terminals across which the voltage is measured. The induction principle can also be explained by using the following simple everyday logic.

The flux lines of the primary current disturb—pass through—the secondary coil that, because "action is equal and opposite to reaction," produces its own current and flux in order to oppose the disturbance. This current and/or flux is associated with the voltage developed. The voltage developed is also referred to as the voltage produced, induced, or generated.

From Eq. (2.5):

$$v = -N\frac{d\phi}{dt} \tag{2.6}$$

The negative sign in the equation signifies that the polarity of the induced voltage opposes the change that produced it. This reaction is a natural one for magnetically coupled coils and follows the principle commonly known as Lenz's law.

As shown in Fig. 2-3, the polarity of the induced voltages is usually identified by the dot symbol. The physical significance of the dot symbol is explained in detail in Chapter 1 (see Section 1.2.14).

Equation (2.6) is a basic law of electromagnetism and one of the most important mathematical relationships of electrical engineering. It governs the operation of motors and generators, and it demonstrates the coexistence of, and the quantitative relationship between, electric and magnetic fields.

The flux within the structure of the transformer changes because it is produced by the alternating voltage supplied to the transformer's input winding.

As Eq. (2.6) makes clear, for sinusoidal input voltages the flux is at a maximum when the voltage is at the zero point of its cycle. The flux and the voltage phasors must be 90° out of phase with each other. Typical voltage and flux waveforms for a transformer are shown in Fig. 2-4.

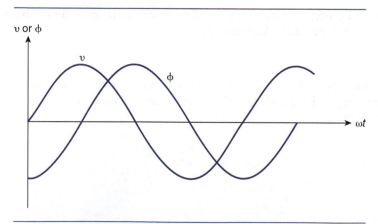

FIG. 2-4 Voltage and flux waveforms in a single-phase transformer.

The equation that relates the induced voltage to the flux produced is derived in Chapter 1 [Eq. (1.148)]. For convenience, it is rewritten here:

$$V = 4.44 N f \phi_m \qquad (2.7)$$

where V is the rms value of the input voltage, f its frequency of oscillation in hertz, and ϕ_m the maximum value of the flux within the magnetic material in webers.

Thus, the rms value of the voltage induced in the secondary winding of the transformer, or in any other coil wound on the same core, will be given by Eq. (2.7). In each case, the appropriate number of turns must be used.

Equation (2.7) is called the fundamental transformer equation, and it is often used in laboratories to calculate the flux level within any shape—toroidal, rectangular, and so on—of magnetic circuit. This equation gives accurate results only if the leakage impedance of the coil is negligible.

The flux produced by the primary winding is divided into two parts: leakage flux and mutual flux. Leakage flux links only the windings of the primary coil and is associated with the transformer leakage impedance. Mutual flux links the windings of the primary and secondary coils and is associated with the magnetizing impedance of the transformer.

2.1.2 Ideal Transformer Relationships

In this section we derive general equations that relate the parameters of the primary winding to the parameters of the secondary winding. We will consider only ideal transformers in order to provide the basis for analysis of nonideal transformers in follow-up sections.

An ideal transformer has zero core loss, no leakage flux, and negligible winding resistances. Refer to Fig. 2-5. Using the induction principle, we have

$$v_1 = -N_1 \frac{d\phi}{dt} \qquad (2.8)$$

$$v_2 = -N_2 \frac{d\phi}{dt} \qquad (2.9)$$

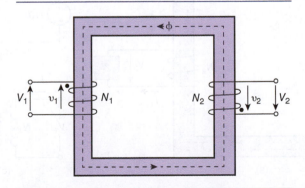

FIG. 2-5 An illustration of the concept of voltage induced. Only the mutual flux is shown.

From Eqs. (2.8) and (2.9), we obtain:

$$v_2 = v_1 \frac{N_2}{N_1} \qquad (2.10)$$

Under the assumed ideal conditions, the induced voltages (v_1, v_2) are equal to their corresponding terminal voltages. Then using rms values, we obtain

$$V_2 = V_1 \frac{N_2}{N_1} \qquad (2.11)$$

That is, voltages of different magnitudes can be obtained by winding coils with different numbers of turns around a magnetic circuit. The effective flux (ϕ) within the magnetic material is dependent on the applied voltage and is essentially independent of the flux of the output current. Thus, under ideal conditions, the voltage induced in the output winding is independent of the load current.

Figure 2-6 shows a schematic of a two-winding transformer, its magnetic equivalent circuit, and its ideal electrical equivalent circuit.

Figure 2-6(a) is the physical representation of an ideal transformer. From Kirchhoff's voltage law (KVL) for magnetic circuits [Fig. 2-6(b)], we have

$$\Sigma \, (NI)_{\text{loop}} = 0 \qquad (2.12)$$

That is, the sum of the magnetic potentials or ampere-turns within a closed magnetic circuit is equal to zero. In mathematical symbols,

$$(N_1 I_1)_{\text{input}} - (N_c I_c)_{\text{core}} - (N_2 I_2)_{\text{output}} = 0 \qquad (2.13)$$

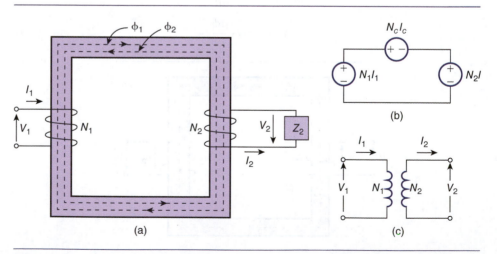

FIG. 2-6 An ideal transformer. **(a)** Physical representation. **(b)** Equivalent magnetic circuit. **(c)** Equivalent electrical circuit.

The magnetic potential $(N_c I_c)_{core}$ is the amount of magnetomotive force (mmf) required to magnetize the core of the transformer. For an ideal transformer ($\mu = \infty$), this mmf is equal to zero. That is,

$$(N_c I_c)_{core} = 0.$$

Thus, from Eq. (2.13), we have

$$I_2 = I_1 \frac{N_1}{N_2} \qquad (2.14)$$

From Ohm's law, we obtain

$$Z_1 = \frac{V_1}{I_1} \qquad (2.15)$$

and

$$Z_2 = \frac{V_2}{I_2} \qquad (2.16)$$

where Z_1 and Z_2 are not the winding impedances, but the impedances as seen from the terminals of winding 1 and winding 2, respectively.

From Eqs. (2.11), (2.14), (2.15), and (2.16), we obtain the impedance transformation property of transformers. That is,

$$Z_1 = Z_2 \left(\frac{N_1}{N_2}\right)^2 \qquad (2.17)$$

It should be emphasized that Eqs. (2.11), (2.14), and (2.17) are applicable only to ideal transformers. An ideal transformer is represented by the equivalent circuit shown in Fig. 2-6(c).

The 5 kVA, 480–120 V, single-phase transformer shown in Fig. 2-7 delivers rated current to a 120 volt load. Neglecting losses, determine the transformer currents and the supply voltage.

EXAMPLE 2-1

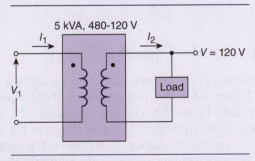

5 kVA, 480-120 V

FIG. 2-7

SOLUTION

The magnitude of the current through the secondary winding is

$$I_2 = \frac{|S|}{V} = \frac{5000}{120} = \underline{41.67 \text{ A}}$$

The windings turns ratio is given by the ratio of the windings voltages:

$$\frac{N_2}{N_1} = \frac{V_2}{V_1} = \frac{120}{480} = \frac{1}{4}$$

From Eq. (2.14), the magnitude of the current through the primary winding is

$$I_1 = \frac{1}{4}(41.67) = \underline{10.42 \text{ A}}$$

The primary voltage is given by the rating of the transformer. That is,

$$\underline{V_1 = 480 \text{ volts}}$$

In an actual transformer, the primary voltage and current would be slightly higher than calculated here because of the effect of the transformer's impedances.

Exercise 2-1

A 5 kVA, 240–120 volt, single-phase transformer supplies rated current to a load at 120 volts. Determine the magnitude of the load impedance as seen from the input terminals of the transformer.

Answer   11.52 Ω

2.1.3 Derivation of the Equivalent Circuit

The analysis of transformers is greatly simplified by using a model, called the *equivalent circuit*. Figure 2-8 shows various forms of the equivalent circuit of a single-phase transformer. Figure 2-8(a) is the schematic of a single-phase transformer, and its approximate equivalent circuit is shown in Fig. 2-8(b).

The resistance R_1 and the reactance X_1 represent the copper losses and the leakage flux of the primary winding, respectively. Similarly, the resistance R_2 and the reactance X_2 represent the copper losses and the leakage flux of the secondary winding, respectively. Then $R_1 + jX_1$ and $R_2 + jX_2$ represent the impedances

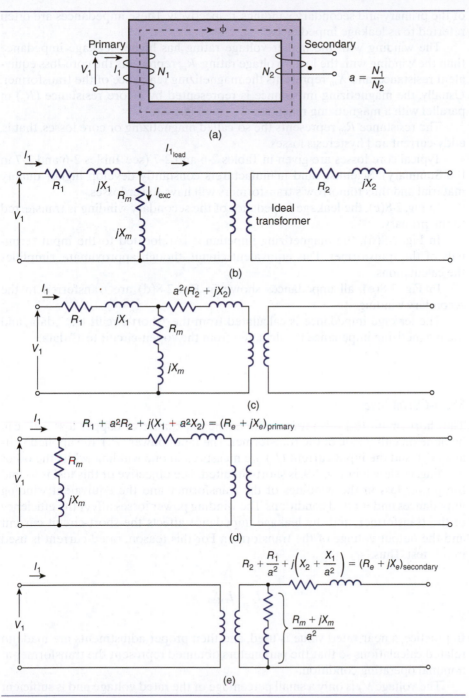

FIG. 2-8 Various forms of the equivalent circuit of a single-phase transformer.
(a) Elementary schematic. **(b)** Individual winding impedances and the magnetizing impedance. **(c)** All impedances transferred to the primary. **(d)** An approximation of part (b). **(e)** All impedances referred or transferred to the secondary.

of the primary and secondary windings, respectively. These impedances are often referred to as leakage impedances.

The winding with the higher voltage rating has higher leakage impedance than the winding with the lower voltage rating. R_m represents the core-loss equivalent resistance, and X_m represents the magnetizing reactance of the transformer. Usually, the magnetizing impedance is represented by a core resistance (R_c) in parallel with a magnetizing reactance X_ϕ [see Fig. 2-8(c)].

The resistance R_m represents the so-called magnetizing or core losses, that is, eddy-current and hysteresis losses.

Typical core losses are given in Tables 2-6 and 2-7 (see Tables 2-6 and 2-7 in the Summary). Scientists and manufacturers constantly develop higher quality material and thus tomorrow's transformers will have lower losses.

In Fig. 2-8(c), the leakage impedance of the secondary winding is transferred to the primary.

In Fig. 2-8(d), the magnetizing impedance is relocated to the input terminals of the transformer. This equivalent circuit, though approximate, simplifies the calculations.

In Fig. 2-8(e), all impedances shown in Fig. 2-8(d) are transferred to the secondary winding.

The leakage impedance is calculated from the "short-circuit test" data, and the magnetizing impedance is calculated from the "open-circuit test" data.

Short-Circuit Test

The short-circuit test (also referred to as the impedance or copper-loss test) can be done on either side of the transformer. The input power (P_z), the applied voltage (V_z), and the input current (I_z) are measured in one winding, while the other winding, as shown in Fig. 2-9, is short-circuited. The objective of this test is to find the power loss in the windings of the transformer and the equivalent winding impedances under rated conditions. The winding power losses affect the efficiency of the transformer, and the leakage impedance affects the short-circuit current and the output voltage of the transformer. For this reason, rated current is used in this test. Thus,

$$I_z = I_{\text{rated}}$$

In practice, a near rated value is used, and then proper adjustments are made to related calculations so that the parameters obtained represent the transformer at nominal operating condition.

The voltage V_z is only a small percentage of the rated voltage and is sufficient to circulate rated current in the windings of the transformer. Usually,

$$V_z = (2\% \longrightarrow 12\%)V_{\text{rated}}$$

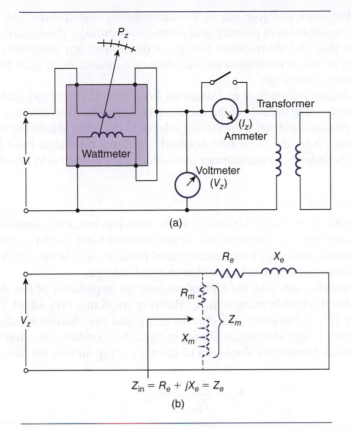

FIG. 2-9 Short-circuit test. **(a)** Laboratory connection schematic. **(b)** Equivalent circuit. (Notice that the magnetizing impedance Z_m is considered infinite.)

As a result, the transformer is magnetized at a relatively low flux density, as shown in Fig. 2-11. In performing this test, the magnetizing impedance is assumed to be open because, in relative terms, it is very large.

Neglecting the losses of the measuring instruments, we have

$$R_e = \frac{P_z}{I_z^2} \text{ ohms} \qquad (2.18)$$

$$|Z_e| = \frac{V_z}{I_z} \text{ ohms} \qquad (2.19)$$

and

$$X_e = \sqrt{Z_e^2 - R_e^2} \text{ ohms} \qquad (2.20)$$

This test does not give the individual winding impedances, but rather the combined impedances of primary and secondary windings. The subscript *e* is used to indicate that the short-circuit test gives the equivalent transformer leakage impedance, or the combination of the winding impedances as seen from one of the transformer windings.

The winding parameters as calculated from Eqs. (2.18), (2.19), and (2.20) are said to be referred to the side of the transformer where the instruments were placed. However, you can easily transfer this impedance to the other side of the transformer by multiplying it by the turns ratio squared. The turns ratio used must be the one seen from the side of the transformer to which this impedance is to be referred.

Open-Circuit Test

The open-circuit test (also referred to as the core-loss test, the magnetization test, the excitation test, the iron-loss test, or the no-load test) furnishes the core loss and the magnetizing impedance under rated conditions. It is usually done on the side of the transformer that has the lower rated voltage.

In conducting this test, the equivalent leakage impedance of the transformer is considered negligible because it is, relatively speaking, very small. The excitation power (P_{exc}), the excitation current (I_{exc}), and the excitation voltage (V_{exc}) are measured in one winding, as shown in Fig. 2-10(a), while the other winding is open-circuited. Neglecting the losses of the measuring meters, we have

$$R_m = \frac{P_{exc}}{I_{exc}^2} \text{ ohms} \tag{2.21}$$

$$|Z_m| = \frac{V_{exc}}{I_{exc}} \text{ ohms} \tag{2.22}$$

and

$$X_m = \sqrt{Z_m^2 - R_m^2} \text{ ohms} \tag{2.23}$$

The excitation current is a small percentage of the transformer's nominal current. Usually,

$$I_{exc} = (3\% \longrightarrow 10\%) \, I_{rated}$$

In order to obtain the core loss that corresponds to rated conditions, the flux on open-circuit test must be equal to the flux within the transformer when the transformer delivers rated current. This is insured, as can be seen from Eq. (2.7), when the excitation voltage is equal to the rated voltage. That is,

$$V_{exc} = V_{rated}$$

Thus, as shown in Fig. 2-11, on an open-circuit test the transformer is energized at approximately the same flux level as under normal operating conditions.

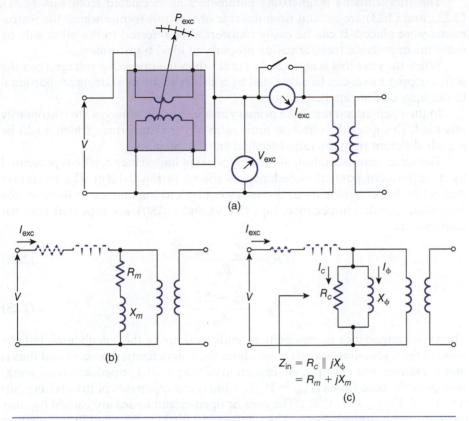

FIG. 2-10 Open-circuit test. **(a)** Laboratory connection schematic. **(b)** Series representation of the magnetizing impedance (notice that $R_1 + jX_1$ are considered negligible). **(c)** Parallel representation of the magnetizing impedance.

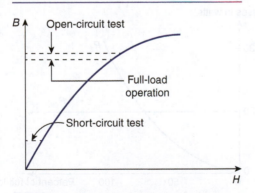

FIG. 2-11 *B-H* curve, showing the relative magnetization levels for a transformer during a short-circuit test, open-circuit test, and under full-load operating conditions.

The transformer's magnetizing parameters as calculated from Eqs. (2.21), (2.22), and (2.23) are as seen from the side of the transformer where the instruments were placed. It can be easily transferred or referred to the other side by using the impedance transformation property of ideal transformers.

When the core loss is measured at other than the operating voltage, then the actual copper losses can be calculated by considering them as being proportional to the square of the applied voltage.

In the open-circuit test, both primary and secondary voltages are customarily recorded. This gives the effective turns ratio of the transformer, which might be slightly different from the ratio specified on its nameplate.

The series representation of the magnetizing impedance can be represented by its equivalent parallel impedance, as shown in Fig. 2-10(c). The equations that relate the components of the series equivalent impedance to those of the equivalent parallel impedance, Eqs. (1.179) and (1.180), are repeated here for convenience:

$$R_c = \frac{R_m^2 + X_m^2}{R_m} \tag{2.24}$$

$$X_\phi = \frac{R_m^2 + X_m^2}{X_m} \tag{2.25}$$

The series impedance representation tends to simplify the calculations, but the parallel representation reveals more about the magnetization process and thus is more common. For example, an inspection of the parallel impedance representation gives the core losses ($P_{exc} = V^2/R_c$) and the components of the exciting current ($I_\phi = V/jX_\phi$, $I_c = V/R_c$). The core or open-circuit losses are caused by eddy currents and hysteresis losses. These losses are further discussed in the sections that follow. Typical winding and iron losses for a 5 kVA single-phase transformer as a function of load current are shown in Fig. 2-12.

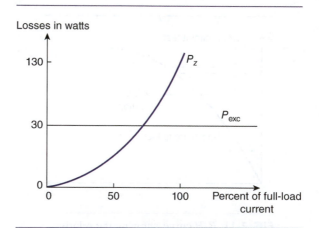

Losses in watts

FIG. 2-12 Typical winding and iron losses as a function of load current for a single-phase transformer (5 kVA, 480–120 V).

In concluding this section on transformer tests, it must be emphasized that Z_e is the input impedance of the transformer when its secondary is shorted, and Z_m is the input impedance of the transformer when its secondary is open-circuited. These statements are, of course, approximate because of the underlying assumptions.

The following results were obtained from testing a 10 kVA, 480–120 V, single-phase transformer.

EXAMPLE **2-2**

Test	Voltage in Volts	Current in Amperes	Power in Watts
Open-circuit	120	2.5	60
Short-circuit	26	20.83	200

Determine:

a. The equivalent leakage impedance of the transformer windings referred to the high-voltage (HV) and low-voltage (LV) winding.
b. The series and parallel components of the magnetizing branch referred to the LV winding.
c. The equivalent circuit of the transformer referred to the HV winding.

SOLUTION

a. The rated current of the transformer through the primary and secondary windings is:

$$I_H = \frac{|S|}{V} = \frac{10,000}{480} = 20.83 \text{ A}$$

$$I_L = \frac{10,000}{120} = 83.33 \text{ A}$$

By comparing the above with the given test data, it is clear that the short-circuit test was done on the HV winding.

Thus,

$$R_{eH} = \frac{P_z}{I_z^2} = \frac{200}{(20.83)^2} = 0.46 \text{ } \Omega$$

and

$$|Z_{eH}| = \frac{V_z}{I_z} = \frac{26}{20.83} = 1.25 \text{ } \Omega$$

$$X_{eH} = \sqrt{Z_{eH}^2 - R_{eH}^2} = \sqrt{(1.25)^2 - (0.46)^2}$$

$$= 1.16 \text{ } \Omega$$

Thus,

$$Z_{eH} = \underline{0.46 + j1.16 \ \Omega}$$

Using the impedance transformation property of the transformers, we obtain the equivalent impedance referred to the low-voltage side:

$$a = \frac{N_2}{N_1} = \frac{120}{480}$$

Thus,

$$Z_{eL} = \left(\frac{120}{480}\right)^2 (0.46 + j1.16) = \underline{0.029 + j0.073 \ \Omega}$$

b.   The equivalent series elements of the magnetizing impedance are

$$|Z_{mL}| = \frac{V_{\text{exc}}}{I_{\text{exc}}} = \frac{120}{2.5} = 48 \ \Omega$$

$$R_{mL} = \frac{P_{\text{exc}}}{I_{\text{exc}}^2} = \frac{0.60}{(2.5)^2} = 9.6 \ \Omega$$

and

$$X_{mL} = \sqrt{Z_{mH}^2 - R_{mH}^2} = \sqrt{48^2 - 9.6^2}$$
$$= \underline{47.03 \ \Omega}$$

The parallel components of the magnetizing impedance, as seen from the *LV* winding, are calculated from the series equivalent components by using Eqs. (2.24) and (2.25):

$$R_{cL} = \frac{R_m^2 + X_m^2}{R_m} = \frac{9.6^2 + 47.03^2}{9.6} = \underline{240 \ \Omega}$$

and

$$X_{\phi L} = \frac{R_m^2 + X_m^2}{X_m} = \frac{9.6^2 + 47.03^2}{47.03} = \underline{48.99 \ \Omega}$$

c.   Referring to the resistance and the reactance of the magnetizing branch to the HV winding, we obtain

$$R_{cH} + jX_{\phi H} = \left(\frac{480}{120}\right)^2 (240 + j48.99)$$

$$= (3840 + j783.84) \ \Omega$$

The equivalent circuit referred to the HV winding is shown in Fig. 2-13.

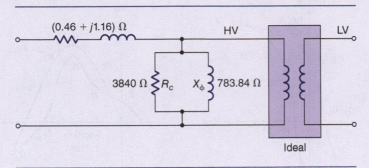

FIG. 2-13

2.1.4 Waveform of Excitation Current

When one of the transformer's windings is open-circuited while the other is connected to rated voltage, the resulting current represents a small percentage of the transformer's rated current. This is due to the transformer's high magnetization impedance and to the infinite impedance seen by one winding while the other is open-circuited. This current is referred to as the excitation current (I_{exc}). In practice,

$$I_{exc} = (3\% \rightarrow 10\%)\, I_{rated} \tag{2.26}$$

The theoretical derivation of the exciting current's waveform is obtained from the transformer's *B-H* characteristic as follows: Draw the waveform of the sinusoidal flux on an ωt-axis. Then, on the same coordinate system, draw the waveform of the exciting current. Refer to Fig. 2-14(a). The waveform of the flux can easily be obtained, because, it will be noted, the points $(\alpha_1, 0)$, (d_1, d_2), and $(g_1, 0)$ are $\pi/2$ radians apart and correspond, respectively, to zero, maximum, and zero values of the flux. In other words, the critical points of the flux are known during half the period of its sinusoidal function. The resulting flux waveform is sketched in Fig. 2-14(b).

The magnitude of the exciting current at each of the coordinate points $[(\alpha_1, 0)$, $(b_1, b_2), (d_1, d_2)$, etc.] of the *B-H* curve is given by the abscissa of these points. This is shown in Fig. 2-14(b). The transformer's voltage, flux, and exciting current are shown in Fig. 2-14(c).

The exciting current is not sinusoidal because of the nonlinearities of the *B-H* curve, and so a conventional phasor diagram of the exciting current cannot be drawn. However, its harmonics content, or its equivalent sinusoidal functions, can be found by using the Fourier series analysis. The Fourier series method is a mathematical tool used to determine the harmonic content of nonsinusoidal waveforms.

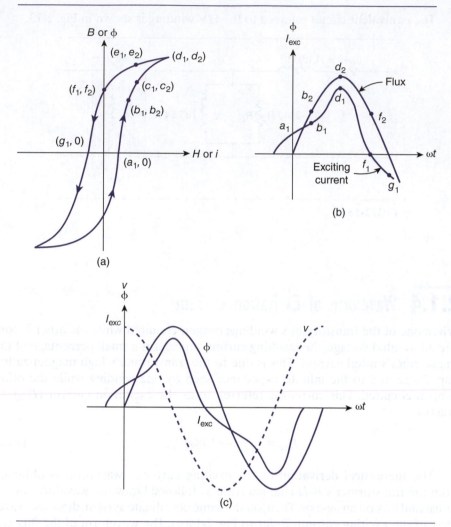

FIG. 2-14 Typical magnetic characteristics of transformers. **(a)** *B-H* curve. **(b)** Instantaneous values of flux and exciting current. **(c)** Instantaneous values of induced voltage, flux, and exciting current.

It can be shown that the exciting current is made up of "odd harmonics,"* that is, a fundamental term, a third harmonic, a fifth harmonic, and so on. The third harmonic can be as high as 40% of the fundamental term.

The exciting current is very important in the operation of all transformers, particularly three-phase transformers. It is discussed in greater detail in Section 2.2.4.

---

*According to Fourier series analysis, any function $f(t)$ of period T has only *odd* harmonics if $f(t)$ satisfies the following relationship: $f(t) = -f[t + T/2]$. The function that has such a characteristic is said to have half-wave *odd* symmetry.

2.1.5 Components of Primary Current and Corresponding Fluxes

When a transformer is connected to its primary supply voltage while its secondary is open-circuited, it draws (as stated previously) a small percentage of its rated current (I_{exc}). Although this current is small, it produces rated flux within the core of the transformer.

As soon as an impedance is connected to the transformer's secondary (see Fig. 2-15(a)), a load current will flow. This current will produce the flux (ϕ_2) that opposes the flux of the primary current. This cannot be tolerated because it would be accompanied by a reduction of the voltage induced in the primary [$v_i = -N(d\phi/dt)$], which would result in violation of KVL ($V_1 = v_i$). To overcome this opposition, the current drawn by the primary winding *increases* in such a way as to completely cancel the magnetic opposition of the secondary current.

The quantity of primary current needed to produce flux sufficient to completely neutralize the opposition of the secondary current is called the load component of the primary current. It is designated by I_{1L}. Alternatively, the mmf of the secondary winding ($N_2 I_2$) is equal and opposite to the increase in the mmf of the primary winding ($N_1 I_{1L}$). As a result, the effective mmf of the transformer at negligible leakage flux is $N_1 I_{exc}$.

The increase in the primary current that accompanies the flow of current in the secondary could also be justified by considering the following principle of energy conservation:

$$\begin{pmatrix} \text{energy drawn by} \\ \text{the transformer} \end{pmatrix} = \begin{pmatrix} \text{energy delivered} \\ \text{to the load} \end{pmatrix} + \begin{pmatrix} \text{energy consumed within} \\ \text{the transformer} \end{pmatrix}$$

The energy demand of the load is met by an increase in the magnitude of the primary current and its power factor relative to the no-load condition.

From the previous discussion it is evident that, under nominal operating conditions, the primary current is made up of two components. One, the exciting current, is required to magnetize the transformer; the other, its load component (L_{1L}), is needed to cancel the opposition of the secondary current. In mathematical form, and as shown in Fig. 2-15(b),

$$I_1 = I_{exc} + I_{1L} \tag{2.27}$$

The primary current I_1 is almost sinusoidal because its nonsinusoidal component, the exciting current, is negligible in comparison to its sinusoidal load component.

The flux of the primary current is shown in Fig. 2-15(a). Its mathematical representation is

$$\phi_1 = \phi_{exc} + \phi_{1L} \tag{2.28}$$

where ϕ_{1L} is the load component of the primary flux.

Phasor Diagram

Refer to Fig. 2-15(b). The transformer delivers power to a lagging-power-factor load. The windings turns ratio and the load voltage V_L normally are known. The load current I_2 is obtained from the load characteristics. The basic circuit equations are

$$V_1 N_2 = V_{nl} N_1 \tag{2.29}$$

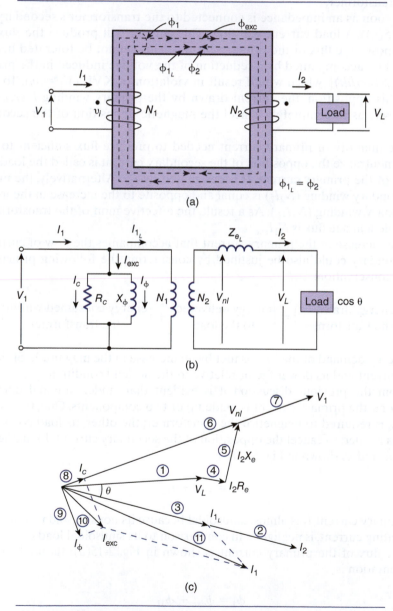

(a)

(b)

(c)

FIG. 2-15 Transformer. **(a)** Polarity and component fluxes. **(b)** Equivalent circuit. **(c)** Phasor diagram.

and

$$V_{nl} = V_L + I_2 Z_{eL} \qquad (2.30)$$

These equations, together with Eq. (2.27), are used to construct the phasor diagram shown in Fig. 2-15(c). The encircled numbers indicate the sequence of steps that you may follow in order to simplify its construction.

2.1.6 Transformer Characteristics

Inrush Current

When switching ON a transformer, the input or the inrush current is often many times larger than its rated current. This occurs when the switching takes place at the instant the input voltage is at the zero point on its time cycle. This results in a maximum flux that corresponds to a near saturation point on the *B-H* curve. At this point, the permeability of the magnetic material, the corresponding inductance, and the magnetizing impedance are at a minimum. Thus, the resulting current is at a maximum.

The magnitude and the waveform of the inrush current also depend on the leakage impedance of the transformer, on its magnetization characteristic, on the residual flux, and on the magnitude of the applied voltage at the instant of switching. The current waveform is such as to satisfy KVL. It cannot be accurately described because of the variable and nonlinear circuit characteristics. You can reduce the magnitude of the inrush current by selecting the transformer's protective circuit breaker with a microprocessor. The microprocessor switches ON the breaker when the supply voltage is zero.

The magnitude of the inrush current is comparable to the current that results from external transformer shorts. However, the short-circuit current is of the

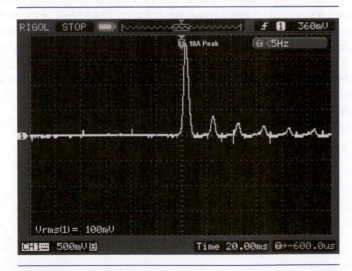

FIG. 2-16 The inrush waveform that is captured when power is applied at the main zero crossing point. *Courtesy of ESP*

fundamental frequency, while the magnetizing current has a large content (up to 40%) of the third harmonic. The magnitude of the inrush current, as per standard industrial practice, is taken as being between 8 and 12 times the rated current of the transformer, and its duration is about 100 milliseconds. As such, it is important in the selection of the upstream protective devices.

Efficiency

An unloaded transformer when connected to its voltage source, draws only the magnetization current on the primary side, the secondary current being zero. As the load is increased, the primary and secondary currents increase as per the load requirements. The volt amperes and wattage handled by the transformer also increase. Due to the presence of no load losses and I^2R losses in the windings, a certain amount of electrical energy gets dissipated as heat inside the transformer. This gives rise to the concept of efficiency (http://nptel.iitm.ac.in/courses/IIT-MADRAS/Electrical_Machines_I/pdfs/1_10.pdf).

By definition, the efficiency (η) of a transformer is given by:

$$\eta = \frac{P_{\text{out}}}{P_{\text{in}}} = \frac{P_{\text{in}} - P_{\text{loss}}}{P_{\text{in}}} = 1 - \frac{P_{\text{loss}}}{P_{\text{in}}} \qquad (2.31)$$

where P_{in} and P_{out} correspond, respectively, to the input and output power of the transformer expressed in watts. The power loss (P_{loss}) is the sum of the core and copper losses of the transformer at the operating conditions under consideration. The copper loss depends on the kVA delivered. When the kVA drawn by the load varies over a 24-hour period, then the so-called 24-hour transformer efficiency may be calculated by using energy instead of power. Generally speaking, the efficiency of transformers increases with their rating.

Industry continuously tries to minimize core loss by developing higher-quality magnetic materials from which the transformers can be economically manufactured. Where economics permit, copper windings are used instead of aluminum because the aluminum gives higher winding loss (I^2R).* The efficiency of the transformer is of primary importance because the cost of energy loss within the transformer over its lifetime is usually greater than that of the transformer itself. Typical transformer losses are shown in Tables 2-5 through Table 2-8 (see Summary). In the absence of winding condensation, idle transformers should be disconnected from their supply lines; otherwise, their energy consumption, due to core loss, is wasted.

Refer to Fig. 2-17. From basic definitions, we have

$$\text{efficiency} = \frac{\text{output power}}{\text{output power} + \text{losses}} \qquad (2.32)$$

or

$$\eta = \frac{V_L I_L \cos\theta}{V_L I_L \cos\theta + P_{\text{exc}} + R_e I_L^2} \qquad (2.33)$$

---

* The resistance of an aluminum conductor with the same current capacity as that of a copper conductor is about 10% higher.

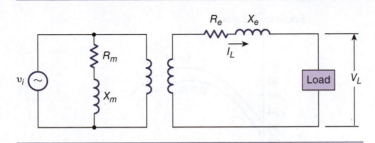

FIG. 2-17 Transformer's equivalent circuit and secondary load.

and

$$\eta = \frac{KI_L}{KI_L + P_{\text{exc}} + R_e I_L^2} \tag{2.34}$$

where

$$K = V_L \cos \theta \tag{2.35}$$

Taking the derivative of the efficiency with respect to the variable parameter I_L, we obtain

$$\frac{d\eta}{dI_L} = \frac{(KI_L + P_{\text{exc}} + R_e I_L^2)K - KI_L(K + 0 + 2R_e I_L)}{(KI_L + P_{\text{exc}} + R_e I_L^2)^2} \tag{2.36}$$

Setting the last equation equal to zero gives

$$P_{\text{exc}} = R_e I_L^2 \tag{2.37}$$

or

$$P_{\text{exc}} = P_z \tag{2.38}$$

Thus, the efficiency of a transformer at a constant load power factor is maximum when the iron loss is equal to its winding loss.

A transformer's efficiency depends on the load current and its power factor. The first condition is obvious, while the second becomes evident when you consider that the primary current, being the vector sum of the exciting current and the component of the load current (see Fig. 2-15(b)), depends on the power factor of the load. The magnitude of the primary current, in turn, controls the primary winding's copper losses. When you speak of a transformer's efficiency, then you must always identify the corresponding current and the power factor.

The variation of efficiency versus load current, as a function of the load current's power factor, is shown in Fig. 2-18.

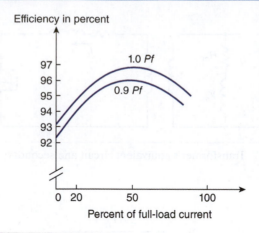

FIG. 2-18 Efficiency as a function of load current and power factor.

The variation of iron loss, winding loss, and efficiency versus load current for a 25 kVA transformer is shown in Fig. 2-19.

Regulation

Voltage regulation is a measure of the change in the magnitude of the output voltage while the load current varies from zero up to its rated value. In mathematical form, regulation is given by

$$\text{regulation \%} = \frac{|V_{nl}| - |V_{fl}|}{|V_{fl}|}(100) \qquad (2.39)$$

where V_{nl} and V_{fl} are, respectively, the magnitudes of the voltages at the output terminals of the transformer at no load and at full load. Full-load voltage is also

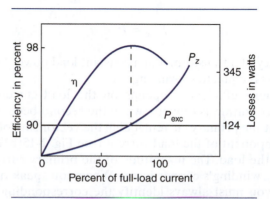

FIG. 2-19 Efficiency, iron loss, and winding loss as a function of load current for a 25 kVA, 480-120 V, single-phase transformer.

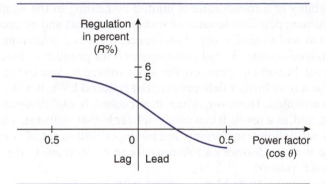

FIG. 2-20 Regulation as a function of load power factor for constant rated current.

referred to as nameplate voltage, rated voltage, or nominal transformer voltage. No-load voltage can be obtained from the input voltage and the nominal voltage ratio of the transformer. Equation (2.39) can be used for any operating condition. In such cases, the full-load voltage is replaced by its actual value.

Regulation depends on the leakage impedance of the transformer and on the power factor of the load. At a particular value of load power factor and leakage impedance, voltage regulation becomes ideal or zero. Depending on the power factor of the load, the regulation could be positive or negative. Usually, under steady-state operating conditions, it is less than plus or minus 5% (±5%). Regulation as a function of power factor is shown in Fig. 2-20. The effects of the load's power factor and the transformer's leakage impedance are demonstrated in Example 2-4.

At leading power factors, regulation is usually negative; that is, the voltage at the secondary terminals of a transformer is larger at full load than it is when the load is disconnected. In such cases, the equipment connected to a transformer's secondary may be subjected to higher than rated voltages. This may occur when the power-factor-correction capacitor banks remain on the network while the plant operates at a reduced load.

The voltage across the load is of primary importance because many operating characteristics of various pieces of equipment depend on it. For example, a 5% reduction of the rated voltage of an incandescent lamp noticeably reduces its output, and the light appears dimmer. Furthermore, the torque delivered by the motors, as explained in Chapter 3, is reduced by a factor of 10% ($T \alpha V^2$). For these reasons, most power distribution transformers, as per manufacturers' standards, are equipped with "off-load" voltage tap changers through which the voltage can be changed by ±2.5%, or by ±5% in relation to their nominal voltage. When the voltage is to be changed, the load of the transformer is disconnected and the number of turns of the primary winding is changed according to the requirements.

kVA Rating

The apparent power, or the kVA rating, of a transformer is always inscribed on its nameplate. It indicates the transformer's transformation capacity in terms of the volt-amperes a transformer is designed to deliver.

The capability of a transformer is limited by heating in the windings (hence, there is a maximum permissible value of sustained current) and by excessively high exciting current and excessive core loss (hence, there is a maximum permissible value for sustained voltage). Rated volt-amperes is the product of rated voltage and nominal current. In normal operation, the input voltage is close to the rated value.

As long as a transformer delivers rated or reduced kVA, it will operate without being overheated. However, when it is cooled, it can dissipate more than nominal heat, and, as a result, it can safely deliver higher-than-rated kVA. For this reason, most substation transformers are equipped with a set of ventilating fans, which enable the transformer's kVA capacity to be increased proportionally to the ventilation furnished.

A transformer of 1000 kVA supplied with fans, for example, can deliver 1333 kVA without being overheated. Such a transformer is identified as 1000/1333 kVA. Transformers with higher kVA ratings can transform higher magnitudes of voltage and current; that is, their windings have, on a relative basis, larger diameters and more insulation. As a result, the *physical* size and the *cost* of a transformer depend on its kVA rating.

The importance of this parameter is further explained in Section 2.6, which provides information about all the important parameters that appear on the nameplate of a transformer. A familiarity with nameplate data facilitates the solution of practical and theoretical transformer problems.

2.1.7 Per-Unit Values

When engineering parameters such as power, torque, speed, voltage, current, and impedance are expressed in their corresponding per-unit values, many engineering concepts and calculations are simplified. For this reason, almost all design and problem solving in the power-distribution field is implemented by expressing all pertinent parameters in their equivalent per-unit values.

Actual engineering parameters are changed to their equivalent per-unit values as follows. The per-unit value of a parameter K is equal to its actual value divided by the base value. That is,

$$K = \frac{\text{actual value of } K}{\text{base value of } K} \text{ pu} \qquad (2.40)$$

Base parameters are normally obtained from the nameplate data of the transformer, as follows:

Base power (S_b) = volt-ampere rating of the transformer $\qquad$ **(2.40a)**

Base voltage (V_b) = rated voltage of the winding under consideration $\quad$ **(2.40b)**

Base current (I_b) = rated current of the winding under consideration

$$= \frac{S_b}{V_b} \qquad (2.40c)$$

$$\text{Base impedance } (Z_b) = \frac{\text{base value of voltage}}{\text{base value of current}} \qquad (2.40d)$$

From Eqs. (2.40c) and (2.40d), we obtain

$$Z_b = \frac{V_b^2}{S_b} \qquad (2.41)$$

The base power is the same in both the high- and the low-voltage windings ($S_{bH} = S_{bL}$). In contrast, the base voltage, base current, and base impedance are different for each winding of the transformer, and thus the subscript H or L should be included in the above expressions to represent parameters at the high- and low-voltage windings. The base value is the value of the parameter expressed in standard engineering units.

When the ohmic value of an impedance is $15\underline{/40°}$ ohms and the value of the base impedance is 300 ohms, then the per-unit value of the impedance is

$$Z = \frac{15\underline{/40°}}{300} = 0.05\underline{/40°} \text{ pu}$$

Similarly, when the actual value of a voltage is 250 volts, and the base voltage is 200 volts, then the per-unit value of the voltage is

$$V = \frac{250}{200} = 1.25 \text{ pu}$$

The per-unit value of a parameter, multiplied by 100, gives the value of the parameter in percent. Thus, when the per-unit value of a parameter is 0.04, its percentage value is 4%.

One advantage of the per-unit system is that when the voltage, current, and impedance of a transformer are expressed in per-unit , they have the same values regardless of the side of the transformer to which they refer. This greatly simplifies the understanding and solution of various transformer problems.

An additional advantage is that when the parameters of a machine are expressed as per-unit, they all fall within a known range, regardless of the rating of the machine. The per-unit values of the equivalent leakage impedance of most single-phase transformers, for example, are within the range of 0.035–0.055 pu. Knowing this is of tremendous importance, greatly simplifying the solution of problems and the conceptual understanding of various transformer operating conditions.

It can be shown that the per-unit value of the transformer's leakage impedance is equal to the voltage required on a short-circuit test, expressed as per-unit of the rated value. That is,

$$Z_{e\text{pu}} = V_{z\text{pu}} \qquad (2.42)$$

This is left as a student problem (see Exercise 2-3).

Because the per-unit system is used in most of the chapters of this text, it is described in detail in the Appendix.

One-Line Diagrams

Transformers, cables, and electric machines are conventionally identified in electrical drawings by their equivalent one-line representations. As the name implies, one-line diagrams represent electrical loads as being supplied through one wire, regardless of how many wires are used in the actual setup.

The main parameters of a system—per-unit impedances, kVA rating of transformers, kW rating of motors, size of cables, protective devices, the short-circuit MVA of the supply voltage source, and so on—are written on one-line diagrams.

For a single-phase system, the short-circuit apparent power ($|S|$) in volt-amperes is

$$|S| = VI_{sc} \tag{2.43}$$

where V and I_{sc} are, respectively, the nominal voltage and the short-circuit current. When the voltage is expressed in kV and the short-circuit current in kA, then Eq. (2.43) gives the short circuit apparent power in MVA.

From Eq. (2.43) and Ohm's law, we obtain

$$|S| = V\frac{V}{Z} \tag{2.44}$$

where Z is the impedance of the supply network in ohms/phase. From Eqs. (2.43) and (2.44), we obtain

$$Z = \frac{V^2}{|S|} \tag{2.45}$$

For a three-phase system, the short-circuit MVA is

$$S = \sqrt{3}\, V_{L\text{-}L} I_{sc} \tag{2.46}$$

where $V_{L\text{-}L}$ and I_{sc} are, respectively, the line-to-line nominal voltage expressed in kV and the short-circuit current expressed in kA. Thus, the short-circuit MVA of a given voltage source can be used to find the internal impedance of the source and/or the short-circuit current.

Protective equipment, such as circuit breakers and fuses, must be capable of safely withstanding the forces produced by the short-circuit currents. Since the short-circuit currents are given indirectly in terms of short-circuit MVA, it is customary in the industry to rate the short-circuit capacity of equipment in terms of short-circuit MVA.

Typical short-circuit MVA's for a 25 kV and 46 kV distribution system are as follows:

kV	MVA
25	500
46	1500

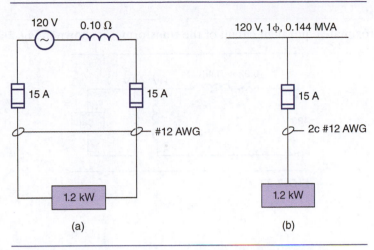

FIG. 2-21 Representation of a distribution system. **(a)** Two-wire. **(b)** One-line.

Under ideal conditions, the source impedance is equal to zero and its frequency is constant. Such a system is identified on the one-line diagram as an "infinite bus." In more general terms an infinite bus is a power source whose voltage and frequency remain constant independent of the active or reactive power being supplied.

Since all electrical drawings represent electrical loads by their one-line diagram, an attempt is made throughout this book to use the one-line diagram as often as possible. Figure 2-21(a) shows a two-wire diagram of a single-phase load. The 120 V, 60 Hz voltage supply has an internal reactance of 0.1 ohm. The 15 A fuses protect the 1.2 kW load from undesirable high currents. The two-conductor (2c) cable is size #12 AWG* (American Wire Gauge). When the load is shorted, the resulting short-circuit MVA is

$$|S| = V \, |I|$$

$$= 120 \left(\frac{120}{0.1} \right) \times 10^{-6}$$

$$= 0.144 \text{ MVA}$$

The one-line diagram of Fig. 2-21(a) is shown in Fig. 2-21(b).

A 5 kVA, 480–120 V single-phase transformer delivers rated current at 0.9 pf lagging. The transformer's equivalent leakage impedance referred to the high-voltage winding is 0.92 + j1.84 ohms. Find the per-unit values of the voltage, current, and leakage impedance on the HV and LV winding.

EXAMPLE **2-3**

---

*3.31 mm²

SOLUTION

The approximate equivalent circuit of the transformer is shown in Fig. 2-22.

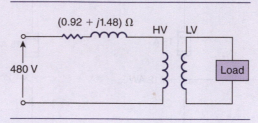

FIG. 2-22

HV Side

The base parameters are

$$S_b = 5000 \text{ VA}$$

$$V_{bH} = 480 \text{ V}$$

$$I_{bH} = \frac{5000}{480} = 10.42 \text{ A}$$

$$Z_{bH} = \frac{(480)^2}{5000} = 46.08 \text{ ohms}$$

The per-unit values are

$$V_H = \frac{\text{actual value}}{\text{base value}} = \frac{480}{480} = \underline{\underline{1.0 \text{ pu}}}$$

$$I_H = \frac{10.42\angle-25.8°}{10.42} = \underline{\underline{1.0\angle-25.8° \text{ pu}}}$$

$$Z_H = \frac{0.92 + j1.84}{46.08} = \underline{\underline{0.02 + j0.04 \text{ pu}}}$$

LV Side

The base parameters are

$$S_b = 5000 \text{ VA}$$

$$V_{bL} = 120 \text{ V}$$

$$I_{bL} = \frac{5000}{120} = 41.67 \text{ A}$$

$$Z_{bL} = \frac{(120)^2}{5000} = 2.88 \text{ ohms}$$

The per-unit values are

$$V_L = \frac{120}{120} = \underline{1.0 \text{ pu}}$$

$$I_L = \frac{41.67\underline{/-25.8°}}{41.67} = \underline{1.0\underline{/-25.8°} \text{ pu}}$$

The leakage impedance referred to the LV side is

$$Z_e = (0.92 + j1.84)\left(\frac{120}{480}\right)^2 = 0.06 + j0.12 \ \Omega$$

$$= \frac{0.06 + j0.12}{2.88} = \underline{0.02 + j0.04 \text{ pu}}$$

The 100 kVA, 440–220 V, 60 Hz single-phase transformer shown in the one-line diagram representation (Fig. 2-23(a)) has negligible magnetizing current and a leakage impedance of 0.03 + j0.040 pu. The transformer is connected to a 440 V source through a feeder whose impedance is 0.04 + j0.08 ohms.

EXAMPLE **2-4**

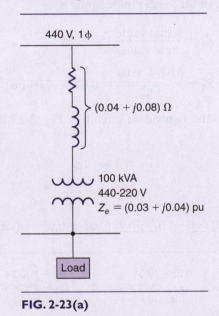

440 V, 1φ

(0.04 + j0.08) Ω

100 kVA
440-220 V
$Z_e = (0.03 + j0.04)$ pu

Load

FIG. 2-23(a)

Determine:

a. The voltage across the load when the transformer is connected to a 440 V source and delivers rated current to a load of 0.80 power factor lagging.

b. The voltage regulation.

c. The primary and secondary short-circuit MVA when the secondary of the transformer is shorted.

d. The rating of the capacitor in kVAR which, when connected across the load, will improve the power factor from 0.80 lagging to unity.

SOLUTION

a. The base impedance in the HV winding is

$$Z_{bH} = \frac{(440)^2}{100,000} = 1.94 \ \Omega$$

The per-unit value of the voltage is

$$V_{pu} = \frac{\text{actual value}}{\text{base value}} = \frac{440}{440} = 1.0 \ \text{pu}$$

Similarly,

$$I_{pu} = 1.0\underline{/-36.9°} \ \text{pu}$$

The per-unit value of the feeder impedance is

$$Z_f = \frac{\text{actual value}}{\text{base value}}$$

$$Z_f = \frac{0.04 + j0.08}{1.94} = 0.02 + j0.04 \ \text{pu}$$

Applying KVL to the equivalent circuit of Fig. 2-23(b) and using per-unit values, we have

$$V_L = V_1 - IZ$$

or

$$V_L\underline{/0°} = 1.0\underline{/\beta_1°} - 1.0\underline{/-36.9°} \ [0.03 + 0.02 + j(0.04 + 0.04)] \qquad \textbf{(I)}$$

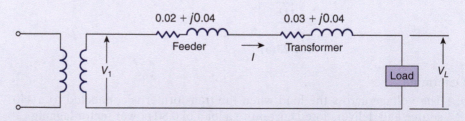

FIG. 2-23(b)

The given power factor of the load establishes the reference for the phase angles.

In other words, the voltage across the load has a phase angle equal to zero, while the input voltage will be at an angle of β_1 degrees to the reference.

From Eq. (I), and after some algebraic calculations, we obtain

$$V_L\underline{/0°} = 1.0\underline{/\beta_1°} - 0.096\underline{/21.2°}$$

Equating the imaginary components of the above equation, we get

$$\beta_1 = \arcsin(0.096 \sin 21.2°) = 2°$$

Equating the real components, we obtain

$$V_L = 1.0 \cos 2° - 0.096 \cos 21.2°$$
$$= 0.91 \text{ pu}$$
$$= 0.91 (220)$$
$$= \underline{200.21 \text{ V}}$$

Alternatively, using *ohmic values*, from KVL we have

$$Z_t = 1.94 (0.03 + j0.044) = 0.06 + j0.08 \ \Omega$$

The total impedance referred to the HV winding is

$$Z = (0.06 + j0.08) + (0.04 + j0.08) = 0.19\underline{/58°} \ \Omega$$

The magnitude of the current in the HV is

$$|I| = \frac{100,000}{440}$$
$$= 227.27 \text{ A}$$

Thus,

$$V_L = \frac{1}{2}(440\underline{/\beta_1°} - IZ)$$

or

$$V_L\underline{/0°} = \frac{1}{2}[440\underline{/\beta_1°} - 227.27\underline{/-36.9°}(0.19\underline{/58°})]$$
$$= \frac{1}{2}[440\underline{/\beta_1°} - 42.156\underline{/21.2°})$$

Equating the imaginary components of the last equation, we obtain

$$\beta_1 = \arcsin \frac{15.25}{440} = 2°$$

Equating the real components, we obtain

$$V_L = \frac{1}{2}(440 \cos 2° - 39.3) = \underline{200.21 \text{ V}}$$

b.　Using Eq. (2.39), the regulation in percent is

$$\text{regulation \%} = \frac{220 - 200.21}{200.21}(100) = \underline{9.9\%}$$

c.　The short-circuit volt-amperes are given by

$$(VA)_{sc} = V_{\text{rated}}I_{sc}$$

$$I_{sc} = \frac{V}{|Z|} = \frac{440}{0.19}$$

$$= 0.2372 \text{ A}$$

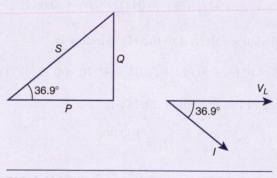

FIG. 2-23(c)

The short-circuit VA expressed in MVA is

$$(MVA)_{sc} = 0.440(2.372)$$

$$= \underline{1.044 \text{ MVA}}$$

Owing to the transformation properties of the transformer, the short-circuit MVA is the same whether it is referred to the high- or low-voltage winding. The power triangle and the phasor diagram of the load are shown in Fig. 2-23(c).

d.   The required reactive power (Q_c) of the capacitor is

$$Q_c = P \tan \theta$$

from which

$$Q_c = 100(0.8)(\tan 36.9°)$$

$$= \underline{60 \text{ kVAR}}$$

The approximate equivalent circuit of a single-phase 5 kVA, 480–240 V transformer is shown in Fig. 2-24.

Exercise

2-2

a.   Draw the equivalent circuit as seen from the high-voltage side.
b.   Draw the equivalent circuit as seen from the low-voltage side.
c.   Find the per-unit value of the transformer's leakage impedance.
d.   Determine the primary voltage when the transformer delivers rated current to a load at 240 volts and 0.80 power factor lagging.

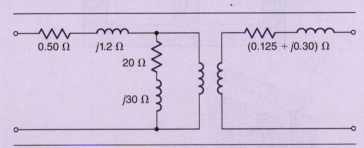

FIG. 2-24

Answer   (a) $Z_{eH} = 1 + j2.4 \ \Omega$;   (b) $Z_{eL} = 0.25 + j0.6 \ \Omega$;
(c) $Z_e = (0.022 + j0.052)$ pu;   (d) $503.52\underline{/1.56°}$ V.

Show that the per-unit value of a transformer's equivalent leakage impedance is equal to the per-unit value of the voltage employed on the short-circuit test.

Exercise

2-3

2.2 Three-Phase, Two-Winding Transformers

2.2.1 Introduction

Three-phase, two-winding transformers are used to interconnect two distribution systems of different voltages, such as a 25 kV transmission line and a 480 V plant. The majority of industrial transformers are of this type. They are called two-winding transformers because for each primary phase winding, there is only one secondary phase winding. The transformers actually have six coils—three for the incoming three-phase power and three for the output three-phase power.

Three-phase transformers can be either of the core type (Fig. 2-25) or of the shell type as shown in Fig. 2-26.

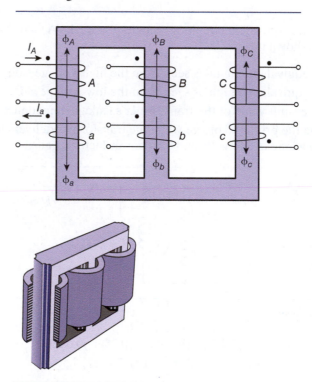

FIG. 2-25 Core-type, 3-ϕ, two-winding transformer.
Based on EPSTN, www.sayedsaad.com

Most transformers are of the core type. Their main advantage is that they prevent (ideally, if the phase core branches are identical and leakage is neglected) the presence of the third-harmonic flux, and hence, avoid inducing third-harmonic voltages. They are also less expensive than shell-type transformers of equivalent rating.

The prerequisites of understanding three-phase transformers are:

- Three-ϕ systems
- One-line diagrams

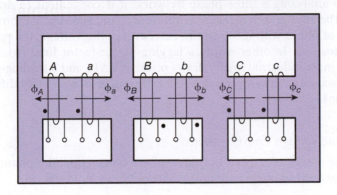

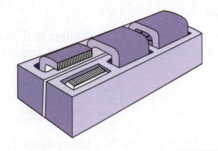

FIG. 2-26 Shell-type, 3-ϕ, two-winding transformer.
Based on EPSTN, www.sayedsaad.com

- Per-unit values
- Magnetic and electric circuit concepts
- Harmonics of the exciting current.

2.2.2 Review of Three-Phase Systems

Although three-phase systems are extensively discussed in Appendix A, they are briefly reviewed here for convenience.

Power Considerations

The active (P), reactive (Q), and complex power (S) drawn by a balanced three-phase load are given by

$$P = \sqrt{3}V_{L\text{-}L}I_L \cos \theta$$

$$Q = \sqrt{3}V_{L\text{-}L}I_L \sin \theta$$

$$S = \sqrt{3}V_{L\text{-}L}I_L{}^*$$

where $V_{L\text{-}L}$ is the rms value of the line-to-line voltage, I_L is the rms value of the line current, and $I_L{}^*$ is its conjugate in phasor form. θ is the phase angle of the load impedance.

When analyzing a three-phase network, it is convenient to draw its power triangle. The real power is taken as reference, and the reactive power is 90 degrees out of phase with it. When the current lags the voltage, the power triangle is drawn leading. In other words, a lagging-power-factor load (which is characteristic of most industrial loads) draws positive VAR, and a leading-power-factor load (capacitive) draws negative VAR (references occasionally differ from this convention).

Current and Voltage Considerations

Depending on the type (Δ, Y) of three-phase connection, the phase and line parameters of a three-phase load are, as shown below, interrelated.

Delta-Connected Load

In a delta-connected load (Fig. 2.27), the line and phase voltages are equal, while the line current lags[†] the phase current by 30° and is larger by a factor of $\sqrt{3}$. Thus,

$$V_{L\text{-}L} = V_{p\text{-}p} \tag{2.47}$$

$$I_L = \sqrt{3}I_p \underline{/-30°} \tag{2.48}$$

where the subscript p stands for the phase parameters.

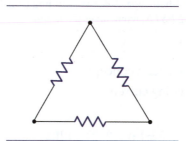

FIG. 2-27 Delta-connected load.

Star-Connected Load

In a star-connected load (Fig. 2-28), the line and phase currents are the same, while the line voltage leads[†] the phase voltage by 30° and is larger by a factor of $\sqrt{3}$. Thus,

$$I_L = I_p \tag{2.49}$$

$$V_{L\text{-}L} = \sqrt{3}V_{L\text{-}N} \underline{/30°} \tag{2.50}$$

[†]The phase shift between the line and phase parameters depends on the order of rotation of the supply voltages (see Appendix A, Fig. A-2).

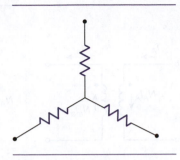

FIG. 2-28 Star-connected load.

Equations (2.48) and (2.50) are derived from a voltage phase sequence ABC.

The analysis of three-phase, two-winding transformers is simplified when their per-phase equivalent electric or magnetic circuits are used. The equivalent circuit of one phase is drawn, and the computations are identical to those used for single-phase transformers. Then the standard three-phase system relationships are employed to find the overall transformer parameters, such as currents, voltages, or power.

Phase Shift between Primary and Secondary Voltages

The phase shift, or the angular displacement, between the primary and the corresponding secondary voltage depends on the type of transformer connection. In a delta–delta or a star–star transformer, the angular displacement is zero; in a delta–star or a star–delta transformer, it is 30 degrees. According to the ASA,[*] the voltages on the HV side lead the corresponding voltages on the LV side by 30 degrees, regardless of which side is star or delta.

Magnetic Circuit Analysis

The magnetic circuit approach to the solution of three-phase transformer problems has the advantage of giving a physical insight into what is happening within the transformer. It can be of great assistance in computations involving the unbalanced operation of transformers.

A per-phase magnetic equivalent circuit is shown in Fig. 2-29(c). Neglecting the effects of the exciting current and applying KVL, we obtain

$$N_A I_A = N_a I_a \qquad (2.51)$$

Also, from KCL, we get

$$\phi_A = \phi_a \qquad (2.52)$$

As before, the upper-case subscripts represent the parameter of the primary side, while the lower-case subscripts represent the parameter of the secondary side.

*American Standard Association, C57.

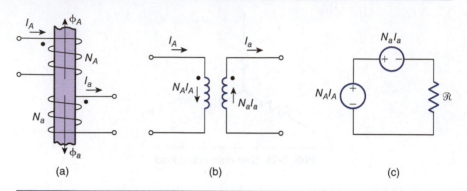

FIG. 2-29 Schematic for one phase of a two-winding, three-phase transformer. **(a)** Polarity of terminals and direction of flux. **(b)** Relative direction of mmf's. **(c)** Magnetic equivalent circuit per phase.

In analyzing problems, it is helpful if the coupled coils, with their actual magnetic volts, are drawn as shown in Fig. 2-29(b).

Electric Circuit Analysis

When electric circuit analysis is used to solve three-phase transformer problems, the following procedure is recommended:

1. Draw the per-phase electrical equivalent circuit.

2. Determine the turns ratio. This is given by the ratio of the phase to neutral voltages of the primary and secondary windings. This ratio is established by taking into consideration the delta- or star-type winding connection.

3. Draw the equivalent star–star connection of the given transformers for delta–star (Δ-Y) or delta–delta (Δ-Δ) type of winding connection. This is easily accomplished by using the standard delta–star equivalent impedances. That is,

$$Z_y = \frac{Z_\Delta}{3} \tag{2.53}$$

where Z_Δ and Z_y represent, respectively, the equivalent impedances connected in delta and star. The above relationship is applicable only when the per-phase winding impedances are identical. When the winding impedances are unequal, changing a delta-connected load to its equivalent star is accomplished by using Eq. (A.6), found in Appendix A.

4. Select the reference voltage from the given data of the problem. Recall that when the transformer is connected delta–star or star–delta, the HV line-to-line voltages lead the corresponding LV line-to-line voltages by 30 degrees.

2.2.3 One-Line Diagram

Three-phase transformer coils can be connected in either star (Y) or delta (Δ) configurations. Figure 2-30(a) shows the industry's standard terminal identification for two-winding, three-phase transformers. The letters H and X represent high and low voltage, respectively, and the subscripts 1, 2, and 3 correspond to phases 1, 2, and 3, or phase rotation ABC.* The subscript 0 is used to designate the grounded terminal of a star-connected transformer. The neutral conductor of the system is also connected to this point when a four-wire distribution system is required. A star–star connected three-phase, two-winding transformer has, on the HV and LV windings, four terminals to which as many as five conductors can be connected.

The three conductors correspond to the 3-ϕ power cables, the fourth conductor constitutes the neutral of the system, and the fifth conductor is the ground of the transformer tank. The neutral is required when single-phase loads are to be supplied with line-to-neutral voltage. The ground and the neutral conductors are shown dotted in order to differentiate them from the other three main cables. The neutral is an insulated cable and carries current under normal operating conditions, while the ground conductor is usually not insulated and carries no current under normal operating conditions. However, when one of the phases of a grounded equipment is accidentally grounded, the resulting high short-circuit current will flow through the ground cable.

In the one-line diagram (Fig. 2-30(b)), the type of winding connections (Δ or Y) is shown. The first symbol indicates the primary or incoming terminal connection, and the second the outgoing or secondary terminal coil connection. The kVA, voltage, and impedance ratings are also marked on the one-line diagram.

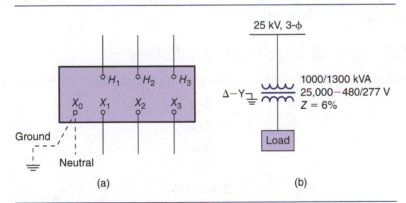

(a)   (b)

FIG. 2-30 Three-phase, two-winding transformers. **(a)** Terminal identification. **(b)** One-line diagram representation of a 1000/1300 kVA, 25,000–480/277 V.

---

*In practice, the phase rotation ABC is often designated as BRB. The latter expression identifies the black, red, and blue colors of the actual cables. When the ground wire is insulated, its color is green. The insulation of the neutral wire is normally white.

FIG. 2-31 Transformer with ventilation fans.
© *Markuso53|Dreamstime.com*

When the transformer is manufactured with special insulation, or equipped with fans, it can transform higher kVA without having its life expectancy reduced. Transformers equipped with one set of ventilating fans cost about 5% more but can deliver 33% more current at nominal voltages. Fig. 2-31 shows a substation transformer equipped with ventilating fans.

The per-unit impedance shown in the one-line diagram is calculated using the base kVA power of the transformer.

2.2.4 Types of Three-Phase Transformers

The following type of three-phase, two-winding transformers will be analyzed:

$$\Delta\text{-Y}$$
$$\Delta\text{-}\Delta$$
$$\text{Y-}\Delta$$
$$\text{Y-Y}$$
$$2\text{-}\phi \text{ to } 3\text{-}\phi$$
$$1\text{-}\phi \text{ to } 3\text{-}\phi$$

Delta–Star Transformers

Typical Δ-Y transformers are shown in Fig. 2-32 and Fig. 2-33. The Δ-Y type of three-ϕ transformers are very popular because they:

- Provide three-phase (L-L-L), one-phase (L-L), and L-N loads. The latter are necessary for computers, general usage receptacles, and the like.

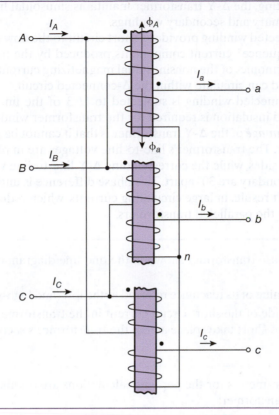

FIG. 2-32 Three single-phase transformers connected Δ-Y to form a two-winding, three-phase transformer bank.

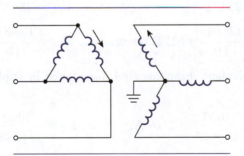

FIG. 2-33 Delta–star transformer.

- Provide means of grounding the neutral terminal which stabilize the reference voltage and can be used—through an impedance—to limit the L-G short-circuit current and thus contribute to mitigation of fires, danger to personnel, and cost of upstream current protective devices.

- Prevent the load's harmonic current components to circulate to the transformer's supply power lines.

- Are relatively more economical and have faster delivery.

Relatively speaking, the Δ-Y transformer maintains sinusoidal line-to-line wave-forms in the primary and secondary windings.

The Δ-connected winding provides a short circuit to the flow of third harmonics. The zero-sequence* current components produced by the unbalanced loads, and the third harmonics of the nonsinusoidal magnetizing current are in phase for each winding and can circulate within the Δ-connected circuit.

The star-connected winding is subjected to $1/\sqrt{3}$ of the line-to-line voltage, and thus reduced insulation is required for the transformer windings.

One *disadvantage* of the Δ-Y transformer is that it cannot be paralleled with a Y-Y transformer. The transformer's line-to-line voltages are in phase in the high- and low-voltage sides, while the corresponding Δ-Y line-to-line voltages between primary and secondary are 30° apart. This phase difference is undesirable for parallel operation. It results in large circulating currents, which reduce the kVA output capability of the paralleled transformers.

EXAMPLE **2-5**

For the three-phase transformer shown in the one-line diagram of Fig. 2-34(a), determine:

a. The ohmic value of its reactance referred to the low- and high-voltage windings.

b. The magnitude of the short-circuit current in the transformer windings when a three-phase short takes place across the transformer's secondary terminals.

SOLUTION

a. The base parameters for the per-unit calculations are obtained from the rating of the transformer:

$$S_b = 1000 \text{ kVA}$$

$$V_{bH} = 25 \text{ kV}, \qquad\qquad\qquad V_{bL} = 480 \text{ V}$$

$$Z_{bH} = \frac{(25.0)^2}{1.0} = 625 \text{ } \Omega/\text{phase}, \qquad Z_{bL} = \frac{(0.48)^2}{1.0} = 0.23 \text{ } \Omega/\text{phase}$$

Z_{bH} is the equivalent star-connected impedance in Ω/phase. The base currents are

$$I_{bH} = \frac{1000}{\sqrt{3} \times 25} = 23.09 \text{ A}, \qquad I_{bL} = \frac{1000}{\sqrt{3}(0.48)} = 1202.81 \text{ A}$$

Then the ohmic values of the reactances are

HV Winding

$$X_H = 0.05(625) = \underline{31.25 \text{ } \Omega/\text{phase, star equivalent}}$$

---

*The zero-sequence currents are due to unbalanced supply voltages and to nonidentical load impedances. They are similar to third harmonic currents.

LV Winding

$$X_L = 0.05(0.23) = \underline{0.012 \ \Omega/\text{phase, star equivalent}}$$

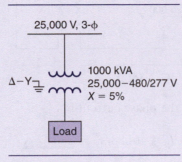

25,000 V, 3-ϕ

Δ–Y

1000 kVA
25,000–480/277 V
X = 5%

Load

FIG. 2-34(a)

b. The short-circuit current can be found by using any of the following methods.

LV Winding

1. Applying Ohm's law in the circuit of Fig. 2-34(b), we find that the magnitude of the current in the LV winding is

$$I = \frac{V}{X} = \frac{480/\sqrt{3}}{0.012} = \underline{24{,}056.26 \text{ A}}$$

2. Using per-unit values:

$$I = \frac{1.0}{0.05} = 20 \text{ pu}$$

$$= 20(1202.81) = \underline{24{,}056.26 \text{ A}}$$

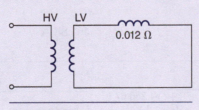

HV   LV

0.012 Ω

FIG. 2-34(b)

HV Side

1. Applying Ohm's law in the star equivalent circuit of Fig. 2-34(c), we obtain the magnitude of the short-circuit current in the HV windings:

$$I = \frac{V}{X} = \frac{25{,}000/\sqrt{3}}{31.25} = 461.88 \text{ A line current}$$

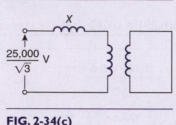

FIG. 2-34(c)

The magnitude of the phase current is

$$I_p = 461.88/\sqrt{3} = \underline{266.67\ \text{A}}$$

2. Using Ohm's law, for the delta-connected equivalent circuit, we obtain

$$X_\Delta = 3X_Y = 3(31.25) = 93.75\ \Omega$$

$$I_p = \frac{25{,}000}{93.75} = \underline{266.67\ \text{A}}$$

and

$$I_L = \sqrt{3}(266.67) = 461.88\ \text{A}$$

3. Refer to Fig. 2-34(d). By using the current transformation property of transformers, we obtain

$$I_p = 24{,}056.26\ \frac{480/\sqrt{3}}{25{,}000} = \underline{266.67\ \text{A}}$$

and

$$I_L = \sqrt{3}I_p = 461.88\ \text{A}$$

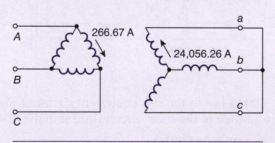

FIG. 2-34(d)

Three identical single-phase transformers, each of 500 kVA, 4160–480 V, 60 Hz, are connected Δ-Y to form a 3-ϕ transformer bank. Data for the short-circuit test on one of the single-phase transformers is as follows:

EXAMPLE **2-6**

Low-voltage winding shorted:

$$V_H = 250 \text{ V}, \qquad I_H = 120.19 \text{ A}, \qquad P = 10 \text{ kW},$$

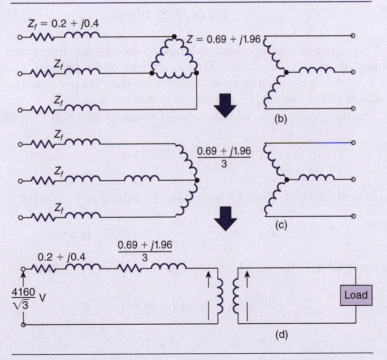

FIG. 2-35(a)

FIG. 2-35 (b), (c), and (d)

The transformer, as shown in Fig. 2-35(a), is connected to an upstream substation of 4160 volts through a feeder whose impedance is $(0.2 + j0.4)$ ohms per phase. When the transformer delivers rated current at 0.9 power factor lagging, determine:

a. The line-to-line voltage across the load.

b. The power supplied by the upstream substation.

c. The current through the transformer windings when a 3-ϕ short takes place at the secondary terminals of the transformer.

SOLUTION

a. From the short-circuit test data, we get

$$|Z_{eH}| = \frac{250}{120.19} = 2.08 \ \Omega/\text{phase}$$

$$R_{eH} = \frac{10{,}000}{(120.19)^2} = 0.69 \ \Omega/\text{phase}$$

$$X_{eH} = \sqrt{2.08^2 - 0.69^2} = 1.96 \ \Omega/\text{phase}$$

Thus,

$$Z_{eH} = 0.69 + j1.96$$

$$= 2.08 \underline{/70.6°} \ \Omega/\text{phase}$$

This is the per-phase impedance referred to the HV winding, which is delta connected. In order to simplify the calculations of the primary, as shown in Fig. 2-35(c), it is changed to an equivalent star. Then the per-phase equivalent circuit, as seen from the HV winding, is shown in Fig. 2-35(d).

The magnitude of the line current in the primary of the Δ-Y transformer is

$$I = \frac{1500}{\sqrt{3}(4.160)} = 208.18 \ \text{A}$$

Refer to Fig 2-35(d). The total impedance in the primary circuit is

$$Z = 0.2 + j0.4 + \frac{0.69 + j1.96}{3} = 1.14 \underline{/67.8°} \ \Omega/\text{phase}$$

Applying KVL to the per-phase equivalent circuit of Fig 2-35(d), we obtain

$$V_1 \underline{/0°} = \frac{4160}{\sqrt{3}} \underline{/\beta_1} - 208.18 \underline{/-25.8°} \ (1.14 \underline{/67.8°})$$

$$= \frac{4160}{\sqrt{3}} \underline{/\beta_1} - 237 \underline{/41.9°}$$

Equating the imaginary parts, we obtain

$$\beta_1 = \arcsin \frac{\sqrt{3}}{4160}\,(237 \sin 41.9°) = 3.8°$$

Equating the real parts, we get

$$V_1 = \frac{4160}{\sqrt{3}}\cos 3.8° - 237\,(\cos 41.9°) = 2220.21 \text{ V/phase}$$

The voltage across the secondary of the transformer is

$$V_2 = V_1\left(\frac{480}{4160}\right) = 2220.21\frac{480}{4160} = 256.18 \text{ V/phase}$$

$$= (256.18)\sqrt{3} = \underline{443.71 \text{ V}_{L\text{-}L}}$$

b. The power supplied is

$$P = \sqrt{3}V_{L\text{-}L}I_L \cos\theta$$

$$= \sqrt{3}(4160)(208.18)\cos(3.8° + 25.8°)$$

$$= 1303.96 \text{ kW}$$

c. The magnitude of the short-circuit current through the feeder is

$$I_L = \frac{V}{Z} = \frac{4160}{\sqrt{3}(1.14)} = 2109.60 \text{ A}$$

The current in the Δ-connected windings is

$$I_P = \frac{I_L}{\sqrt{3}} = \frac{2109.60}{\sqrt{3}} = \underline{1218 \text{ A}}$$

The current in the Y-connected windings is

$$I_s = (1218)\frac{4160}{480/\sqrt{3}} = \underline{\underline{18{,}283.20 \text{ A}}}$$

Alternatively, using per-unit values,

$$Z_{bH} = \left(\frac{4160}{1.5}\right)^2 = 11.54 \ \Omega/\text{phase}$$

Thus,

$$Z_{e\ pu} = \frac{1.14}{11.54} = 0.099 \text{ pu}$$

$$I_{sc} = \left(\frac{1}{0.099}\right) = 10.13 \text{ pu}$$

In the LV side:

$$I_{sc} = 10.13 \frac{1500}{\sqrt{3}(0.480)} = \underline{\underline{18,283.20 \text{ A}}}$$

In the HV side:

$$I_{sc} = 10.13 \frac{1500}{\sqrt{3}(4.16)} = \underline{\underline{2109.60 \text{ A}}}$$

Exercise

2-4

The three-phase transformer shown in the one-line diagram of Fig. 2-36 is connected to a 4.16 kV source. The magnetizing current is negligible.

1. If the transformer delivers its base kVA at the rated voltage, determine:
   a. The primary and secondary line and phase currents.
   b. The ohmic values of the leakage reactance referred to the HV and LV windings.
   c. The primary and secondary line and phase currents when the secondary terminals of the transformer are shorted.
2. Repeat (a), (b), and (c) when the transformer operates with the fans and delivers 2000 kVA at rated voltage.

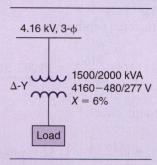

FIG. 2-36

Answer   1. (a) 1804.2 A, 208.18 A, 120.2 A
          (b) 2.08 Ω, 0.0092 Ω/ph
          (c) 30,070.3 A, 2003.2 A, 3469.6 A
      2. (a) 2405.6 A, 277.6 A, 160.3 A
          (b) and (c) no change

The three-phase, three-winding transformer shown in Fig. 2-37 delivers 25 A to a single-phase load. The tertiary winding (T) is shorted, and the coils have an equal number of turns.

a. Mark the polarity of each coil and show the direction of its flux.
b. Determine the current of each winding. (*Hint*: The net flux through each leg must be equal to zero.)

Exercise
2-5

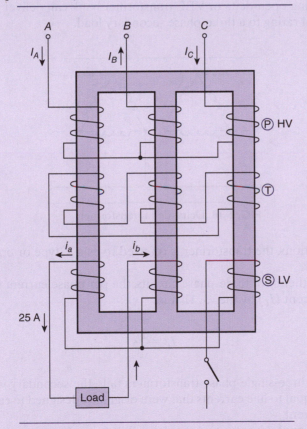

FIG. 2-37

Answer $I_T = 0$, $i_A = i_B = 25$ A, $i_C = 0$

Δ-Δ Transformer

A Δ-Δ winding connection is shown in Fig. 2-38. This type of transformer has the following advantages:

1. When a line-to-ground fault—the most common type of fault—occurs in the secondary distribution of the transformer, the resulting current is very small and therefore the power flow is not interrupted. This is of tremendous importance to industries whose processes—such as the melting of metals—cannot be interrupted.

2. This type of transformer connection maintains sinusoidal voltages at its output terminals. This is because the delta–delta winding connection provides a path for the flow of the third harmonic component of the magnetizing current and for the circulation of the zero-sequence component of the unbalanced load currents.

3. When three single-phase transformers are connected in Δ-Δ to form a 3-ϕ transformer bank, one of the single-phase transformers can be removed and the resulting open-delta or VEE transformer bank can deliver about 58% of its original rating to a three-phase secondary load.

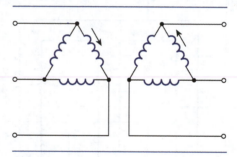

FIG. 2-38 Delta–delta transformer.

In such operations, the transformer is referred to as a V-type or open-delta type of transformer.

From the theory of three-phase circuits, the per-phase current (I_P) is related to the line current (I_L) by the $\sqrt{3}$. That is,

$$I_P = \frac{I_L}{\sqrt{3}}$$

When one of three-single-phase transformers fails, the secondary winding phase currents are equal to line currents that were originally designed to carry only 58% of the line current.

The additional advantage of the delta–delta transformer is that it will continue to operate while there is a line-to-ground short circuit.

The disadvantage of this type of winding configuration is that it cannot supply L–N single-phase loads.

Star–Delta Transformer

The Y-Δ transformer is of limited use because it does not provide a grounded neutral and thus does not support L-N, single-phase loads.

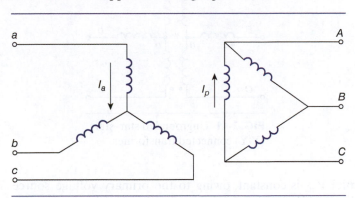

FIG. 2-39 Y-Δ Transformer.

Star–Star Transformer

The Y-Y transformer, when the star-point is grounded, can provide L–L–L, L–L, and L-N loads. However, it permits the flow of harmonic current through the upstream and downstream network.

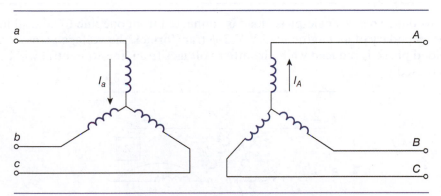

FIG. 2-40 Y-Y Transformer.

An ungrounded Y-Y transformer is shown in Fig. 2-41. When the transformer's load is slightly unbalanced, the neutral points (n) may not be at a fixed potential with respect to ground. This is referred to as the "floating" neutral. The potentially unstable neutral is due mainly to unbalanced load currents.

Considering Fig. 2-41, from KVL, in the primary side, we have

$$V_{Cn} + V_{nA} + V_{AC} = 0$$

or

$$V_{AC} = V_{nC} + V_{An}$$

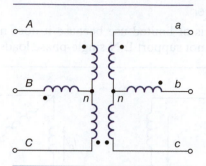

FIG. 2-41 Ungrounded star–star (Y-Y) connected transformer.

The potential V_{AC} is constant, owing to the primary voltage source. The other voltages (V_{nC} and V_{An}) will vary in a way that satisfies the previous two equations.

To stabilize the potential of the neutral point, a delta-connected winding—the so-called tertiary—is wound on the core of the transformer. In this case, the transformer becomes a three-phase, three-winding transformer. Such transformers have high MVA ratings and are normally used by the utilities.

EXAMPLE **2-7** Show that, when a single-phase load is connected from one line to neutral in the secondary of an ungrounded Y-Y, 3-ϕ transformer, the voltage across the loaded phase is reduced while the other voltages from line-to-neutral will be increased.

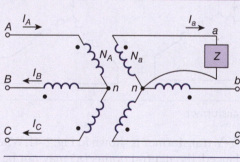

FIG. 2-42

SOLUTION

Assume a load is connected from line "a" to neutral, as shown in Fig. 2-42. The load impedance will draw a current I_a because of the voltage induced in the secondary winding:

$$I_a = \frac{V_a}{Z}$$

Since there is a current on the secondary means, there must also be a current I_A in the primary, because of the current-transformation property of transformers, namely:

$$N_A I_A = N_a I_a$$

Applying KCL, in the primary winding, we have

$$I_A = I_B + I_C$$

I_A exists, and, since $I_b = I_c = 0$, then I_B and I_C must represent increases in the exciting currents in phases B and C, respectively. Thus, their corresponding voltages V_{Bn} and V_{Cn} must also be *increased*. This leads to the reduction of V_{AB} ($V_{AB} = V_{nB} + V_{An}$) and V_{ab}.

Compare and contrast the Δ-Δ and Δ-Y type of three-phase, two-winding transformers. Consider insulation levels, exciting currents, and output voltage waveforms.

Exercise

2-6

Special Type of Transformers

In this section we will describe the transformers that change a 2-ϕ and 1-ϕ voltage supply to a 3-ϕ balanced system. These special transformers may be analyzed by drawing their corresponding magnetic equivalent circuits in terms of their mmf's.

Scott-Connection

A scott-transformer connection, as shown in Fig. 2-43, is used to interconnect a 2-ϕ system to a 3-ϕ distribution network. The 2-ϕ voltages are equal in magnitude and at 90 electrical degrees to each other.

Such transformers have two cores, and in the 3-ϕ side, one of the windings has 0.866 N number of turns and is connected to the center tap of the other winding, which has N number of winding turns. This winding arrangement insures a 120-degree phase shift among the voltages of the 3-winding.

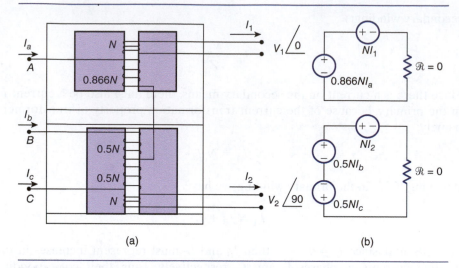

(a)   (b)

FIG. 2-43 Scott-type of transformer. (a) Winding configuration. (b) Equivalent magnetic circuits.

Refer to Fig. 2-43. Assuming an equal number of winding turns (N), we have:

Winding One

Neglecting the core losses and applying KVL in the magnetic circuit of winding 1, we obtain

$$NI_1 = 0.866\, NI_a$$

or

$$NI_1 = \underline{0.866\, NI_a \underline{/0}}$$

Similarly, for winding 2, we obtain

$$NI_2 + 0.5\, NI_b - 0.5\, NI_c = 0$$

or

$$NI_2 = 0.5N(I_c - I_b)$$

The magnitudes of the line currents I_a, I_b, and I_c are equal:

$$NI_2 = 0.5NI_a\,(1\underline{/120} - 1\underline{/-120})$$

$$= \underline{0.866 NI_a \underline{/90}}$$

Thus the mmf's on the two windings are equal in magnitude and at 90 electrical degrees to each other.

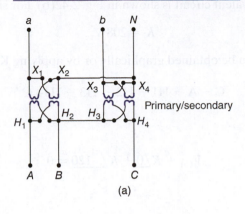

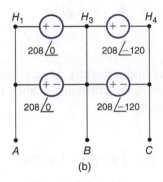

FIG. 2-44 One-phase to three-phase transformation. (a) Winding configuration. (b) Equivalent electrical circuit (ideal).

(The mmf's are proportional to magnetic flux whose derivative is proportional to a corresponding voltage).

1-φ to 3-φ Transformation

Figure 2-44 shows the winding configuration of a special transformer that is used to change a single-phase line-to-line voltage to a three-phase balanced voltage supply.

These types of transformers are for economical reasons used within a plant to supply a remote 3-φ panel. The savings result by using two wires instead of three wires.

For a 208 V line-to-line supply voltage, we have

$$V_{ab} = 208\underline{/30}\ \text{V}, \qquad V_{an} = 120\underline{/0} \quad \text{and} \quad V_{bn} = 120\underline{/-120}\ \text{V}$$

The transformer's winding turns ratio is such as to develop 208 V across the secondary windings. That is,

$$V_{AB} = 208\underline{/0}, \qquad V_{BC} = 208\underline{/-120}\ \text{V}$$

The electrical equivalent circuit is shown in Fig. 2-42(b). For simplicity,

$$K = 208 \text{ V}$$

The voltage V_{CA} can be obtained graphically or by applying KVL in the loop:

$$C - A - H1 - H2 - H3 - H4 - C$$

Thus,

$$V_{CA} + K\underline{/0} + K\underline{/-120} = 0$$

from which

$$V_{CA} = K\underline{/120}$$
$$= 208\underline{/120} \text{ V}$$

Due to leakage impedances and a slightly unequal turns ratio, there would be circulating currents. As a result, the efficiency of such transformers is lower than that of their equivalent standard units.

2.2.5 Harmonics of the Exciting Current

This section discusses the harmonics of the exciting current and the advantages and disadvantages of the various transformer winding connections.

Harmonics of the Exciting Current

Consider the exciting current of one phase only. Neglecting its fourth and higher-order harmonics, we have

$$i_{\text{exc}} = i_1 + i_2 + i_3$$
$$= I_{m1} \cos \omega t + I_{m2} \cos 2\omega t + I_{m3} \cos 3\omega t \qquad \textbf{(2.54)}$$

where I_{m1}, I_{m2}, and I_{m3} are the maximum values of the fundamental, second, and third harmonics.

Second Harmonics

Designating I_{m2} as the maximum value of the second harmonic in one phase, we see that the vectorial sum for all three phases is

$$i_2 = I_{m2} \cos 2\omega t + I_{m2} \cos 2(\omega t + 120°) + I_{m2} \cos 2(\omega t + 240°) \qquad \textbf{(2.55)}$$

From the above:

$$i_2 = 0$$

Thus, the sum of the second, or even, harmonics of a three-phase system is equal to zero, regardless of the transformer winding connections.

Third Harmonic

The third harmonic components of the exciting current in the lines of a three-phase system do not add up to zero, as in the case of the fundamental and second harmonics. Mathematically,

$$i_3 = I_{m3} \cos 3\omega t + I_{m3} \cos 3(\omega t + 120°) + I_{m3} \cos 3(\omega t + 240°) \qquad \textbf{(2.56)}$$

From the above,

$$i_3 = 3I_{m3} \cos 3\omega t \qquad \textbf{(2.57)}$$

When third harmonics are permitted to flow, the exciting current is nonsinusoidal in waveform. This waveform—as can be verified by the *B-H* characteristic of high-permeability magnetic materials—is accompanied by sinusoidal flux. Sinusoidal flux produces sinusoidal secondary voltages, and nonsinusoidal flux produces nonsinusoidal secondary voltages. These nonsinusoidal voltages are undesirable because they may cause electrical apparatuses—motors, computers, and the like—to malfunction. Thus, it is desirable that the waveform of the exciting current be nonsinusoidal.

The nonsinusoidal waveform of the exciting current is ensured when the transformer connections permit the flow of third harmonics. Star-connected windings with a grounded neutral provide a physical link so that the harmonics can flow from the system's lines down to ground, then eventually return to the generator's grounded neutral. Therefore, star-grounded windings permit the flow of nonsinusoidal excitation currents, and so their secondary phase voltages are sinusoidal. Delta-connected windings also produce sinusoidal fluxes and voltages because they permit the third harmonics to circulate within their windings.

In contrast, ungrounded star-connected windings do not permit the flow of third harmonics of exciting currents, and thus the induced voltages are nonsinusoidal.

2.3 Autotransformers

This section covers the principles of operation, equivalent circuits and losses. Figure 2-45(a) shows the schematic representation of a 10 kVA, 200–100 V, single-phase transformer. Its autotransformer-type connection can be obtained as shown in Figs. 2-45(b) and (c), by a special interconnection of its primary and secondary windings.

In autotransformers, one single winding is used as the primary winding as well as secondary winding, as against two distinctly separate windings in a conventional power transformer. Autotransformers are smaller in size and more economical than two-winding transformers of the same rating. They are normally used for voltage-transformation ratios close to 1:1 and in variacs, which provide a variable secondary voltage.

Understanding and analyzing autotransformers is simplified by noting the following:

1. The ampere-turns of each coil are the same whether the transformer is connected as an autotransformer or as a two-winding transformer. That is,

$$(NI) = \text{constant} \qquad\qquad (2.60)$$

In other words, the magnitude of the current through each coil remains the same, regardless of whether the transformer operates as an autotransformer or as a conventional two-winding transformer.

Referring to Fig. 2-45, the ampere-turns of each coil are the same in each of the three situations shown, but the complex power transferred from the supply to the load differs from one case to the next.

2. Since the current through each autotransformer winding is the same as in the conventional two-winding transformer, the winding loss remains the same. However, the efficiency of the autotransformer is increased if the output power is increased.

3. Neglecting losses, the complex power (kVA) at the input is equal to the complex power delivered to the output:

$$(\text{kVA})_{\text{input}} = (\text{kVA})_{\text{output}} \qquad\qquad (2.61)$$

The kVA transformation capability of the autotransformer is the same as that of the two-winding transformer. An autotransformer, however, delivers higher kVA than the conventional transformer because of the direct electrical connection between the primary and secondary windings. In other words, part of the output kVA is conducted from the primary to the secondary winding. The conducted kVA is referred to as untransformed kVA.

4. The two-winding conventional transformer has its primary and secondary circuits electrically isolated, while in the autotransformer electrical disturbances in the primary can be easily passed to the secondary through their direct electrical connection.

A two-winding transformer, as shown in Fig. 2-45, can be connected in two ways to supply a load as an autotransformer. The two different connections are identified as A and B. For reasons of comparison, the highlights of type A and B connections are summarized in Table 2-1. In both cases, the output voltage is smaller than the input voltage. However, when one of the voltage windings is reversed, the output voltage will be larger than the input voltage.

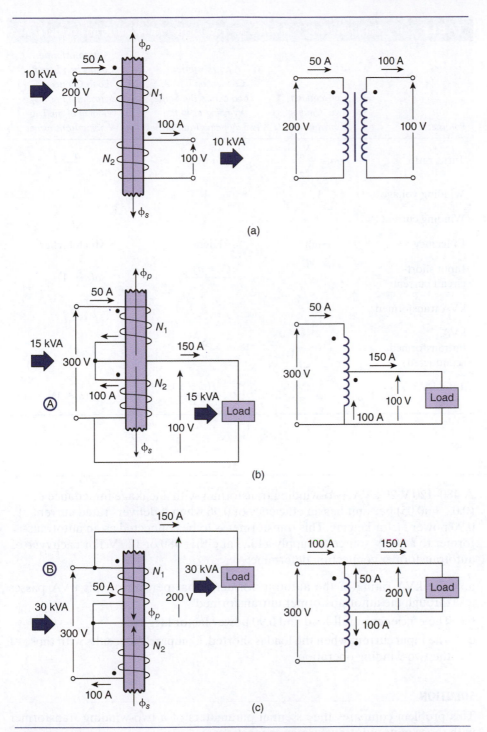

FIG. 2-45 Transformer representations and their schematics. (a) A two-winding transformer. (b) Type A autotransformer connection. (c) Type B autotransformer connection.

		Autotransformer	Autotransformer Connection Type B:
		Connection Type A:	Load Connected
	Conventional	Load across the Same	across the Primary
	Two-Winding	Winding as in the	Winding of the Two-
Parameter	Transformer	Two-Winding Transformer	Winding Transformer
TABLE 2-1 Parameters of two-winding transformers and autotransformers			
Turns ratio	$a = \dfrac{N_1}{N_2}$	$a + 1$	$\dfrac{a + 1}{a}$
Winding voltage	V	V	V
Winding current	I	I	I
Efficiency	High	Higher	Much higher
Input short-circuit current	I_{sc}	$\dfrac{(1 + a)}{a} I_{sc}$	$a(a + 1)I_{sc}$
kVA transformed	S	S	S
kVA untransformed (conducted)	0	$\dfrac{S}{a}$	aS
Total kVA Output	S	$S\dfrac{(a + 1)}{a}$	$S(a + 1)$

EXAMPLE **2-8**

A 480–120 V, 20 kVA, two-winding transformer with a leakage impedance of $(0.02 + j0.05)$ per unit has an efficiency of 0.96 when it delivers rated current at 0.90 power factor lagging. This transformer is to be connected as an autotransformer to a 600 V source to supply a load at either 480 or 120 V. For each *type* of autotransformer connection determine:

a. The kVA rating of the autotransformer. What percent of this kVA passes through the autotransformer untransformed?

b. The efficiency at full-load and 0.90 power factor lagging.

c. The input current when the load is shorted. Compare the results with those of the two-winding operation.

SOLUTION

This problem compares the essential parameters of a two-winding transformer with those of its autotransformer connections.

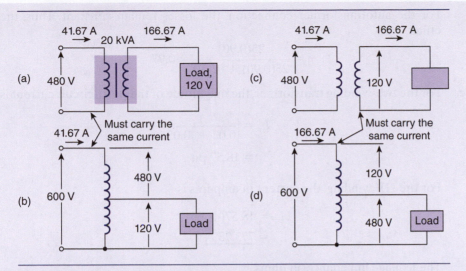

FIG. 2-46 (a) and (b): Two-winding and autotransformer connection (type A). (c) and (d): Two-winding and autotransformer connection (type B).

Type (A) Autotransformer Connection (Load across the 120 V Winding)

a. Referring to Fig. 2-46(b), the magnitude of the current through the input will be the current rating of the 480 V winding. That is,

$$I = \frac{20,000}{480} = 41.67 \text{ A}$$

Therefore, the input apparent power to the autotransformer is

$$S = 41.67(600)$$

$$= 25 \text{ kVA}$$

Of these 25 kVA, 5 kVA pass unaltered, or untransformed, from input to output; the remaining 20 kVA are transformed, as in the case of the two-winding transformer connection. Thus,

$$\text{percentage of untransformed kVA} = \frac{5}{25}(100)$$

$$= 20\%$$

b. The losses of the two-winding transformer are

$$\text{losses} = \text{power input} - \text{power output} = \frac{P_{\text{out}}}{\eta} - P_{\text{out}}$$

$$= 20,000(0.9)\left(\frac{1}{0.96} - 1\right) = 750 \text{ W}$$

For the autotransformer connection, the losses remain constant. Thus, the efficiency is

$$\eta = \frac{25(0.90)}{25(0.90) + 0.750} = 0.97$$

c. For the two-winding transformer, the magnitude of the short-circuit current is

$$I_{sc} = \left| \frac{1}{0.02 + j0.05} \right|$$

$$= 18.57 \text{ pu}$$

For the HV winding, the current in amperes is

$$I_{sc} = 18.57(41.66)$$

$$= \underline{773.73 \text{ A}}$$

The leakage impedance, in ohms, is

HV side

$$Z_{bH} = \frac{(480)^2}{20,000} = 11.52 \ \Omega$$

$$Z_H = (0.02 + j0.05)(11.52) = 0.62 \underline{/68.2°} \ \Omega$$

LV side

$$Z_{bL} = \frac{(120)^2}{20,000} = 0.72 \ \Omega$$

$$Z_L = (0.02 + j0.05)(0.72) = 0.04 \underline{/68.2°} \ \Omega$$

In the case of the autotransformer, when the 120 V winding is shorted, the source of 600 V is applied across the 480 V winding. Thus, the magnitude of the short-circuit current is

$$I_{sc} = \frac{600}{0.62} = \underline{967.16 \text{ A}}$$

or, by using the appropriate relationship from Table 2-1.

$$I_{sc} = 773.73 \frac{(1 + a)}{a} = 773.73 \left(\frac{5}{4} \right)$$

$$= \underline{967.16 \text{ A}}$$

Type (B) Autotransformer Connection (Load across the 480 V Winding)

a. Referring to Fig. 2-46(d), the current through the 600 V source will be the rated current of the 120 V winding. That is,

$$I = \frac{20 \times 10^3}{120} = 166.67 \text{ A}$$

This current must pass through the input terminals. Therefore, the apparent input power for this type of autotransformer connection is

$$S = 166.67(600)$$
$$= \underline{100}\ \text{kVA}$$

Of these 100 kVA, 80 kVA pass unaltered from the input to the output, while the remaining 20 kVA are transformed as in the case of the two-winding connection.

Expressing the untransformed kVA in percentage of the total output kVA, we have

$$(kVA)_{\text{untr.}} = \frac{80}{100}(100) = \underline{80\%}$$

b. Since the losses are constant at 0.75 kW, the efficiency is

$$\eta = \frac{100(0.9)}{100(0.9) + 0.75}$$
$$= \underline{0.99}$$

c. By shorting the load, the 600 V are applied across the 120 V winding. Thus, the magnitude of the short-circuit current is

$$I_{sc} = \frac{600}{0.04} = \underline{15,474.6\ \text{A}}$$

For comparison purposes, the results are summarized in Table 2-2. Check the results by using the applicable relationships given in Table 2-1.

TABLE 2-2	Summary of results of Example 2-8		
		Autotransformer Connection	
Parameter	Two-Winding Transformer Connection	Type A Load across 120 V	Type B Load across 480 V
Output kVA	20	25	100
Efficiency	0.96	0.97	0.99
Short-circuit current through the source in A	773.7	967.16	15,474.6

Exercise

2-1

A single-phase, 1000 VA, 60 Hz autotransformer has four coils, as shown in Fig. 2-47. The voltage rating of the HV windings is 240 V/coil, and that of the LV windings is 120/coil. Show the field connections, and determine the maximum VA that can be delivered to the load when:

a. The transformer supplies 120 V from a 240 V source.
b. The transformer supplies 240 V from a 480 V source.
c. Two of these transformers are connected in an open-delta configuration to supply a 3-ϕ, 480 V load from a 3-ϕ, 600 V source.

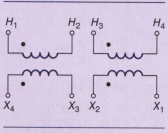

FIG. 2-47

Answer   (a) 1000 VA;   (b) 1000 VA;   (c) 8660 VA

Disadvantages of Autotransformers

The disadvantages of autotransformers are as follows:

- They are not economical for voltage rations larger than 2.
- They require primary protective devices of higher capacity.

   1. Because of the electrical conductivity of the primary and secondary windings, the lower voltage circuit is liable to be impressed upon by higher voltage. To avoid breakdown in the lower voltage circuit, it becomes necessary to design the low-voltage circuit to withstand higher voltage.

   2. The autotransformer has a common terminal between the primary and the secondary windings, and when the secondary is shorted, the voltage applied to the primary is much higher than its rated voltage. This results in higher short-circuit currents and thus requires more expensive protective devices.

   3. The connections on primary and secondary sides must necessarily be the same, except when using interconnected starring connections. This introduces complications due to changing primary and secondary phase angles, particularly in the case-by-case of the delta–delta connection.

   4. Because a common neutral in a star–star connected autotransformer, it is not possible to ground the neutral of one side only. Both of its sides must have its neutrals either grounded or isolated.

5. It is more difficult to preserve the electromagnetic balance of the winding when voltage adjustment tappings are provided. It should be known that the provision of adjusting tapping on an autotransformer increases the frame size of the transformer considerably. (Source: http://www.electrical4u.com/electrical-transformer/auto-transformer.php)

2.4 Parallel Operation of Transformers

Transformers are sometimes paralleled in order to meet the increased kVA demands of a particular load. Transformers that are to be paralleled satisfactorily should satisfy the following conditions:

a. Equal impedances.

b. Equal turns ratio.

c. Equal phase shift between the primary and secondary open-circuited voltages.

d. The same phase rotation.

When these conditions are not satisfied, the transformers—as is demonstrated below—will be damaged.

The one-line representation of parallel transformers, as well as their corresponding equivalent circuits, are shown in Fig. 2-48. For the analysis of paralleled

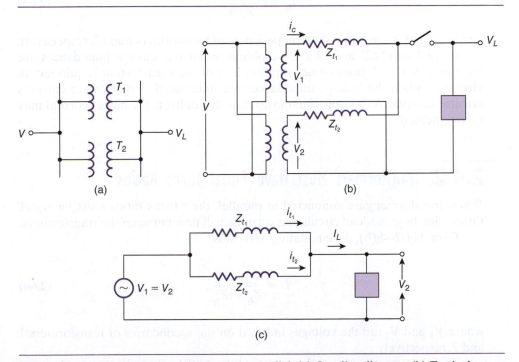

(a)   (b)

(c)

FIG. 2-48 Two single-phase transformers in parallel. (a) One-line diagram. (b) Equivalent circuit for determining the circulating current. (c) Equivalent circuit for determining the load on each transformer. (The turns ratio for all units are assumed to be equal.)

transformers, the per-phase equivalent circuits of Fig. 2-48(b) and Fig. 2-48(c) can be used.

In Fig. 2-48(b), the load is disconnected from the common secondary transformer terminals, and thus this circuit can be used to find the transformer's common circulating current. When the transformers have an equal number of winding turns, then Fig. 2-48(c) may be used to find each transformer's load as a function of the power delivered to the common load.

2.4.1 Transformers Must Have Equal Leakage Impedances

Assuming equal transformer voltage ratios, then from Fig. 2-48(c), and by applying the current-divider concept, we obtain each transformer's current as a function of the load current I_L:

$$I_{t_1} = I_L \frac{Z_{t_2}}{Z_{t_1} + Z_{t_2}} \tag{2.62}$$

and

$$I_{t_2} = I_L \frac{Z_{t_1}}{Z_{t_1} + Z_{t_2}} \tag{2.63}$$

where Z_{t_1} and Z_{t_2} are the leakage impedances of transformers 1 and 2, respectively.

From Eqs. (2.62) and (2.63), we obtain, when the leakage impedances are the same, that each transformer supplies half of the load current requirements. However, when the leakage impedances are unequal, the transformer currents are also unequal. As a result, the transformer that delivers the higher current may be overheated.

2.4.2 Transformers Must Have Equal Turns Ratios

When transformers are connected in parallel, their turns ratios must be equal. Otherwise, large no-load circulating currents will flow between the transformers.

From Fig. 2-48(b), the circulating current is

$$I_c = \frac{V_1 - V_2}{Z_{t_1} + Z_{t_2}} \tag{2.64}$$

where V_1 and V_2 are the voltages induced on the secondaries of transformers 1 and 2, respectively.

To minimize the circulating current, the "off-load" voltage tap changers should, if possible, be properly adjusted to equalize the voltages induced in the secondaries of the transformers.

If the transformers supply a load when their turns ratios are different, ordinary circuit-analysis techniques can be used to find the individual transformer currents.

2.4.3 Equal Phase Shift Between the Voltages of Paralleled Units

The three-phase units must also produce secondary line-to-line voltages in phase when they are paralleled. Otherwise, large circulating currents will flow.

For this reason, a Δ-Y transformer cannot be paralleled with a Y-Y transformer. As previously mentioned, the line-to-line voltages on the delta side of a Δ-Y transformer lead the corresponding line-to-line voltages on the star side by 30°, while the line-to-line voltages on the primary of a star-star (Y-Y) transformer are in phase with the line-to-line voltages on the secondary side.

The line-to-line voltages for one phase only of a Δ-Y and Y-Y three-phase transformer are shown in Figs. 2-49(a) and (b), respectively.

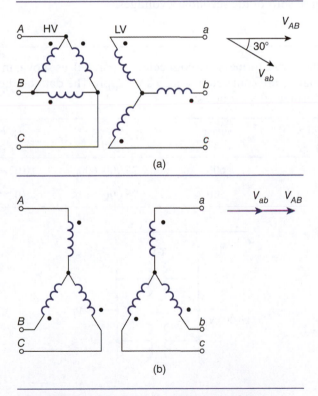

FIG. 2-49 Two-winding, three-phase transformers and a partial voltage phasor diagram. (a) Delta–star. (b) Star–star.

2.4.4 Same Phase Rotation

When two transformers are paralleled, the primary and secondary voltages must have the same phase rotation. This condition is easily verified in the field, and if necessary, the connections of the two phases can be interchanged. This will result in the same phase rotation of the three-phase voltages. When the phase rotation is not the same, large circulating currents will flow through the transformer windings, and unbalanced voltages will be delivered to the load.

The paralleling of transformers, in general, results in higher rated and short-circuit currents through the common primary feeders. This, in turn, would require more expensive primary protective devices.

Circulating Currents

To prevent large circulating currents within the windings of the paralleled transformers, the following conditions must be satisfied.

- Equal winding leakage impedances.
- Equal winding turns ratio per phase.
- Equal phase shift of the secondary voltages.

EXAMPLE **2-9**

Two single-phase transformers are connected in parallel, as shown in Fig. 2-50(a), and their primaries are connected to a 25 kV supply. The data on the nameplate of each transformer are as follows:

Transformer	Rating in kVA	Voltage Ratio	Leakage Impedances in Per-Unit
1	600	25,000–600	$0.02 + j0.05$
2	500	25,000–610	$0.02 + j0.06$

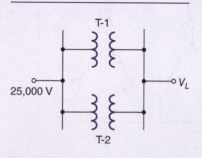

FIG. 2-50(a)

Determine:

a. The no-load circulating current.

b. The power loss due to the no-load circulating current.

SOLUTION

a. Change the per-unit values of the transformer impedances to the equivalent ohmic values of the low-voltage winding.

For Transformer 1:

$$Z_{b_L} = \frac{(600)^2}{600 \times 10^3} = 0.6 \ \Omega$$

and

$$Z_{t_1} = (0.02 + j0.05)(0.6) \ \Omega$$
$$= 0.012 + j0.03 \ \Omega$$

For Transformer 2:

$$Z_{bL} = \frac{(610)^2}{500 \times 10^3} = 0.74 \ \Omega$$

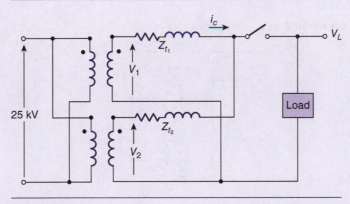

FIG. 2-50(b)

and

$$Z_{t_2} = (0.02 + j0.06)(0.74) \ \Omega$$
$$= 0.015 + j0.045 \ \Omega$$

Thus, from the equivalent circuit of Fig. 2-50(b), we have

$$I_c = \frac{V}{Z} = \frac{610 - 600}{0.012 + 0.015 + j(0.03 + 0.045)} = \frac{10}{0.027 + j0.075}$$

$$= 125.95\underline{/-70.2°}\ \text{A}$$

b. The power loss due to circulating current is

$$I^2R = (125.95)^2(0.027) = \underline{426.73\ \text{W}}$$

Exercise 2-8

Two single-phase transformers are connected in parallel, as shown in Fig. 2-51, to supply a load of 500 kVA at 480 volts and 0.90 power factor lagging. The data on the transformer's nameplate are as follows:

Transformer	Rating in kVA	Voltage Ratio	Leakage Impedance in Percent
1	300	4160–480	2 + j3.5
2	250	4160–480	1.8 + j4.0

For each transformer, neglect the magnetizing impedances, and determine:
a. The current.
b. The kVA loading.

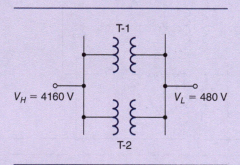

FIG. 2-51

Answer (a) 590.4$\underline{/-23.4°}$ A, 452.3$\underline{/-29°}$ A
(b) 283.4$\underline{/23.4°}$ kVA, 217.1$\underline{/29°}$ kVA

2.5 Instrument Transformers and Wiring Diagrams

This section discusses instrument transformers and wiring diagrams. Wiring diagrams illustrate how the instrument transformers are interconnected to the various measuring and indicating meters. These topics not only supplement the basic transformer concepts but also give an overview of the practicing engineer's work.

2.5.1 Instrument Transformers

Instrument transformers are used to measure and control electrical parameters (voltage and currents). There are many different classes, each with a specific range of measuring accuracy.

Instrument transformers used for relaying and indicating purposes have a larger margin of error than those used for measuring the power and energy on which the billing demand of a particular plant is based.

There are two types of instrument transformers: potential transformers, used for measuring voltage, and current transformers, used for measuring current (see Fig. 2-52). The equipment connected to their secondaries constitutes the so-called load, or burden. In normal operation, the burden of the current transformer is of very low impedance, while the load on the potential transformer is of very high impedance. An elementary representation of instrument transformers and some of the meters connected to their secondaries is shown in Fig. 2-53.

The polarity and the grounding of the potential and current transformers are as shown in the figure. Correct polarity identification will lead to correct instrument connections; proper device grounding will ensure the safety of personnel and the protection of equipment.

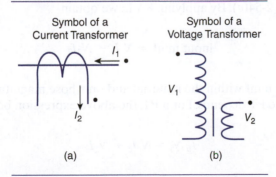

FIG. 2-52 Symbols of instrument transformers.
(a) Current transformer. (b) Voltage transformer.

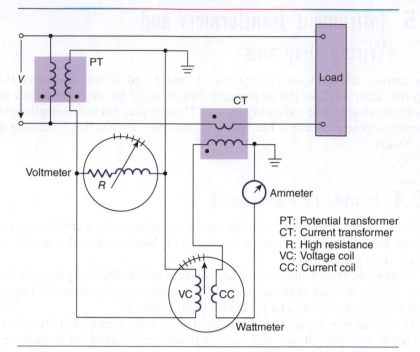

FIG. 2-53 An elementary representation of instrument transformers and measuring meters.

Potential Transformers

Potential transformers are used for stepping down voltages to nonhazardous and conventional voltage levels—normally 120 volts—for measuring and controlling purposes.

As shown in Fig. 2-54, their primary winding is connected in parallel to the high-voltage circuit whose voltage is to be measured. The voltage across the secondary winding is applied to potential coils of meters, relays, and other instruments, depending on what is desired.

Refer to Fig. 2-54(c). By applying KVL, we obtain

$$\text{input mmf} = N_c I_c + N_2 I_2 \tag{2.65}$$

where $N_c I_c$ is the mmf within the material and on whose magnitude the magnetization level of the PT depends. For a PT, the above expression becomes

$$I_p N_1 = N_c I_c + N_2 I_2 \tag{2.66}$$

where I_p is the current through the PT's primary winding. The magnitude of the primary voltage of a potential transformer is nearly constant; therefore, the magnitude of the flux within the magnetic material is also constant (see Eq. (2.7)). For

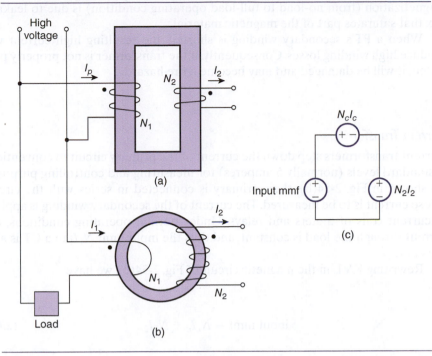

(a)

(b)

(c)

Input mmf

FIG. 2-54 Instrument transformers. (a) Schematic diagram for a PT. (b) Schematic diagram for a CT. (c) Magnetic equivalent circuit for instrument transformers.

all practical purposes, this results in a constant-magnitude mmf within the material, regardless of the magnitude of the load current.

The variation of the PT's magnetization level from a no-load ($I_2 = 0$) to a full-load operating condition is shown in Fig. 2-55(a). The small change in the levels of

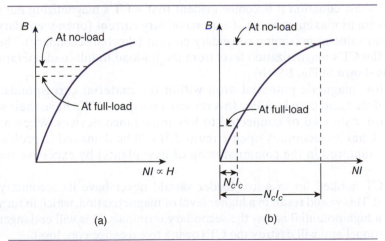

(a)

(b)

FIG. 2-55 Levels of magnetization from no-load to full-load for instrument transformers. (a) Potential transformers. (b) Current transformer.

magnetization (from no-load to full-load operating condition) is due to leakage flux that saturates part of the magnetic material.

When a PT's secondary winding is shorted, the resulting high current will produce high winding losses. Consequently, if the transformer is not properly protected, it will be damaged and may become a fire hazard.

Current Transformers

Current transformers step down the current of the primary circuit to conventional standard levels (normally 5 amperes) for measuring and controlling purposes. As shown in Fig. 2-54(b), their primary is connected in series with the circuit whose current is to be measured. The current of the secondary winding is applied to current coils of meters and relays. Under normal operating conditions, the current through the load is constant, and thus the input mmf $N_1 I_1$ to a CT is also constant.

Rewriting KVL in the magnetic circuit of Fig. 2-54(c), we have

$$\text{input mmf} = N_c I_c + N_2 I_2 \tag{2.67}$$

The above expression for a CT becomes

$$N_1 I_1 = N_c I_c + N_2 I_2 \tag{2.68}$$

From the last equation it becomes evident that a CT's magnetizing mmf ($N_c I_c$) varies from its maximum value at zero secondary current (open secondary) to its minimum value at maximum secondary current (shorted secondary). The variation of the CT's magnetization level from the no-load to full-load operating condition is shown in Fig. 2-55(b).

A low magnetic potential drop within the material corresponds to low levels of flux and core loss. For this reason, energized CT's have their secondaries either shorted or connected to low impedance devices. When an energized CT has its secondary open-circuited, it will be damaged (a problem usually encountered in the commissioning of new plants) by excessive magnetic losses.

A CT hooked up to a live feeder should never have its secondary open-circuited. This would result in a higher level of magnetization, which in turn would induce a high potential across the secondary terminals. This will endanger working personnel and will destroy the CT (owing to excessive core loss).

For the purposes of comparison, the characteristics of CT's and PT's are summarized in Table 2-3.

TABLE 2-3 Main characteristics of PT's and CT's		
Description	PT	CT
Input winding connection, with respect to plant's load	Parallel	Series
Step down the magnitude of the voltage	Yes	
Step down the magnitude of the current		Yes
Levels of magnetization depends on	Supply line voltage	Current through its secondary
Damaged when its secondary is	Short-circuited	Open-circuited
Cause of damage	High copper loss	High core loss

Exercise

2-9

When the single-phase load of Fig. 2-54 is accidentally shorted, what will be the effect on the potential and current transformers?

Optical Instruments

Optical instrument transformers refer to potential and current transformers, which contrary to classical PT's, measure the voltage and current to a distribution system without being in contact with the live conductors. They are very expensive and are presently used in high-voltage substations. Their usage, however, will be extended to lower voltages because of their great advantages, such as security of personnel, very low maintenance, high accuracy, no magnetic or copper losses, prevention of fires, no ferroresonance, and small size.

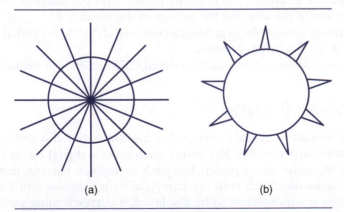

(a)   (b)

FIG. 2-56 Light bulb. (a) Unpolarized light. (b) Light (polarized) as seen with reading glasses.

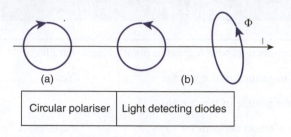

Circular polariser	Light detecting diodes

FIG. 2-57 Schematic diagram of an optical CT.
(a) Polarized circular light encircling a wire and
magnetic field of cable's current. (b) Generation
and detection of polarized light.

*In general, a light polarizing device, depending on its configuration, transmits
part of the light. This process of polarization can be easily observed at night
when you use reading glasses and look at the street incandescent lights and/or
when you look toward the lights of incoming cars when driving (Fig. 2-56).

Optical Current Transformers

The optical CT consists of two sections, one of which produces opposite rotating
circular polarized lights and the other incorporates light detecting diodes.

For a given wire to a distribution network, two circular polarized lights are
generated (Fig. 2-57), one of which is aided by the magnetic field of the conductor's current ($\phi \, \alpha \, I$) and the other is opposed. The change in the illumination level
is detected by the light-detecting diodes. The diode's current is further processed,
and the output of the optical sensor is transmitted to the control room by wireless
technology or by optic fibers.

PT's

The sensor of an optical PT detects the intensity of the electric field (E) produced
by the conductor's current. The intensity depends on the distance of the sensor
from the section of the wire and the voltage of the wire's field.

The corresponding voltage produced ($v = -\int E \, dx$) is the weighted average of
the output of the detecting elements.

Some manufacturers combine the optical CT's and PT's in the same unit.

2.5.2 Wiring Diagrams

Wiring diagrams are electrical drawings that use conventional symbols to indicate
the actual hook-up of meters. The wiring diagram of a digital meter is shown in
Fig. 2-58. It measures many parameters such as voltage, current, power, power
factor, and harmonics. Such units are equipped with a display and a set of push
buttons. When a parameter is to be displayed, a corresponding push button is
depressed. These meters can also send "measuring pulses" to computers, which in
turn can print out the instantaneous variations of the system parameters.

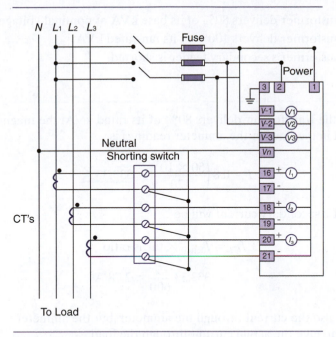

FIG. 2-58 Wiring diagram of a digital multimeter.

EXAMPLE

2-10

For the single-phase transformer shown in the one-line diagram in Fig. 2-59, determine the ammeter reading for the following operating conditions:

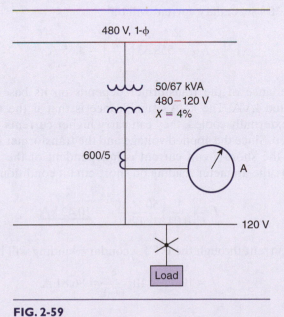

FIG. 2-59

a. The transformer delivers 80% of its base kVA at nominal voltage.
b. The transformer delivers 100% of its fan-rated kVA.
c. The transformer's secondary feeder is shorted.

SOLUTION

a. When the transformer delivers 80% of its rated kVA, the magnitude of the current in the line (or the ammeter reading) is

$$I = 0.8 \left(\frac{50 \times 10^3}{120} \right) = \underline{333.33 \text{ A}}$$

The CT's secondary current will be

$$I_{CT} = I_{\text{actual}} \times \text{turns ratio}$$

$$= 333.33 \, \frac{5}{600} = 2.78 \text{ A}$$

This is also the current through the ammeter, but the ammeter is always calibrated to give the actual current through the load.

b. Similarly, the ammeter reading will be

$$I = \left(\frac{67 \times 10^3}{120} \right) = \underline{558.33 \text{ A}}$$

The CT's secondary current will be

$$I_{CT} = 558.33 \left(\frac{5}{600} \right) = 4.65 \text{ A}$$

c. The impedance of the transformer depends on its base kVA and not on its fan-rated kVA. The kVA's significance is that if the transformer windings are externally cooled, they can carry higher currents without becoming overheated. Since the applied voltage and the transformer impedance remain constant, the short-circuit current is independent of the level of operating kVA. Thus, the ammeter reading on short-circuit conditions is

$$I = \left(\frac{1}{0.04} \right) \frac{50 \times 10^3}{120} = \underline{10.42 \text{ kA}}$$

The current through the CT's secondary winding will be

$$I = 10.42 \times 10^3 \, \frac{5}{600} = 86.81 \text{ A}$$

EXAMPLE

2-11

a. Refer to Fig. 2-59. When the ammeter reads 1500 A and has an accuracy of 99%, what is the apparent power drawn by the inductive load?

b. What is the power factor of the load when the kW meter reads 2000 kW and the reactive power meter reads 1500 kVAR?

SOLUTION

a. The scales on switchboard-type meters are calibrated in terms of the line-side parameters. Thus, the current through the 4.16 kV lines is

$$I_L = \frac{1}{0.99}(1500) = 1515.15 \text{ A}$$

By definition, the apparent power is

$$|S| = \sqrt{3}V_{L\text{-}L}I_L = \sqrt{3}(4.16)(1515.15) = \underline{10{,}917.17 \text{ kVA}}$$

b. The real and reactive powers are given, respectively, by

$$P = \sqrt{3}V_{L\text{-}L}I_L \cos\theta$$

$$Q = \sqrt{3}V_{L\text{-}L}I_L \sin\theta$$

Thus,

$$\theta = \arctan\frac{Q}{P} = \arctan\frac{1500}{2000} = 36.9°$$

The power factor is

$$\cos 36.9° = \underline{0.80 \text{ lagging}}$$

Exercise

2-10

a. Why is the secondary of a PT normally supplied with a fuse but not the secondary of the CT?

b. Explain why a voltmeter never measures the voltage and an ammeter never measures the current.

Ferroresonance

Human Resonance

Resonance in general is a condition that results in maximum response (love, hate, indifference, and so forth). Some people instinctively love red colors, while others hate them or yet some others are indifferent to them. The future of an individual most likely will be strongly influenced by his or her inner inclination (resonance) in conjunction with that of fellow humans (boss, partner, friend, and such). It has been reported that an international bank, in order to select the most suitable employees to a managerial position, subjects all candidates to a questionnaire (searching for resonance) and from this, a decision is made that is apparently 100% accurate.

Linear Resonance

In the linear resonance, the overvoltages and overcurrents are of predictable magnitude. Such a resonance takes place when the circuit's inductive reactance is equal to the capacitive reactance. Furthermore, the circuit's power factor is equal to 100%, and the magnitude of the resulting overvoltages and overcurrents are sinusoidal and of predicable frequency. Similarly, a building is characterized with linear resonance whose frequency is in the range of 4–6 Hz.

Nonlinear Resonance

In nonlinear resonance the voltages and current are very high and of unpredictable frequency. The nonlinear resonance is known as a ferroresonance. It was observed almost 100 years ago that ferroresonance was taking place in distribution systems whose transformers were made from ferrous materials (iron laminations).

Ferroresonance depends on the circuit's initial conditions, the applied voltage, the nonlinear circuit inductance, and on the capacitance between cables and/or cables and ground. In ungrounded systems, the voltages developed could be up to 4 pu, while in the grounded systems, voltages could reach 2.5 pu.

The inductance is inversely proportional to the permeability, which in turn is the slope of the material's B-H diagram. At higher than nominal voltages, the transformer is driven into saturation, the permeability is reduced, and thus the inductance is increased.

Many causes can initiate ferroresonance, but the main one is the variable inductive reactance of the power, potential, and measuring potential transformers that are driven into saturation. The initiation of this phenomenon could be attributed to the lightning strokes, breaker switching, or malfunctioning and single-phase operation of a three-phase distribution system.

One way to mitigate the effects of resonance is to oversize the transformers so that an accidental increase in the supply voltage will not drive them into saturation and/or add a resistance (damping effect) on the secondary windings of the transformers.

It has been well publicized that if a small step-down PT and a small capacitor in a control system initiated ferroresonance, that can cause the failure (insulation damage) of dozens of downstream motors.

K-Factor of Transformers

Introduction

A transformer designated as *k*-type (*k*-4, *k*-9, *k*-13, etc.) indicates that it will not be overheated when the *k*-factor of the line currents is lower than that of the transformer.

When $k = 1$, it means that the transformer will not be overheated when it supplies rated power to linear load—that is, when the load does not draw any harmonic current. The rate of heat loss within the transformer is the sum of the core loss (eddy-current and hysteresis loss) plus the stray losses and I^2R of its windings.

The eddy-current losses and hysteresis losses are nonlinear functions of the currents' frequency.

By far, the largest effects of the harmonic components of the currents are on the eddy-current losses.

Calculation of the *K*-Factor

The *k*-factor is defined as follows:

$$k = \sum_{h=1}^{h_{max}} I_h^2 h^2 \qquad (2.69)$$

where h stands for the number of the harmonic and I_h is the rms current of harmonic h, in per unit of the line current.

EXAMPLE

2-12

The line current (ι) in a power distribution system is

$$\iota = 10 \sin \omega t + 4 \sin 3 \omega t + 2 \sin 5 \omega t....$$

Determine the *k*-factor.

Solution

The rms value (I) of the line current is

$$I = \sqrt{\left(\frac{10}{\sqrt{2}}\right)^2 + \left(\frac{4}{\sqrt{2}}\right)^2 + \left(\frac{2}{\sqrt{2}}\right)^2} = 7.746 \text{ A}$$

and the per-unit values of the current components are

$$I_1 = \left(\frac{10/\sqrt{2}}{7.746}\right) = 0.9129$$

$$I_3 = \left(\frac{4/\sqrt{2}}{7.746}\right) = 0.3651 \text{ pu}$$

and

$$I_5 = \left(\frac{2/\sqrt{2}}{7.746}\right) = 0.1826 \text{ pu}$$

From Eq. (2.69), we obtain

$$k = [(0.9129)(1)]^2 + [(0.3651)(3)]^2 + [(0.1827)(5)]^2$$

$$= \underline{2.8664}$$

2.6 Transformer's Nameplate Data

Some of the essential markings on a transformer's nameplate are as follows.

Apparent Power Rating in kVA

This parameter indicates the designed kVA base transformation capability of the transformer. As long as the transformer operates at or below its marked kVA, it will not overheat. This is also the base power for the per-unit calculations of the transformer impedances. When transformers are equipped with cooling fans, they can deliver higher kVA than their rated value without being overheated.

For example, a transformer designated as

$$1500/2000 \text{ kVA}$$

has a base transformation capacity of 1500 kVA. However, with the use of cooling fans, the transformer can deliver 2000 kVA (33% increase) without being overheated. When a second set of fans is used, the transformer can deliver an additional 25% kVA without being overheated. Also, when a special winding insulation is used, the transformer can safely deliver an additional 12% kVA. For example, a transformer with nameplate data of 12/16/20/22.4 MVA should be interpreted as follows:

12 MVA	is the base power
16 MVA	is its capacity with one set of fans (1.33)
20 MVA	is its capacity with two sets of fans (1.25)
22.4 MVA	is its capacity due to upgraded winding insulation (1.12)

The ohmic value of the transformer's impedance remains the same, regardless of whether the transformer delivers its base or its fan-rated kVA. The resistance of the windings depends on the temperature of the conductors and will change somewhat with increases or decreases in load current.

Impedance in Percent

This is the magnitude of the equivalent winding or leakage impedance. The transformer's magnetizing impedance is not included on the nameplate data. The winding impedance (expressed in percent) indirectly gives the equivalent transformer's impedance I ohms and also indicates what percent of the rated voltage is required to circulate rated current through the windings of the transformer when the secondary is shorted.

The leakage impedance varies with the size of the transformer, class voltage, type of winding (copper or aluminium), and frequency. For the standard voltages and in the range of 1000 to 2000 kVA, the transformer's impedance is about 6%.

Efficiency in Percent

This is the efficiency of the transformer when it delivers rated kVA at unity power factor. Generally, the higher the kVA capacity of the transformer, the higher its efficiency. In the range of 100 to 2000 kVA, the efficiency is usually between 96 and 99%.

"OFF-Load" Voltage Tap Changers

Normally, transformers are equipped with four "OFF-load" voltage tap changers ($\pm$ 2.5% and $\pm$ 5%). The positive tap settings are used to compensate for voltage drop through the transformer's upstream feeders, while the negative tap settings are used to compensate for generated voltages that are slightly above nominal values.

Winding Connection—Rated Voltages

The nameplates of three-phase transformers also indicate whether their windings have a delta (Δ) or a star (Y) connection and give their corresponding rated voltages. The latter may be used in establishing the line-to-line or the phase-to-neutral turns ratio. Nameplate data also indicate whether provisions have been made for grounding the neutral terminal of the star-connected windings.

2.7 Conclusion

Transformers change the voltage and the current from one level to another and, as such, constitute the main interconnecting link between two systems of different voltages.

The design, operation, and analysis of all types of transformers are based on the induction principle, which says that a voltage is induced in coils whose flux linkages ($N\phi$) change as a function of time. Under ideal conditions, the induced voltage and the resulting current in a secondary coil are related to the corresponding parameters of the source coil (primary coil) by the windings turns ratio.

Each type of transformer can be analyzed by using either the magnetic or the electric equivalent circuits. Although using the magnetic circuit is simpler, the trend throughout the industry is to employ the electric equivalent circuits.

The core and copper losses of transformers—even though they are a small fraction of the power-transformation capability of any transformer—often constitute the criteria for selecting transformers. This is because the purchasing cost of a transformer is much lower than the cost of the energy losses within the transformer over its operating life. (A kilowatt consumed continuously over a 20-year period has a present worth of about $7215 when the annual interest rate is 6% and the cost of energy is 15¢/kWh and increases 4% per year. In comparison, the cost of a 25-kVA transformer that has 1 kW of power loss is about $2000.)

In the analysis of circuits that include transformers, it is much more convenient to use per unit values because the impedances, voltages, or currents, when expressed in per-unit, are the same in either side of the transformer.

Throughout the various sections of this chapter, the short-circuit MVA of transformers is calculated because of its importance in equipment selection (circuit breakers, fuses, etc.). When properly selected, this equipment safeguards personnel and downstream apparatuses.

Single-phase power transformers have relatively small ratings, usually in the range of 1–100 kVA.

The coils of the transformer are magnetically coupled, and the relative polarity of the induced voltages is normally identified by a dot.

By convention, the dotted terminal is at higher potential than the undotted terminal. This leads to determination of the direction of current flow and flux. This is of extreme practical importance for proper field connections of measuring equipment and of protective relays. Misinterpretation of the dotted terminals will result in incorrect meter indication, a problem that usually arises during the commissioning of industrial plants.

The analysis of *three-phase*, two-winding transformers is simplified by using their electric or magnetic per-phase equivalent circuits. An additional simplification results when the per-unit values of the transformer parameters are used. The latter remove the difficulty of long calculations and the confusion that may arise from the various three-phase transformer connections.

Transformers of 5–20 MVA rating have a nominal impedance in the range of 5% to 8% and an efficiency of 0.97 to 0.99.

The phase shift between the primary and secondary line-to-line voltages depends on the transformer's connections. For the delta–star connection, the voltages on the HV side lead their corresponding LV side by 30 degrees. The majority of three-phase transformers are of Δ-Y type. They tend to maintain sinusoidal secondary voltages, and the secondary star windings can be used to ground the distribution network and to connect line-to-neutral loads.

Three-phase transformers are normally equipped with a set of ventilating fans (the additional cost is 5% above nominal). As a result, the transformer's transformation capacity is increased by 33%.

Autotransformers are used in large power systems and in testing. Standard transformers are paralleled in order to increase the reliability of a substation (two smaller units are more reliable than a larger unit) and to provide additional transformed kVA that meets a plant's increased demand. In order to parallel two transformers, they must have equal impedances, equal turns ratios, the same phase rotations, and equal phase shifts between the primary and secondary open-circuited voltages.

Instrument transformers step down the voltage and the current to more standard and less dangerous levels.

Usually, the secondary of a potential transformer is rated at 120 V, while the secondary of a current transformer is rated at 5 A.

The burden, or the load (voltmeters, voltage coils of kW-meters, etc.), of potential transformers is of very high impedance; conversely, the burden of the current transformers is of very low impedance.

Energized potential transformers are damaged—owing to high copper loss—when their secondary is shorted. Current transformers are damaged—owing to high core loss—when their secondary is open-circuited.

Wiring diagrams are electrical drawings that use conventional symbols to indicate the actual field connections of the measuring equipment. As such, wiring diagrams are of fundamental importance to power-distribution engineers, to those who participate in the commissioning of industrial plants, or to those whose responsibility it is to maintain the plant's electrical equipment.

In ferroresonance, high and repetitive voltages are developed within the distribution system. It results from circuit breaker switching, lightning strokes, and the like that drive the plant's coils and transformers into saturation.

The *k*-rating of the transformers reveals the level of harmonics at which it will not be overheated. The sources of harmonics are lighting fixtures, microprocessors, and variable speed drives.

2.8 Summary

The following tables summarize the main concepts on transformers and their technical characteristics and parameters. The main concepts of the chapter are condensed in Table 2-4, and typical manufacturer's data are given in Table 2-5 through Table 2-8.

TABLE 2-4 Summary of Main Transformer Concepts

Item	Description	Remarks
1	Principle of operations of transformers: $$v = -N\frac{d\phi}{dt}$$	Eq. (2.6)
2	Fundamental transformer equation: $$V = 4.44Nf\phi_m$$	Eq. (2.7)
	Ideal Transformer Relationships	
3	$$V_2 = V_1\frac{N_2}{N_1}$$	Eq. (2.11)
4	$$I_2 = I_1\frac{N_1}{N_2}$$	Eq. (2.14)
5	$$Z_1 = Z_2\left(\frac{N_1}{N_2}\right)^2$$	Eq. (2.17)
6	*Short-circuit test* $$I_z = I_{\text{rated}}$$ $$V_z = (2\% \rightarrow 12\%)V_{\text{rated}}$$ Measure I_z, V_z, and P_z. Calculate: $$R_e = \frac{P_z}{I_z^2}, \quad |Z_e| = \frac{V_z}{I_z}, \quad X_e = \sqrt{Z_e^2 - R_e^2}$$	Section 2.1.3
7	*Open-circuit test* $$I_{\text{exc}} = (3\% \rightarrow 10\%)I_{\text{rated}}$$ $$V_{\text{exc}} = V_{\text{rated}}$$ Measure: $$I_{\text{exc}}, V_{\text{exc}}, \text{ and } P_{\text{exc}}$$ Calculate: $$R_m = \frac{P_{\text{exc}}}{I_{\text{exc}}^2}, \quad |Z_m| = \frac{V_{\text{exc}}}{I_{\text{exc}}}, \quad X_m = \sqrt{Z_m^2 - R_m^2}$$	Section 2.1.3
8	Transformer operates at maximum efficiency when $$P_z = P_{\text{exc}}$$	Eq. (2.38)
9	$$\text{Regulation}\% = \frac{|V_{nl}| - |V_{fl}|}{|V_{fl}|}(100)$$	Eq. (2.39)

(Continued)

TABLE 2-4	(Continued)	
Item	Description	Remarks
10	$Z_{e_{pu}} = V_{z_{pu}}$	Eq. (2.42)
	3-ϕ Transformers	
11	$N_A I_A = N_a I_a$	Eq. (2.51)
12	Star–delta impedance: $$Z_y = \frac{Z_\Delta}{3}$$	Eq. (2.53)
13	Third harmonic component of exciting current: $$i_3 = 3I_{m_3} \cos 3\omega t$$	
13a	Evaluation of a transformer's k-rating: $$k = \sum_{h=1}^{h_{\max}} I_h^2 h^2$$	Eq. (2.69)
	Autotransformers	
14	The ampere-turns of each winding are constant, regardless of the type of transformer connection. That is, $$(NI) = \text{constant}$$	Eq. (2.60)
15	Neglecting losses, $$(\text{kVA})_{\text{input}} = (\text{kVA})_{\text{output}}$$	Eq. (2.61)
	Parallel Operation of Transformers	
16	To parallel two transformers, the following conditions must be satisfied: a. equal impedances b. equal turns ratios c. equal phase shift between the primary and the secondary open-circuited voltages d. the same phase rotations	Section 2.4
	PT's and CT's	
17	When short-circuited, energized potential transformers are damaged because of high copper loss	Section 2.5.1
18	When open-circuited, energized current transformers are damaged because of high core loss	Section 2.5.1

TABLE 2-5 Typical parameters of dry-type transformers with copper windings at 60 Hz

a. Single phase (480–240/120 V)

kVA	Losses in Watts at 170°C		Efficiency in Percent at 170°C		Impedance in Percent at 170°C
	Core	Copper	Full-Load	¼ Full-Load	
10	70	275	96.7	96.6	3.6
25	110	825	96.4	97.5	4.1
50	150	2000	95.9	97.8	5.7
100	280	3350	98.5	98.1	4.7

b. Three-phase (480–208/120 V)

kVA	Core	Copper	Full-Load	¼ Full-Load	Impedance
15	130	410	96.5	96.0	3.6
75	300	3100	95.7	97.4	5.7
150	540	5000	96.4	97.8	4.7
500	1120	11,531	97.5	98.5	5.0

Based on data from Westinghouse Canada Inc.

TABLE 2-6 Parameters of 4160–480/277 V, dry-type transformers with copper windings

kVA	Resistance in Ohms at 170°C		Exciting Current in Percent	Losses in kW		Efficiency at Full-Load	Impedance in Percent
	HV	LV		Core	Copper		
500	2.9	0.011	1.5	1.2	7.6	98.3	5.5
1000	1.09	0.004	1.12	3.8	13.1	98.3	5.9
1500	0.75	0.003	0.70	4.0	20	98.4	6.1
2000	0.48	0.016	0.65	4.8	23	98.6	6.5

Based on data from Westinghouse Canada Inc.

TABLE 2-7 Parameters of a 15/20/22.4 MVA, 60–4.16/2.4 kV, liquid-type transformer with copper windings

Exciting Current in Percent	Core Loss in kW	Copper Loss in kW	Impedance in Percent
0.41	15.0	65	7.0

Based on data from Westinghouse Canada Inc.

TABLE 2-8 Parameters of 4160–480/277 V, 60 Hz, dry-type transformer with copper and/or aluminum windings

kVA	Cost in 1988 Windings		Core Loss in kW at 170°C Windings		Copper Loss in kW at 170°C Windings		Total Weight in kg		Estimated Impedance in Percent
	Copper	Aluminum	Copper	Aluminum	Copper	Aluminum	Copper	Aluminum	
750	$22,800	$19,000	3.0	3.4	8.1	8.8	2100	2050	6
1000	26,400	22,000	3.8	4.0	13.1	14.2	2500	2425	6
1500	30,600	25,000	4.2	4.2	20	21.5	3300	3100	6

Based on data from Westinghouse Canada Inc.

2.9 Review Questions

1. In general, what fraction of the transformer's output power is its iron and winding losses?

2. Show that 1 kW consumed continuously over a 20-year period has a present-worth value of $7215. Assume the interest rate is 6% per year, the cost of energy is 5 cents/kWH, and the increase in the cost of energy is 4% per year.

3. Explain why the magnetizing impedance is not a constant parameter but depends on the level of magnetization.

4. How would you measure the magnetic potential drop (N_cI_c) in a magnetic material? How is this potential related to the permeability of the material?

5. Explain why the per-phase analysis of a three-phase balanced system is analogous to the per-pole analysis of a magnetic circuit.

6. For maximum efficiency, how is winding loss related to core loss?

7. Why does the selection of the transformer's primary protective devices depend on the time-current waveform of the inrush current?

8. Sinusoidal voltages produce sinusoidal fluxes, which in turn produce nonsinusoidal exciting currents. Explain.

9. What is the predominant harmonic of the exciting current and of the inrush current of a transformer?

10. Show that the per-unit value of the transformer's leakage impedance is the same whether it is referred to as the HV or the LV winding.

11. Draw the power triangle for a balanced three-phase inductive load and give the formulas for real, imaginary, and complex power.

12. Nonsinusoidal exciting currents produce sinusoidal secondary voltages. What are the two essential requirements for the existence of nonsinusoidal exciting currents?

13. Explain why the most important characteristic of an autotransformer is that the current in its windings is the same as when the transformer operates as a conventional two-winding transformer.

14. What four conditions must be satisfied before any two transformers are paralleled?

15. What is the essential difference between a CT and a PT?

16. Explain the nameplate data marked on your laboratory's transformer.

17. Compare the following: exciting current, inrush current, load component of primary current and short circuit current of a transformer.

2.10 Problems

2-1 A 50 kVA, 2300–230 V, 60 Hz, single-phase step-down transformer has primary and secondary leakage impedances of $(0.5 + j2.6)$ and $(0.005 + j0.026)$ ohms, respectively. Neglecting the magnetizing impedance, determine:

   a. The equivalent circuit, as seen from the high-voltage and the low-voltage windings.

   b. The secondary voltage, when the transformer is connected to a 2300 V source and delivers rated current to a load at 0.9 Pf lagging.

2-2 Data from the open- and short-circuit test on a 10 kVA, 2400–240 V, 60 Hz step-down transformer are as follows:

Test	Voltage in Volts	Current in Amperes	Power in Watts
Open circuit	2400	0.35	150
Short circuit	12	41.67	320

Determine:

   a. The equivalent circuit, as seen from the high- and low-voltage winding.

   b. The primary voltage, when the transformer delivers rated current at 0.9 power factor to a 240 V inductive load.

   c. The voltage regulation and efficiency.

   d. The per-unit value of the transformer's equivalent leakage impedance.

2-3 A single-phase 50 kVA, 2400–120 V, 60 Hz transformer has a leakage impedance of $(0.023 + j0.05)$ per-unit and a core loss of 600 watts at rated voltage. Determine:

   a. The load current for maximum efficiency.

   b. The efficiency for the condition in (a) and the efficiency when the transformer delivers rated current. In both cases, assume unity power factor.

   c. The power factor of the load that will result in the best regulation (0%). Assume that the load current and the primary voltage remain constant at their rated values.

2-4 A 100 kVA, 480–120 V, 60 Hz, single-phase transformer has an efficiency of 95.75% at rated conditions and unity power factor. Its leakage impedance is $(2.5 + j5.0)$ percent. The transformer's open-circuit test data are as follows:

Voltage in Volts	Power in Watts	Frequency in Hertz
100	1400	50

a. Determine the eddy-current and hysteresis losses at 60 Hz and 120 V.

b. Determine the rate at which heat is produced within the structure of the transformer at 100 V and 50 Hz due to the friction caused by the motion of the magnetic domains.

c. Can the transformer be used to deliver its rated power at rated voltage and 50 Hz? Explain.

2-5 A 15 kVA, 480–240/120 V, 60 Hz, single-phase transformer has equal primary and secondary leakage impedances of $(0.01 + j0.02)$ per unit. Determine:

a. The no-load voltage across the secondary terminals at rated flux.

b. The voltage regulation when the transformer is connected to a 480 V source and delivers rated current at a power factor of:

   1. Unity.

   2. 0.80 leading.

   3. 0.80 lagging.

2-6 A transformer has a leakage impedance of $(0.015 + j0.06)$ per unit. When the transformer delivers rated current at 0.9 Pf lagging, its efficiency is 97%. Determine:

a. The nominal core loss in per-unit and the voltage regulation.

b. The no-load voltage tap setting that will minimize the voltage regulation.

2-7 Figure P2-7 shows a core-type 150 kVA, 4160–480/277 V, three-phase transformer. Connect the primary in delta and the secondary in star. The neutral is to be solidly grounded. Show coil polarities and direction of flux lines and find the coil currents for the following load conditions:

a. A load drawing 60 A is connected from line a to neutral.

b. A load drawing 60 A is connected from line a to line b.

c. A 3-ϕ delta-connected load of 100 kW, 480 V, 0.90 efficient, and 0.8 power factor lagging is connected across the secondary.

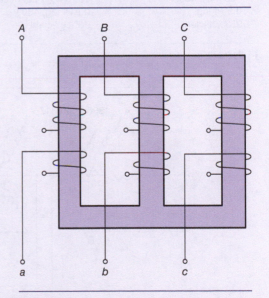

FIG. P2-7

2-8 A three-phase, 2000 kVA, 25,000–480/277 V, Δ-Y transformer supplies two three-phase loads through feeder B as shown in Fig. P2-8. The Δ-connected high-voltage transformer windings are connected

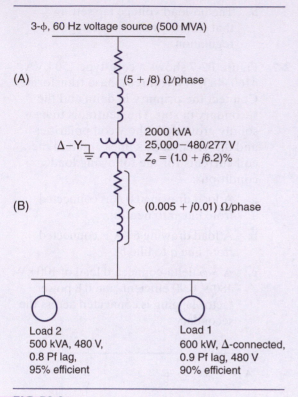

3-φ, 60 Hz voltage source (500 MVA)

(A) (5 + j8) Ω/phase

2000 kVA
Δ–Y 25,000–480/277 V
$Z_e = (1.0 + j6.2)\%$

(B) (0.005 + j0.01) Ω/phase

Load 2
500 kVA, 480 V,
0.8 Pf lag,
95% efficient

Load 1
600 kW, Δ-connected,
0.9 Pf lag, 480 V
90% efficient

FIG. P2-8

through cable A to a 3-φ, 60 Hz power source whose short-circuit MVA is 500. The ratings of various equipment are as shown in the diagram. If the three-phase loads draw rated currents at rated voltage, determine:

a. The per-unit values of the impedances of cables A and B, using the ratings of the transformer as base values.

b. The total losses of the transmission system.

c. The voltage required at the source.

2-9 Figure P2-9 shows a three-phase, five-wire (3-φ, 5W) distribution system. If the connected loads draw rated currents, determine:

a. The minimum required kVA capacity of the transformer.

b. The phase currents through each of the six transformer coils.

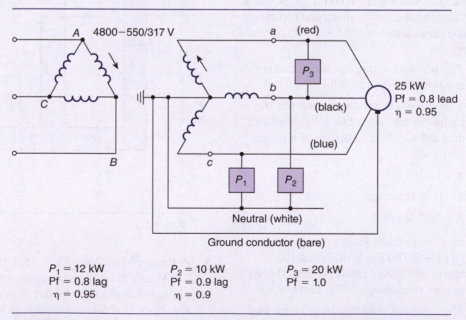

A 4800–550/317 V a (red)

P_3

b

(black)

c

(blue)

25 kW
Pf = 0.8 lead
η = 0.95

B

P_1 P_2

Neutral (white)

Ground conductor (bare)

$P_1 = 12$ kW
Pf = 0.8 lag
η = 0.95

$P_2 = 10$ kW
Pf = 0.9 lag
η = 0.9

$P_3 = 20$ kW
Pf = 1.0

FIG. P2-9

2-10 Figure P2-10 shows a three-phase transformer whose neutral point is grounded through a 1.0 ohm resistor. The transformer delivers rated voltage to the three-phase heater. When the incoming live part of cable *C* (phase *ABC*) accidentally makes contact with the grounded metallic enclosure of the heater, estimate (assuming negligible transformer leakage impedance) the following:

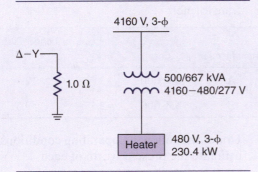

4160 V, 3-φ

Δ–Y

1.0 Ω

500/667 kVA
4160−480/277 V

Heater | 480 V, 3-φ
230.4 kW

FIG. P2-10

a. The line and ground currents.

The voltage from the heater's enclosure to ground.

The reaction of the upstream protective devices.

b. Repeat (a), assuming that the enclosure is not grounded.

c. Repeat (a), assuming that the neutral of the transformer is solidly grounded.

d. Repeat (a), assuming that the neutral of the transformer is not solidly grounded.

2-11 A 5 kVA, 480–120 V, two-winding transformer is to be connected as an autotransformer to supply a 480 V load from a 600 V supply. Determine:

a. The autotransformer kVA.

b. The untransformed kVA delivered to the load.

2-12 A two-winding, 2400–1000 V, 200 kVA transformer with a leakage impedance of $(0.02 + j0.06)$ per unit has an efficiency of 0.96 when it delivers rated current at 0.85 power factor lagging. This transformer is to be connected as an autotransformer in order to supply a 2400 V load from a 3400 V supply. Determine:

a. The kVA rating of the autotransformer. What percentage of this kVA passes through the transformer unaltered?

b. The efficiency at full load and at 0.85 power factor.

c. The primary current when the secondary of the transformer is shorted.

2-13 Two single-phase transformers are connected in parallel, as shown in Fig. P2-13, to supply rated voltage and current to a 1000 kVA, 480 V, 0.95 power-factor-lagging load. The nameplate information for each transformer is as follows:

Transformer	Rating in kVA	Voltage Ratio	Leakage Impedances in Per-Unit
1	550	4160–480	$(0.02 + j0.04)$
2	550	4160–468	$(0.02 + j0.05)$

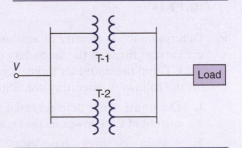

T-1

V

Load

T-2

FIG. P2-13

Assuming constant primary voltage and negligible magnetizing current, determine:

a. The primary voltage.

b. The current through each transformer.

c. The output kVA of each transformer.

d. The no-load circulating current.

e. The current in the common primary feeder if the load is shorted.

2-14 A transformer is connected to a 25 kV substation through a cable, as shown in the one-line diagram in Fig. P2-14. The characteristics and parameters of the system's components are as shown.

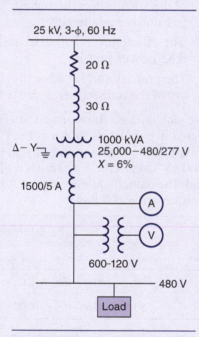

25 kV, 3-ϕ, 60 Hz

20 Ω

30 Ω

Δ–Y 1000 kVA
25,000–480/277 V
X = 6%

1500/5 A

Ⓐ

Ⓥ

600–120 V

480 V

Load

FIG. P2-14

a. Determine the voltmeter indication, the current through the secondary of the CT, and the ammeter indication for the following operating conditions:

1. The transformer delivers rated current at 0.90 power factor lagging.

2. A sustained three-phase short takes place at the 480 V terminals of the transformer.

3. The transformer delivers rated current at 0.9 Pf leading. Assume the short-circuit MVA of the source to be 500.

b. Explain why "a PT must not be energized with its secondaries short-circuited, and a CT must not be energized with its secondaries open-circuited."

2-15 Two manufacturers quoted the following on a request for the purchase of one (1) 1000 kVA, three-phase, 4160–480 V transformer:

Manufacturer	Cost	Core Loss in kW	Copper Loss in kW
A	$30,000	3	12
B	$35,000	2.5	9.5

Under the following operating conditions, estimate the present worth of each manufacturer's transformer over a five-year operating period.

a. Continuous operation at 75% of full-load.

b. Continuous operation at 50% of full-load.

c. Operation at full-load, 50% of the time.

Assume that the interest rate is 10%. The cost of energy is 6¢/kWh, and its annual increase is 8%.

2-16 The current (ι) drawn from a power distribution transformer is given by

$$\iota = 100 \sin \omega t + 30 \sin 3 \omega t \\ + 15 \sin 5 \omega t + 10 \sin 7 \omega t \ldots$$

Determine the required k-type of transformer.

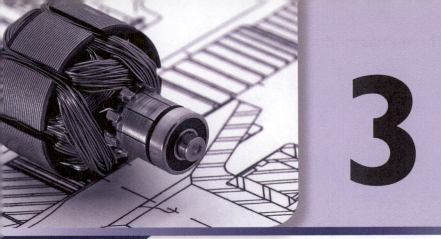

3

Three-Phase Induction Machines

3.0   Introduction

3.1   Three-Phase Induction Motors

3.2   Industrial Considerations

3.3   Measurement of Equivalent Circuit Parameters

3.4   Asynchronous Generators

3.5   Controls

3.6   Conclusion

3.7   Tables

3.8   Review Questions

3.9   Problems

What You Will Learn in This Chapter

A

1 Stator and rotor windings
2 Principles of operation
3 Rotating magnetic fields
4 Mechanical and electrical radians
5 Equivalent circuit
6 Torque and power relationships
7 Industrial considerations
8 Mechanical loads
9 Voltage, efficiency, and power factor
10 Actual motor characteristics
11 Measurement of equivalent circuit parameters
12 Special applications
13 Asynchronous generators
14 Electromechanical transients
15 Plugging
16 Y-Δ starters
17 Variable frequency drives
18 Summary of formulas used
19 Manufacturers' data

B Outstanding Practical Highlights

1 Principles of operation
2 Equivalent circuit
3 Ohmic and per-unit analysis
4 Governing equations and their derivations
5 Torque-speed characteristics of motors
6 Torque-speed characteristics of mechanical loads
7 Speed control and the corresponding energy savings
8 Soft start
9 Y-Δ start
10 Starting, accelerating time, and corresponding energy consumption
11 Asynchronous generators
12 Manufacturers' motor parameters

C Additional Students' Aid on the Web

1 Additional solved and unsolved problems on VFD
2 Plugging
2.1 Rotor winding losses as a function of the rotor's polar moment of inertia
3 Speed as a function of time
4 Electromagnetic transients
5 Frequency and voltage multipliers

3.0 Introduction

Three-phase induction motors are by far the most widely used motors in industry, accounting for about 80% of the total number of motors used in the average plant.* In many small and medium-sized industries, all motors above 3 kW are three-phase induction motors. Three-phase induction motors are popular because they are more economical, last longer, and require less maintenance than other types of motors. Because of their importance, this chapter provides a detailed analysis of three-phase motors and a thorough discussion of their applications. The following subjects are discussed: principles of operation, rotating magnetic fields, equivalent circuits, governing mathematical relationships, industrial considerations, induction generators, and solid-state control.

Manufacturers' data for three-phase induction machines are included in the tables in Section 3.7 at the end of this chapter. Their control schematics are extensively covered in Chapter 7. The physical distribution of the machine windings and their unbalanced voltage operation are covered in more specialized textbooks.

3.1 Three-Phase Induction Motors

3.1.1 Stator and Rotor

Like all motors, three-phase induction motors transform electrical power into mechanical power by means of a stationary part called the stator and a rotating part called the rotor. Stator or *armature windings* are housed on the stator, whereas *rotor windings* are installed on the rotor. For an elementary representation of stator and rotor windings, see Fig. 3-1.

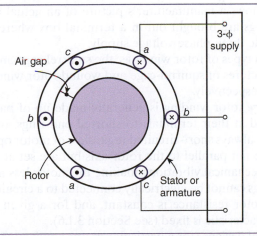

FIG. 3-1 An elementary representation of a three-phase induction motor. (In an actual machine, the armature and the field windings are placed in the stator and rotor slots, respectively.)

---

*Only about 50% of the energy used by a plant is consumed by 3-ϕ induction motors. Roughly 35% is absorbed by synchronous and dc machines, and the remaining 15% is consumed by single-phase motors, heating, cooling, lighting, and miscellaneous.

FIG. 3-2 Stator windings. *Photo courtesy of Siemens Industry, Inc.*

Stator windings can be either star- or delta-connected. Their main functions are to receive three-phase ac power and to produce a *single rotating* magnetic field that has an approximately sinusoidal space distribution. This rotating field completes its magnetic path through the stator, two air gaps, and the rotor structure. Figure 3-2 shows a manufacturer's picture of an actual armature winding. The stator windings are brought out in a terminal box where they can be connected to a suitable three-phase voltage supply.

There are two types of rotor windings: the squirrel cage and the wound type. Manufacturer's pictures of squirrel-cage and wound-rotor windings are shown in Figs. 3-3 and 3-4, respectively.

A *squirrel-cage* rotor winding is generally made up of bare aluminum bars that are connected at their terminals to shorted end rings. In other words, the rotor windings are always short-circuited regardless of motor operating condition. The rotor bars are not parallel to the rotor axis but are set at a slight skew. This feature reduces mechanical vibrations, making the motor less noisy.

Rotor windings cannot be electrically connected to a circuit outside the rotor. As a result, the rotor resistance is constant, and for a given stator voltage, its torque-speed characteristic is fixed (see Section 3.1.6).

In some special designs (for a particular motor application, for instance), the rotor may have double squirrel-cage windings, each with a different resistance. This construction gives higher starting torque, lower starting current, and higher full-load power factor.

FIG. 3-3 Squirrel-cage rotor. *Photo courtesy of Siemens Industry, Inc.*

FIG. 3-4 Wound-rotor winding. *Courtesy of General Electric*

FIG. 3-5 Three-phase induction motors: 200 kW, 460 V. *Courtesy of General Electric*

FIG. 3-6 Induction motor rotors: 6300 kW, 4160 V. *Courtesy of General Electric*

A very small percentage of induction machines have a wound rotor. *Wound-rotor windings* terminate at the slip rings on which the brushes rest. The brushes can then be connected to a three-phase variable resistor, and the resistance of the rotor winding can be externally controlled. This variable resistor, as demonstrated in this chapter, controls the torque-speed characteristic of the motor. Figures 3-5 and 3-6 show, respectively, manufacturer's pictures of three-phase induction motors and rotors.

3.1.2 Principles of Operation

The operation of three-phase induction machines is based on the generation of a revolving field, the transformer action, and the alignment of the magnetic field axes.

When balanced three-phase currents are injected into the stator windings, they produce a *rotating magnetic field* that, unless the rotor is revolving at the same speed as the magnetic field, will *induce* voltages* in the rotor windings. This results in rotor current and therefore rotor flux. The magnetic fields of the stator and rotor try to *align their magnetic axes*—a natural phenomenon—and in so doing, a torque is developed.

Refer to Fig. 3-7. The flux of the synchronously rotating stator field (ϕ_s) and the flux of the rotor currents (ϕ_f) combine vectorially to produce the net or resultant flux (ϕ_R) within the structure of the stator and rotor.

From Ampere's Law (Eq. (1.161)), the force and thus the torque (T) developed by two interacting fields are proportional to the product of the

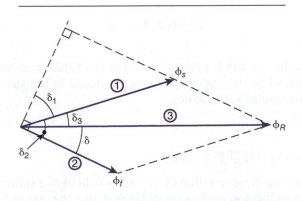

FIG. 3-7 Stator and rotor flux in a 3-ϕ induction machine at standstill.

*For these reasons, three-phase induction motors are sometimes referred to as rotating transformers.

strength of the two fields times the sine of the smallest angle between the two fields. Mathematically,

$$T = K\phi_f\phi_s \sin \delta_2 \tag{3.1a}$$

Since

$$\delta_1 + \delta_2 = 90° $$

Then

$$T = K\phi_f\phi_s \cos \delta_1 \tag{3.1b}$$

where the constant of proportionality K is a function of the physical parameters of the machine, and the factor $(\phi_s \cos \delta_1)$ is the quadrature component of the stator flux with respect to the rotor field.

From basic trigonometric concepts, we have

$$\phi_s \cos \delta_1 = \phi_R \cos (\delta_1 + \delta_3) \tag{3.2}$$

and

$$\phi_R \cos (\delta_1 + \delta_3) = \phi_R \sin \delta \tag{3.3}$$

where δ is the phase angle between the resultant field and the rotor field. This angle is called the torque angle. From Eqs. (3.1b), (3.2), and (3.3), we obtain

$$T = K\phi_f\phi_R \sin \delta \tag{3.4}$$

Under ideal conditions, the torque produced in the rotor structure is delivered to the shaft load, while the opposing torque produced in the stator structure is transmitted to the motor's foundation.

3.1.3 Rotating Magnetic Field

This section analyzes the generation of a rotating field by the stator windings of a 3-ϕ induction machine and derives formulas that give the strength and speed of the rotating field.

Consider the elementary two-pole, three-phase machine shown in Fig. 3-8(a). The dots in the center of the three conductors indicate that the current's direction is toward the reader, while the crosses indicate that the assumed positive direction of current is away from the reader. The stator windings carry balanced three-phase currents as shown in Fig. 3-8(b).

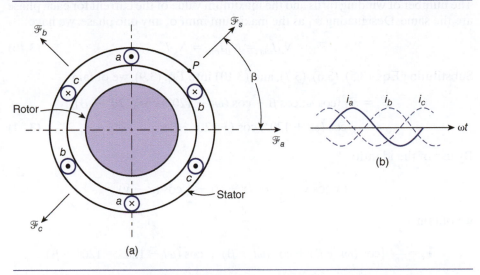

FIG. 3-8 Three-phase induction machine: **(a)** an elementary two-pole induction machine, and **(b)** balanced three-phase stator currents of sequence ABC.

For a phase sequence ABC, the phase magnetomotive forces (mmf's) as functions of time are as follows:

$$\mathscr{F}_a = N_A I_{ma} \cos \omega t \tag{3.5}$$

$$\mathscr{F}_b = N_B I_{mb} \cos (\omega t - 120°) \tag{3.6}$$

$$\mathscr{F}_c = N_C I_{mc} \cos (\omega t + 120°) \tag{3.7}$$

where I_{ma}, I_{mb}, and I_{mc} are the maximum values of the phase currents, and N_A, N_B, and N_C are the number of turns of the phase windings. The quantity v is the angular frequency of oscillation of the stator currents, which by definition is

$$\omega = 2\pi f \text{ electrical radians per second} \tag{3.8}$$

where f is the frequency of the stator currents in hertz.

At the instant under consideration, the mmf of phase a coincides with the horizontal axis, as shown in Fig. 3-8(a). The directions of phase mmf's $\mathscr{F}_a$, $\mathscr{F}_b$, and $\mathscr{F}_c$ are obtained by using the right-hand rule.

The resultant stator mmf ($\mathscr{F}_s$) along an axis at an angle β to the horizontal is found by summing up the projections of the phase mmf's along this line:

$$\mathscr{F}_s = \mathscr{F}_a \cos \beta + \mathscr{F}_b \cos (120° - \beta) + \mathscr{F}_c \cos (-120° - \beta) \tag{3.9}$$

The number of winding turns and the maximum value of the current for each phase are the same. Designating $\mathcal{F}_1$ as the maximum mmf of any one phase, we have

$$\mathcal{F}_1 = N_A I_{ma} = N_B I_{mb} = N_C I_{mc} \tag{3.10}$$

Substituting Eqs. (3.5), (3.6), (3.7), and (3.10) into Eq. (3.9), we obtain

$$\mathcal{F}_s = \mathcal{F}_1 \left[\cos \omega t \cos \beta + \cos (\omega t - 120°) \cos (120° - \beta) \right.$$
$$\left. + \cos (\omega t + 120°) \cos (-120° - \beta) \right] \tag{3.11}$$

By use of the identity

$$\cos x \cos y = \frac{1}{2} \cos (x + y) + \frac{1}{2} \cos (x - y)$$

we obtain

$$\mathcal{F}_s = \frac{\mathcal{F}_1}{2} \left[\cos (\omega t + \beta) + \cos (\omega t - \beta) + \cos (\omega t - 120° + 120° - \beta) \right.$$

$$+ \cos (\omega t - 120° - 120° + \beta) + \cos (\omega t + 120° - 120° - \beta)$$

$$\left. + \cos (\omega t + 120° + 120° + \beta) \right] \tag{3.12}$$

Simplifying, we get

$$\mathcal{F}_s = \frac{\mathcal{F}_1}{2} \left[\underline{\cos (\omega t + \beta)} + \cos (\omega t - \beta) + \cos (\omega t - \beta) + \underline{\cos (\omega t + \beta - 240°)} \right.$$

$$\left. + \cos (\omega t - \beta) + \underline{\cos (\omega t + \beta + 240°)} \right] \tag{3.13}$$

The sum of the three underlined terms is equal to zero because these phasors are displaced from each other by 120° and because their magnitudes are equal. Therefore, Eq. (3.13) becomes

$$\mathcal{F}_s = \frac{3}{2} \mathcal{F}_1 \cos (\omega t - \beta) \tag{3.14}$$

Equation (3.14) describes a revolving field that rotates counterclockwise with an angular velocity of ω radians per second. The speed of the revolving field is normally designated by ω_s and is referred to as synchronous speed ($\omega_s = \omega$). The flux at a point P that is at β degrees to the horizontal will vary sinusoidally with the same frequency as that of the stator currents.

The revolving field may be visualized as being equivalent to the field generated by a permanent magnet rotated about an axis that coincides with the rotor of the machine.

The effective mmf through the stator structure is equal to 1.5 times the mmf produced by one phase alone. The resulting effective flux, at the absence of saturation and rotor current, is directly proportional to this mmf and inversely proportional to the reluctance of the path through which the flux completes its magnetic circuit.

In general, a rotating field of constant amplitude is produced by an m-phase winding wound $2\pi/m$ electrical radians apart and excited by balanced m-phase currents. The magnitude of the rotating field is $m/2$ times the field produced by any one phase, and its speed of rotation is given by Eq. (3.8). The direction of rotation of the field depends on the phase sequence of the applied currents. When the phase sequence of the supply voltages is reversed, the direction of rotation of the stator field and the speed of the motor are also reversed. The revolving fields of single- and two-phase ac motors are developed in Chapter 4.

Mechanical and Electrical Radians

Consider the elementary four-pole machine shown in Fig. 3-9(a). Starting at point A and going counterclockwise (2π mechanical radians) around the periphery of the machine's stator, the following field polarities are encountered: N_1S_1–S_2N_2–N_3S_3–S_4N_4. This corresponds to two complete electrical cycles (4π electrical radians), as can be seen in Fig. 3-9(b).

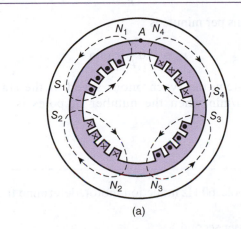

(a)

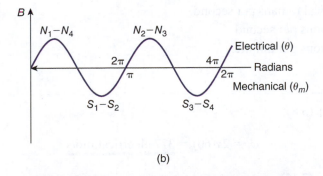

(b)

FIG. 3-9 **(a)** Elementary four-pole machine with stator slots and windings omitted; and **(b)** space distribution of field illustrating the relationship between mechanical (θ_m) and electrical (θ) radians.

For a four-pole machine, then, one mechanical or physical revolution will correspond to two complete cycles of the field. In general, for a p-pole machine,

$$\theta = \frac{p}{2}\,\theta_m \tag{3.15}$$

where θ and θ_m are the electrical and mechanical radians, respectively.

From Eqs. (3.8) and (3.15), we get the speed of the rotating field in mechanical radians per second:

$$\omega_m = \frac{2}{p}\,(2\pi f) = 4\pi\frac{f}{p} \text{ mechanical radians/s} \tag{3.16}$$

Since one revolution is equivalent to 2π mechanical radians, the speed in revolutions per second (n_s) is

$$n_s = 2\frac{f}{p} \text{ r/s} \tag{3.17}$$

The speed in revolutions per minute is

$$n_s = 120\frac{f}{p} \text{ r/min} \tag{3.18}$$

Thus, for a 60 Hz system, the synchronous speed of the stator field will be 3600, 1800, or 1200 r/min when the number of poles is two, four, or six, respectively.

EXAMPLE **3-1**

For a 480 V, 3-ϕ, four-pole, 60 Hz induction motor, determine the speed of the stator field in:

a.   Electrical radians per second.
b.   Mechanical radians per second.
c.   Revolutions per second.
d.   Revolutions per minute.

SOLUTION

a.   From Eq. (3.8),

$$\omega_s = 2\pi(60) = 377 \text{ electrical rad/s}$$

b.   From Eq. (3.16),

$$\omega_m = 377\left(\frac{2}{4}\right) = 188.5 \text{ mechanical rad/s}$$

c. From Eq. (3.17),

$$n_s = 2\left(\frac{60}{4}\right) = \underline{30 \text{ rps}}$$

d. From Eq. (3.18),

$$n_s = 120\left(\frac{60}{4}\right) = \underline{1800 \text{ r/min}}$$

Prove that the speed of rotation of the stator field is reversed when the order of rotation of the stator currents is reversed.

Exercise

3-1

3.1.4 Slip

Under normal operating conditions, the rotor rotates in the same direction as the magnetic field of the stator but at a reduced speed. The difference between the synchronous speed of the stator field and the actual rotor speed (n_a) defines the slip (s) of the motor. The slip of a motor is an important parameter, used extensively in the design and analysis of induction machines. The per-unit value of the slip is given by

$$s = \frac{n_s - n_a}{n_s} \tag{3.19}$$

where n_s is the synchronous speed of the rotating field, given by Eq. (3.18).

To achieve higher efficiency, the majority of three-phase induction motors are designed to operate at a very small slip (usually less than 5%) when delivering rated power.

When three-phase induction machines operate as induction generators, their actual rotor speed is higher than their synchronous speed, and their velocity is in the same direction as the synchronously rotating stator field. Thus, their slip is *negative*. Induction generators are used to connect small generating stations to large utility networks. (For a detailed analysis of induction generators, see Section 3.4.2.)

When an induction machine is driven by another motor in such a way that its actual rotor speed is in the direction opposite to that of the synchronously rotating stator field, then n_a in Eq. (3.19) becomes negative, and thus the slip is *greater than unity*. Induction machines operating at a slip greater than unity are used as frequency and voltage multipliers. They are covered in Section 3.4.

EXAMPLE **3-2**   A 75 kW, 60 Hz, 480 V, 1176 r/min, three-phase induction motor has, at full-load, a power factor of 80 lagging and an efficiency of 0.90. Under rated operating conditions, determine,

a. The magnitude of the current drawn by the motor.
b. The real, reactive, and complex power drawn by the motor.
c. The output torque at full-load.
d. The operating slip.

SOLUTION

a. By definition,

$$\text{efficiency} = \frac{\text{output power}}{\text{input power}}$$

In mathematical symbols,

$$\eta = \frac{P_{\text{out}}}{P_{\text{in}}}$$

The output power is 75 kW. The input power is

$$P_{\text{in}} = \sqrt{3}\,V_{L\text{-}L}I_L \cos\theta$$

From the above, the magnitude of the line current is

$$I_L = \frac{P_{\text{out}}}{\eta(\sqrt{3}\,V_{L\text{-}L}\cos\theta)} = \frac{75{,}000}{0.90\,(\sqrt{3}(480)(0.80))} = \underline{125.29\ \text{A}}$$

The current phasor is shown in Fig. 3-10(a).

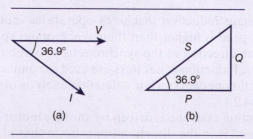

(a)   (b)

FIG. 3-10

b. The real or consumed power is

$$P_{in} = \sqrt{3}V_{L\text{-}L}I_L \cos \theta$$

$$= \sqrt{3}(480)(125.29)(0.80) = \underline{83.33 \text{ kW}}$$

or

$$P_{in} = \frac{P_{out}}{\eta} = \frac{75}{0.9} = \underline{83.33 \text{ kW}}$$

The reactive power is

$$Q = \sqrt{3}V_{L\text{-}L}I_L \sin \theta$$

$$= \sqrt{3}(480)(125.29)(\sin 36.9°) = \underline{62.5 \text{ kVAR}}$$

or

$$Q = P \tan \theta = 83.33 \tan 36.9° = \underline{62.5 \text{ kVAR}}$$

The complex power is

$$S = \sqrt{3}V_{L\text{-}L}I^*$$

$$= \sqrt{3}(480)(125.29)\underline{/36.9°} = \underline{104.17\underline{/36.9°} \text{ kVA}}$$

Alternatively,

$$S = P + jQ$$

$$= 83.33 + j62.5 = 104.17\underline{/36.9°} \text{ kVA}$$

The motor's power triangle is shown in Fig. 3-10(b).

c. The full-load output torque is

$$T = \frac{\text{power}}{\text{speed}} = \frac{75{,}000}{1176\dfrac{2\pi}{60}} = \underline{609 \text{ N} \cdot \text{m}}$$

d. By inspection of the given data, it can be determined that the motor's synchronous speed is 1200 r/min. The operating slip is

$$s = \frac{n_s - n_a}{n_s} = \frac{1200 - 1176}{1200} = \underline{0.02 \text{ per unit (pu)}}$$

Exercise

3-2

A 480 V, 50 kW, delta-connected three-phase, 60 Hz, six-pole induction motor operates at a slip of 1.5% and has a power factor of 82% and an efficiency of 88%. Determine the following:

a. The magnitude of the line and phase currents.
b. The actual rotor speed.
c. The real, reactive, and complex power drawn by the motor.

Answer (a) 83.34 A, 48.12 A; (b) 1182 r/min;
(c) 56.82 kW, 39.66 kVAR, $69.29\underline{/34.9°}$ kVA

Exercise

3-3

In a laboratory, a four-pole, 60 Hz induction machine is operated as follows:

a. As an induction motor at a speed of 1760 r/min.
b. As an induction generator at a speed of 1850 r/min.

In each of these three cases, determine the operating slip and the direction of the rotor's rotation relative to the clockwise rotation of the stator field.

Answer (a) 2.22%, cw; (b) −2.78%, cw.

3.1.5 Equivalent Circuit

The analysis of polyphase electric machines is simplified by using their per-phase equivalent circuits. The derivation of an equivalent circuit is based on the consideration of the electromagnetic coupling between the stator and rotor when the rotor is stationary and when it is running.

Rotor Stationary

When the rotor of the motor is stationary, the time rate of change of the flux linkages between the stator and the rotor depends on both the number of winding turns in the rotor and the speed and magnitude of the rotating mmf of the stator. This is similar to the transformer action, where the alternating flux of the primary current sets up flux linkages between the primary and secondary windings. The voltage induced in the secondary winding of a transformer is proportional to the time rate of change of its flux linkages. Similarly, the voltage induced in the rotor winding of a three-phase induction motor is proportional to the time rate of change of the winding's flux linkages.

The frequency (f_r) of the voltage induced in the rotor windings is equal to that of the stator windings. That is,

$$f = f_r \qquad\qquad (3.20)$$

where f is the frequency of the stator currents.

The per-phase equivalent circuit of a three-phase induction motor at standstill, then, is similar to that of a transformer. Refer to Fig. 3-11(a). The impedance $R_1 + jX_1$ represents the per-phase leakage impedance of the stator windings. The impedance $R_c /\!/ jX_\phi$ represents the per-phase magnetizing impedance of the motor. Because of the air gap between the stator and rotor structures, this

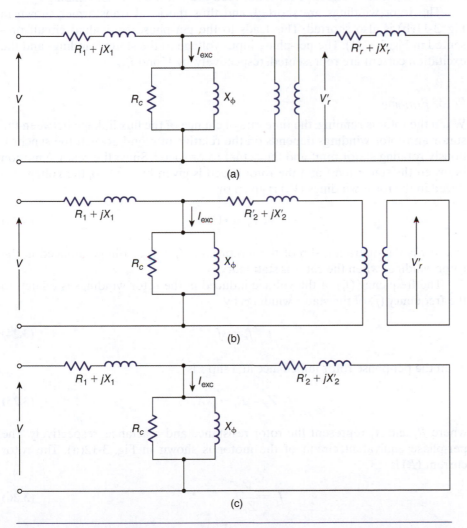

FIG. 3-11 Per-phase equivalent circuit of a three-phase motor at standstill: **(a)** stator and rotor circuits, **(b)** rotor impedance referred to the stator, **(c)** equivalent of (b).

impedance is substantially larger than that of a static transformer of equivalent voltage and volt-ampere rating. The impedance $R_r' + jX_r'$ represents the per-phase leakage impedance of the rotor windings at standstill.

The rotor impedance as seen from the stator (Fig. 3-11(b)) is

$$Z_2' = a^2(R_r' + jX_r') \tag{3.21}$$

or

$$Z_2' = R_2' + jX_2' \tag{3.22}$$

where Z_2' is the rotor impedance at standstill as seen from the stator, and a represents the number of effective turns between the stator and rotor windings.

The rotor windings are shorted, and thus the ideal transformer shown in Fig. 3-11(b) is also shorted. This leads to the per-phase equivalent circuit presented in Fig. 3-11(c). The per-phase input voltage to the stator windings and the excitation current are represented, respectively, by V and I_{exc}.

Rotor Running

When the rotor is running, the time rate of change of the flux linkage between the stator and rotor windings depends on the relative motion between the synchronously rotating stator mmf and the actual rotor speed. Since the relative motion between the stator mmf and the rotor speed is given by the slip, the voltage induced in the rotor windings (V_r) is given by

$$V_r = sV_r' \tag{3.23}$$

where s is the operating slip of the motor and V_r' is the voltage induced in the rotor windings when the rotor is stationary.

The frequency (f_r) of the voltage induced in the rotor windings is related to the frequency (f) of the stator windings by

$$f_r = sf \tag{3.24}$$

Then the per-phase rotor impedance at a slip s is

$$Z_r = R_r + jsX_r \tag{3.25}$$

where R_r and X_r represent the rotor resistance and reactance, respectively. The per-phase equivalent circuit of the motor is shown in Fig. 3-12(a). The rotor current (I_r) is

$$I_r = \frac{V_r}{Z_r} \tag{3.26}$$

$$= \frac{sV_r'}{R_r + jsX_r} \tag{3.27}$$

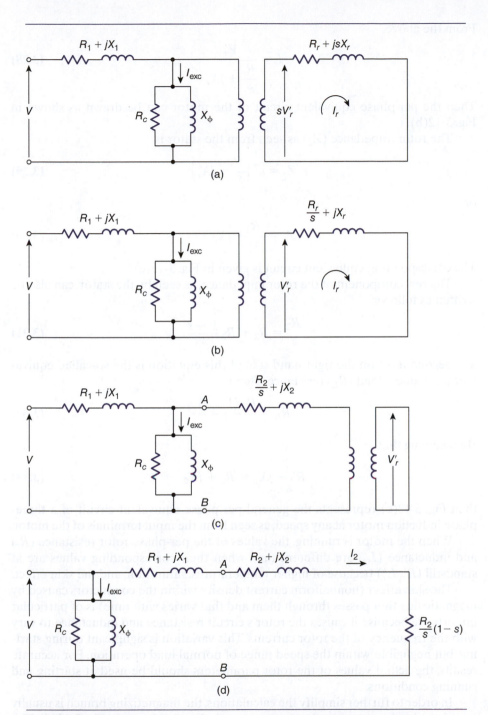

FIG. 3-12 Per-phase equivalent circuit of a three-phase induction motor:
(a) stator and rotor circuits, **(b)** equivalent of (a), **(c)** equivalent of (b), and
(d) approximation of (c).

From the above,

$$I_r = \frac{V'_r}{\dfrac{R_r}{s} + jX_r} \tag{3.28}$$

Then the per-phase equivalent circuit of the motor can be drawn as shown in Fig. 3-12(b).

The rotor impedance (Z_2) as seen from the stator is

$$Z_2 = a^2\left(\frac{R_r}{s} + jX_r\right) \tag{3.29}$$

or

$$Z_2 = \frac{R_2}{s} + jX_2 \tag{3.30}$$

The corresponding equivalent circuit is given in Fig. 3-12(c).

The real component of the rotor impedance, as seen by the stator, can also be written as follows:

$$\frac{R_2}{s} = R_2 + R_2\frac{1 - s}{s} \tag{3.31}$$

The second term on the right-hand side of this equation is the so-called equivalent mechanical load (R_L) resistance. That is,

$$R_L = \frac{R_2(1 - s)}{s} \tag{3.32}$$

By assuming that

$$R'_2 + jX'_2 = R_2 + jX_2 \tag{3.33}$$

then Fig. 3-12(c) represents the general per-phase equivalent circuit of a three-phase induction motor at any speed, as seen from the input terminals of the motor.

When the motor is running, the values of the per-phase rotor resistance (R_r) and inductance (L_r) are different than when their corresponding values are at standstill (R'_r, L'_r) because of higher temperatures, saturation, and the skin effect.

The skin effect (nonuniform current density within the conductors caused by magnetic flux that passes through them and that varies with time) is of particular importance because it causes the rotor's circuit resistance and inductance to vary with the frequency of the rotor currents. This variation is significant during starting, but negligible within the speed range of normal load operation. For accurate results, the actual values of the rotor parameters should be used at starting and running conditions.

In order to further simplify the calculations, the magnetizing branch is usually transferred to input terminals of the stator winding, as shown in Fig. 3-12(d). The magnetizing impedance could also be represented, as in the case of transformers, by its equivalent series-connected components ($R_m + jX_m$). The measurement of the parameters of the equivalent circuit is discussed in Section 3.3.

3.1.6 Torque and Power Relationships

Torque and power relationships will be derived by using the equivalent circuit shown in Fig. 3-12(d). In deriving the general expression for the torque of an induction motor, the effects of the magnetizing impedance will be neglected. Because this magnetizing impedance is relatively large, its effect on the torque developed by the motor at full-load is negligible.

From basic definitions, the per-phase torque developed is

$$T = \frac{\text{power developed}}{\text{actual speed}} = \frac{I_2^2 R_L}{\omega_a} \tag{3.34}$$

The equivalent mechanical load resistance (R_L) is

$$R_L = R_2 \frac{1 - s}{s}$$

and the actual speed is

$$\omega_a = \omega_s (1 - s)$$

Substituting the above into Eq. (3.34), we obtain

$$T = \frac{I_2^2 R_2 \frac{1 - s}{s}}{\omega_s (1 - s)} \tag{3.35}$$

$$= \frac{I_2^2 R_2}{\omega_s s} \ \text{N} \cdot \text{m/phase} \tag{3.36}$$

From the equivalent circuit, the magnitude of the current I_2 is

$$I_2 = \left| \frac{V}{Z} \right| = \frac{V}{\sqrt{\left(R_1 + \frac{R_2}{s}\right)^2 + X^2}} \tag{3.37}$$

where $X = X_1 + X_2$.

From the above, the torque developed in terms of the motor's parameters is

$$T = \frac{V^2}{\left[\left(R_1 + \frac{R_2}{s}\right)^2 + X^2\right]} \left(\frac{R_2}{\omega_s s}\right) \ \text{N} \cdot \text{m/phase} \tag{3.38}$$

where V is the phase-to-neutral voltage.

The variation of torque versus slip is shown in Fig. 3-13. The abscissa of this curve is the motor's speed or slip. The symbols T_{fl}, T_{st}, and T_m correspond to the full-load, starting, and maximum torque of the motor, respectively. The same diagram also shows the torque-speed characteristic of the mechanical load. The intersection of these characteristics gives the operating torque and speed of the motor. For this reason, it is extremely important that, before selecting a suitable motor, you accurately calculate the torque-speed characteristic of the driven load.

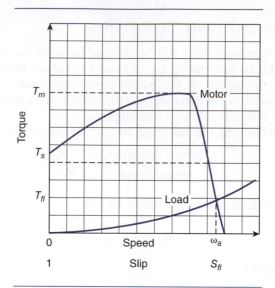

FIG. 3-13 Typical torque-speed characteristics.

The motor and load torque-speed characteristics are also used (as explained on the website) to find the accelerating and decelerating times of the motor. These parameters are important for the transient analysis of machines and for the proper selection of the motor's protective devices.

From Eq. (3.38), it is evident that, for a constant slip, the torque is proportional to the applied voltage squared:

$$T \propto V^2 \tag{3.39}$$

In practice, you may have to evaluate the starting torque capability of a motor because its high starting currents—depending on the impedance of the supply network—reduce the voltage delivered to the stator, and as a result the starting torque is also reduced. When the Thévenin impedance $(R_{th} + jX_{th})$ of the supply network is known, R_1 in Eq. (3.38) should be replaced by $R_1 + R_{th}$, and X should be replaced by $X + X_{th}$.

From Eq. (3.38), after expanding and simplifying, we obtain

$$T = \frac{V^2 s R_2}{\omega_s[(sR_1)^2 + 2sR_1R_2 + R_2^2 + (sX)^2]} \quad \text{N} \cdot \text{m/phase} \tag{3.40}$$

For slips in the full-load range, the predominant term in the denominator of the above equations is R_2^2. Therefore,

$$T \approx \frac{V^2}{\omega_s} \frac{s}{R_2} \tag{3.41}$$

or

$$T \approx K \frac{s}{R_2} \tag{3.42}$$

where K is the constant of proportionality given by

$$K = \frac{V^2}{\omega_s} \qquad (3.43)$$

From Eq. (3.42), it is evident that in the full-load operating range, the higher the actual speed of the motor, the lower its output torque. It can also be shown that for a certain range of rotor resistance, the starting torque (T_{st}) of the motor is proportional to its rotor resistance. Mathematically,

$$T_{st} \propto R_2 \qquad (3.44)$$

The higher the rotor resistance, the higher the motor's starting torque, and the lower its inrush current. The derivation of Eq. (3.44) is left as a student problem (see Problem 3-2).

Because of the effects of rotor resistance on starting torque and starting current, the high-power motors supplied through weak power systems (i.e., a supply network with relatively high impedance) are of the wound-rotor type.

Maximum or Breakdown Torque

Refer again to the approximate equivalent circuit of Fig. 3-12(d). For maximum power transfer from stator to the rotor-load resistance, the following relationship must be satisfied:

$$|R_1 + jX| = \frac{R_2}{s} \qquad (3.45)$$

Equating the magnitudes of this relationship, we get the equation for the slip of the motor at maximum torque (s_{mt}):

$$s_{mt} = \frac{R_2}{\sqrt{R_1^2 + X^2}} \qquad (3.46)$$

where

$$X = X_1 + X_2$$

Substituting Eq. (3.46) into Eq. (3.38), after simplification, we find the following expression for the maximum torque developed by the motor:

$$T_m = \frac{V^2}{\left[\left(R_1 + \sqrt{R_1^2 + X^2}\right)^2 + X^2\right]} \times \frac{\sqrt{R_1^2 + X^2}}{\omega_s} \quad \text{N} \cdot \text{m/phase} \qquad (3.47)$$

Thus, the maximum torque is independent of rotor resistance. The rotor resistance controls the slip or the speed where the maximum torque occurs. Typical torque-speed characteristics, as a function of rotor resistances, are shown in Fig. 3-14.

With negligible stator resistance $(R_1 = 0)$, it can be shown that the torque (T) at any slip (s) is related to the maximum torque (T_m) and its corresponding slip (s_{mt}) by

$$\frac{T}{T_m} = \frac{2ss_{mt}}{s_{mt}^2 + s^2} \qquad (3.48)$$

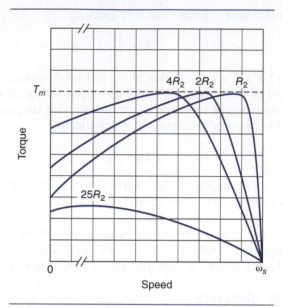

FIG. 3-14 Torque-speed characteristics as a function of rotor resistance.

If the torque-speed characteristic of the motor is not available, then the above equation can be used, in conjunction with some of the motor parameters (T_{st}, T_m, T_{fl}, s_{fl}) to sketch it. The derivation of Eq. (3.48) is left as a student problem (see Problem 3-2).

The torque developed by a motor must be sufficient to provide the motor's rotational losses and the torque requirements of the mechanical load. The torque requirements of a mechanical load may include any or all of the following torque components: the torque component $[(J(d\omega_a/dt)]$ required to accelerate the rotating masses up to the steady-state speed; the torque component $(B\omega_a)$ that represents the viscous friction of the rotating parts; and any other torque (T_L) requirements of the load. In mathematical form,

$$T = J\frac{d\omega_a}{dt} + B\omega_a + T_L \tag{3.49}$$

where

J = the polar moment of inertia of the rotating masses in Newton-meter-second squared

B = the coefficient of viscous friction in N · m/(rad/s)

T_L = a component of load torque in N · m

ω_a = the actual speed of the motor in rad/s.

Equation (3.49) is derived from Newton's second law of motion for rotating bodies. This law states that the sum of the torques (ΣT) about an axis of rotation is equal to the polar moment of inertia (J) times the angular acceleration (α) of the rotating masses. Mathematically,

$$\Sigma T = J\alpha \tag{3.50}$$

This law is of primary importance to the dynamic analysis of all machines, and its applications are discussed on the accompanying website.

In developing the torque relationships, the magnetizing impedance of the motor was neglected in order to simplify the calculations. The magnetizing impedance, as already mentioned, does not affect the torque of the motor at full load; however, at light loads, it has significant effects. For accurate torque calculations, the magnetizing impedance should be taken into consideration. You can easily do so by replacing, in Eqs. (3.38) and (3.47), the voltage source (V) and stator impedance (Z_1) by the equivalent Thévenin's voltage and impedance. Referring to Fig. 3-12(c), and looking toward the input from terminals A–B, we obtain

$$V_{th} = \frac{V(R_c \mathbin{/\mkern-5mu/} jX_\phi)}{R_1 + jX_1 + R_c \mathbin{/\mkern-5mu/} jX_\phi} \tag{3.51}$$

and

$$Z_{th} = (R_1 + jX_1) \mathbin{/\mkern-5mu/} (R_c \mathbin{/\mkern-5mu/} jX_\phi) \tag{3.52}$$

From Eqs. (3.37) and (3.46), we find that the current I at a slip s is related to the current at maximum torque (I_{mt}) and its corresponding slip (S_{mt}) by

$$\left(\frac{I}{I_{mt}}\right)^2 = \frac{2s^2}{s_{mt}^2 + s^2} \tag{3.53}$$

The equations developed in this section are based on the approximate equivalent circuit of Fig. 3-12(d). They are *not* applicable to motors with deep-bar rotor construction or to double squirrel-cage rotors. In such cases, the proper equivalent circuit should be drawn, then from basic principles, the desired relationships can be derived.

Power Considerations

The power across the air gap (P_{ag}) provides the power developed and the rotor copper loss. That is,

$$P_{ag} = I_2^2 R_2 \frac{1 - s}{s} + I_2^2 R_2 \tag{3.54}$$

Thus,

$$P_{ag} = I_2^2 \frac{R_2}{s} \text{ W/phase} \tag{3.55}$$

As Eq. (3.55) shows, the rotor-winding losses are directly proportional to the air-gap power and the operating motor slip. Thus, the smaller the slip, the lower the rotor copper losses. For this reason, the slip, at full-load, is usually kept below 5%.

The power developed (P_d) furnishes the output power as well as the rotational losses. Mathematically,

$$P_d = I_2^2 R_2 \frac{1-s}{s} \text{ W/phase} \tag{3.56}$$

$$= P_{\substack{\text{rotational} \\ \text{losses}}} + P_{\text{output}}$$

The rotational losses include windage, friction, and stray load losses.

EXAMPLE 3-3

Figures 3-15(a) and (b) show, respectively, the one-line diagram of a three-phase induction motor and its per-phase equivalent circuit referred to the stator. Assuming constant rotor impedance, and neglecting the effects of the magnetizing current and rotational losses, determine:

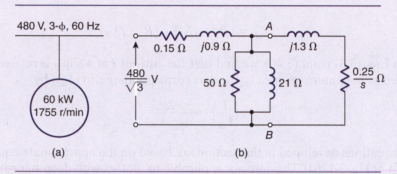

(a)   (b)

FIG. 3-15

a.   The number of poles.
b.   The synchronous speed of the stator field.
c.   The per-unit slip.
d.   The line current at starting.
e.   The starting torque.
f.   The full-load current, power factor, torque, and efficiency.
g.   The slip and current at maximum torque.
h.   The maximum torque.

SOLUTION

a.   Under normal operating conditions, the slip of a three-phase induction motor is kept small because of its direct relationship to the copper losses. Thus, the actual speed of the motor should be very close to the speed of the synchronously rotating stator field. Since the actual speed is given as 1755 r/min, the closest synchronous speed is 1800 r/min. From Eq. (3.18), the number of poles is

$$p = \frac{120 \times 60}{1800} = \underline{4 \text{ poles}}$$

b. The speed of the stator field is

$$n_s = \underline{1800 \text{ r/min}}$$

or

$$\omega_s = 1800 \frac{2\pi}{60} = \underline{188.5 \text{ rad/s}}$$

c. Using Eq. (3.19), we find that the slip in per unit is

$$s = \frac{1800 - 1755}{1800} = \underline{0.025 \text{ pu}}$$

d. The line current at starting is found by applying Ohm's law in the equivalent circuit of Fig. 3-15(c):

$$I_{st} = \frac{480/\sqrt{3}}{0.15 + 0.25 + j2.2} = \underline{123.94 \underline{/-79.7°} \text{ A}}$$

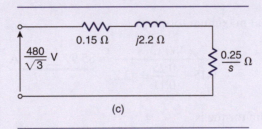

(c)

FIG. 3-15(c)

e. Since the motor is star-connected and the magnetizing impedance is neglected, the rotor current as seen from the motor terminals is equal to the stator or line current. Substituting the known data into Eq. (3.36) gives the motor's starting torque:

$$T_{st} = 3(123.94)^2 \frac{(0.25)}{188.5} = \underline{61.12 \text{ N} \cdot \text{m}}$$

f. The current at full-load is

$$I_{fl} = \frac{480/\sqrt{3}}{0.15 + \dfrac{0.25}{0.025} + j2.2} = \underline{26.68 \underline{/-12.2°} \text{ A}}$$

By definition, the power factor is $\cos 12.2° = 0.977$ lagging. The power factor is unrealistically high because the magnetizing impedance was assumed to be negligible.

The full-load torque is

$$T_{fl} = \frac{3(26.68)^2}{188.5}\left(\frac{0.25}{0.025}\right) = 113.32 \text{ N} \cdot \text{m}$$

The efficiency of the motor at full-load is

$$\eta = 1 - \frac{P_{loss}}{P_{in}} = 1 - \frac{3I^2R}{\sqrt{3}V_{L\text{-}L}I_L \cos\theta} = 1 - \frac{3(26.68)(0.40)}{\sqrt{3}(480)(0.977)}$$

$$= 0.96$$

g. From Eq. (3.46), the slip at maximum torque is

$$s_{mt} = \frac{0.25}{\sqrt{0.15^2 + 2.2^2}} = 0.11 \text{ pu}$$

The corresponding motor's speed is

$$n_a = 1800(1 - 0.11) = 1596 \text{ r/min}$$

The current at maximum torque is

$$I = \frac{480/\sqrt{3}}{0.15 + \dfrac{0.25}{0.11} + j2.2} = 85.99\underline{/-43°} \text{ A}$$

h. The maximum torque is

$$T_m = \frac{3(85.99)^2(0.25)}{188.5(0.11)} = 259.44 \text{ N} \cdot \text{m}$$

For purposes of comparison, the results are summarized in Table 3-1.

TABLE 3-1	Summary of the results of Example 3-3			
Operating Condition	Speed in r/min	Current in A	Power Factor Lagging	Torque in N · m
Starting (s = 1)	0	123.94	0.18	61.12
Full-load (s = 0.025)	1755	26.68	0.977	113.32
Maximum torque ($s_{mt} - 0.11$)	1596	85.99	0.73	259.44

The per-phase equivalent circuit of a 480 V, 60 Hz, 1176 r/min induction motor is shown in Fig. 3-16. Neglecting the magnetizing impedance, determine the starting, full-load, and maximum torque of the motor.

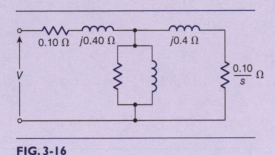

FIG. 3-16

Answer   269.63 N · m, 343.99 N · m, 1011.6 N · m

3.2 Industrial Considerations

This section discusses the industrial classification of squirrel-cage induction motors and mechanical loads, the change in a motor's input parameters that accompany variations in the load requirements, a motor's efficiency and power factor, and the effects of the supply voltage on the torque-speed characteristic.

3.2.1 Classification of Induction Motors

The National Electrical Manufacturers Association (NEMA) has developed a code-letter system by which a letter (A, B, C, or D) designates a *particular class of motors* with specific characteristics. The various characteristics are mainly obtained by a unique rotor design. Table 3-2 shows the general characteristics and the corresponding applications of each class of motor. Typical torque-speed characteristics for the various code-letter designations of squirrel-cage induction motors are shown in Fig. 3-17. Before a motor is selected, the characteristics of the load must be known. The load characteristics include power requirements, starting and full-load torque, speed variation, acceleration characteristics, and the environment in which the motor is to operate.

To select a particular type of motor for a given mechanical load, the torque-speed characteristic of the motor must be compared to that of the mechanical load. The intersection of the motor-load characteristics must correspond not only to the speed and torque requirements of the load, but also to the highest efficiency for the motor. The *starting torque* of the motor must be larger than the

TABLE 3-2 Characteristics and applications of 3-ϕ Induction Motors

Design	Starting Torque, in Per Unit	Starting Current, in Per Unit	Maximum Torque, in Per Unit	Range of Full Load Slip	Applications
A	1.5 to 1.75	5 to 8	2 to 25	2% to 5%	General-purpose motors, fans, blowers, most machinery tools
B	1.5 to 1.75	4.5 to 5	2 to 3	3% to 6%	Same as Class A
C	2 to 2.5	3.5 to 5	1.9 to 2.25	4% to 8%	Used in conveyors, compressors
D	2.75 to 3	3 to 8	2.75	7% to 17%	Used for high-inertia loads, low efficiency

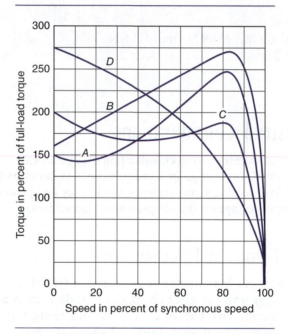

FIG. 3-17 Typical torque-speed characteristics of squirrel-cage induction motors.

load requirement, or the motor will not be able to start rotating. As a result, it will draw its locked-rotor current, perhaps damaging the motor.

The minimum value of the torque in the region between starting and maximum torque is called the *pull-in torque*. As can be seen from the general torque-speed characteristics (Fig. 3-17), only class A and C motors have a meaningful pull-in torque of the motor, which must be lower than that of the motor itself; otherwise, the rotor will not accelerate smoothly.

The *maximum torque* developed by a motor indicates the capability of the machine to overcome high transient-load torques. A motor with a maximum torque that is relatively low may stall when a sudden load torque exceeds the motor's breakdown torque.

The *accelerating torque* of a motor can be obtained from its torque-speed characteristics by comparing the differences in the areas under the torque-speed curves of the motor and the load. (For more detail, see Chapter 3W on the accompanying website.)

A motor's accelerating time (the time it takes a motor, starting from rest, to reach its steady-state speed) is inversely proportional to its accelerating torque. For all practical purposes, during acceleration, a motor draws locked-rotor current. The longer a motor's accelerating time, the greater the possibility of thermal damage to its windings. For this reason, the maximum number of permissible starts during a specific time interval should be carefully determined, and the motor's manufacturer should — as is usually the practice — be consulted.

For any given mechanical load, a class D motor is best equipped to reduce acceleration time (see Fig. 3-17). The cost of a motor, however, increases progressively from class A to class D.

3.2.2 Mechanical Loads

The selection of a particular motor for a specific load requirement is based on the torque-speed characteristic of the mechanical load. Manufacturers of motors will provide the characteristics of a motor: its starting current and torque, maximum torque, starting and full-load power factors, and so on. Similarly, manufacturers of fans, pumps, and the like can provide the torque-speed characteristics of their equipment.

When the torque-speed characteristic of a particular load is not available, it can be derived by drawing the free-body diagram of the system under consideration and by applying Newton's laws of motion.

Load Classification

Mechanical loads are classified according to their torque-speed characteristics. There are actually many such characteristics, but all may be grouped into one of four general categories: constant torque; torque proportional to speed; torque inversely proportional to speed; and torque proportional to the square of the speed (see Fig. 3-18). The mathematical descriptions and the actual torque-speed characteristics of mechanical loads are usually evaluated by mechanical process engineers and can generally be found in standard mechanical engineering handbooks.

Constant Torque

A constant-torque load is a load whose torque remains constant within the operating speed range of the load. Examples of constant-torque loads are conveyors, grinding mills, and crane-hoist systems.

Torque Proportional to Speed

A load whose torque requirement is proportional to its speed is the calender — a machine used for pressing and smoothing cloth or paper between rollers. The smoothness of this page depends on the calender of the paper machine that produced it.

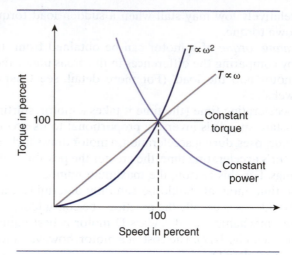

FIG. 3-18 General torque-speed characteristics of mechanical loads.

Torque Inversely Proportional to the Speed

A load whose torque is inversely proportional to its speed is also referred to as the constant-power load. Examples of constant-power loads are circular saws and lathe drives.

Torque Proportional to the Square of the Speed

Examples of loads whose torque requirements are proportional to the square of their speed (power is proportional to the third power of the speed) are centrifugal fans and blowers. All the loads of air-handling fans used in heating, ventilating, and air conditioning systems (HVAC) are of this type.

Exercise

3-5

Prove the following:

a. The mechanical load on a motor that drives a crane hoist is of the constant torque type. (*Hint*: Torque is equal to the force times its moment arm.)

b. The torque of a motor that drives a fan in an air-handling system is directly proportional to the square of the motor's speed. (*Hint:* From the energy equation of hydraulics, the frictional losses are proportional to the square of the speed of the air flow.)

c. The torque of a motor that drives rollers between which paper is pressed and smoothed is directly proportional to the speed of the rollers.

d. The torque required by a circular saw is inversely proportional to its speed.

3.2.3 Mechanical Load Changes and Their Effects on a Motor's Parameters

This section details how changes in load requirements affect the current, power factor, and efficiency of a motor. A load characteristic may be changed when the dampers of a ventilating fan are adjusted, for example, or when a pump's control valve is reset. Such adjustments may be necessary in order to satisfy a particular rate-of-flow condition, which in turn enables the successful completion of a process.

Figure 3-19 shows the torque-speed characteristics of a motor and of a variable mechanical load. Under *nominal* operating conditions, the speed and the torque developed by the motor are given by the intersection of the motor and load characteristics, identified by the symbol Q_1. Under this load requirement, the motor operates at full-load condition. Its losses (copper and rotational) are minimum in comparison to the output power; its stator and rotor currents are of nominal magnitude, and thus the motor—all other conditions being equal—will operate satisfactorily during its entire lifetime. The motor's power factor, efficiency, torque, and speed under nominal operating conditions are all inscribed on its nameplate.

When the torque requirements of the load are *reduced*, the new operating point is given, as before, by the intersection of the torque-speed characteristics. This point is identified by Q_2. As shown, the speed of the motor will increase and its slip will be reduced. Then the impedance* of the motor, as can be seen from the motor's equivalent circuit, will also increase. As a result, the current—at constant terminal voltage—will be reduced.

The reduction in the stator current will reduce the copper losses of the motor but will not appreciably affect the rotational losses. Although the total losses

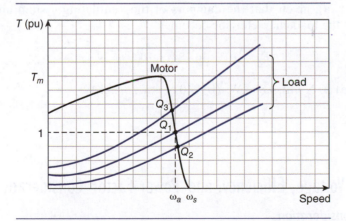

FIG. 3-19 Torque-speed characteristics of an induction machine and its mechanical load.

*$Z = R_1 + (R_2/s) + jX$ ohms.

decrease—as does the output power—the efficiency of the motor also *decreases*. The power factor will increase because the reduction in slip increases the real component of motor impedance. The motor, as seen from its supply lines, appears more resistive.

When the torque requirements of the load *increase*, the motor's torque-speed characteristic is shifted upward. The new operating point of the motor-load system is given by Q_3. The speed of the motor is reduced, and its torque is increased in order to satisfy the new load requirements. The slip of the motor increases [see Eq. (3.19)], and as a result, the impedance of the motor decreases and becomes more inductive. This increase in the operating slip is accompanied by an increase in the current drawn by the motor; consequently, the motor's copper losses increase. The rotational losses remain about constant, and the efficiency of the motor normally decreases, although the output power of the motor increases.

Exercise 3-6

A six-pole, 60 Hz, 3-ϕ induction motor drives a ventilating fan. The motor torque at full-load, as a function of slip, is

$$|T_{mfL}| = 4000s \text{ N} \cdot \text{m}$$

The torque of the load as a function of the speed is

$$T_L = 10 + 4.616 \times 10^{-3} \omega^2 \text{ N} \cdot \text{m}$$

a. Sketch the torque-speed characteristics of the motor and load, and determine the torque and power of the motor.

b. Repeat (a), given that the dampers of the ventilating system are adjusted so that the load characteristic becomes

$$T_L = 10 + 16.484 \times 10^{-3} \omega^2 \text{ N} \cdot \text{m}$$

Answer (a) 80 N · m, 9.85 kW; (b) 240 N · m, 28.35 kW

3.2.4 Voltage, Efficiency, and Power-Factor Considerations

Voltage Considerations

Induction motors whose nominal capacity is less than 200 kW are designed to operate at voltages less than 1 kV. These are referred to as low voltages (LV). In contrast, motors with a power rating of more than 200 kW are generally designed to operate at voltages within the 1 kV to 15 kV range. These are referred to as medium voltages (MV). The most popular MV levels are 2.3 kV, 4.16 kV,

6.6 kV, and 13.2 kV. The 4 MW grinding mills used in the cement industry, for example, are rated at 4.16 kV, while the 8 MW motors used in the pulp and paper industry are designed to operate at 13.2 kV. MV induction machines are certainly more expensive than LV motors of the same kW rating, but they draw reduced inrush and full-load currents, require less expensive cables, and decrease the losses throughout the upstream electrical network.

In general, the torque developed by a motor at a given speed is proportional to the square of the applied voltage, as demonstrated by Eq. (3.38). Refer to Fig. 3-20. For a given voltage, while all other parameters remain constant, the motor's torque-speed characteristic is fixed. Changes in the supply voltage will change the motor's characteristic accordingly. By reducing the voltage from V' to V volts, the motor's torque-speed characteristic is shifted downward. As a result, the motor's torque—at any speed—is reduced.

For small variations in the load requirements and for a constant ac voltage supply, the motor's *speed variation is limited*. A moderate reduction in the speed will be accompanied by a drastic increase in the current drawn by the motor (see manufacturer's data, Fig. 3-23). If the current protective devices do not remove the motor from its supply voltage, it will overheat. For a given mechanical load, on the other hand, a substantial increase in the speed of the motor is not possible. Under normal conditions, the motor operates very close to synchronous speed, regardless of supply voltage. This effect is evident from the torque-speed characteristics of any 3-ϕ induction motor.

Thus, squirrel-cage induction machines supplied through a constant ac voltage source have a very small speed-variation range; consequently, they cannot be used for mechanical loads that require even a ±2% variation in speed. Another disadvantage of the constant ac voltage supply is that the inrush current of the motor—about six times its nominal value—may cause a significant voltage drop through the upstream impedances. As a result, the motor's *starting torque* is reduced, and so the motor may not be able to start rotating its mechanically coupled load.

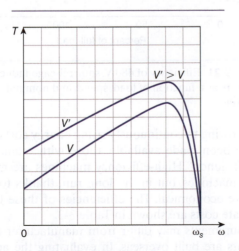

FIG. 3-20 Torque-speed characteristics as a function of voltage.

Furthermore, large inrush currents produce excessive heat through the rotor-stator structure, so successive starts within seconds or even within minutes may damage the motor's windings. For this reason, large motors that draw intolerably high starting currents are provided with controls—see Section 3.5—through which their starting current is reduced during the starting of the machine.

Exercise

3-1

Under normal operating conditions, the windings of a 3-ϕ induction motor are connected in delta (Δ); at starting, they are connected in star (Y). Determine the motor's torque and line current at starting in per unit.

Answer Answer $T = 0.33$ pu

Efficiency Considerations

As illustrated in Fig. 3-21, the efficiency of the motor depends on its speed, its power rating, and the driven load. These values are all normally selected by mechanical process engineers, and not by electrical engineers, who handle all other aspects of the motor—its specifications, purchasing, protection, and so on.

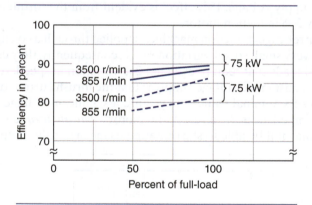

FIG. 3-21 Efficiency of 480 V, squirrel-cage induction motors as a function of load, speed, and nominal power.

In recent years, owing to continuously rising energy costs, two types of induction machines have been made available: those with "standard efficiency" and those with "high efficiency." High-efficiency machines are more expensive than standard-efficiency machines, but in the long run, thanks to their lower energy losses, they are more economical. The efficiencies of these two types of motors and their approximate costs are shown in Table 3-3.

Efficiency measurements may differ from manufacturer to manufacturer, especially when motors are built overseas. In evaluating the advantages of higher-efficiency machines, then, it is important to know how these efficiency ratings are determined.

TABLE 3-3   Squirrel-cage, 480 V, 3-ϕ, four-pole induction machines: typical "standard" and "high" efficiencies and their corresponding costs.

Rating in kW	Efficiency in Percent		Cost Per Unit*	
	Standard	High	Standard Efficiency	High Efficiency
7.5	85.9	90.2	1	1.4
50	91.3	95	8	9.8
75	93	95	10	12
100	95	95.8	16	19

*The cost per unit, in 2010 dollars, is $500.
Based on data from General Electric Canada, Inc.

Power-Factor Considerations

As illustrated in Fig. 3-22, the power factor of the motor depends on its speed, its kW rating, and the operating load. When the power factor of a plant is low, it can be improved by installing sufficient capacitor banks. Selecting a capacitor bank requires consideration and analysis of the following:

- Optimum kVAR and voltage ratings.
- Optimum connection within the power distribution network.
- Resonance conditions and the effects of harmonics.

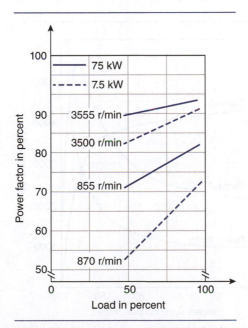

FIG. 3-22 Power factor of 480 V induction machines as a function of load, speed, and nominal power.

When a motor and its power-factor improvement capacitor are switched through a common disconnect device, the rating of the capacitor must be such as not to magnetize the motor each time the motor is switched OFF. In order to avoid this condition, the rated current of the capacitor must be less than the reactive no-load magnetizing current of the motor. If the capacitor bank is oversized, it may, during motor stopping, magnetize (source of excitation) the stator windings. Therefore, owing to the motor's decelerating rotor (source of rotation), the motor may operate as a generator. This condition may be accompanied by high transient voltages and torques, and as a result, the normal life expectancy of the motor may be shortened.

Exercise

3-8

Explain the following:

a. The higher the nominal speed of a motor, the higher its power factor and efficiency.
b. The designing of a plant with a low power factor is due to oversized equipment.

Actual Motor Characteristics

Figure 3-23 shows the torque, current, and power factor versus speed characteristic for a particular 2400 kW, 2300 V, 3-ϕ, 60 Hz, squirrel-cage induction machine. It should be noted that a small decrease in the operating speed will be accompanied by a large increase in the current. For this reason, the characteristics of the mechanical load should be accurately evaluated; otherwise, the motor will operate at larger slips and thus higher motor currents. In practice, all ac machines can withstand a continuous current overload of 10% without overheating.

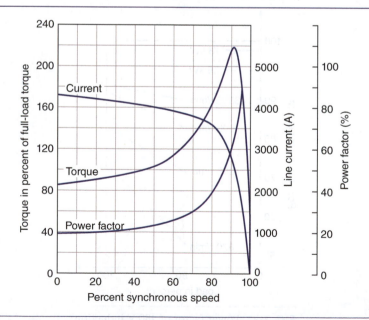

FIG. 3-23 Output torque, line current, and power factor versus speed for a four-pole, 60 Hz, 2400 kW, 2300 V squirrel-cage induction motor. Rated speed: 1765 r/min; rated current: 705 A; power factor: 90%. *Based on data from Westinghouse Canada Inc.*

EXAMPLE 3-4

The three-phase induction motor shown in the one-line diagram of Fig. 3-24(a) has the following starting characteristics:

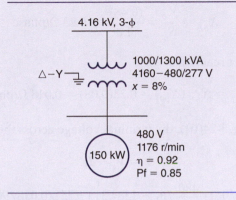

4.16 kV, 3-φ

1000/1300 kVA
4160−480/277 V
x = 8%

Δ−Y

480 V
1176 r/min
η = 0.92
Pf = 0.85

150 kW

FIG. 3-24(a)

$$T_{st} = 2.2 \text{ pu}, \quad I_{st} = 6\underline{/-75°} \text{ pu}$$

The motor receives its voltage supply through a transformer whose nameplate data are shown on the diagram. Neglecting the impedance of the cables and that of the 4.16 kV source, determine, at motor starting:

a. The voltage at the motor terminals.

b. The starting torque of the motor in per unit.

c. The size of the capacitor bank that, when connected in parallel to the motor, will minimize the current through the upstream network.

d. The torque developed by the motor after the capacitor bank has been added.

SOLUTION

The magnitude of the rated current of the motor is

$$I = \frac{P}{\sqrt{3}V_{L\text{-}L} \cos \theta(\eta)} = \frac{150{,}000}{\sqrt{3}(480)(0.92)(0.85)} = 230.72 \text{ A}$$

The magnitude of the starting current of the motor is

$$I_{st} = 230.72(6) = 1384.31 \text{ A}$$

In phasor form,

$$I_{st} = 1384.31\underline{/-75°} \text{ A}$$

Taking the nameplate data of the transformer as base parameters, we have

Base impedance:

$$X_{bL} = \frac{V_b^2}{S_b} = \frac{(0.48)^2}{1.0} = 0.23 \ \Omega/\text{phase}$$

Transformer reactance:

$$X_{aL} = X_{bL} \ (X_{\text{pu}}) = 0.23(0.08) = 0.018 \ \Omega/\text{phase}$$

a. From KVL (Fig. 3-24(b)), the terminal voltage across the motor at starting is

$$V = V_s - IZ$$

$$V\underline{/0°} = \frac{480}{\sqrt{3}} \ \underline{/\theta_1} - 1384.31\underline{/-75°} \ (0.018\underline{/90°})$$

from which

$$\theta_1 = 1.4°$$

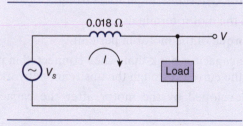

FIG. 3-24(b)

and

$$V = 252.40 \ \text{V/phase}$$

$$= \underline{437.18 \ \text{V, L-L}}$$

Since the motor's terminal voltage is reduced from 480 V to 437.18 V, the starting current must also be proportionately reduced. By repeating the calculations with a new starting current, one will find that the motor's terminal voltage remains close to 437 V.

b. The torque developed by the motor is proportional to the square of the voltage. That is,

$$T \propto V^2$$

from which

$$\frac{T_2}{T_1} = \left(\frac{V_2}{V_1}\right)^2$$

where subscript 1 identifies the parameters at starting, as given by the data of the motor, and subscript 2 identifies the corresponding motor parameters at the actual starting conditions. Substituting the known values, we obtain

$$T_{st} = 2.2 \left(\frac{437.18}{480}\right)^2 = \underline{1.82 \text{ pu}}$$

The starting-torque capability of the motor has been reduced from 2.2 pu to 1.82 pu, owing to the voltage drop on the upstream network. If the driven mechanical load requires a starting torque of 2.0 pu, for example, the motor will not be able to start rotating.

c. One method of reducing the voltage drop at starting is to connect a capacitor bank in parallel to the motor. The capacitor bank must be of a magnitude that will cancel the reactive component of the starting current. Thus, the capacitor's current (I_c) should be

$$I_c = 1384.31 \sin 75° = 1337.14 \text{ A}$$

The size of the capacitor bank is

$$Q = \sqrt{3} V_{\text{L-L}} I \sin \theta$$
$$= \sqrt{3}(480)(1337.14)(1)$$
$$= 1111.68 \text{ kVAR}*$$

d. The upstream current is

$$I = I_m + I_c$$
$$= 1384.31\underline{/-75°} + 1337.14\underline{/90°}$$
$$= 358.29\underline{/0°} \text{ A}$$

The voltage across the motor is

$$V\underline{/0°} = \frac{480}{\sqrt{3}}\underline{/\theta_2} - 358.29\underline{/0°} (0.018\underline{/90°})$$

from which

$$\theta_2 = 1.4°$$

---

*The cost of this capacitor is about 40% of the cost of the motor.

and

$$V = 277.05 \text{ V/phase}$$
$$= 479.86 \text{ V, L-L}$$

The torque developed at starting is

$$T = 2.2 \left(\frac{479.86}{480}\right)^2 = \underline{2.2 \text{ pu}}$$

The results are summarized in Table 3-4.

TABLE 3-4 Summary of the results of Example 3-4

Operating Condition	Starting Torque (pu)	Feeder Current at Starting (A)
Nominal	2.2	1384.31
Without the capacitor	1.82	1384.31
With the capacitor	2.2	358.29

Exercise 3-9

For the motor whose characteristics as a function of speed are shown in Fig. 3-23, the nominal speed is 1765 r/min. Estimate the current, power factor, and torque of the motor when its speed varies by ±2% of its nominal value. Tabulate the results.

3.3 Measurement of Equivalent-Circuit Parameters

The parameters of the approximate equivalent circuit* of a three-phase induction machine can be obtained from the equivalent standard transformer tests. The actual measurements are carried out on the stator of the machine. Consequently, the calculated parameters and the equivalent circuit are as seen from the stator.

---

*For a more accurate and complete discussion of tests on induction machines, see IEEE Test Code for Polyphase Induction Motors and Generators No. 112A (New York: Institute of Electrical and Electronic Engineers, 1964).

Measurement of Stator Resistance

Per-phase stator resistance (R_1) is measured by any of the standard dc methods. This resistance is usually adjusted in order to compensate for the effects of higher operating motor temperatures. Usually $R_{ac} \approx 1.3 \, R_{dc}$.

Locked-Rotor Test (s = 1)

The locked-rotor test on an induction machine—often referred to as the blocked-rotor test—corresponds to the short-circuit test on a transformer. Keeping the rotor stationary, we find that the slip is equal to unity, and thus the equivalent mechanical load is shorted.

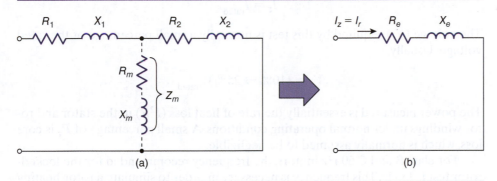

FIG. 3-25 Equivalent circuit corresponding to locked-rotor test $(s = 1)$: **(a)** actual $(Z_m$ large); **(b)** approximation of (a).

Refer to Fig. 3-25(b). The locked-rotor test gives the equivalent impedance of the stator and rotor windings. That is,

$$Z_e = R_e + jX_e \tag{3.57}$$

These parameters are calculated by measuring the stator's 3-ϕ power (P_z), the line voltage (V_z), and the line current (I_z).

From basic considerations, we have

$$Z_e = \frac{V_z}{I_z} \tag{3.58}$$

$$R_e = \frac{P_z}{I_z^2} = R_1 + R_2 \tag{3.59}$$

and

$$X_e = \sqrt{Z_e^2 - R_e^2} = X_1 + X_2 \tag{3.60}$$

where V_z and P_z are the phase-to-neutral voltage and per-phase power, respectively.

The empirical relationships between the stator and rotor reactances are given in Table 3-5.

TABLE 3-5 Empirical relationships for induction motor leakage reactances (from IEEE Test Code)

Type of Motor	Class A	Class B	Class C	Class D	Wound-Rotor
X_1	$0.5X_2$	$0.67X_2$	$0.43X_2$	$0.5X_2$	$0.5X_2$

Based on data from IEEE Test Code

The current for the locked-rotor test (I_z) is usually the rated current of the motor:

$$I_z = I_{rated}$$

The voltage (V_z) required by this test is normally a small percentage of the rated voltage. Usually,

$$V_z \approx (10\% \rightarrow 25\%)V_{rated}$$

The power measured is essentially the rate of heat loss (I^2R) in the stator and rotor windings under normal operating conditions. A small percentage of P_z is core loss, which is normally assumed to be negligible.

For class B and C 60 Hz motors, the frequency recommended for the locked-rotor test is 15 Hz. This frequency is necessary in order to simulate a rotor-heating effect equivalent to running conditions. The reactances measured at 15 Hz are only one-quarter of their actual values.

For class A motors, the recommended frequency is the same as that required under normal operating conditions.

No-Load Test (s = 0)

The no-load test of an induction motor corresponds to the open-circuit test on a transformer. The machine runs unloaded at a speed very close to synchronous, while the excitation power (P_{exc}), the voltage (V_{exc}), and the current (I_{exc}) are measured.

The excitation voltage should be equal to the rated voltage of the motor so that rated flux conditions prevail. That is,

$$V_{exc} = V_{rated}$$

The excitation current is a large percentage of the rated current:

$$I_{exc} = (25\% \rightarrow 40\%)I_{rated}$$

The excitation current is larger than that of a transformer with the same kVA and voltage ratings because of the larger air gap between the rotor and stator structures and because of the rotational losses that accompany the no-load test.

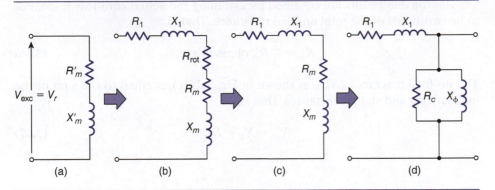

FIG. 3-26 Equivalent circuit obtained from the no-load test: **(a)** measured parameters ($Z'_m = R'_m + jX'_m$); **(b)** detailed representation of the measured parameters, $Z'_m = (R_1 + R_{rot} + R_m) + j(X_1 + X_m)$; **(c)** the resistance (R_{rot}) that represents the rotational losses; **(d)** parallel representation of the magnetizing impedance shown in (c).

The data of the no-load test gives the so-called gross open-circuit motor impedance (Z'_m), as shown in Fig. 3-26(a).

$$Z'_m = R'_m + jX'_m \tag{3.61}$$

This impedance includes the stator and magnetization impedances as well as the rotational losses in their equivalent ohmic values. From basic definitions, we have

$$Z'_m = \frac{V_{exc}}{I_{exc}} \tag{3.62}$$

$$R'_m = \frac{P_{exc}}{I^2_{exc}} \tag{3.63}$$

and

$$X'_m = \sqrt{Z'^2_m - R'^2_m} \tag{3.64}$$

The no-load resistance (R'_m) is equal to the stator resistance (R_1) plus the equivalent resistances of the core loss (R_m) and the rotational losses. (See Fig. 3-26(b).)

$$R'_m = R_1 + R_m + R_{\text{rotational losses}} \tag{3.65}$$

The rotational losses* are normally incorporated in the mechanical load and are considered part of the power consumed by the equivalent load resistance $R_2(1 - s)/s$.

---

*The rotational losses are made up of the stray-load losses and the mechanical losses. The mechanical losses can be accurately measured by plotting, at no-load, the power versus voltage squared. The intercept with zero voltage gives the mechanical losses.

Reasonable results are obtained by assuming the actual core-loss resistance to be two-thirds of the total no-load resistance. That is,

$$R_m = \frac{2}{3} R'_m \text{ ohms/phase} \tag{3.66}$$

The no-load reactance (X'_m), as shown in Fig. 3-26(c), is equal to the sum of the magnetizing and stator reactances. That is,

$$X'_m = X_1 + R_m \tag{3.67}$$

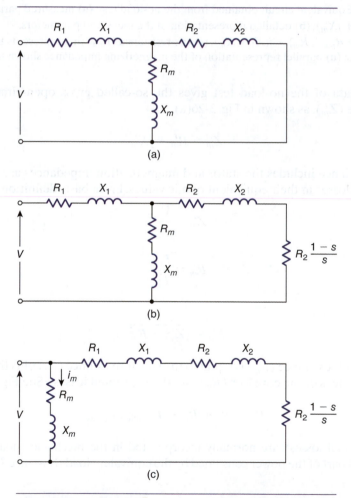

(a)

(b)

(c)

FIG. 3-27 Per-phase approximate equivalent circuits as seen from the stator: **(a)** rotor stationary, **(b)** rotor running, **(c)** a further approximation of (b); the magnetizing impedance has been relocated in order to simplify the calculations.

The stator reactance is relatively small, and for all practical purposes,

$$X'_m \approx X_m \tag{3.68}$$

The series magnetizing impedance is often represented by an equivalent parallel impedance, as shown in Fig. 3-26(d).

The parallel representation of the magnetizing impedance is more common and can easily be found from its equivalent series representation, as discussed in Chapter 2 (Eqs. (2.24) and (2.25)).

The parameters of the motor's equivalent circuit are shown in Fig. 3-27. The most commonly used approximate equivalent circuit for the engineering analysis of induction machines is that of Fig. 3-27(c).

EXAMPLE 3-5

A 100 kW, three-phase, 60 Hz, 480 V, star-connected, 1146 r/min, Class C induction motor yielded the following results when tested:

a. Average value of dc resistance between stator terminals: 0.20 ohms
b. No-load test:

$$V_{exc} = 480 \text{ V, L-L}$$

$$P_{exc} = 1920 \text{ W, three-phase}$$

$$I_{exc} = 40 \text{ A}$$

c. Locked-rotor test:

$$V_z = 32 \text{ V, L-L}$$
$$P_z = 2560 \text{ W, three-phase}$$
$$I_z = 65.6 \text{ A}$$
$$f = 15 \text{ Hz}$$

Determine the per-phase equivalent circuit of the motor.

SOLUTION

No-Load Test (Fig. 3-28)

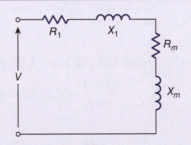

FIG. 3-28 Circuit corresponding to no-load test.

Using Eq. (3.63),

$$R'_m = \frac{P_{exc}}{I^2_{exc}} = \frac{1920}{3(40)^2} = 0.40 \text{ } \Omega/\text{phase}$$

From Eq. (3.66),

$$R_m = \frac{2}{3}(0.40) = 0.27 \text{ } \Omega/\text{phase}$$

The per-phase stator resistance is

$$R_1 = \frac{0.2}{2} = 0.10 \text{ } \Omega/\text{phase}$$

The rotational loss equivalent resistance is found by using Eq. (3.65).

$$R_{r.l.} = 0.40 - 0.10 - 0.27 = 0.033 \text{ } \Omega/\text{phase}$$

This resistance is not explicitly shown in the equivalent circuit, but its power consumption will be part of the power developed by the motor.

From Eq. (3.62), the gross magnetizing impedance is

$$Z'_m = \frac{V_{exc}}{I_{exc}} = \frac{480/\sqrt{3}}{40} = 6.92 \text{ } \Omega/\text{phase}$$

The gross magnetizing reactance is

$$X'_m = \sqrt{6.92^2 - 0.40^2} = 6.92 \text{ } \Omega/\text{phase}$$

Locked-Rotor Test (Fig 3-29)

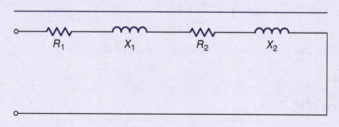

FIG. 3-29 Locked-rotor test.

From the data of the locked-rotor test and Eqs. (3.58) and (3.59), we obtain

$$Z_e = \frac{V_z}{I_z} = \frac{32/\sqrt{3}}{65.6} = 0.28 \text{ } \Omega/\text{phase}$$

$$R_e = \frac{P_z}{I^2_z} = \frac{2560}{3(65.6)^2} = 0.2 \text{ } \Omega/\text{phase}$$

The rotor resistance, as seen from the stator, is found by using Eq. (3.59).

$$R_2 = R_e - R_1$$
$$= 0.20 - 0.10$$
$$= 0.10 \ \Omega/\text{phase}$$

The equivalent motor reactance is found by using Eq. (3.60).

$$X_e = \sqrt{Z_e^2 - R_e^2} = \sqrt{0.28^2 - 0.20^2} = 0.20 \ \Omega/\text{phase}$$

The stator and rotor reactances can be found by using Tables 3-5 and Eq. (3.60), as follows:

$$X_e = X_1 + X_2$$

Substituting, we get

$$0.2 = X_1 + \frac{X_1}{0.43}$$

From the last two equations,

$$X_1 = 0.06 \ \Omega/\text{phase}$$

and

$$X_2 = 0.14 \ \Omega/\text{phase}$$

The above reactances were measured at a frequency of 15 Hz. Thus, at the rated frequency of 60 Hz,

$$X_1 = 4(0.06) = 0.24 \ \Omega/\text{phase}$$

and

$$X_2 = 4(0.14) = 0.56 \ \Omega/\text{phase}$$

The magnetizing reactance is found by using Eq. (3.67):

$$X_m = X_m' - X_1$$
$$= 6.93 - 0.24 = 6.69 \ \Omega/\text{phase}$$

The per-phase equivalent circuit, as obtained from the data of the given tests, is shown in Fig. 3-30.

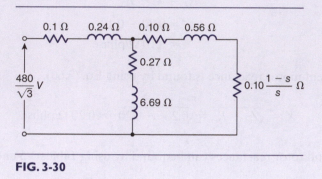

FIG. 3-30

3.4 Asynchronous Generators

When the rotor of a 3-ϕ induction machine is driven above synchronous speed and in the same direction as the synchronously rotating field, the slip—as can be seen from the general slip definition—becomes negative. The current and the voltage of the rotor windings are reversed in polarity, and the torque developed is in a direction opposite to the rotation. In this mode of operation, the machine delivers power to the stator terminals; that is, the motor acts as an induction generator.

An ordinary 3-ϕ induction motor, driven in the proper direction at hypersynchronous speeds, becomes an induction generator. The *excitation* of the generator is provided by the source of the stator voltage, and the *shaft's rotation* is derived from a water, wind, solar, steam, or gas-driven turbine. In some cases, a capacitor bank provides the excitation voltage.

Figures 3-31(a) and (b) show an elementary representation of an induction motor and an induction generator. In many respects, the induction generator is

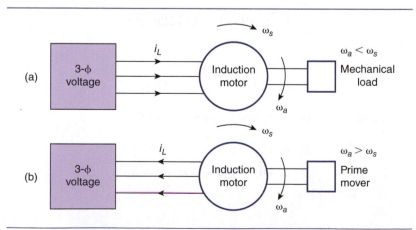

FIG. 3-31 Induction machines: **(a)** induction motor, **(b)** induction generator.

identical to the induction motor. The equivalent circuits are the same, but the energy flows in opposite directions. An induction machine that drives a hoist operates as an induction generator for descending loads and as a motor for ascending loads.

The induction generator finds increasing applications because it provides the most economical method of interconnecting a small power station to a large power-distribution network. Since the late 1970s, small power-generating stations have been operating in many places because the law requires utilities to purchase, at reasonable rates, the electricity produced by their own consumers.

The induction generator is not *self-excited*; it requires magnetizing kVAR (kilovolt-amperes reactive), which is drawn from the power-distribution network to which the generator is connected. If they are the right size, capacitor banks connected in parallel to the generator can compensate for the magnetizing or reactive power drawn by the induction generator. However, the power factor of the network—as seen from the terminals of the generator—depends on the impedance of the other loads that are connected to the same power grid. The magnetization of the machine can also occur through:

- The residual magnetism of the machine's metallic structure.
- A permanent magnet in the rotor.

The prime movers of induction generators are equipped with speed-control governors that limit the no-load speed of the machines to safe levels. This safeguard is necessary because, when the electrical load of the generator is suddenly disconnected, its rotor—owing to the constant prime mover's power and diminished electrical counter-torque—will overspeed.

Overspeeds are undesirable because the resulting centripetal and centrifugal forces may damage the rotor if the rotor is not designed to accommodate them. The level of overspeed depends on the ability of the control apparatus to instantly reduce the power developed by the prime mover. It is not always possible to instantly reduce the power, however. If a hydropower station were to attempt this operation, for example, destructive pressure rises would be produced within upstream water conduits (penstocks).

High overspeeds are also undesirable because they may induce high stator voltages. The magnitude of the induced voltage depends on the machine's electrical time constant (L/R) and on the prime mover's speed. For example, if an induction generator has an open-circuit time constant of 6 seconds, then its terminal voltage after 2 seconds—at a constant prime-mover speed—will be 72% ($e^{-2/6}$) of nominal voltage.

If, however, the speed of the prime mover is increased at the same time by 200%, then the terminal stator voltage will be about 144%. This overvoltage has undesirable effects, not only on the generator but also on the auxiliary equipment (potential transformers, relays, etc.) that is part of the generator's control apparatus.

The power rating of installed induction generators is usually in the range of 20 kW to 50 MW. If a generator's voltage rating is different from that of the distribution network, a transformer is used to interconnect the two voltage levels.

Equivalent Circuit

The per-phase equivalent circuit of an asynchronous generator is shown in Fig. 3-32. The generated voltage (V_g) is directly proportional to the machine's magnetic field (ϕ) and the rotor's speed (ω). That is,

$$V_g = K\phi\omega$$

The terminal voltage (V_t) and frequency are slightly higher than those of the utility's transmission lines. The stator (Z_1), rotor (Z_2), and magnetizing impedances (Z_m) are slightly higher than those corresponding to motor operation because the frequency of the generated voltage is larger than that when the machine operates as a motor.

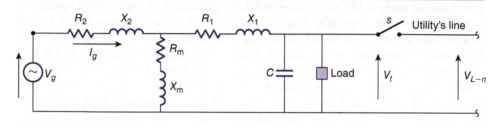

FIG. 3-32 Per-phase equivalent circuit of an asynchronous generator.

EXAMPLE **3-6**   If a 100 kW, 480 V windmill asynchronous generator operates at 1225 rpm and delivers 115 A at 0.9 Pf to a three-phase 60 Hz system, estimate:

a. The generator's operating slip.

b. The maximum value of the capacitor in order to provide the machine's magnetizing field.

c. The ratio of the induction generator/induction motor reactances.

d. What should be the line-to-line voltage of the utility's voltage for proper interconnection to the asynchronous generators?

SOLUTION

a. The closest synchronous speed of the rotating field is

$$n_s = 120\frac{f}{P} = 120\frac{(60)}{P}$$

For a six-pole machine,

$$n_s = 120\frac{(60)}{6} = 1200 \text{ rpm}$$

and the machine's corresponding slip is

$$s = \frac{1200 - 1225}{1200} = \underline{-0.02 \text{ pu}}$$

b. The apparent power of the machine is

$$|S| = \sqrt{3}(480)(115)$$
$$= 95.61 \text{ kVA}$$

and the nominal reactive power (Q) of the machine is

$$Q = P \tan \theta$$
$$= 95.61(0.9) \tan (\arccos 0.9)$$
$$= 41.68 \text{ kVAR}$$

From which

$$Q = VI \sin \theta_c = VI \sin 90° = V\frac{V}{Z} = V^2 \, \omega C$$

and

$$C = \frac{41.68 \times 10^3}{(480)^2(377)} = 480 \text{ } \mu\text{F}$$

c. The motor's nominal operating slip is

$$s = 0.02 \text{ pu}$$

and the ratio of the reactances is

$$\frac{X_g}{X_m} = \frac{2\pi f g(L)}{2\pi f m \,(L)}$$

$$= \frac{f_g}{f_m} = \frac{f(1.02)}{f(0.98)} = 1.04$$

d. The terminal voltage and frequency of the asynchronous generator must be slightly larger than those of the utility's interconnecting systems.

Briefly elaborate on the following:

Exercise

3-10

a. On interconnecting an asynchronous generator to a utility's network, the utility's terminal voltage must be slightly smaller than that of the asynchronous generator.

b. The impedances of the induction motor are smaller than those of the asynchronous generator.

c. How can the phase sequence of the voltages be changed?

d. Under what circumstances can a capacitor furnish the magnetizing field of an asynchronous generator, and what are its adverse effects?

e. The maximum kVAR of the capacitor must be equal to the motor's reactive power as registerd at the no-load test.

f. Why do the stator and rotor magnetic fields in a synchronous motor repel each other?

3.5 Controls

This section discusses the following techniques used to control the starting, stopping, and operating torque-speed characteristics of 3-ϕ induction motors.

- Reduction of the motor's high starting current and torque
- Variable frequency drives
- Coasting a motor to a stop

3.5.1 Reduction of the Motor's High Starting Current and Torque

Induction motors, upon starting at nominal line voltages, draw high starting currents and also develop high torques. The starting current could be about six times rated and the torque developed two to three times rated.

High starting currents produce high copper losses (I^2R) and may slightly increase the premises' power demand and cause voltage sags (momentary reductions in the voltage of the upstream network) that may affect the operation of equipment. In addition, the utilities may prohibit such a condition. Higher than nominal torques overstress the structure of the machine. The methods used to reduce a motor's high starting current and torque are electromechanical and electronic. In the first case, one can use in-line resistors or inductors, autotransformers, or Y-Δ starters. Electronic control includes the so-called soft start. (The soft start is discussed in Section 3.5.3.)

In-line Resistors or Inductors

The in-line resistors or inductors incorporate a start-stop push button, a contactor, and a timer that sequentially switches the in-series elements ON or OFF so that the motor's voltage and current are kept at a desirable level. The main disadvantage of such a setup is the energy dissipated upon starting.

Autotransformers

Autotransformers and in-line resistors have a similar function. The desirable voltage level of the autotransformer is selected through the secondary voltage taps of the secondary winding by using a start-stop push button, a time relay, and a contactor. The momentary change in the voltage level is, however, accompanied by high currents, which is true of all inductive circuits.

Y-Δ Starters

The Y-Δ starter incorporates simple controls through which, at starting the motor's windings, are connected in a Y-configuration; when the motor reaches about 80% of rated speed, a time delay relay switches OFF the Y-winding configuration and

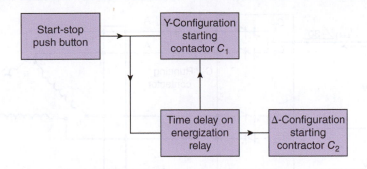

FIG. 3-33 Block diagram representation of a Y-Δ starter.

switches ON the Δ-winding configuration. This scheme reduces the motor's high starting current and torque and thus mitigates their adverse effects. The block diagram configuration for the Y-Δ starter is shown in Fig. 3-33.

Y-Winding Configuration

Refer to Fig. 3-34. From the starter to the motor's terminals, there are six wires: three for the line currents and the other three for the windings' neutral connection. Through a start-stop push button, the contactor's coil (C_1) is energized (not shown in the diagram), its contacts close (C_1), and a time delay is also energized. The motor's winding receives 0.58 of the line voltage $\frac{1}{\sqrt{3}}$, the shaft rotates, and

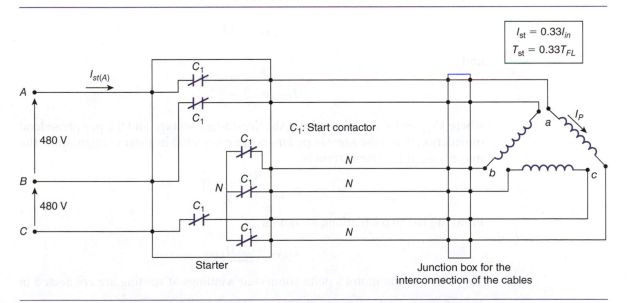

FIG. 3-34 Starting a three-phase motor. Start-connected schematic.

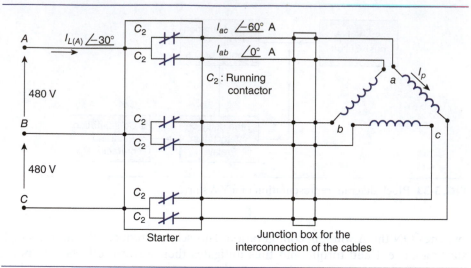

FIG. 3-35 Operating a 3-ϕ motor. Delta-connected schematic.

after a predetermined time, the coil of contactor C_1 is deenergized and the coil of contactor C_2 is energized. These steps lead to operation of the motor in the Δ-winding configuration (Fig. 3-35).

Current Considerations

At starting, the motor's windings are star-connected. The magnitude of the phase (I_P) and line currents (I_L) of a delta-connected load is

$$I_{L(\Delta)} = \sqrt{3}\, I_P$$

and

$$I_{L(\Delta)} = \sqrt{3}\, \left| \frac{V_{\text{L-L}}}{Z} \right|$$

where $V_{\text{L-L}}$ and Z are, respectively, the line-to-line voltage and the per-phase load impedance. When the *same* impedances are connected in a star configuration, the magnitude of the line current is

$$I_{L(Y)} = \left| \frac{V_{\text{L-L}}/\sqrt{3}}{Z} \right|$$

From the last two equations, we obtain

$$I_{L(Y)} = \underline{0.33 I_{L(\Delta)}}$$

That is, when the motor's delta-connected windings at starting are connected in Y-configuration, the starting current is reduced by a factor of $\frac{1}{3}$.

Torque Considerations

The torque of a three-phase induction motor is proportional to the voltage squared. That is,

$$T \propto V^2$$

For the Y-winding and delta configuration, we obtain

$$\frac{T_{st\,(Y)}}{T_{st\,(\Delta)}} = \frac{\left(\frac{V_{L\text{-}L}}{\sqrt{3}}\right)^2}{(V_{L\text{-}L})^2}$$

from which

$$\underline{T_{st\,(Y)} = 0.33\,T_{st\,(\Delta)}}$$

That is, when at starting the motor's windings are star-connected, the torque developed is also one-third that of the delta-winding configuration. In a particular application, the reduction in the torque must be verified to ensure that it is sufficient to meet the starting requirements of the load. At the instant of switching, the current and the torque of the motor increase and the windings' insulation and the motor's structure are overstressed.

Δ-Winding Configuration

Figure 3-35 presents the same motor starter for the Δ-winding configuration. Only the contacts of the contactor C_2 are shown. The line current to the starter is the nominal current of the motor. Each line current feeds two phases of the motor's windings. That is, for line "A", we have

$$I_A = I_{ab} + I_{ac}$$

$$I_A = I_{ab}\,\underline{/0} + I_{ac}\,\underline{/-60°}$$

The 60 degree phase shift is due to the standard 120 degree phase shift between voltages and currents in a three-phase system. The phase currents are of equal magnitude. That is,

$$I_P = I_{ab} = I_{ac}$$

From the last two relationships, the magnitude of the phase current is

$$I_P = \frac{I_A}{\sqrt{3}}$$

which is the standard relationship between the phase and line current in a delta-connected system. Occasionally, the problem arises regarding the size of the cables from the starter—several meters away—to the motor. These cables carry not $\frac{1}{2}$ of the line current but $\frac{1}{\sqrt{3}}$ to that of the supply line to the starter.

EXAMPLE **3-7** | A 200 kW, 480 V, three-phase induction motor has a starting and a nominal operating current, respectively, of 1800 A and 300 A. The motor's winding-configuration is star-connected on starting and delta-connected on steady-state operation. Determine the six-line currents from the starter to the motor at:

a. Starting.
b. Nominal operation.

SOLUTION

a. At starting, the magnitude of the line currents (I_L) is equal to the phase current (I_P)

$$I_L = \frac{1}{3}(1800) = 600 \text{ A}, \quad \text{line “a”}$$

The currents in the lines "b" and "c" are equal to the current of line "a" and 120 degrees to it.

The neutral wires carry no current for balanced voltage supply and equal load impedances.

b. At nominal operation, the motor's winding configuration is delta, and the line currents to the starter are nominal at 300 A.

From the starter, however, to the motor's terminals,

$$I_A = I_{ab} + I_{ac}$$
$$= K\underline{/0} + K\underline{/-60}$$
$$= \sqrt{3}K\underline{/-30°}$$

From which

$$K = \frac{300}{\sqrt{3}}\underline{/30}$$
$$= 173.21\underline{/30} \text{ A}$$

Thus, the magnitude of the current through each of the six wires to the motor is

$$173.21 \text{ A.} \quad \text{That is,} \left(I_P = \frac{I_L}{\sqrt{3}}\right)$$

Briefly explain the following:

a. The phase angle of the starting current of an induction motor is very high (about 70°) relative to that of a nominal operation (about 25°).

b. For a given number of magnetic poles, the higher the speed of the motor, the higher is its power factor.

c. A partially loaded motor operates at a low-power factor.

d. Each of the lines to Y-Δ starter supplies current to two-motor lines on a delta-winding configuration. Each of these lines carries 0.58 of the corresponding line current to the starter.

e. Draw in block diagram from the elementary control diagram of the Y-Δ motor starter.

3.5.2 Variable Frequency Drives

The variable frequency drives (VFD) refer to a control system through which the voltage and its frequency applied to an induction motor vary in such a way as to meet the torque-speed requirements of a particular mechanical load and to minimize the corresponding energy consumed. As shown in Fig. 3-36, the VFDs have four distinct sections: the rectifier, the filter, the inverter, and the controller.

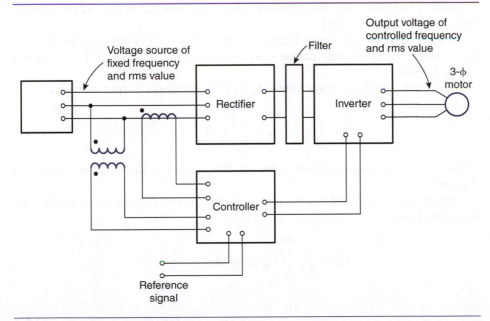

FIG. 3-36 VFD. Block diagram representation.

The rectifier is usually a six-pulse diode bridge and changes the incoming ac voltage to a dc waveform. The filter, as its name implies, blocks the transmission of high-frequency harmonics. The inverter changes the rectifier's output dc voltage waveform to a three-phase ac signal of desired voltage and frequency. The controller incorporates all the devices necessary for the protection and automatic control of the inverter's output. These components of VFD are discussed in some detail in the following sections.

Rectifiers

A rectifier receives a three-phase ac waveform at its input terminals and produces a dc voltage waveform at its output terminals. A typical six-pulse, diode-bridge rectifier is shown in Fig. 3-37(a). It is called a six-pulse rectifier because, during a complete cycle of the input voltage, its output consists of six voltage pulses. The diodes act as switches that are either open- or short-circuited, according to the relative polarity of the voltage from their anode (A) to cathode (K) (see diode D_1). When the diode's anode is at a higher potential than its cathode, the diode is shorted (forward-biased or conducting). When the diode's cathode is at higher potential than its anode, the diode is open-circuited (reverse-biased or nonconducting).

The ON-OFF condition of the diodes (such as D_1 and D_2) connected to terminals a and b depends on the magnitude of the voltage between a and $b(v_{ab})$ relative to the other line-to-line voltages. At the particular time interval when the voltage v_{ab} is the highest, then diodes D_1 and D_2 conduct. Refer to Fig. 3-37(b). The conducting period of these two diodes is from $\pi/3$ to $2\pi/3$ radians.

The conducting period of each pair of diodes can be determined in this manner. The output current pulses for a resistive load are shown in Fig. 3-37(c). According to Ohm's law, these current pulses are similar to the output voltage pulses given in Fig. 3-37(b).

Each ac supply line carries four of the six output current pulses in an alternating symmetry, thus the line currents do not have a dc component. This is shown in Fig. 3-37(d).

By definition, the average value (V_{av}) of the rectifier's output voltage is

$$V_{av} = \frac{1}{T}\int_0^T v_0 \, dt \tag{3.69}$$

where v_0 is the instantaneous value of the output voltage whose period is T seconds.

Using symmetry, and considering only the output voltage pulse that is due to v_{ab}, we have

$$V_{av} = \frac{6}{2\pi}\int_{\pi/3}^{2\pi/3} V_m \sin \omega t \, d\omega t$$

$$= -\frac{3V_m}{\pi}(\cos 120° - \cos 60°) \tag{3.70}$$

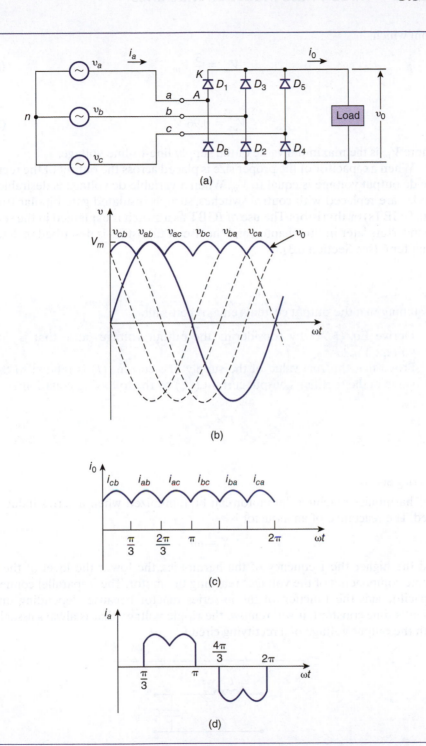

FIG. 3-37 Rectifier: **(a)** typical circuit; **(b)** input-output voltage waveforms; **(c)** output current; **(d)** line current (i_a).

from which

$$V_{av} = \frac{3}{\pi} V_m \tag{3.71}$$

or

$$V_{av} = 0.96 \, V_m \tag{3.72}$$

where V_m is the maximum value of the supply line-to-line voltage.

When a capacitor of the proper size is placed across the output of the rectifier, the dc output voltage is equal to V_m. When a variable dc voltage is desirable, the diodes are replaced with control switches, such as insulated gate bipolar transistors (*IGBT*s) or thyristors. The use of *IGBT* as a switch is explained in the section on inverters later in this chapter, and the use of thyristors is described in detail in Chapter 6 (see Section 6.3).

Exercise 3-12

Assuming that the output current remains constant,

a. Derive Eq. (3.71) by considering the output voltage pulse that is due to voltage V_{cb}.
b. Prove that the rms value of the supply line current (I) is related to the dc value of the rectifier's output current (I_{av}) by the following equation:

$$I = \sqrt{\frac{2}{3}} \, I_{av}$$

Filtering Section

The harmonics reaching the motor can be minimized when a series inductor is used. The reactance of an inductor is

$$X = \omega L$$

and the higher the frequency of the harmonics, the lower the level of the harmonic components of the voltages reaching the motor. The in-parallel connected capacitor aids the function of the in-series reactor because, depending on the circuit's time constant, it will remove the ripple voltage that is always associated with the output voltage of a rectifying circuit.

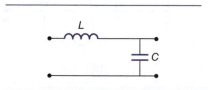

FIG. 3-38 Filtering section of a VFD.

| Input section | Analog-to-digital converter (A/D) | Microprocessor | Digital-to-analog converter (D/A) |

FIG. 3-39 Block-diagram representation of a digital controller.

Controller

The controller is the brain of a frequency converter. Not only does it provide a closed-loop control for the motor, but it also monitors the motor's performance and protects it from abnormal operating conditions.

A controller in block-diagram form is shown in Fig. 3-39. The input section of the regulator receives the reference signal and the outputs from the potential and the current transformers.

The output of the potential transformers (PT) is used to produce the dc voltages required to operate the various digital circuits. The output of the current transformers (CT) is directly proportional to the current drawn by the motor. When properly evaluated, it reveals the magnitudes of the motor's torque and speed, as can be seen in Fig. 3-23.

In some applications, a dc type of CT may be connected to the dc lines of the rectifier. The operation of such a CT is based on the Hall effect, not on the electromagnetic coupling of two coils. The incoming section of the controller subtracts the CT's signal from the reference signal, and the resulting signal is converted to an equivalent digital signal. This digital signal is fed into the microprocessor, which compares it to a set of stored instructions and produces an output accordingly.

The output of the microprocessor is converted to an analogue signal that controls the ON-OFF status of the *IGBT*. (See Fig. 3-40(a).)

Inverters

The function of an inverter is to produce a three-phase ac waveform of controlled frequency from a single dc voltage supply. A typical IGBT three-phase inverter is shown in Fig. 3-40(a).

Depending on the voltage from gate to emitter, each *IGBT** in an inverter is either open-circuited or short-circuited; that is, the transistor operates as a switch and not as an amplifier. When the V_{GE} is sufficiently high, the transistor is short-circuited ($v_{CE} \approx 0$), and when the V_{GE} is zero, the transistor is open-circuited ($I_c = 0$).

When a transistor is short-circuited, it is said to be operating in the saturation region. When a transistor is open-circuited, it is said to be operating in the cut-off region.

---

* The *IGBT* operation is discussed on the website.

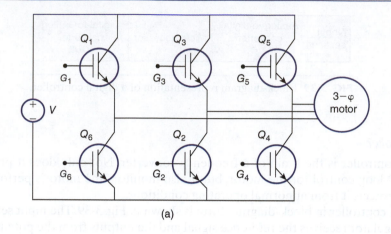

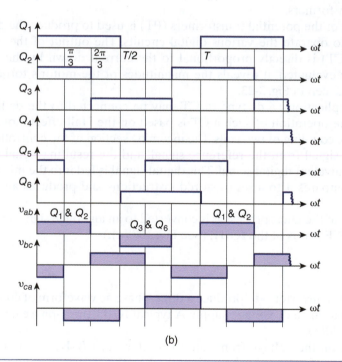

FIG. 3-40 3-ϕ Inverter: **(a)** circuit, **(b)** pertinent waveforms.

The ON and OFF status for each *I*GBT is equal to half the period ($T/2$) of the output voltage. The switching ON of a transistor is delayed $\pi/3$ radians relative to the transistor that was previously conducting. For the current to flow, two *I*GBTs must conduct simultaneously.

Consider the output voltage waveform v_{ab}. As shown in the diagram and in the following table, for each component of this output voltage, a unique pair of *I*GBTs conduct.

Time Interval of v_{ab}	Conducting IGBTs
$0 - 2\dfrac{\pi}{3}$	Q_1 and Q_2
$2\dfrac{\pi}{3} - \pi$	None
$\pi - 5\dfrac{\pi}{3}$	Q_3 and Q_4
$5\dfrac{\pi}{3} - 2\pi$	None

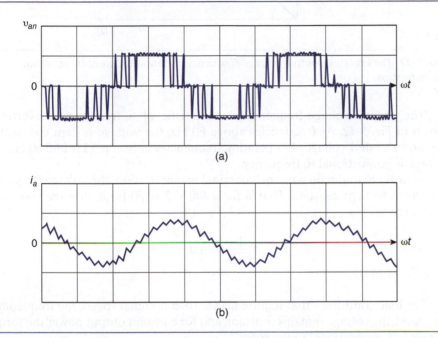

FIG. 3-41 Actual output waveforms of an inverter: **(a)** voltage, **(b)** current.

The pair of conducting transistors of the other line-to-line voltages can be determined in the same manner. The 3-ϕ voltages produced have a phase sequence ABC.

Normally, the operation of an inverter's switches produces an output voltage that, over a complete voltage cycle, is made up of a series of pulses whose width, as seen in Fig. 3-41(a), is variable. This is referred to as pulse-width modulation (PWM). Although the output voltage is made up of a series of pulses, the resulting motor current resembles a sine function, as shown in Fig. 3-41(b).

The width of the voltage pulses controls the rms value of the output voltage, and the number of pulses within a voltage cycle controls the frequency of the output voltage. The switching of the *IGBTs* produces high transient voltages due to the inductance of the load; these voltages are usually bypassed by connecting a diode across the collector-emitter terminals.

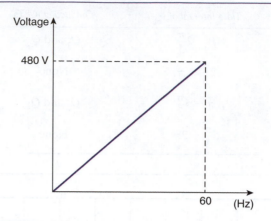

FIG. 3-42 The inverter's output voltage-frequency variation for constant machine magnetization.

The output voltage-frequency characteristic of a frequency converter is shown in Fig. 3-42. At frequencies above 60 Hz, the voltage is kept constant at the motor's rated voltage. At operating frequencies between 5 Hz and 60 Hz, the voltage is proportional to frequency.

In order to maintain a motor's nominal magnetization, the voltage/frequency ratio must be kept constant. That is, for a 480 V, 3-ϕ, 60 Hz motor, the voltage to frequency ratio is

$$\frac{480}{60} = 8.0$$

(See Eq. (1.148))

For load conditions that require higher than nominal speed, the frequency is increased, the voltage remains constant, and for constant output power the torque is inversely proportional to the speed.

EXAMPLE **3-8** The speed, power, and torque of a 460 V, 60 Hz, 3-ϕ induction motor are controlled through a VFD.

a. Draw the voltage-frequency output characteristic of the inverter.

b. When the output frequency of the inverter is reduced from 60 Hz to 30 Hz, what is the rms value of the inverter's output voltage?

c. What is the reason that up to nominal speed, the voltage/frequency ratio should be kept constant?

d. Explain why, as a result of increasing the inverter's output above 60 Hz while the power to the motor is kept constant, the motor's torque is inversely proportional to the speed and the motor's magnetization is reduced.

e. Why, for a constant load torque, is the motor's power proportional to the frequency?

SOLUTION

a. See Fig. 3-42.

b. The voltage to frequency ratio is

$$\frac{V}{f} = \frac{460}{60} = 7.67$$

Thus,

$$V = 7.67f$$

$$V = 7.67(30)$$

$$V = 230 \text{ V, line-to-line}$$

(The motor's magnetization did not change.)

c. The voltage to frequency ratio is maintained at a constant in order to operate the motor at a nominal magnetization level.

d. I) $P = T\omega$ and $T = \dfrac{P}{\omega}$

II) From Eq. (1.148), by increasing the frequency, the level of magnetic field is reduced.

e. $P = T\omega = TK_1 f = K_2 f$

EXAMPLE 3-9

A 10 HP, 480 V, 1770 rpm, 3-ϕ, induction motor drives a ventilator that is equipped with inlet damper control. The ventilator at the operating region has the pressure (h_2) versus air flow (Q_2) characteristic given by

$$h_2 = 2Q_2^2$$

and at nominal operating condition

$$h_1 = (1)Q_1^2$$

The fan is to operate for six months per year continuously at 80% of rated air flow. Determine the energy consumed by using:

a. Damper control

b. VFD

(**Note:** The pressure versus flow characteristic is derived by the process or mechanical engineers).

SOLUTION

a. The system parameters designated with subscripts 1 and 2 are

$$h_2 = 2Q_2^2$$
$$h_1 = (1)Q_1^2$$

and

$$h_2 = 2\left(\frac{Q_2}{Q_1}\right)^2$$
$$= 2\left(\frac{0.8}{1.0}\right)^2$$
$$= 1.28$$

The air power is equal to pressure times rate of air flow.

$$HP_2 = h_2 Q_2 \quad \text{or} \quad HP_2 = HP_1\left(\frac{h_2}{h_1}\left(\frac{Q_2}{Q_1}\right)\right)$$

Substituting for the ratio of the pressures, we obtain

$$HP_2 = HP_1 \frac{2Q_2^3}{Q_1^3} = (HP_1)\, 2\left(\frac{0.8}{1}\right)^3 = \underline{1.024 HP_1}$$

Corresponding energy is

$$W_1 = 6 \times 30 \times 24(1.024 \times 0.746) = \underline{3300\ \text{kWh}}$$

b. Based on the fan laws commonly known as affinity laws, we have

$$P \propto Q^3$$

Thus,

$$\frac{P_2}{P_1} = \left(\frac{Q_2}{Q_1}\right)^3$$
$$P_2 = P_1\left(\frac{0.8}{1}\right)^3$$
$$= 0.512 P_1$$
$$= 0.512(10)(0.746)$$
$$= 3.82\ \text{kW}$$

and energy (W_2):

$$W_2 = 6 \times 30 \times 24(3.82) = \underline{16,502 \text{ kWh}}$$

For comparison purposes, see the following table:

No.	Method of Rate of Air Flow Control	Energy Consumption in kWh
1	Use of dampers	16,502
2	VFD	3300

3.5.3 Soft Start

Soft start refers to the gradual increase in the motor's speed by reducing the time interval of the voltage applied to the motor. This increase is accompanied by a reduction in the motor's starting current and torque. Although the peak value of the applied voltage does not change, its rms value is reduced accordingly. A soft start control circuit could be part of a VFD or a much simpler circuit incorporating *IGBTs* and a timer.

Reduction of the voltage's rms value by using 2-*IGBT* per phase is illustrated in Fig. 3-43. For simplicity, only one phase is shown. One *IGBT* controls the duration

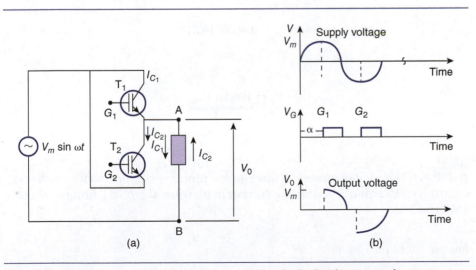

FIG. 3-43 *IGBT* control of starting current: **(a)** Circuit. **(b)** Pertinent waveforms.

of the time interval of the positive part of the voltage cycle, and the other controls that of the negative cycle. The starting point of the voltage cycle depends on the application of the gating pulse, which is commonly known as the firing angle (α).

In Fig. 3-43(a), when the supply voltage is positive and T-1 is triggered, current flows to the load from terminal A to B. When the supply voltage is negative and T-2 is triggered, current flows to the load from terminal B to A.

In Fig. 3-43(b), the firing angles for T-1 and T-2 are $\frac{\pi}{6}$ and $\frac{7\pi}{6}$ radians, respectively. The voltage to the load (V_0) is as shown.

EXAMPLE **3-10** | The rms value of the current through a 208 V, 60 Hz, 3-ϕ resistive load is controlled through one *IGBT* per phase. The *IGBT*'s firing angles for phase "a" are $\frac{\pi}{3}$ and $\frac{4\pi}{3}$ radians. Estimate the rms load voltage.

SOLUTION

$$V_0^2 = \frac{2}{2\pi} \int_{\pi/3}^{\pi} (V_m \sin \omega t)^2 \, d\omega t$$

$$= \frac{V_m^2}{2\pi} \int_{\pi/3}^{\pi} (1 - \cos 2\omega t) \, d\omega t$$

$$= \frac{V_m^2}{2\pi} \left[\omega t \Big|_{\pi/3}^{\pi} - \frac{1}{2} \sin 2\omega t \Big|_{\pi/3}^{\pi} \right]$$

$$= \frac{V_m^2}{2\pi} \left[\frac{2\pi}{3} - \frac{1}{2} (\sin 360 - \sin 120) \right]$$

$$= 0.40 \, V_m^2 = 0.4(208)\sqrt{(2)^2}$$

and

$$V_0 = 186.56 \, V_{\text{rms}}$$

Exercise
3-13

A 480 V, 60 Hz, delta-connected, three-phase motor is equipped with a soft start controller. Determine the starting current in pu when the *IGBT* firing angle is 150° and 240°.

Answer   0.12 pu

3.5.4 Plugging

A motor can be stopped by any of the following methods:

1. Removing the supply voltage.
2. Reversing the phase sequence of the applied voltages.
3. Applying a dc field to a stator phase winding while the ac voltage supply is removed.
4. Using a mechanical brake.

With each method, a specific amount of time is required to bring the motor to a complete stop.

3.6 Conclusion

The operation of induction machines is based on the development of the rotating field, on the induction principle, and on the natural tendency of magnetic fields to align their axes.

When balanced 3-ϕ currents are circulated through properly distributed coils, a rotating field is produced. This rotating field completes its loop by cutting through the rotor and stator structures. As a result, a voltage is induced in the rotor coils. This voltage, in turn, causes current to flow through the rotor windings, and thus a rotor field is produced. The stator and rotor fields try to align their magnetic axes, and as a result, the motor produces torque.

The rotor of an induction machine is of either the squirrel-cage or the wound-rotor type. Induction machines with squirrel-cage rotors are much more popular than those with wound rotors because they are less expensive to both purchase and operate. The National Electrical Manufacturers Association (NEMA) groups motors with squirrel-cage rotors into four categories—class A, B, C, or D—according to their torque-speed characteristics. Each class of motor is suited to a particular application. Class B motors, for example, are for general-purpose use and can be found in fans and blowers.

Induction machines with wound rotors are used for special applications, but because they are less efficient at higher slips, they are gradually being replaced by variable-speed drives. Wound-rotor induction machines can be found in the following special applications:

1. To operate variable torque-speed characteristics for loads that operate over a wide range of speed and torque.
2. To limit the starting current of a motor in a distribution system where the supply power system is weak (large upstream impedance).
3. To limit the starting current of motors that are required to start and restart frequently, such as those used in the crushing machines of the mining industry. The reduction in the starting current increases the number of permissible starts during any given time interval, so the motor's life is not reduced.
4. To give high torque at high slip.

A motor's slip indicates the relative motion between the speed of the synchronously rotating stator field and the speed of the shaft of the motor. The slip can be *positive* and less than unity (in an induction motor); *positive* and larger than unity (in voltage and frequency multipliers); or *negative* and less than unity (in induction generators).

For all practical purposes, the real part of the impedance of a motor as seen from its stator is inversely proportional to the slip. Thus, at starting (s = 1), the impedance is small, whereas under full-load conditions (s ≈ 0.015), the impedance is very high.

Consequently, a motor's current at this operating condition—known as the rated condition—is about six times smaller than it is at starting. The large starting currents of a motor not only lower its starting-torque capability—owing to a high voltage drop in the upstream impedances—but also develop extremely high temperatures in the rotor and stator windings. These large currents may limit the number of motor starts permissible within a given time interval.

The motor's equivalent circuit is of paramount importance. It simplifies understanding of the motor's operation and facilitates analysis of a particular problem. In practice, however, the circuit parameters are often not known, and an application engineer has to make decisions based on other data (such as starting and full-load torque and currents) readily available from the manufacturers.

Induction generators are gaining in popularity because they provide a simple and the most economical method of connecting a wind turbine or a small hydropower station to a distribution network. Power stations driven by wind, solar, water, or process heat are becoming increasingly common because of recent legislation requiring utilities to buy, at reasonable rates, electricity produced by their consumers.

Variable frequency drives (VFDs) change the frequency and the voltage applied to a motor. As a result, a motor's torque-speed characteristic can be modified to meet diverse load requirements. Additionally, VFDs help reduce energy consumption per-unit output more than the other methods of speed control.

The following tables summarize the basic concepts of this chapter in their mathematical form and also furnish typical manufacturers' data for three-phase induction machines.

3.7 Tables

TABLE 3-7.1 Summary of important equations

Item	Description	Remarks
1	Magnitude of rotating field $$F_s = \frac{3}{2} F_1 \cos(\omega t - \beta)$$	Eq. (3.14)
2	Mechanical and electrical degrees or radians $$\theta = \frac{p}{2} \theta_m$$	Eq. (3.15)

TABLE 3-7.I (*Continued*)

Item	Description	Remarks
3	Speed of rotating field in r/min $$n_s = 120\frac{f}{p}$$	Eq. (3.18)
4	Per-unit slip $$s = \frac{n_s - n_a}{n_s}$$	Eq. (3.19)
5	Voltage induced in rotor windings $$V_r = sV'_r$$	Eq. (3.23)
6	Frequency of voltage induced in the rotor windings $$f_r = sf$$	Eq. (3.24)
7	Mechanical load in equivalent electrical resistance $$R_L = R_2\frac{(1-s)}{s}$$	Eq. (3.32)
8	Torque developed $$T = \frac{V^2}{\left[\left(R_1 - \dfrac{R_2}{s}\right)^2 + X^2\right]\left(\dfrac{R_2}{\omega_s s}\right)}$$	Eq. (3.38)
9	Torque in the full-load region $$T \approx K\frac{s}{R_2}$$	Eq. (3.42)
10	Slip at maximum torque $$s_{mt} = \frac{R_2}{\sqrt{R_1^2}} + X_2$$	Eq. (3.46)
11	Motor's output torque $$T = J\frac{d\omega_a}{dt} + B\omega_a + T_L$$	Eq. (3.49)
12	Power developed $$P_d = I_2^2 R_2\frac{1-s}{s}$$	Eq. (3.56)
13	Starting current $$I_{st} = (5 \rightarrow 6)I_{fl}$$	

Variable Frequency Drives (VFDs)

Item	Description	Remarks
14	Output voltage of a six-pulse bridge rectifier $$V_{av} = \frac{3}{\pi}V_m$$	Eq. (3.71)

TABLE 3-7.2   Typical Data for three-phase 60 Hz, 460 V, NEMA Design B, service factor 1.0, class insulation B, standard squirrel-cage induction motors

Power in kW	Full-Load Speed in r/min	Approximate Amperes		Starting Torque in Percent	Maxi- mum Torque in Percent	Efficiency in Percent		Power Factor in Percent	
		Full- Load	Locked- Rotor			Full- Load	50% Load	Full- Load	50% Load
0.75	1730	1.64	15	275	300	78.5	72	73	51
	870	2.3	15	135	215	70.5	61	58.5	39.5
7.5	3500	12.23	81.25	135	200	85.5	82.5	90	82
	880	15.86	81.25	125	200	83	79.5	71.5	52.5
20	3540	32.2	182	130	200	86.5	83	90	82.5
	885	35.2	182	135	200	88.5	87.5	80.5	64.5
40	3545	62.3	362.5	125	250	89	87	90.5	83
	880	69.3	362.5	145	230	90.5	88.5	80	65
50	3560	75.4	542.5	115	250	90.5	88	92	85
	885	86.2	542.5	145	210	91	90	80	66.5
75	3555	112.4	725	110	200	90.5	88.5	92.5	88.5
	885	128.4	725	125	200	90.5	89.5	81	70
100	3565	149.9	1085	110	240	91	88	92	88.5
	1775	151.6	1085	125	210	92	90.5	90	86

Based on data from Siemens Electric Limited

TABLE 3-7.3   Data for three-phase 60 Hz, 4000 V, totally enclosed, fan-cooled, squirrel-cage induction motors

kW	Full-Load Speed in r/min	Full-Load Amperes	Locked- Rotor Amperes	Starting Torque in Percent	Maxi- mum Torque in Percent	Efficiency in Percent		Power Factor in Percent	
						Full- Load	50% Load	Full- Load	50% Load
150	1764	28	162	100	200	91.8	88.7	84.1	71
	1178	27.1	162	120	200	91.8	88.7	87	75.5
200	589	32.4	162	60	175	90.8	88.6	73.7	53.5
	589	42.6	202	60	175	91.1	89	74.3	54.5
225	1180	40.4	242	100	175	92.3	89.2	87	76.4
	589	47.5	242	60	175	91.4	89.3	74.8	55.5
300	1182	53.8	323	60	175	92.7	89.6	87	76
400	1184	71.4	404	60	175	93	89.9	87	77.1
1500	1175	259	1510	140	210	95	94.5	88	77

Based on data from Siemens Electric Limited

TABLE 3-7.4 Typical test results and measurements of a 260 kW, 4000 V, 0.75 Pf, 0.88 efficient, 10-pole, squirrel-cage induction motor

a. Tests

Test	Frequency (Hz)	Voltage (V)	Current (A)	Power (kW)
No-load test	60	4000	33.1	9.5
Locked-rotor test	60	975	57	27
Locked-rotor test	30	537.8	57	29.2

b. Calculation of Load Characteristics at Rated Voltage and Frequency

Load in percent	25	50	75	100	125
Line current in A	35	40.2	47.9	57.6	68.7
Efficiency in percent	86.6	91.6	92.7	92.7	92.2
Power factor in percent	31.1	51.2	63.7	70.7	74.4
Slip in percent	0.32	0.65	1.01	1.4	1.83

Starting current (locked rotor) in pu: 4.6
Starting torque in pu: 1.5
Maximum torque in pu: 2.2

c. Temperature rise

Hours run: 4
Volts: 4000
Load: 100%

	Frame	Bearings	
		Load Side	Opposite Side
Temperature in °C	50	19	15

d. Winding resistance between terminals

At 20°C   1.69 ohms

e. Air-gap measurement

1.13 mm, average

Based on data from Siemens Electric Limited

TABLE 3-7.5 Typical data for three-phase, 4000 V, 60 Hz, squirrel-cage rotor induction motors

Rating in kW	Number of Poles	Polar Moment of Inertia Referred to Speed of Motor in kg-m²			Torque in Percent			Accelerating Time in Seconds	Maximum Permissible Stall Time in Seconds	
		Rotor	Load	Total	Starting	Pull-In	Maximum		Starting From Cold	Re-Starting Immediately after Operation (Hot-Start)
200	4	6	100	106	110	90	230	20	25	20
200	4	6	40	46	110	90	230	9	25	20
250	6	31	217	248	90	70	220	20	25	20
250	14	68	34	102	135	100	260	2	20	16
300	6	46	125	171	175	130	210	8	20	16
375	6	53	150	203	175	130	210	8	20	16

Based on data from Siemens Electric Limited

3.8 Review Questions

1. What are the essential differences between squirrel-cage and wound-rotor three-phase induction machines?

2. Why are three-phase induction machines often called rotating transformers?

3. Differentiate between a synchronous speed and an actual speed.

4. What is "slip," and under what operating conditions is it:

   a. Positive and smaller than unity?

   b. Positive and larger than unity?

   c. Negative and smaller than unity?

5. What are the frequencies of rotor currents at starting and at full-load? How does frequency affect the impedance of the rotor?

6. Why is rated voltage used in the "no-load test" and near full-load current used in the locked-rotor test?

7. Why, in general, is the starting torque for a three-phase squirrel-cage induction motor 1.5 to 2 times the full-load torque, and the starting current 5 to 6 times the full-load current?

8. For the A, B, C, and D classifications of motors, how does the rotor reactance compare with that of the stator?

9. *Explain*: The torque at the starting region is inversely proportional to the operating slip, while at the full-load region, it is proportional to the slip.

10. Why is the time that it takes a motor, starting from rest, to reach its rated speed of paramount importance?

11. Describe the four methods by which a three-phase induction motor can be stopped.

12. How does the motor of a hoist change the direction of rotation?

13. What provides the excitation power for an induction generator?

14. *Explain:* The maximum rate of heat dissipation of a machine is given by its efficiency.

15. What are the advantages and disadvantages of VFDs?

3.9 Problems

3-1 A 480 V, 60 Hz, 1740 r/min, 0.90 Pf, 7.5 kW, three-phase induction motor has a full-load efficiency of 85%. Determine:

   a. The full-load slip.

   b. The number of poles.

   c. The current drawn by the motor.

3-2 For a three-phase induction motor, show that:

   a. At starting, the motor's torque is directly proportional to its rotor resistance. That is,

$$T_{st} \propto R_2$$

   b. The magnitude of the maximum torque is independent of rotor resistance.

   c. By neglecting stator resistance, the torque (T) at slip (s) is related to the maximum torque (T_m) and its corresponding slip (s_{mt}) by the following equation:

$$\frac{T}{T_m} = \frac{2ss_{mt}}{s_{mt}^2 + s^2}$$

3-3 A 240 V, 3-ϕ, 1140 r/min, three-phase induction motor has the per-phase equivalent circuit shown in Fig. P3-3.

   a. Neglecting the rotational losses and the magnetizing impedance, determine:

     1. The synchronous speed of the stator field.

     2. The per-unit slip.

     3. The line current at starting.

     4. The full-load current, power factor, torque, and efficiency.

     5. The per-unit impedance of the stator and rotor circuits.

     6. The slip and current at maximum torque.

     7. The maximum torque.

   b. Repeat (a3), taking the magnetizing impedance into consideration.

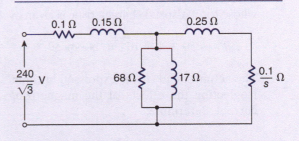

FIG. P3-3

3-4 A 480 V, six-pole, 1140 r/min, three-phase induction motor has the per-phase equivalent circuit shown in Fig. P3-3. Neglecting stator resistance, determine:

   a. The full-load slip.

   b. The slip at maximum torque.

   c. The starting torque in pu.

   d. The maximum torque in pu.

3-5 A 440 V, 60 Hz, 1746 r/min, 60 kW, star-connected, class B, three-phase induction motor yielded the following results when tested:

1. Average value of dc resistance between stator terminals = 0.16 Ω.

2. Locked-rotor test:

   $V_z = 28$ V, line-to-neutral

   $P_z = 2.24$ kW, three-phase

   $I_z = 60$ A

3. No-load test:

   $V_{exc} = 440$ V, line-to-line

   $P_{exc} = 1.6$ kW, three-phase

   $I_{exc} = 12$ A

Determine:

a. The per-phase approximate equivalent circuit.

b. The torque developed at full-load.

3-6 Figures P3-6(a) and (b) show, respectively, the one-line diagram and the per-phase equivalent circuit of a 3-ϕ wound-rotor induction motor. The motor drives a load whose torque-speed characteristic is given by

$$T_L = 50 + 4 \times 10^{-2} n_a \text{ N} \cdot \text{m}$$

The actual speed (n_a) is expressed in r/min. Neglecting the effects of the magnetizing current, determine:

a. The load viscous damping factor (B) in N · m/rad/s.

b. The minimum required additional rotor resistance that must be inserted in each phase in order to start the motor.

c. The operating speed with the external resistance in the rotor circuit reduced to zero.

3-7 A ventilating fan whose torque is proportional to the square of its speed is driven by a squirrel-cage induction motor at nominal conditions.

a. Assuming that the torque requirements of the fan change by ± 10% in comparison to the full-load torque, estimate qualitatively how this will affect (increase or decrease) the speed, efficiency, line current, and power factor of the motor.

b. Repeat (a), assuming that the motor is oversized by 10% above the nominal requirements.

c. Repeat (a), assuming that the motor is undersized by 10% below the nominal requirements.

3-8 The 208 V, 60 Hz, three-phase, four-wire distribution system shown in Fig. P3-8 delivers power to a three-phase induction motor and to single-phase loads. The starting current of the three-phase

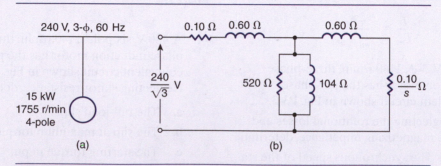

(a)　　　　　　　　　(b)

FIG. P3-6

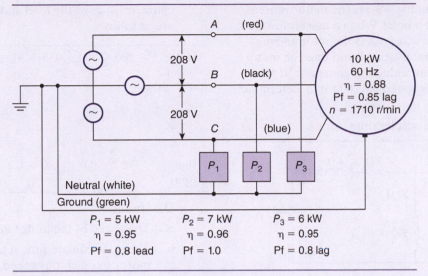

FIG. P3-8

induction motor is four times the full-load value and lags the applied voltage by 84°. Determine the currents through the transmission line and through the neutral wire of the distribution system for each of the following operating conditions:

a. The motor at starting.

b. The motor running at rated condition.

3-9 Referring to Table 3.7.4(a):

a. Determine the maximum permissible size of a capacitor bank that can be used to correct the power factor of a 260 kW motor. Assume that the motor and capacitor bank are switched through the same disconnect device.

b. What is the power factor of the motor and capacitor, as seen from their common disconnect device?

3-10 The three-phase induction motor shown in Fig. P3-10 has a starting torque and current as follows:

$$T_{st} = 1.8 \text{ pu}, \qquad I_{st} = 6\,\underline{/-70°} \text{ pu}$$

The reactances of the cables are as shown. When the 500 kW load draws rated current and the 200 kW motor starts, determine:

a. The torque developed by the motor.

b. The size of the capacitor bank that, when connected in parallel with the motor, will minimize the starting current through the upstream network.

c. The torque developed at minimum starting current.

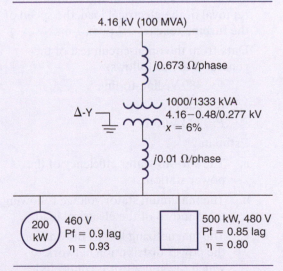

FIG. P3-10

3-11 A 480 V, 3-ϕ, 60 Hz, six-pole induction motor has the per-phase equivalent circuit shown in Fig. P3-11. The motor is used to drive a hoist. When a hoist is lowered, the load accelerates the rotor above synchronous speed, and thus the motor runs as an induction generator. If the resulting speed is 1260 r/min, determine:

a. The line current.

b. The accelerating torque.

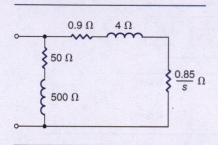

FIG. P3-11

characteristics of the load and the motor are as follows:

Motor torque (N · m)	200	390	455	375	250	0
Load torque (N · m)	0	60	90	120	135	150
Slip (pu)	1	0.6	0.40	0.2	0.1	0

Determine:

a. The speed of the motor at steady state.

b. The approximate time it takes the motor to reach this speed.

c. The approximate time it takes the motor to reach the speed of maximum torque.

3-12 A 3-ϕ, 480 V, three-phase induction generator delivers 375 kW when powered through a hydroturbine. The water's rate of flow and head at the turbine entrance are 4.70 m³/s and 10 m, respectively. The open-circuit time constant of the induction generator is 4 ms, and at the sudden removal of the electrical load, the speed of the turbine reaches 2.2 pu in 1 ms.

Data from the open-circuit test of the generator are as follows:

V: 480 V, line-to-line

I: 332 A

P: 72 kW, 3-ϕ

Estimate:

a. The wire-to-water efficiency of the power station.

b. The maximum stator voltage following the removal of the electrical load.

c. The magnetizing kVAR drawn from the power distribution network to which the generator is connected.

3-13 A three-phase, 75 kW, 600 V, 60 Hz, 1160 r/min induction motor drives a pure-inertia load of 5 kg . m². The torque-slip

3-14 A six-pole, three-phase, star-connected, 60 Hz, 460 V induction motor has a stator leakage impedance of 0.5 + j1.0 ohms per phase, a rotor leakage impedance of 0.6 + j1.2 ohms per phase, and a magnetizing impedance of 4 + j40 ohms per phase.

a. Draw the per-phase equivalent circuit at 60 Hz and at 15 Hz.

b. Estimate the motor torque when operating at 15 Hz and a slip of 10%. Neglect the magnetizing current.

3-15 The motor in Fig. P3-15 drives a pure-inertia load and has a per-phase equivalent circuit as shown. Neglecting stator resistance and the effects of the magnetizing current, determine:

a. The energy stored in the rotating masses while the motor is running unloaded.

b. The time it takes to plug the motor to a complete stop, and the associated energy loss in the rotor windings.

c. The time it takes to reverse the speed of the motor by plugging, and the associated energy loss in the rotor windings.

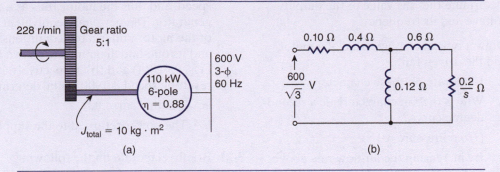

FIG. p3-15

d. The value of the rotor resistance that will furnish the minimum time for (b).

e. The number of maximum reversals per minute. Assume total motor loss to be 1.8 times the rotor copper loss.

3-16 The torque-speed characteristic of a 60 Hz, 3-ϕ squirrel-cage induction motor is controlled by a VFD. Making reasonable assumptions, prove the following:

a.
$$\frac{T_1}{T_2} = \frac{s_1}{K_1 s_2}$$

where T_1 is the torque at a slip s_1 at 60 Hz, and T_2 is the torque at slip s_2 at a frequency of 60 K_1 Hz ($K_1 < 1$).

b. For equal torques, the corresponding winding currents are equal.

3-17 The voltage output of an inverter to an inductive load is as shown in Fig. P3-17.

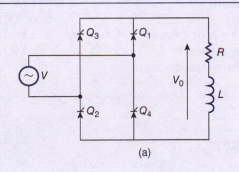

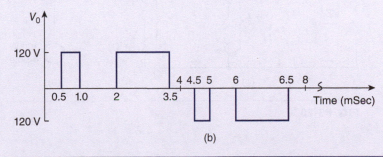

FIG. P3-17

Determine the rms value of the output voltage and its frequency.

3-18 Draw a typical characteristic of a motor and the driven fan.

   a. Elaborate on these characteristics.

   b. Why is the fan characteristic a second-degree curve?

   (*Hint*: Pressure $\alpha\ V^2$)

   c. By increasing the air-flow rate above nominal using a VFD, what are the adverse effects?

3-19 Referring to the data in Table 3-3, compare the present cost of the energy losses of the 50 kW squirrel-cage induction motor for the two commercially available efficiencies. Assume that:

   The motor operates at full-load 16 hours a day, 46 weeks a year, over a five-year period.

   The energy cost is 10¢/kWh and increases by 8% per year.

   The nominal interest rate is 12% and is compounded annually.

3-20 A three-phase induction motor drives a hoist. During lowering, the load accelerates the motor above synchronous speed, and thus the motor runs as a generator. The per-phase equivalent circuit of the motor under rated conditions, and its one-line diagram, are shown in Figs. P3-20(a) and (b), respectively. If the resulting speed is 1900 r/min, determine:

   a. The line current.

   b. The power returned to the supply.

3-21 Briefly elaborate on the following:

   a. On interconnecting an asynchronous generator to a utility's network, the utility's terminal voltage must be slightly smaller than that of the asynchronous generator.

   b. The impedances of the induction motor are smaller than those of the asynchronous generator.

   c. How can the phase sequence of an asynchronous generator be changed?

   d. Under what circumstances can a capacitor furnish the magnetizing field of an asynchronous generator, and what are its adverse effects?

   e. The reactive power of the capacitor (kVAR) must be at maximum equal to the motor's reactive power.

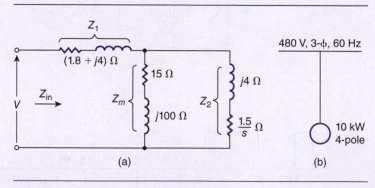

(a)   (b)

FIG. P3-20

4

Single-Phase Motors

4.0   Introduction

4.1   Revolving Fields

4.2   Equivalent Circuit

4.3   Torque Developed

4.4   Methods of Starting

4.5   Magnetic Fields at Starting

4.6   Types of 1-ϕ Motors

4.7   Conclusion

4.8   Problems

What You Will Learn in This Chapter

A   Theoretical Aspects
1   Single-phase motors
2   Revolving fields
3   Equivalent circuit and torque developed
4   Effects of auxiliary winding
5   Start capacitor
6   Magnetic fields at starting

B   Outstanding Practical Highlights
1   One-phase motors
2   Torque developed
3   Effects of auxiliary motor winding
4   How to cancel the noise-producing electromagnetic field
5   Typical motor parameters

C   Additional Students' Aid on the Web
Two-phase motors
1   Balance—Voltage operation
2   Unbalance—Voltage operation
3   Two-phase voltages; Symmetrical components
4   Equivalent circuit and torque developed

4.0 Introduction

Single-phase motors are used mainly in domestic applications for the operation of vacuum cleaners, washing machines, ventilating fans, compressors, and the like. About 90% of all motors manufactured are of the single-phase type. Because of their low efficiency and power factor, their current rating, depending on the kW rating of the motor, may be relatively high.

Single-phase motors have low-power ratings and are generally noisy because the instantaneous power drawn from a single-phase system (the only kind available in an ordinary home) is pulsating in nature (it has a positive and negative value over a complete voltage cycle). In contrast, three-phase motors are less noisy because the instantaneous power drawn from a balanced three-phase system is constant. Nevertheless, more expensive single-phase motors, such as those used for residential room ventilators, incorporate circuits that minimize the negative component of the pulsating power.

The most commonly used types of motors that may be operated from a single-phase supply are as follows:

1. Induction motors, which are, internally, two-phase motors, at least during starting. This class of motors includes the so-called split-phase motor,

the capacitor-start motor, the capacitor-run motor, and the capacitor-start-capacitor-run motor.

2. Shaded-pole motors.

3. Single-phase synchronous motors.

4. Series ac/dc motors.

The theory of operation behind a single-phase motor is based on the fact that a single stator winding produces two spacewise revolving fields* that rotate at synchronous speeds but in opposite directions. The next section discusses the development of these two rotating fields.

4.1 Revolving Fields

In Fig. 4-1, the mmf of the main winding is

$$\mathscr{F}_{main} = \mathscr{F}_m \cos (\omega t - \theta) \tag{4.1}$$

where $\mathscr{F}_m$ is the maximum value of the mmf produced by the main winding, ω is the angular frequency of the stator voltage, and θ is the phase angle of the winding's impedance.

The field, at an angle of β degrees from the axis of the main winding's mmf, is

$$\mathscr{F} = \mathscr{F}_{main} \cos \beta \tag{4.2}$$

From Eqs. (4.1) and (4.2), we obtain

$$\mathscr{F} = \mathscr{F}_m \cos \beta \cos (\omega t - \theta) \tag{4.3}$$

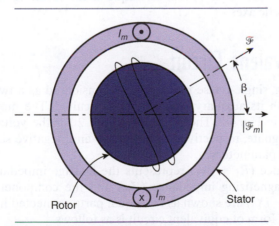

FIG. 4-1 Single-phase induction motor.

---

*The development of two fields in 1-ϕ induction motors further increases their noise in comparison to 3-ϕ induction motors in which just a single field is produced.

From the above, by using the following trigonometric identities,

$$\cos A \cos B = \frac{1}{2} [\cos (A + B) + \cos (A - B)] \qquad (4.4)$$

we obtain (after rearranging and simplifying),

$$\mathscr{F} = \frac{\mathscr{F}_m}{2} [\cos (\omega t + \beta - \theta) + \cos (\beta - \omega t + \theta)] \qquad (4.5)$$

Thus, a single-phase motor produces two *spacewise* oscillatory fields that rotate in opposite directions. The angular speed of the stator fields is designated with ω_s and is referred to as synchronous speed ($\omega_s = \omega$).

4.1.1 Rotor Stationary

When the rotor is stationary, the two revolving fields have equal amplitudes and, relative to the rotor, rotate at the same speed but in opposite directions. As a result, no effective field is induced in the rotor windings, and thus the motor does not have a starting-torque capability. For this reason, some domestic fan motors will not start without an external "kicking." This kicking increases the torque in the direction of the external disturbance, and so the equilibrium of the mmf's is not maintained. As a result, the motor starts to rotate.

4.1.2 Rotor Turning

When the rotor is turning, the relative angular velocities of the positive and negative fields, with respect to the rotor, are $s\omega_s$ and $(2 - s)\omega_s$, respectively. S is the slip of the motor. The two fields induce unequal voltages in the rotor winding. This condition leads to the development of torque as the resulting mmf's attempt to align their magnetic axes.

4.2 Equivalent Circuit

A single-winding, single-phase motor can be considered as a two-winding, two-phase motor with its auxiliary winding open-circuited. The motor's equivalent circuit is shown in Fig. 4-2. The subscripts 1 and 2 for the voltage, current and impedances designate, respectively, the positive and negative sequence components of the said parameters.

The impedance ($R_1 + jX_1$) represents the leakage impedance of the main winding. The magnetizing impedance (Z_m) and the components of the rotor impedance ($R_2 + jX_2$) are shown as two equal parts connected in series. The justification for this form of equivalent circuit is as follows.

The positive- and negative-sequence components of the currents through the main winding are

$$I_{m_f} = I_{m_b} = I_m \qquad (4.6)$$

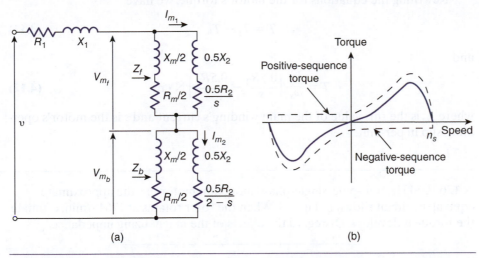

FIG. 4-2 Single-phase motor: **(a)** equivalent circuit, **(b)** torque-speed characteristic.

Disregarding the magnetizing impedance, the rotor's impedances are given by

$$Z_1 = 0.5X_2 + \frac{0.5R_2}{s} \tag{4.7}$$

$$Z_2 = 0.5X_2 + \frac{0.5R_2}{s - 2} \tag{4.8}$$

where s is the motor's operating slip. At standstill, the positive-sequence impedance is equal to the negative-sequence impedance.

The positive- and negative-sequence components of the voltage across the rotor impedances are

$$V_{m_1} = I_{m_1}Z_1 \tag{4.9}$$

$$V_{m_2} = I_{m_2}Z_2 \tag{4.10}$$

At standstill, no voltage is induced in the auxiliary winding. However, when the rotor is turning, Z_1 is not equal to Z_2, and thus a voltage is induced in the auxiliary winding.

For an accurate analysis, the variation of the rotor's parameters with the operating slip should be considered.

4.3 Torque Developed

The net torque of the motor, as well as its positive- and negative-sequence components, are shown in Fig. 4-2(b).

Rewriting the equations for the motor's torque, we have

$$T = T_f - T_b \tag{4.11}$$

and

$$T = \frac{I_m^2}{\omega_s}\left(\frac{0.5R_2}{s} - \frac{0.5R_2}{s-2}\right) \text{N} \cdot \text{m} \tag{4.12}$$

where I_m is the rms value of the main winding's current and s is the motor's operating slip in per-unit.

EXAMPLE **4-1**

A 120 V, 60 Hz, four-pole, single-phase induction motor has the approximate equivalent circuit shown in Fig. 4-3. When the rotor rotates at 1764 r/min, estimate the torque it develops. Disregard the effects of the magnetizing impedance.

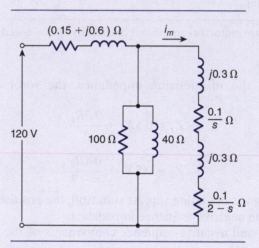

FIG. 4-3

SOLUTION

At 1764 r/min, the slip is

$$s = \frac{1800 - 1764}{1800} = 0.02 \text{ pu}$$

The impedance of the circuit is

$$Z = 0.15 + \frac{0.1}{s} + \frac{0.1}{2-s} + j1.2 \ \Omega$$

Substituting the operating slip into the above, we obtain

$$Z = 0.15 + \frac{0.1}{0.02} + \frac{0.1}{2-0.02} + j1.2$$

$$= 5.34 \ \underline{/13°} \ \Omega$$

The current is

$$I = \frac{V}{Z} = \frac{120\underline{/0^\circ}}{5.34\underline{/13^\circ}}$$

$$= 22.48 \underline{/-13^\circ} \text{ A}$$

The torque is calculated by substituting the known parameters into Eq. (4.12), as follows:

$$T = \frac{22.48^2}{1800 \frac{2\pi}{60}} \left(\frac{0.1}{0.02} - \frac{0.1}{2 - 0.02} \right)$$

$$= 13.41 - 0.135 = \underline{13.27 \text{ N} \cdot \text{m}}$$

Exercise

4-1

For the motor of Example 4-1, estimate:

a. The voltage induced in the rotor windings and its positive- and negative-sequence components, all as seen from the stator.

b. The motor's copper losses.

Answer (a) $V_f = 56.31\underline{/-9.6^\circ}$ V, $V_b = 3.42\underline{/67.5^\circ}$ V, $V = 57.18\underline{/-6.2^\circ}$ V
(b) 176.9 W

4.4 Methods of Starting

A single-phase induction motor can be started in several ways. In addition to the main winding, the motor's stator includes a starting, or auxiliary, winding. The motor's cost and its method of starting depend on the configuration of this auxiliary winding. The starting winding is always connected in parallel with the main winding. Because it has higher impedance than the main winding, the starting winding draws less current. The motor's starting torque depends on the impedance of its windings at standstill and on the effective turns ratio between the stator and rotor windings. Motors equipped with a starting winding are often referred to as "split-phase induction motors."

The various types of split-phase induction motors are described briefly in the next section.

4.4.1 Auxiliary Winding, Permanently Connected

A split-phase motor with a permanently connected auxiliary winding is shown in Fig. 4-4(a). The auxiliary winding may—or may not—include a capacitor. The capacitor, depending on its relative size, will further increase the starting

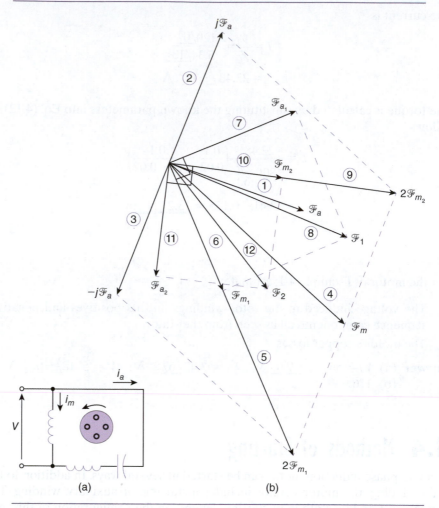

FIG. 4-4 Single-phase motor: **(a)** auxiliary winding, permanently connected, **(b)** phasor diagram.

and full-load torques of the motor. A motor with a permanently connected auxiliary winding does not need any maintenance and is used where a low starting torque is required, such as in the operation of a small ventilating fan. (The torque-speed characteristic of this motor is shown later in this chapter in Fig. 4-7.)

Phasor Diagram

The phasor diagram of a split-phase motor with a permanently connected auxiliary winding is shown in Fig. 4-4(b). The phasor diagram is constructed by following the sequence identified by the encircled numbers.

Phasors (1), (2), and (3)

The mmf of the auxiliary winding ($\mathcal{F}_a$) is arbitrarily drawn as shown. When this phasor is rotated by plus and minus 90°, it gives the phasors ($j\mathcal{F}_a$) and ($-j\mathcal{F}_a$), respectively.

Phasors (4), (5), and (6)

The mmf of the main winding ($\mathcal{F}_m$), relative to that of the auxiliary winding, is larger and more inductive. Its phasor is drawn as shown.

Phasors (7) and (8)

The positive-sequence components of the main and auxiliary windings are equal to each other in both magnitude and phase quadrature. The algebraic summation of the positive-sequence components yields the motor's net positive-sequence component ($\mathcal{F}_1$).

Phasors (9) and (10)

The algebraic sum of the main winding's mmf and the phasor $j\mathcal{F}_a$ yields the negative-sequence component of the main winding's mmf.

Phasors (11) and (12)

The negative-sequence component of the auxiliary winding's mmf lags the corresponding one of the main winding by 90°. The algebraic sum of the negative-sequence components of the winding's mmf gives the motor's net negative-sequence component ($\mathcal{F}_2$).

4.4.2 Auxiliary Winding with Capacitor Start

The starting torque of the motor increases substantially when a properly sized capacitor is connected in series with the auxiliary winding as shown in Fig. 4-5(a). The auxiliary winding is normally disconnected from the circuit by a centrifugal switch when the motor reaches about 75% of its rated speed. When the shaft of the

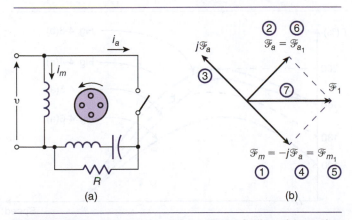

FIG. 4-5 Single-phase motor: **(a)** auxiliary winding with starting capacitor, **(b)** phasor diagram—ideal case.

motor is not accessible, as in hermetically enclosed compressors, then the contact of a voltage or current relay is used to disconnect the auxiliary winding.

Phasor Diagram

Figure 4-5(b) shows the phasor diagram of a capacitor-start motor in an ideal case. Here, the mmf's are arbitrarily assumed to be equal in magnitude and in space quadrature to each other.

4.4.3 Auxiliary Winding, Removed After Starting

Figure 4-6(a) presents a split-phase induction motor with an auxiliary winding that is removed after starting. The majority of motors, such as those used in ventilating fans and clothes washers/dryers, are of this type. The auxiliary winding can be removed through a centrifugal switch or through the switching action of a relay. The auxiliary winding should be connected in parallel to a properly sized resistor that will dissipate the coils' stored energy following the current interruption. As can be seen in Fig. 4-7,

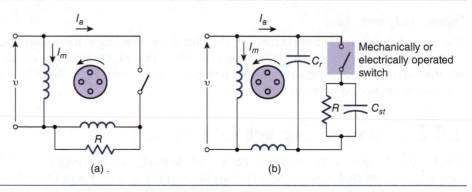

FIG. 4-6 Single-phase motor: **(a)** auxiliary winding, removed after starting, **(b)** with starting and running capacitors.

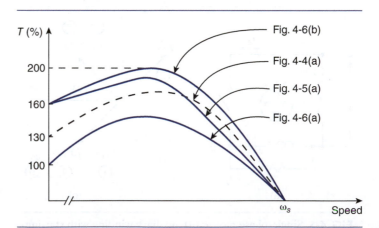

FIG. 4-7 Torque-speed characteristic of split-phase induction motors.

the starting and running torques of this motor are smaller than those of a motor with a permanently connected auxiliary winding that includes a capacitor.

4.4.4 Auxiliary Winding, with a Starting and a Running Capacitor

The highest starting and running torque of a single-phase induction motor is obtained when its split-phase winding is equipped, as shown in Fig. 4-6(b), with a starting and a running capacitor. For a 3 kW compressor motor, typical values for a starting and a running capacitor are, respectively, 200 μF and 50 μF.

4.5 Magnetic Fields at Starting

As mentioned previously, at standstill, the currents through the main and auxiliary windings produce two magnetic fields that rotate synchronously in opposite directions. Designating the mmf's of the main and auxiliary windings $N_m I_m$ and $N_a I_a$, we have

$$\begin{bmatrix} \mathscr{F}_{m_1} \\ \mathscr{F}_{m_2} \end{bmatrix} = \frac{1}{2} \begin{bmatrix} 1 & -j \\ 1 & j \end{bmatrix} \begin{bmatrix} N_m I_m \\ N_a I_a \end{bmatrix} \tag{4.13}$$

where $\mathscr{F}_{m_1}$ and $\mathscr{F}_{m_2}$ represent, respectively, the positive- and the negative-sequence components of the motor's mmf, and I_m and I_a are the rms values of the current through the main and auxiliary windings, respectively.

From the matrix equation, we obtain

$$\mathscr{F}_{m_1} = \frac{1}{2}(N_m I_m - jN_a I_a) \text{ ampere-turns} \tag{4.14}$$

and

$$\mathscr{F}_{m_2} = \frac{1}{2}(N_m I_m + jN_a I_a) \text{ ampere-turns} \tag{4.15}$$

From Eq. (4.15), the negative field is zero when the following condition is satisfied.

$$N_m I_m = -jN_a I_a \tag{4.16}$$

Replacing the currents by their corresponding voltage/impedance ratios (see Fig. 4-5(a) and assume R = ∞), we obtain

$$N_m \frac{V}{R_m + jX_m} = -jN_a \frac{V}{R_a + jX} \tag{4.17}$$

where V is the supply voltage and X is equal to the reactance of the auxiliary winding (X_a) minus the reactance of the capacitor (X_c). That is,

$$X = X_a - X_c \tag{4.18}$$

From Eq. (4.17), we have

$$N_m R_a + jN_m X = N_a X_m - jN_a R_m \tag{4.19}$$

Equating the real parts, we obtain

$$R_a = \frac{N_a}{N_m} X_m \tag{4.20}$$

Equating the imaginary parts, we obtain

$$X = -\frac{N_a}{N_m} R_m \tag{4.21}$$

or

$$X_a - X_c = \frac{N_a}{N_m} R_m \tag{4.22}$$

From the above,

$$X_c = \frac{N_a}{N_m} R_m + X_a \tag{4.23}$$

Equations (4.20) and (4.23) can be used to design the stator and rotor windings in a way that optimizes the starting torque of the motor. In other words, when these equations are satisfied, a single rotating field is developed at starting.

EXAMPLE **4-2**

A 120 V, 60 Hz, single-phase induction motor of the capacitor-start type has a main winding with 180 effective turns and an auxiliary starting winding with 250 effective turns. With the rotor stationary, the input impedance of the main winding is $5 + j10$ ohms and that of the auxiliary winding is $13.89 + j15.50$ ohms. Determine at standstill:

a. The magnitude of the forward- and backward-rotating fields.

b. The size of the starting capacitor necessary to yield a single rotating magnetic field at starting.

SOLUTION

a. A schematic of the single-phase induction motor is shown in Fig. 4-8. The forward- and backward-rotating magnetic fields of the main winding are obtained from Eqs. (4.14) and (4.15), respectively.

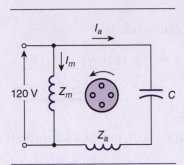

FIG. 4-8

$$\mathscr{F}_{m_1} = \frac{1}{2}\left(N_m I_m - j N_a I_a\right)$$

$$= \frac{1}{2}(120)\left(\frac{180}{5 + j10} - j\frac{250}{13.89 + j15.50}\right)$$

$$= 1349.10\,\underline{/-94.5^\circ}\ \text{A rms}$$

$$\mathscr{F}_{m_2} = \frac{1}{2}\left(N_m I_m + j N_a I_a\right)$$

$$= \frac{1}{2}(120)\left(\frac{180}{5 + j10} + j\frac{250}{13.89 + j15.50}\right)$$

$$= 1041.7\,\underline{/-21.6^\circ}\ \text{A rms}$$

The positive-sequence component of the auxiliary winding's mmf is

$$\mathscr{F}_{a_1} = j\mathscr{F}_{m_1} = 1349.10\,\underline{/-4.5^\circ}\ \text{A rms}$$

and the negative-sequence component is

$$\mathscr{F}_{a_2} = -j\mathscr{F}_{m_2} = 1041.7\,\underline{/-111.6^\circ}\ \text{A rms}$$

Thus, the net forward-rotating field is

$$\mathcal{F}_1 = \mathcal{F}_{a_1} + j\mathcal{F}_{m_1} = 1{,}349.1(1 \underline{/-4.5°} + 1 \underline{/-4.5°})$$

$$= 2{,}698.2 \underline{/-4.5°} \text{ A rms}$$

The net backward-rotating field is

$$\mathcal{F}_2 = \mathcal{F}_{a_2} + j\mathcal{F}_{m_2} = 1041.7(1 \underline{/-111.6°} + 1 \underline{/-21.6°})$$

$$= \underline{\underline{90.9 \underline{/155.9°} \text{ A rms}}}$$

b. To have a single rotating field, the backward-rotating mmf must be equal to zero. That is,

$$\frac{1}{2}(120)\left(\frac{180}{5 + j10} + j\frac{250}{Z_a}\right) = 0$$

The impedance of the auxiliary winding is

$$Z_a = 13.89 + j(15.5 - X_c)$$

where X_c is the reactance of the capacitor. From the last two relationships,

$$Z_a = -j\left(\frac{250(5 + j10)}{180}\right) = 13.89 + j(15.5 - X_c)$$

Solving for X_c, we obtain

$$X_c = 22.44 \ \Omega$$

Thus,

$$C = \frac{1}{\omega X_c} = \frac{1}{2\pi(60)(22.44)}$$

$$= \underline{\underline{118.2 \ \mu\text{F}}}$$

Alternatively, using Eq. (4.23),

$$X_c = X_a + \frac{N_a}{N_m} R_m = 15.5 + \frac{250}{180}(5) = 22.44 \ \Omega$$

This reactance will give the same capacitance as that obtained previously.

Exercise

4-2

a. A 120 V, 60 Hz, four-pole, single-phase induction motor operates at 1764 r/min. Determine the angular velocity of the rotating fields with respect to the rotor.

b. The net positive- and negative-sequence mmf's of a single-phase motor are as follows:

$$\mathscr{F}_1 = 100 \underline{/-30°} \text{ A rms}$$

$$\mathscr{F}_2 = 80 \underline{/-60°} \text{ A rms}$$

Find the mmf's of the main and auxiliary windings.

Answer (a) 36 r/min cw, 3564 r/min ccw

(b) 64.8 $\underline{/-34.1°}$ A rms; 110.5 $\underline{/-48.7°}$ A rms

Refer to Fig. 4-6(b). Explain why the motor must not be restarted within a time interval that is substantially less than four times the time constant (RC) of the capacitor circuit.

Exercise

4-3

4.6 Types of I-ϕ Motors

4.6.1 Split-phase

The currents in the two stator windings of a split-phase motor are displaced from each other phasewise by either (1) making the resistance of one winding higher than that of the other, or (2) connecting a capacitor *in series* with one of the windings. The resulting magnetic field may be resolved into two oppositely revolving components, one larger than the other, thereby producing a net forward torque. To avoid overheating, the higher-resistance winding of the split-phase motor, or the winding with the capacitor in a capacitor-start motor, is disconnected when running speed is approached by either a relay's contact or a centrifugal switch. Once the rotor is turning, excitation of a single stator winding will yield forward torque. In the same class, with regard to operation while running, is the hand-started induction motor. This motor has only a single stator winding, but it will run in whichever direction it is started.

4.6.2 Shaded-Pole Motors

A shaded-pole, single-phase induction motor has a squirrel-cage rotor and a salient-pole stator (see Fig. 4-9). Wound on the stator are the main multi-turn winding and the auxiliary single-turn winding. (Note that this differs from the distributed windings of split-phase motors.) The auxiliary winding is a short-circuited copper band commonly referred to as a shaded-pole winding.

When the main winding is connected to the voltage supply, the flux through the shaded part of the pole lags the flux through the unshaded part because of the opposition to the flux generated by the induced currents in the auxiliary winding. As a result, the stator flux has two components, one of which lags the other. Each of these fluxes completes its loop through the stator poles, air gaps, and rotor.

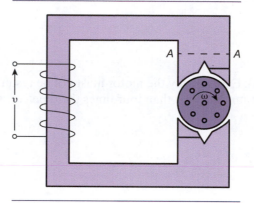

FIG. 4-9 Shaded-pole motor.

4.6.3 Revolving Field

The flux through the stator at section A-A (ϕ_A) is the vector sum of the fluxes through the shaded (ϕ_{sh}) and unshaded (ϕ_{un}) parts of the pole. Thus,

$$\phi_A = \phi_{un} + \phi_{sh} \qquad (4.24)$$

or

$$\phi_A = \phi_{un} \, (1\underline{/0°} + K\underline{/-\theta}) \qquad (4.25)$$

where: $K = \dfrac{\phi_{sh}}{\phi_{un}}$ and θ is the phase angle by which the flux through the shaded part of the pole lags the flux through the unshaded part of the pole. The parameters ϕ_{un}, K, and θ are variables, and, at a particular point in the voltage waveform, have a unique value. As a result, the stator flux has two rectangular components that change as a function of time, which leads to the development of a kind of spacewise revolving field. The rotor is subjected to two magnetizing forces whose vector sum induces a rotor current and a corresponding rotor flux. The interaction of stator and rotor fluxes produces torque and rotation whose direction is from the unshaded to the shaded part of the pole.

Shaded-pole motors have a low power factor (0.25 to 0.50) and efficiency (0.25 to 0.40), but are very economical to manufacture. Their power rating is in the range of 1 mW to 200 W. Almost all motors rated at less than 50 W are of the shaded-pole type. Their starting torque is 0.30 to 0.85 pu, and their maximum torque is 1.2 to 1.5 pu.

Usually, shaded motors have two, four, or six poles. When supplied with integral gear systems, they can drive loads at speeds that are slower than one revolution per day. Shaded-pole motors are used in photocopy machines, humidifiers, fans, toys, phonograph turntables, and many other products.

4.6.4 Stepper Motors

Stepper motors rotate in increments or steps that are determined by rotor construction and the type of stator excitation. The stator winding is often a multiphase winding (sometimes up to four phases are used), and the excitation voltage is a series of current pulses (digital signal). The higher the number of stator phases, the smaller the increment of angular rotation per input pulse.

The rotor may be one of two types: permanent magnet or reluctance. A four-pole reluctance motor is shown in Fig. 4-10(a). The shaft of this kind of motor has one salient rotor configuration for each stator phase. That is, for a four-phase stator winding, the rotor will have along its axis four distinct pole configurations.

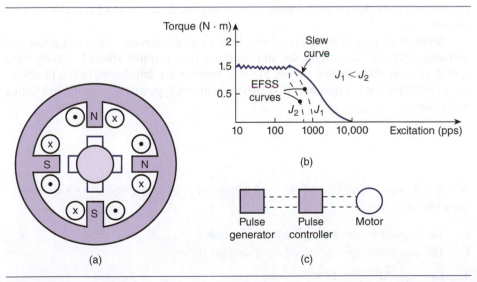

FIG. 4-10 Stepper motors: **(a)** a four-pole reluctance motor, **(b)** torque: excitation characteristics, and **(c)** control configuration.

When one of the stator windings is excited, the rotor will rotate in such a way that the path of least reluctance is established for the stator flux.

Reluctance motors can operate with a minimum rotation of 1.8°/pulse; that is, to effect a complete revolution, 200 pulses are required. The characteristics of stepper motors [see Fig. 4-10(b)] are expressed in terms of the torque produced as a function of excitation pulses per second (pps).

The torque of the motor is constant up to a specific rate of excitation pulses. At a further increase in the rate of excitation pulses, the torque is decreased either along the solid curve or along the dotted line characteristic.

In Fig. 4-10(b), the dotted lines represent the motor operation without controlled acceleration and deceleration. This operation is referred to as an error-free-start-stop (EFSS) operation. In this mode of operation, the higher the effective moment of inertia, the lower the motor's output torque. The solid line characteristic represents a motor operation with controlled acceleration and deceleration. The solid line characteristic is commonly known as the slew curve.

Permanent magnet stepping motors develop higher torque than reluctance motors and rotate in larger steps per unit pulse excitation.

4.6.5 Control

Since the position of the rotor in a stepper motor depends on the number of pulses it receives, one can easily control the rotation of the shaft by controlling the number of pulses.

Figure 4-10(c) shows the control configuration of a stepper motor. It is an open-loop control where the generated pulses are applied to the motor's windings at a controlled rate through a programmable digital controller. The characteristics of a simple open-loop control system are compatible to those of the closed-loop design.

Stepper motors at steady state accelerate and decelerate with each pulse and consequently are not useful for applications that require smooth steady-state speed, such as phonograph drives. Stepper motors are employed in the positioning of machine tools, computer disk drives, printers, typewriters, sewing machines, and others.

Exercise

4-4

A 5 A, 1.8° pulse reluctance-type stepper motor is to rotate 36° in 0.02 second. Determine:

a. The required number of excitation pulses.

b. The amplitude and width of each current pulse.

Answer (a) 20; (b) 10 A, 0.50 ms

4.6.6 Series AC/DC Motors

Small single-phase motors that can operate with either dc or ac input voltage are called universal or series ac/dc motors. These motors are normally rated at less than 1 kW, and their speed is usually between 2000 and 12,000 r/min. They are generally used in hand tools and household appliances. They operate much like dc series motors. The dc series motors are discussed extensively in Chapter 6; the student is advised to read Section 6.1.7 before attempting to understand the operation and analysis of series ac/dc motors.

The equivalent circuit of series ac/dc motors and its torque-speed characteristics are shown in Figs. 4-11(a) and (b), respectively. The field current (I_f) is equal to the armature current (I_a) because the field and armature are connected in series. The voltage generated in the armature winding, V_g, is given by

$$V_g = K\phi\omega \qquad (4.26)$$

where K is a constant of the motor and ϕ is its effective field. The torque developed by the motor is proportional to the effective flux and to the current drawn from the line. Mathematically,

$$T = K\phi I_a \qquad (4.27)$$

The constant of proportionality is given by Eq. (6.14).

At a specified speed, the motor develops less torque for an ac operation than it does for a dc operation. This situation is due to magnetic losses and to the fact that, for ac operations, the armature current is reduced because of the inductive reactance of the motor's winding.

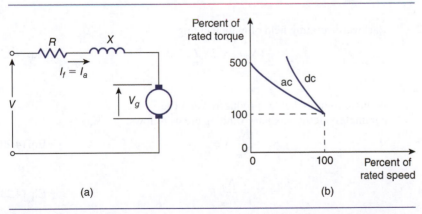

FIG. 4-11 Universal motor: **(a)** equivalent circuit, **(b)** typical torque-speed characteristics.

4.7 Conclusion

Single-phase induction motors are used extensively in hand tools and in almost all home appliances. The power drawn from a single-phase voltage source produces two spacewise rotating electromagnetic fields. When the rotor is *stationary*, these two fields are equal in magnitude and opposite in direction. Consequently, single-phase motors have no starting-torque capability. However, by specially designing the stator windings (split-phase motors), one of the two fields can be reduced, and the resultant field can develop the desired starting torque. It is theoretically possible to design the stator windings in a way that eliminates one of the rotating fields. This would result in maximum torque at starting. When the rotor is *rotating*, one of the revolving fields is greatly reduced. As a result, the motor develops nominal torque without any auxiliary circuit components.

The different nature of the fields at starting and running leads to the development of two equivalent circuits from which the motor characteristics can be derived.

The speed of a single-phase motor can be controlled smoothly through a solid-state frequency converter or, in steps, through a voltage-reducing device (autotransformer, inductor, resistor, etc.).

Table 4-1 summarizes some of the essential mathematical relationships that govern the operation of single-phase and two-phase motors. In addition, manu-

TABLE 4-1 Summary of important equations

Item	Description	Remarks
1	Forward-rotating field of main winding $$\mathscr{F}_{m_1} = \frac{1}{2}(N_m I_m - jN_a I_a)$$	Eq. (4.14)
2	Backward-rotating field of main winding $$\mathscr{F}_{m_2} = \frac{1}{2}(N_m I_m + jN_a I_a)$$	Eq. (4.15)
3	A single rotating field is produced when the parameters of the auxiliary winding are as follows: $$R_a = \frac{N_a}{N_m} X_m$$	Eq. (4.20)
4	$$X_c = X_a + \frac{N_a}{N_m} R_m$$	Eq. (4.23)

facturer's data for a single-phase motor used in household clothes washers are presented in Table 4-2.

TABLE 4-2 Data for a split-phase, 120 V, 250 W, 1800 r/min, single-phase induction motor used in household clothes washers

Parameter	Winding	
	Main	Auxiliary
Resistance in ohms	1.26	3.03
Reactance in ohms	Running condition	Locked-rotor condition
Positive sequence (X_1)	1.58	0.69
Negative sequence (X_2)	1.12	0.52
Zero sequence (X_0)	29.7	16.4
Effective turns ratio	50	65

4.8 Problems

4-1 Figure P4-1 shows the equivalent circuit of a 60 Hz, 120 V, two-pole single-phase induction motor. When the motor is running at a slip of 4%, neglecting the magnetizing current, determine:

  a. The input current and power factor.

  b. The forward torque.

  c. The backward torque.

  d. The torque developed.

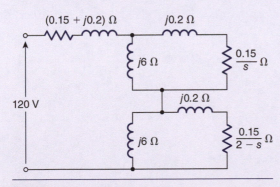

FIG. P4-1

4-2 A 120 V, 60 Hz, single-phase induction motor of the capacitor-start type has the following characteristics:

Parameter	Winding	
	Main	Auxiliary
Effective turns	200	250
Input impedance at standstill in ohms	5 + j10	12.5 + j16

Determine:

  a. The amplitudes of the forward- and backward-rotating fields.

  b. The size of the capacitor bank that, when connected in series with the auxiliary winding, will produce at starting only a forward-rotating field.

4-3 For the shaded-pole motor shown in Fig. P4-3, prove the following:

a. When one of the switches (A or B) is closed, the direction of the rotation is from the unshaded part of the pole toward the shaded part of the pole.

b. The flux through the unshaded part of the pole is maximum at the instant the supply voltage is maximum.

c. The speed of the motor is maximum when the supply voltage is applied across the minimum number of winding turns.

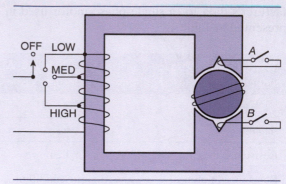

FIG. P4-3

4-4 Draw the torque-speed characteristics and explain the principle of operation of:

a. Split-phase motors.

b. Shaded-pole motors.

c. Stepper motors.

5

Synchronous Machines

5.0   Three-Phase Synchronous Machines

5.1   Three-Phase Cylindrical Rotor Machines: Motors

5.2   Three-Phase Cylindrical Rotor Machines: Generators

5.3   Salient-Pole Synchronous Machines

5.4   Conclusion

5.5   Review of Important Mathematical Relationships

5.6   Manufacturer's Data

5.7   Review Questions

5.8   Problems

What You Will Learn in This Chapter

A. Theoretical Aspects

1 Principle of operation of synchronous machine
2 Equivalent circuit
3 Phasor diagram
4 Power and torque produced
5 Effects of field current on motor characteristics
6 Three-phase cylindrical rotor generator
7 Generator characteristics
8 Measurement of their parameters
9 Three-phase salient-rotor machines
10 Phasor diagrams, torque developed

B. Outstanding Practical Highlights

1 Vee curves
2 How, by adjusting the field current, one can control a distribution system's power factor and reactive power
3 How to draw a step-by-step phasor diagram
4 Generator's voltage regulation
5 Manufacturer's motor-generator parameters
6 Operating speed and torque developed
7 Why the inductive loads in a synchronous generator weaken its field

C. Additional Student Aids on the Web

1 Three-phase faults and selection of breakers
2 Principle of constant-flux linkage
3 Stator currents
4 Field current

5.0 Three-Phase Synchronous Machines

Three-phase synchronous motors are used to drive loads that operate at constant speed. Such loads include, for example, the grinding mills of the mining industry and the rolling mills of the paper and textile industries.

Synchronous motors operating at 1800 r/min with ratings larger than 1000 kW have efficiencies that are as much as 4% higher than those of equivalent three-phase squirrel-cage induction motors.

Synchronous generators produce a balanced three-phase voltage supply at their output terminals by transforming the mechanical energy transmitted to their rotors into electrical energy. The source of the mechanical power commonly known as the prime mover may be wind, hydropower, or steam.

Synchronous generators are used mainly by utility companies to generate their standard, three-phase, 60 Hz power. In fact, most ac power is generated through synchronous generators, the ratings of which sometimes exceed 1000 MVA. In consumers' plants, smaller machines (up to 1 MVA) provide standby emergency power to critical loads in the event the utility's power system fails. Remote locations use synchronous generators with ratings from 150 watts to several kilowatts.

Increasingly, small synchronous generators—down to 20 kW—are being installed where a source of power can produce electricity economically. Such power may be supplied, for example, by small hydroelectric plants, or they may derive from the excess steam from chemical processes, the heat produced by factories burning dried cotton, and the like. Most automobiles are also equipped with three-phase generators.

Synchronous machines are classified as either *cylindrical rotor* or *salient rotor*. (Machines of both types will be discussed in the sections that follow. The three-phase short circuits of the machines are discussed in the Web section.)

The control schematics and the modern electronic control of synchronous machines are explained in Chapter 7.

5.1 Three-Phase Cylindrical Rotor Machines: Motors

Three-phase synchronous motors are externally controlled, variable-power-factor machines. They run at synchronous speeds, and, as shown in Fig. 5-1(a), they require two sources of excitation:

1. A balanced three-phase ac source for the stator windings.
2. A dc source for the rotor windings.*

The structure and winding arrangement of the stator of a synchronous motor is identical to that of a three-phase induction motor. The rotor, however, is of the wound type and may be of either cylindrical or salient construction.

The dc current for the rotor windings may be obtained either from an external dc source through the slip rings of the rotor (classical machines) or from the rectified ac voltage induced in a special set of rotor windings (modern or brushless machines). In the second case, in addition to the dc *field winding,* the rotor carries a three-phase *ac winding* and a set of *solid-state devices* (thyristors and diodes). These solid-state devices are located between the ac and dc windings. When the rotor rotates, the magnetomotive force of an external dc supply induces a voltage in the ac winding.

Solid-state devices rectify the induced voltage and control the timing of the application of the rectified voltage to the dc windings. (A detailed description of the dc voltage source for the field windings of the motor is given in Section 7.3.2.)

Brushless synchronous motors also include those synchronous machines that obtain their field excitation from an auxiliary generator, commonly known as an exciter. The exciter often is driven by the same prime mover as the synchronous machine.

---

*Permanent-magnet rotors may be used in small machines.

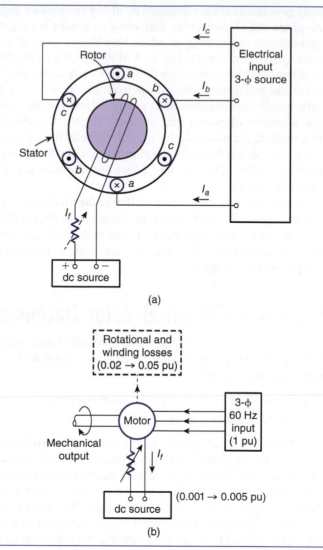

FIG. 5-1 Synchronous motor: **(a)** elementary representation of stator and rotor windings–cylindrical rotor, **(b)** direction of power flow and typical losses for a 100 kW machine.

The motor's mechanical output power comes from transformation of the input ac power. Although the dc supply is essential for operation of the motor, it transfers no power to the output. Figure 5-1(b) shows typical values of the input power and losses for a 100 kW motor.

The electrical input to the rotor windings of large synchronous machines is less than 1% of the total power. In efficiency calculations, therefore, it may be neglected. The field current essentially controls the reactive volt-ampere requirements of the stator circuits and has a pronounced effect on the motor's output torque.

In synchronous machines, the magnetic flux of the armature may aid or oppose the magnetic flux of the field current. This interaction, termed *armature reaction*, depending on its effects, is also called the magnetizing or demagnetizing effect of the armature current. Determining the effects of the armature current is essential in analyzing and understanding synchronous machines.

The main advantage of synchronous motors is that their power factor (cos θ), input line current (I_L), and input reactive power (Q) can be easily controlled by properly adjusting the field current. In other words, synchronous motors, as seen from the network terminals, behave as either variable-shunt capacitors or variable-shunt inductors. This topic is discussed in detail in Section 5.1.9.

Synchronous motors provide variable torque for loads that require a constant speed. For ordinary applications, however, they are neither economical nor practical because they operate at only one speed and require an ac source for stator excitation and a dc current for their rotor windings. Besides this drawback, they have another crucial disadvantage: Sudden or cyclic variations in the mechanical load torque may drive the motor out of synchronism. This problem may cause disturbances and lead both the motor and the supplying electrical system to shut down.

A major engineering problem related to the steady-state operation of synchronous motors involves calculating the *field current* that corresponds to a specific armature current and power factor.

The following sections discuss the cylindrical-rotor motor's principle of operation, equivalent circuit, governing equations, phasor diagrams, and external characteristics, commonly known as Vee curves. The photographs shown in Fig. 5-2 to Fig. 5-5 show some physical characteristics of synchronous motors.

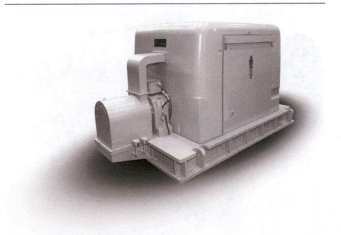

FIG. 5-2 A four-pole, 15 MW, 60 Hz synchronous motor.
Courtesy of General Electric

FIG. 5-3 Stator of a 2.6 MW synchronous motor. *Courtesy of General Electric*

FIG. 5-4 Rotor of a 2.6 MW synchronous motor. *Courtesy of General Electric*

FIG. 5-5 Salient-rotor synchronous motor: Continuous-end rings. *Courtesy of General Electric*

5.1.1 Rotor

An elementary representation of a synchronous motor with a cylindrical rotor is shown in Fig. 5-1(a). In general, motors with rated speeds of 1800 or 3600 r/min have cylindrical rotors, whereas synchronous motors with lower rated speeds are designed, for economical reasons, with salient rotors.

Cylindrical-rotor machines are relatively simple to analyze; they simply cease to operate when their field current is removed. Salient-pole machines, however, are more difficult to analyze. Without any excitation, they may develop a torque that may be up to 40% of the rated torque.

5.1.2 Rotating Fields

Synchronous machines have two synchronously rotating magnetic fields:

1. The dc field of the rotating rotor.

2. The stator or armature field.

The stator field is identical to that produced by the armature of three-phase induction motors. For convenience, its description is repeated here.

Refer to Fig. 5-6(a). The magnitude and speed of rotation of the stator field are given by

$$\mathscr{F}_s = \frac{3}{2}\mathscr{F}_1 \cos{(\omega t - \beta)} \text{ ampere-turns} \tag{5.1}$$

$$n_s = 120\frac{f}{p} \text{ r/min} \tag{5.2}$$

where:

$\mathscr{F}_s$ = is the rotating stator mmf produced by the three-phase armature currents

$\mathscr{F}_1$ = is the maximum value of the mmf produced by one phase only. For balanced armature currents and an equal number of turns per winding, we have

$$\mathscr{F}_1 = N_a I_{am} = N_b I_{bm} = N_c I_{cm} \text{ ampere-turns} \tag{5.3}$$

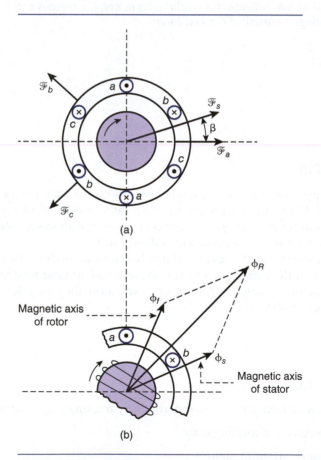

(a)

(b)

FIG. 5-6 Magnetic fields in a synchronous motor: **(a)** stator mmf's, **(b)** stator, rotor, and resultant magnetic flux.

β = an arbitrary angle measured with respect to the horizontal line, as shown in Fig. 5-6(a).

ω = the angular speed of the stator currents

f = the frequency of the stator currents

η_s = the synchronous speed of the rotating field

p = the number of poles of the motor.

Equations (5.1) and (5.2) are derived in Section 3.1.3. The magnetic field of the armature is equivalent to the field of a synchronously rotating permanent magnet. Under normal operating conditions, this field does not induce any voltage in the rotor winding because the rotor winding is rotating at the same speed.

As shown in Fig. 5-6(b), the field of the rotor is along the magnetic axis of the rotor. Since the rotor rotates at synchronous speeds, under normal operating conditions, its field also travels at the same speed.

Under normal operating conditions, then, the stator and rotor fields travel at the same angular speed, which is known as the synchronous speed and is given by Eq. (5.2).

5.1.3 Principle of Operation

The motor's operation, or torque-producing capability, results from the natural tendency of two magnetic fields to align their magnetic axes. The fields under consideration are:

1. The stator field (ϕ_s).
2. The rotor field (ϕ_f).

The flux that accompanies the synchronously rotating stator field (ϕ_s) completes its magnetic circuit by passing through the stator, air gaps, and rotor. The flux of the rotor field (ϕ_f) also passes through the rotor, the air gaps, and the structure of the stator. The vector sum of the armature and field flux, as shown in Fig. 5-6(b), gives the so-called resultant (ϕ_R) air-gap flux. Mathematically,

$$\vec{\phi}_R = \vec{\phi}_s + \vec{\phi}_f \tag{5.4}$$

The magnetic axis of the field flux tries to align itself with the magnetic axis of the stator flux. This process produces motor action. The mathematical expression for the torque developed was derived in Chapter 3 on three-phase induction machines (see Section 3.1.2). This electromagnetic torque is in the direction of rotation of the stator field. When an additional load is applied to the shaft, the rotor or the magnetic axis of the field flux slips in space behind the magnetic axis of the resultant air gap flux. The result is a larger space angle; therefore, a larger torque is developed.

5.1.4 Starting

Starting a three-phase synchronous motor is similar to starting a three-phase induction motor. Balanced voltages are applied to the stator windings, while the rotor winding is short-circuited or connected to a resistance. Synchronous motors are commonly equipped with damper windings, which are similar to the rotor windings of squirrel-cage induction motors. During any nonsynchronous-speed operation, damper windings contribute to the motor's output torque. The stator voltages produce a rotating magnetic field, which induces a voltage in the rotor winding. This voltage will circulate rotor current, which in turn will produce a magnetic field.

The interaction of these two fields produces motor action, which will bring the rotor up to a speed that is very close to synchronous. At this instant, the rotor windings are switched to the normal dc supply, and the rotor pulls in synchronism with the armature rotating field. In practice, this switching often takes place at that particular instant when the flux of the field momentarily reduces the voltage already impressed in the stator. This results in high inrush currents in the armature windings, which are accompanied by a characteristic "thump."

The resistance (R_2) of the rotor circuit limits the rotor current and thus controls the starting torque of the motor. Generally speaking, the higher the resistance of the field windings, the higher the starting torque. Mathematically,

$$T_{st} \propto R_2 \qquad\qquad (5.5)$$

The torque that the motor develops, at the instant the dc field is "switched in," is usually referred to as the pull-in torque. The pull-in torque must be larger than the load requirements at that speed; otherwise, the motor will not be able to synchronize.

At starting, synchronous motors have very low impedance. As a result, their starting current is five to six times greater than their rated current. Depending on the Thévenin's impedance of the supply network, this high current may cause a dip in the voltage delivered to the plant where the motor is operating. This voltage reduction will reduce the starting-torque capability of the motor ($T \propto V^2$) and may adversely affect other electrical loads, such as motors, lights, and magnetic contactors.

The accelerating time, the magnitude of the inrush current, and the torque are identical to those of three-phase induction machines.

Where circumstances do not permit high inrush currents, an autotransformer may be used to reduce the voltage applied to the motor. A more popular method is to design a motor with low starting torque and inrush current. Such a motor is coupled to the load through an air clutch. The motor starts unloaded, and after it is synchronized, the pressurized air—which is conducted through a properly sized opening along the inner section of the motor's shaft—engages or couples the air clutch to the load. Use of an air clutch not only permits the design of motors with lower starting torque and currents, but also develops pneumatic power that is sufficient to provide the high starting torque that some loads require.

The appendix to this chapter gives manufacturers' data for high- and low-starting-torque (soft or air-clutch starting) synchronous motors.

5.1.5 Equivalent Circuits

This section develops and analyzes the approximate equivalent circuit(s) of a synchronous motor. The flow of energy from the ac and dc sources and the electromagnetic coupling between their corresponding fields are shown in Fig. 5-7(a). This electromagnetic coupling leads to the development of torque. From Ampere's law, the torque developed depends on the magnitude and relative position of the two interacting fields.

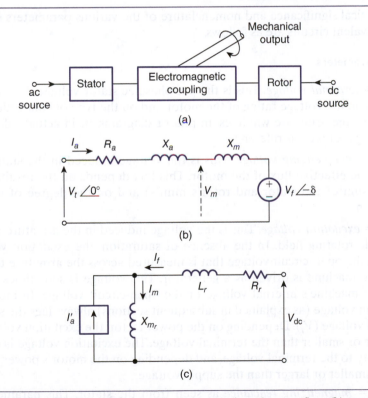

FIG. 5-7 Synchronous motor: **(a)** elementary presentation of energy flow, **(b)** approximate per-phase equivalent circuit as seen from the stator, **(c)** approximate equivalent circuit as seen from the rotor.

The approximate per-phase equivalent circuit of a motor, as seen from the stator (Fig. 5-7(b)), is derived by considering the effects of the two rotating fields on the stator. The resultant field magnetizes the stator (a process represented by X_m). The rotor field induces a voltage (V_f) in the stator windings whose phase angle is at $-\delta°$ to the stator voltage supply (V_t) because the synchronously rotating rotor field lags the stator field by the same angle. The leakage impedance of the stator windings is represented by $R_a + jX_a$.

The approximate equivalent circuit of a motor as seen from the rotor (Fig. 5-7(c)) is derived by considering the effects of the two rotating fields on the rotor. The resultant field magnetizes the rotor (a process represented by the reactance X_{mr}), and the effects of the stator current are represented by a current source of I'_a amperes. The actual resistance and inductance of the rotor windings are as shown in the figure.

The stator and rotor equivalent circuits supplement each other. They give an in-depth view of the machine's conceptual design and behavior, enabling the application engineer to predict the motor's operation.

Equivalent-Circuit Parameters

The physical significance and nomenclature of the various parameters shown in the equivalent circuits are as follows.

Stator Parameters

V_t = *terminal voltage.* This is the per-phase ac stator voltage. Its magnitude depends on the voltage rating of the motor and on the type of connection (delta or star) of the armature windings. In phasor diagrams, or in actual calculations, this voltage is taken as reference.

V_m = *magnetizing voltage.* This is the voltage induced in the stator windings by the effective flux of the motor. This flux depends on the resultant mmf (vector sum of the stator and rotor's mmf's) and on the degree of magnetic saturation.

V_f = *excitation voltage.* This is the voltage induced in the armature windings by the dc rotating field. In the absence of saturation, the excitation voltage is equal to the open-circuit voltage that is measured across the armature terminals when the machine is driven as a generator. This voltage is sometimes referred to as the machine's internal voltage or the open-circuit voltage. In motors, the excitation voltage (as explained in subsequent sections) always lags the supply or terminal voltage (V_t). Depending on the power factor, the excitation voltage may be larger or smaller than the terminal voltage. The excitation voltage is opposite in polarity to the terminal voltage, and depending on the motor's power factor, it may be smaller or larger than the supply voltage.

X_m = *magnetizing reactance* as seen from the stator. This parameter represents the magnetization of the stator that results from the effective field of the armature. It depends on the characteristics of the magnetic material and the degree of saturation.

X_a = *stator leakage reactance.*

R_a = *stator resistance.* This is the effective or ac resistance of the armature windings—that is, the stator dc resistance adjusted to include magnetic losses and the effects of higher operating temperatures.

X_s = *synchronous reactance.* This is the sum of the stator leakage reactance and the magnetizing reactance. That is,

$$X_s = X_a + X_m \tag{5.6}$$

Z_s = *synchronous impedance.*

$$Z_s = R_a + jX_s \tag{5.7}$$

I_a = *armature current.* When the motor is star-connected, this current is equal to the line current.

Power Angle (δ)

The power angle, designated by the Greek letter delta (δ), is the phase angle between the excitation voltage (V_f) and the magnetizing voltage (V_m). It is also called the torque or the load angle. This angle is equal to the phase angle between the resultant air-gap flux (ϕ_R) and the field flux (ϕ_f).

The phase angle between the terminal voltage (V_t) and the excitation voltage (V_f) is approximately equal to the torque angle because the leakage impedance is very small compared to the magnetizing reactance.

The torque angle for a synchronous motor is always negative (V_f lags V_t), regardless of operating power factor. The torque angle used for the steady-state analysis of the motors (and shown in the phasor diagrams) is expressed in electrical degrees. However, the torque angle used for analysis of the machine's electromechanical transient is expressed in mechanical degrees. The relationship between mechanical (δ_m) and electrical degrees (δ) is derived in Section 3.1.3 and is rewritten here for convenience:

$$\delta_m = \frac{2}{p}\delta \tag{5.8}$$

where p is the number of poles of the machine.

Typical ranges of the parameters of synchronous motors are given in Table 5-1.

TABLE 5-1	Typical synchronous machine parameters	
Parameter	*Minimum**	*Maximum**
X_a	0.08	0.30
X_m	0.4	2.4
R_a	0.002	0.04

All values are in per unit, based on the machine's rating.

Rotor Parameters

V_{dc} is the dc voltage applied across the rotor windings. In all modern machines, this voltage is obtained by rectifying the voltage induced in another set of rotor windings from a source that is magnetically coupled to the rotor but not electrically connected to it.

X_{mr} is the machine's equivalent magnetizing reactance as seen from the rotor terminals.

R_r and L_r are the rotor resistance and self-inductance, respectively.

I_f is the dc current of the rotor winding.

I_m is the component of the field current that flows through the magnetizing reactance X_{mr}.

I'_a is the component of the stator current that represents the armature reaction. This is the contribution of the armature current to the magnetization of the machine; it is expressed in equivalent field amperes. Depending on the power factor at which the motor draws power, the armature reaction may aid or oppose the actual field current. (For an illustration, see the Phasor diagrams of Fig. 5-10.)

The ratio of the armature current (I_a) to its armature reaction component gives the effective turns ratio (N_e) between the stator and the rotor windings. Mathematically,

$$N_e = \frac{I_a}{I'_a} \tag{5.9}$$

The measurement of a machine's effective turns ratio and armature reaction is discussed in Section 5.2.3.

5.1.6 Field Current

From the equivalent circuit of Fig. 5-7(c), the motor's magnetizing current (I_m) is given by

$$I_m = I_f + I'_a \tag{5.10}$$

where I_f is the actual field current and I'_a is the contribution of the armature current to the magnetization process. Rewriting the last relationship, we have

$$I_f = I_m - I'_a \tag{5.11}$$

I'_a is in phase with the armature current and is related to it by the effective turns ratio, as previously mentioned. The field current and its magnetizing component lag their corresponding voltages by 90°, as shown in Fig. 5-8.

The field current and its magnetizing component can be obtained graphically by projecting their corresponding voltages on the machine's open-circuit characteristic (OCC). See Fig. 5-9.

The actual values of V_f and V_m at any particular motor operating condition can be found by applying KVL to the equivalent circuit of Fig. 5-7(b). The measurement of the open-circuit characteristic, and further applications, are discussed in detail in Section 5.2.3.

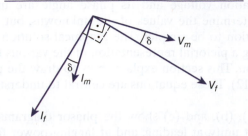

FIG. 5-8 Excitation and magnetization voltage phasors and their corresponding currents.

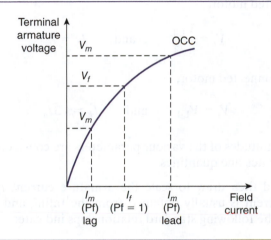

FIG. 5-9 The effects of the power factor on the magnetization voltage and on the magnetizing component of the armature current.

Equation (5.10) may be written in terms of the interacting mmf's. That is, the effective magnetizing mmf is equal to the mmf of the field plus the mmf of the stator. Although this expression would be conceptually advantageous, it would be of limited usefulness because an additional unknown—the number of winding turns—is introduced.

5.1.7 Phasor Diagrams

A graphical representation, such as a phasor diagram, often simplifies the solution of a problem and enhances the understanding of the subject matter under consideration. Note, for example, the following equation, which is obtained by applying KVL to the motor's equivalent circuit of Fig. 5-7(b):

$$V_f \underline{/-\delta} = V_t \underline{/0°} - I_a (R_a + jX_s) \tag{5.12}$$

Usually, the excitation voltage and its phase angle are unknown. You may mathematically determine the values of the unknowns, but you will often find the graphical solution to be simpler. The graphical solution has the additional advantage of giving a pictorial representation of the various terms and concepts under consideration. This section explains how to draw the phasor diagrams of Eqs. (5.11) and (5.12). These equations are central to understanding the motor's operation.

Figures 5-10(a), (b), and (c) show the phasor diagrams of a synchronous motor operating at unity, at leading, and at lagging power factors, respectively. The sequence of steps required to complete the diagrams is outlined below and is identified on the phasor diagrams by the encircled numbers.

1. Draw the terminal voltage V_t to scale and take it as reference. Recall that for a star-connected motor,

$$V_t = \frac{V_{\text{L-L}}}{\sqrt{3}} \qquad \text{and} \qquad I_L = I_a$$

For a delta-connected motor,

$$V_t = V_{\text{L-L}} \qquad \text{and} \qquad I_L = \sqrt{3}I_a$$

Only the magnitudes of the various parameters are considered here. The subscript L indicates line quantities.

2. Calculate and then draw to scale the armature current I_a. The per-phase armature current is usually obtained from the rating and efficiency of the machine, as the following standard relationships indicate:

$$\text{input power} = \frac{\text{output power}}{\text{efficiency}} \qquad\qquad \text{(5.13)}$$

Using mathematical symbols,

$$P_{\text{in}} = \frac{P_{\text{out}}}{\eta} \qquad\qquad \text{(5.14)}$$

$$P_{\text{in}} = \sqrt{3}\, V_{\text{L-L}} I_L \cos\theta \qquad\qquad \text{(5.15)}$$

where η and $\cos\theta$ represent the motor's efficiency and its power factor, respectively.

Note that the output power is also called brake horsepower, nameplate power, or rated power.

3. From the tip of V_t, draw the phasor $I_a R_a$. Because this represents the voltage drop in the stator resistance, it should be in phase with the armature current. However, owing to the negative sign in Eq. (5.12), it must be placed at an angle 180° from its normal direction.

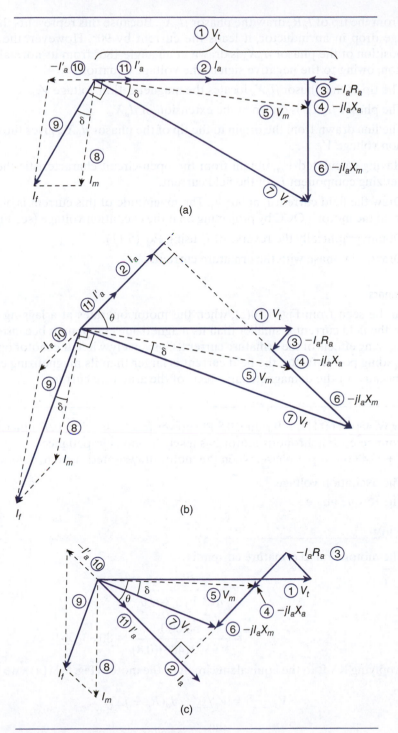

FIG. 5-10 Phasor diagrams for synchronous motors: **(a)** unity power factor, **(b)** leading power factor, **(c)** lagging power factor.

4. From the tip of I_aR_a draw the phasor jI_aX_a. Because this represents the voltage drop in an inductor, it leads the current by 90°. However, the actual position of the phasor jI_aX_a is drawn at an angle 180° from its normal direction, owing to the negative sign of the voltage equation.

5. The tip of the phasor jI_aX_a locates the magnetization voltage V_m.

6. The phasor jI_aX_m is drawn at the extension of jI_aX_a.

7. The line drawn from the origin to the tip of the phasor jI_aX_m gives the excitation voltage V_f.

8. Having calculated V_m, obtain from the open-circuit characteristic the magnetizing component I_m of the field current.

9. Draw the field current I_f at 90° V_f. The magnitude of this current is obtained from the motor's OCC by projecting on it the excitation voltage (see Fig. 5-9).

10. Obtain graphically the reverse of I'_a using Eq. (5.11).

11. Draw I'_a in phase with the armature current.

Comments

As can be seen from Fig. 5-10(c), when the motor operates at a lagging power factor, the field current is smaller than its magnetizing component because of the magnetizing effects of the armature current. Conversely, when the motor operates at a leading power factor, the field current is larger than its magnetizing component because of the demagnetizing effects of the armature current.

EXAMPLE **5-1**

A 50 kW, 480 V, 60 Hz, 900 r/min, 0.8 Pf (power factor) leading, 0.93 efficient, star-connected synchronous motor has a synchronous impedance of $0.074 + j0.48$ ohms per phase. When the motor draws rated current, determine:

a. The excitation voltage.
b. The torque angle.

SOLUTION

a. The motor's rated armature current is

$$I_a = \frac{P_{\text{out}}}{\eta\sqrt{3}\ V_{\text{L-L}}\cos\theta}$$

$$= \frac{50{,}000}{0.93\sqrt{3}(480)(0.8)} = 80.83\underline{/36.9°}\ \text{A}$$

Applying KVL to the equivalent circuit of the motor (Fig. 5-11(a)), we obtain

$$V_f\underline{/-\delta} = V_t\underline{/0°} - I_a(R_a + jX_s)$$

$$= \frac{480}{\sqrt{3}} - 80.83\underline{/36.9°}\ (0.074 + j0.48)$$

$$= 295.62 - j34.63$$

$$= 297.64 \underline{/-6.7°} \text{ V/phase}$$

and

$$V_f = 515.54 \text{ V, L-L}$$

b. The torque angle is

$$\delta = -6.7°$$

The phasor diagram is shown in Fig. 5-11(b).

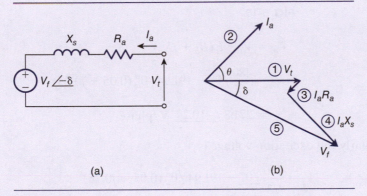

(a)   (b)

FIG. 5-11

A 480 V, 3-ϕ, 60 Hz, 150 kW, 0.94 efficient synchronous motor has a per-phase synchronous impedance of $0.05 + j0.75$ ohms. The leakage reactance is 0.25 ohm per phase. The OCC in the operating region is given by

$$V_f = 10 + 20I_f$$

When the motor draws rated current at unity power factor, determine:

a. The approximate and exact values of the torque angle.
b. The magnetizing component of the field current.
c. The field current.
d. The armature reaction in equivalent field amperes.
e. The effective turns ratio.

EXAMPLE **5-2**

SOLUTION

a. The rated current is

$$I_a = \frac{P_{in}}{\sqrt{3} V_{L\text{-}L} \cos \theta} = \frac{150,000}{0.94} \left(\frac{1}{\sqrt{3}(480)(1)} \right) = 191.94 \underline{/0°} \text{ A}$$

The magnetizing voltage is found by applying KVL to the equivalent circuit of Fig. 5-12.

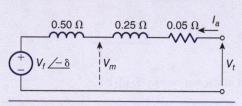

FIG. 5-12

$$V_m = V_t - I_a(R_a + jX_a)$$

$$= \frac{480}{\sqrt{3}} \underline{/0°} - 191.94 \underline{/0°} (0.05 + j0.25)$$

$$= 271.8 \underline{/-10.2°} \text{ V/phase}$$

Similarly, the excitation voltage is

$$V_f = \frac{480}{\sqrt{3}} \underline{/0°} - 191.94 \underline{/0°} (0.05 + j0.75)$$

$$= 303.8 \underline{/-28.3°} \text{ V/phase}$$

Thus, the approximate and exact values of the torque angle in electrical degrees are

$$\delta_{appr} = \underline{-28.3°}$$

$$\delta_{ex} = -28.3° - (-10.2°) = \underline{-18.1°}$$

b. The magnitude of the magnetizing component of the field current can be found from the given open-circuit characteristic:

$$I_m = \frac{V_m - 10}{20} = \frac{271.8\sqrt{3} - 10}{20} = 23.04 \text{ A}$$

This current lags its corresponding voltages by 90°. Thus,

$$I_m = \underline{23.04 \underline{/-90° -10.2°}}$$

$$= \underline{23.04 \underline{/-100.2°}} \text{ A}$$

c. Similarly, the magnitude of the field current is

$$I_f = \frac{303.8\sqrt{3} - 10}{20} = 25.81 \text{ A}$$

In phasor form:

$$I_f = 25.81 \underline{/-90° - 28.3°}$$

$$= 25.81 \underline{/-118.3°} \text{ A}$$

d. The armature reaction is found by using Eq. (5.11).

$$I'_a = I_m - I_f = 23.04 \underline{/-100.2°} - 25.81\underline{/-118.3°}$$

$$= 8.16 \text{ A}$$

e. The effective turns ratio is obtained from Eq. (5.9).

$$N_e = \frac{I_m}{I'_a} = \frac{191.94}{8.16} = 23.51$$

Exercise 5-1

Prove the following:

a. The torque angle is equal to the phase angle between the field current and its magnetizing component.

b. The approximate value of the torque angle of a synchronous motor that operates at a leading power factor of θ degrees and has negligible armature resistance is given by

$$\delta = \arctan \frac{I_a X_s \cos \theta}{V_t + I_a X_s \sin \theta}$$

c. The per-unit copper losses in a motor are equal to the motor's per-phase armature resistance expressed in per unit.

Exercise 5-2

A 375 kW, 2200 V, 60 Hz, 0.80 Pf lagging, 0.966 efficient, 900 r/min, star-connected motor has a synchronous impedance of $0.015 + j0.702$ pu. When the motor draws rated current at a nominal power factor, determine:

a. The excitation voltage.

b. The approximate torque angle.

Answer (a) 1741.5 V, L-L;   (b) $-44.3°$

A synchronous motor has negligible armature impedance. Under rated operating conditions, its power factor is unity and its field current is 30 A. On an open-circuit test, 26 A are required to give the rated terminal voltage. Determine:

a. The armature reaction in equivalent field amperes.
b. The torque angle.

Answer (a) 14.97 A, (b) −29.9°

5.1.8 Power and Torque Developed

In this section, general expressions will be derived for the complex, real, and reactive powers drawn by a synchronous motor that is connected to an infinite bus (see Fig. 5-13(a)). The system's per-phase equivalent circuit is shown in Fig. 5-13(b). The impedance (Z) is equal to the synchronous impedance of the motor plus the impedance of the transmission line between the motor and the infinite bus.

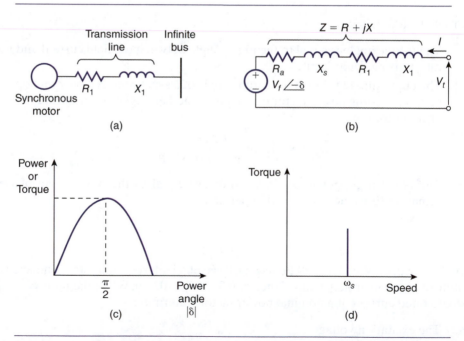

FIG. 5-13 **(a)** Synchronous motor connected to an infinite bus, **(b)** per-phase equivalent circuit, **(c)** power or torque versus load angle, **(d)** torque versus speed.

The complex power drawn from the infinite bus is

$$S = P + jQ = V_t I^* \tag{5.16}$$

where I^* is the conjugate of the current phasor and V_t is now the infinite-bus voltage. Using Ohm's law,

$$I^* = \left(\frac{V_t \underline{/0°} - V_f \underline{/-\delta}}{Z} \right)^* \tag{5.17}$$

By definition,

$$Z = |Z| \underline{/\theta} \tag{5.18}$$

From Eqs. (5.16), (5.17), and (5.18), we obtain

$$S = \frac{V_t}{|Z|} (V_t \underline{/-\theta} - V_f \underline{/-\theta - \delta})^* \tag{5.19}$$

$$= \frac{V_t}{|Z|} [V_t \cos\theta - j\sin\theta) - V_f(\cos(-\theta - \delta) + j\sin(-\theta - \delta))]^* \tag{5.20}$$

$$= \frac{V_t}{|Z|} [V_t \cos\theta - V_f\cos(-\theta - \delta) - j(V_t\sin\theta + V_f\sin(-\theta - \delta))]^* \tag{5.21}$$

Taking the conjugate of the expression within the brackets, we have

$$S = \frac{V_t}{|Z|} [V_t \cos\theta - V_f\cos(-\theta - \delta) + j(V_t\sin\theta + V_f\sin(-\theta - \delta))] \tag{5.22}$$

The active power drawn from the infinite bus is given by the real component of the complex power. Thus,

$$P = \frac{V_t}{|Z|} [V_t \cos\theta - V_f\cos(-\theta - \delta)] \tag{5.23}$$

$$= \frac{V_t}{|Z|} (V_t \cos\theta - V_f(\cos\theta \cos\delta - \sin\theta \sin\delta) \tag{5.24}$$

Also by definition,

$$\cos\theta = \frac{R}{|Z|} \quad \text{and} \quad \sin\theta = \frac{X}{|Z|}$$

Thus,

$$P = \frac{V_t}{|Z|^2} (V_t R - R V_f \cos\delta + X V_f \sin\delta) \text{ watts/phase} \tag{5.25}$$

The torque angle at which the motor draws maximum power is found by applying the calculus theory of minimum and maximum.

$$\frac{\partial P}{\partial \delta} = \frac{V_t V_f}{Z} (R \sin\delta + X \cos\delta) = 0 \tag{5.26}$$

From the above, the torque angle (δ_{m_1}) at which the motor develops maximum power is

$$\delta_{m_1} = -\arctan \frac{X}{R} \tag{5.27}$$

From Eq. (5.25), by neglecting the resistive component of the impedance and considering only magnitudes, we obtain

$$P = \frac{V_t V_f}{X} \sin \delta \text{ watts/phase} \tag{5.28}$$

The power input to the motor will be equal to the power developed. The power developed provides the output power and the rotational losses.

By definition, the torque is

$$T = \frac{\text{power}}{\text{speed}} = \frac{V_t V_f}{\omega_s X} \sin \delta \text{ N} \cdot \text{m/phase} \tag{5.29}$$

For constant excitation, the power is given by

$$P = K \sin \delta \tag{5.30}$$

and the torque by

$$T = K_1 \sin \delta \tag{5.31}$$

The constants K and K_1 can be obtained from Eqs. (5.28) and (5.29), respectively.

As shown in Fig. 5-13(c), the maximum torque of the motor occurs at a power angle of $\pi/2$ radians. Maximum torque is often referred to as breakdown torque or pull-out torque. Since the speed of the motor is constant, the torque-versus-speed characteristic will be as shown in Fig. 5-13(d).

The imaginary part of the complex power of Eq. (5.22) gives the reactive power drawn from the infinite bus. Thus,

$$Q = \frac{V_t}{|Z|} [V_t \sin \theta + V_f \sin (-\theta - \delta)] \tag{5.32}$$

and

$$Q = \frac{V_t}{|Z|} [V_t \sin \theta - V_f (\cos \theta \sin \delta + \sin \theta \cos \delta)] \tag{5.33}$$

Replacing $\sin \theta$ and $\cos \theta$ as before by their equivalent expressions, we get

$$Q = \frac{V_t}{|Z|^2} (XV_t - V_f R \sin \delta - XV_f \cos \delta) \text{ VAR/phase} \tag{5.34}$$

The maximum reactive power drawn from the infinite bus will occur at a torque angle of

$$\delta_{m_2} = \arctan \frac{R}{X} \tag{5.35}$$

By neglecting the resistive component of the impedance in Eq. (5.34), we obtain

$$Q = \frac{V_t}{X}(V_t - V_f \cos \delta) \qquad (5.36)$$

When the quantity within the parentheses of Eq. (5.36) is positive, the reactive power is also positive. Thus, the machine operates at a lagging power factor, and the motor is said to be underexcited. Conversely, when the quantity within the parentheses is negative, the motor operates at a leading power factor, and the motor is said to be overexcited.

When the parameters of Eqs. (5.25), (5.28), (5.34), and (5.36) are expressed in per unit, then the derived formulas give the corresponding total output of the machine in per unit. When a synchronous motor is used only to improve the power factor of a network, its torque angle is approximately zero. Thus, the reactive power drawn by the motor is

$$Q = \frac{V_t(V_t - V_f)}{X_s} \text{ VAR/phase} \qquad (5.37)$$

Equation (5.37) makes it clear that, by varying the dc current, the reactive power of a synchronous motor can be varied smoothly in either direction. In comparison, the reactive power of a capacitor can change only in steps (discontinuously) and only in one direction.

Figure 5-14(a) shows a synchronous motor used to improve the power factor of a load. The phasor diagram of the pertinent parameters is shown in Fig. 5-14(b).

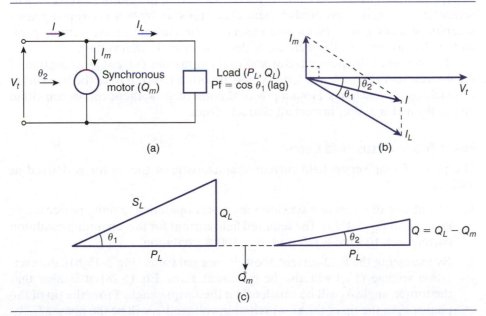

(a)

(b)

(c)

FIG. 5-14 Ideal synchronous motor and its effects on the overall power factor of a plant: **(a)** electrical network, **(b)** phasor diagram, **(c)** power triangles of the load, motor, and source.

The power triangles of the load, motor, and source are shown in Fig. 5-14(c). Note that, in this case, a synchronous motor draws only reactive power.

Speed Control

The speed of a synchronous motor can be controlled either by changing the number of poles or by varying the frequency of the stator voltage. The frequency is varied through "frequency converters," and the number of poles is changed by reconnecting the stator and rotor windings into the desired pole configuration. The technique of frequency conversion is the same for both synchronous and three-phase induction machines, and so it need not be discussed again here. One exercise will be given, however, just to review the essentials.

5.1.9 Effects of Field Current on the Characteristics of the Motor

Previous sections discussed the effects of the field current on the motor's internal parameters, such as torque angle, excitation voltage, and the magnetization of the machine. This section will analyze the effects of the field current on the motor's external parameters. Specifically, qualitative proofs will be developed that give the power factor, the armature current, and the real and reactive powers as functions of the field current. The qualitative proofs are based on two assumptions: that the motor's output power remains constant, and that its stator resistance is negligible. These proofs will lead to the construction of the motor's external characteristics. One can easily predict these characteristics by first drawing a phasor diagram at unity power factor and observing the changes in the various parameters that accompany the increase or decrease of excitation voltage.

The curves of the power factor and of the armature current as a function of the field current resemble the letter V and are often referred to as the Vee curves of synchronous machines. From a practical point of view, these curves constitute one of the motor's most important characteristics.

Power Factor versus Field Current

The power-factor versus field-current characteristic of the motor is derived as follows:

1. The phasor diagram of a synchronous motor operating at unity power factor is shown in Fig. 5-15(a). The required field current for this operating condition corresponds to what is referred to as 100% excitation.

2. By increasing the field current above its normal value (Fig. 5-15(b)), the excitation voltage (V_{f_2}) will also be increased. From Eq. (5.28), it is clear that the torque angle δ_2 will be smaller than the torque angle. From the tip of the phasor V_t to the tip of the new excitation voltage V_{f_2} will be the phasor $I_{a_2} X_s$. This is so because of the following general voltage equation:

$$V_{f_2} = V_t - jI_{a_2} X_s \tag{5.38}$$

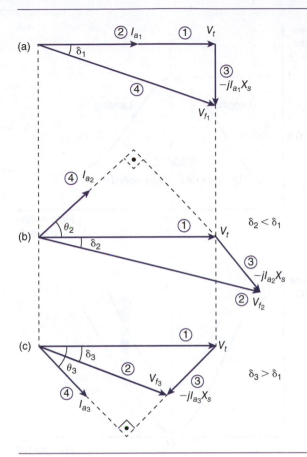

FIG. 5-15 Phasor diagrams for the qualitative derivation of Vee curves: **(a)** unity power factor (normal excitation), **(b)** leading power factor (overexcitation), and **(c)** lagging power factor (underexcitation).

The extension of $I_{a_2}X_s$ (dotted line) must cross the line that represents the armature current at $90°$. This is possible only when the armature current *leads* the terminal voltage, as shown in Fig. 5-15(b).

3. By decreasing the field current below its normal value (Fig. 5-15(c)), the excitation voltage (V_{f_3}) will also be decreased, and the new torque angle δ_3 will be larger than the torque angle δ_1. As before, from the tip of the phasor V_t to the tip of the new excitation voltage will be the phasor $I_{a_3}X_s$. The dotted extension of the phasor $I_{a_3}X_s$ must cross the line of the armature current at $90°$. This is possible only when the armature current *lags* the terminal voltage, as shown in Fig. 5-15(c)).

As this discussion makes clear, a change in the field current will be accompanied by a change in the power factor of the motor. Qualitatively speaking, the

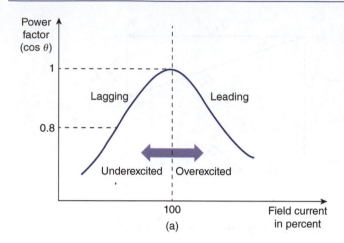

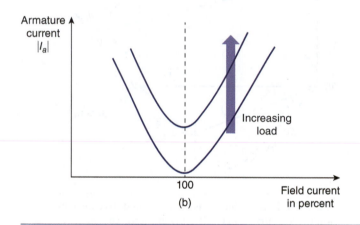

FIG. 5-16 Vee curves for a synchronous motor: **(a)** power factor versus field current, **(b)** magnitude of armature current versus field current.

variation of the power factor as a function of the field current can be represented by a curve that resembles an inverted V, as shown in Fig. 5-16(a).

Magnitude of Armature Current versus Field Current

Neglecting armature resistance, we find that the input power is equal to the power developed, which is assumed to remain constant. Mathematically,

$$V_t I_a \cos \theta = P_d \text{ watts/phase} \qquad (5.39)$$

From Eq. (5.39),

$$I_a = \frac{K}{\cos \theta} \qquad (5.40)$$

where the constant K is given by

$$K = \frac{P_d}{V_t} \qquad (5.41)$$

Qualitatively speaking, the previously derived Vee curves of Fig. 5-16(a) indicate the following:

For variations of the field current up to 100% excitation,*

$$\cos \theta \propto I_f \qquad (5.42)$$

For variations of the field current above 100% excitation,

$$\cos \theta \propto \frac{1}{I_f} \qquad (5.43)$$

From Eqs. (5.40), (5.42), and (5.43), we obtain

For excitation above 100%, $\qquad I_a \propto I_f \qquad (5.44)$

For excitation below 100%, $\qquad I_a \propto \frac{1}{I_f} \qquad (5.45)$

The variation of the armature current as a function of field current may be represented by the curves shown in Fig. 5-16(b), whose shape resembles the letter V. The armature current is minimum when the power factor is maximum. In other words, minimum armature current corresponds to 100% excitation or unity power factor.

Reactive Power versus Field Current

By definition, the reactive power drawn by a motor is

$$Q = V_t I_a \sin \theta \ \text{VAR/phase} \qquad (5.46)$$

The Vee curves of Fig. 5-16(a) indicate that for excitation up to 100%,

$$\cos \theta \propto I_f$$

An increase in the cosine function will be accompanied by a decrease in the sine function. Thus,

$$\sin \theta \propto \frac{1}{I_f} \qquad (5.47)$$

---

*The proportionality symbols indicate the trend in the change of the variables but not true direct proportion or inverse proportion.

From Eqs. (5.45), (5.46), and (5.47), we get

$$Q \propto \frac{1}{I_f^2} \qquad \text{(5.48)}$$

Similarly, for excitation above 100%, from Eqs. (5.43), (5.44), and (5.46), we obtain

$$Q \propto I_f^2 \qquad \text{(5.49)}$$

From Eqs. (5.48) and (5.49), the reactive power versus the motor's field current can be drawn as shown in Fig. 5-17. The leading power factor corresponds to negative VAR and the lagging power factor to positive VAR. At 100% field excitation, the power factor is equal to unity and the reactive power is equal to zero.

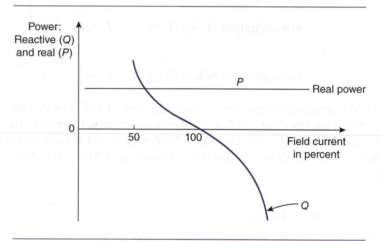

FIG. 5-17 A synchronous motor's real and reactive powers as a function of field current.

Real Power

The real power drawn by a synchronous motor is independent of the field current because the real component of the armature current ($I_a \cos \theta$) does not depend on the excitation. The variation of the real power as a function of the field current is shown in Fig. 5-17. In an actual machine, however, the stator ($I_a^2 R_a$) losses, and hence the input power, will increase whenever the stator current increases as a result of changing the field current.

The curves derived in this section are approximate and are to be used for qualitative analysis only. Exact relationships can be derived from the magnetization characteristic of the particular motor and by use of the appropriate equations discussed in this chapter.

EXAMPLE **5-3**

Explain:

a. The fact that the torque angle of a synchronous motor always lags the terminal voltage.

b. The fact that a sudden increase in the torque requirements of a synchronous motor will momentarily decrease the synchronous speed of the rotor, and thus the torque angle will be enlarged.

c. The magnetizing and demagnetizing effects of armature current. Use phasor diagrams to illustrate the concepts.

SOLUTION

a. The output power of a synchronous motor is produced by the transformation of electrical input power. In other words, before the motor starts to rotate and produce an output, the electrical input (V_t) must first be applied; the rotation of the shaft will then follow (V_f). Since the torque angle is approximately equal to the phase angle between the terminal voltage and the excitation voltage, the torque angle is negative, or lags the applied voltage V_t, as shown in Fig. 5-18. The excitation voltage is produced by the flux of the rotor, which rotates at the same angular speed as the shaft.

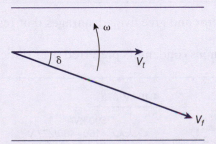

FIG. 5-18

b. As shown by the following equation, a sudden increase in the load torque requirements of a synchronous motor will be accompanied by an increase in the torque angle:

$$P = V_f \frac{V_t}{X_s} \sin \delta$$

Since the torque angle is always lagging, its enlargement will force the shaft to momentarily rotate at a lower speed relative to the stator field. Alternatively, owing to physical considerations, loading the shaft of the motor will momentarily reduce the rotor's angular speed and thus, as depicted in Fig. 5-18, the torque angle will become larger. This phenomenon is only transient. Once the power equilibrium is reestablished, the rotor will attain synchronous speed and the torque angle will settle at a magnitude that will satisfy the new power requirement.

c. Economic considerations dictate that, under normal operating conditions, the motor should operate at a point close to the knee of the magnetization curve. Achieving this level of magnetization requires a certain amount of field current. Since the field current has two components, its magnitude will depend on the relative magnitudes and positions of its components. Rewriting the equation for the field current, we have

$$I_f = I_m - I'_a$$

When the motor operates at a lagging power factor, as shown in Fig. 5-10(c), less field current is required than when it operates at a leading power factor, as shown in Fig. 5-10(b). At a lagging power factor, on the one hand, the flux of the armature current aids the magnetizing effect of the field current. This is known as the *magnetizing effects of the armature current*. At a leading power factor, on the other hand, the armature current produces flux that opposes the magnetization of the machine. This is referred to as the *demagnetizing effects of the armature current*.

EXAMPLE **5-4**

1. Define power factor and give five advantages that result from its improvement.
    Why are synchronous condensers preferred over static capacitors?

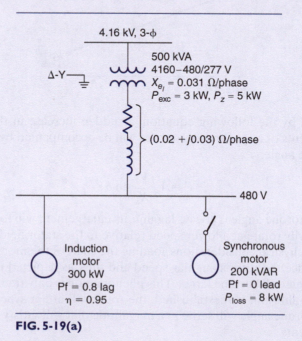

4.16 kV, 3-φ

Δ-Y

500 kVA
4160–480/277 V
X_{e_i} = 0.031 Ω/phase
P_{exc} = 3 kW, P_z = 5 kW

(0.02 + j0.03) Ω/phase

480 V

Induction
motor
300 kW
Pf = 0.8 lag
η = 0.95

Synchronous
motor
200 kVAR
Pf = 0 lead
P_{loss} = 8 kW

FIG. 5-19(a)

2. For the network shown in Fig. 5-19(a), calculate with and without the synchronous motor in the circuit:

   a. The losses in the feeder.
   b. The kVA delivered by the transformer.
   c. The losses in the transformer.
   d. The regulation.

SOLUTION

1. The power factor of a given circuit is equal to the cosine of the phase angle between the circuit voltage and current. The voltage is always taken as reference.

   The ordinary consumer of electrical energy is not interested in the phase angle between the voltage and the current applied to a motor. He or she is concerned not with power factor, but with the actual kW output of his motor. However, for constant voltage and output power, the current through the incoming feeders depends on the power factor, as shown by the following equation:

$$I = \frac{P_{in}}{\sqrt{3} V_{L\text{-}L} \cos \theta}$$

For example, a reduction in the power factor from unity to 0.80 will result in a 25% increase in the current. Conversely, an increase (improvement) in the power factor will result in decreasing current, which will reduce the following:

   a. The losses in the upstream feeders.
   b. The required size of the upstream feeders.
   c. The required size of the transformer.
   d. The losses in the transformer.
   e. The voltage fluctuation across the motors and other electrical loads.

   Synchronous motors can provide output torque and leading or lagging kVAR to the network. Their operation is very smooth and their losses are nominal. Capacitors can provide only kVAR lagging in steps, but they consume negligible power. For simplicity, we assume here that the motor does not provide any torque.

   Furthermore, the effects of harmonics and switching problems are more severe for capacitors than for synchronous motors. Nevertheless, a synchronous machine has maintenance requirements for its bearings, ventilation, and so on, that static capacitors do not have.

2. *Without the synchronous motor in the circuit:*
   a. The line or feeder current is

$$I = \frac{300,000}{0.95\sqrt{3}(0.8)480} = 474.79 \underline{/-36.9°} \text{ A}$$

Thus, the losses in the feeder (P_{lf}) are

$$P_{lf} = 3(0.02)(474.79)^2 = \underline{13.53\text{ kW}}$$

b. The transformer should supply the power required by the load plus the power consumed by the feeder. From the circuit of Fig. 5-19(b), the voltage

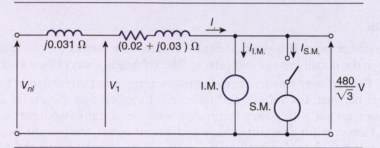

FIG. 5-19(b)

at the output terminals of the transformer is

$$V_1 = V_L + IZ = \frac{480}{\sqrt{3}}\underline{/0°} + 474.79\underline{/-36.9°}(0.02 + j0.03)$$

$$= 293.32\underline{/-1.1°}\text{ V/phase},\quad\text{and}\quad V_1 = 508.1\text{ V L-L}$$

Thus, the apparent power delivered by the transformer is

$$S = \sqrt{3}\,V_{\text{L-L}}I_L$$

$$= \sqrt{3}(508.1)(474.79) = \underline{417.8\text{ kVA}}$$

c. Assuming that the core loss is proportional to the square of the terminal voltage, the transformer's losses are

$$P_{l_{(t)}} = 3\left(\frac{508.1}{480}\right)^2 + 5\left(\frac{474.79}{601.4}\right)^2 = \underline{6.48\text{ kW}}$$

d. The no-load voltage at the secondary of the transformer is

$$V_{nl} = \frac{480}{\sqrt{3}}\underline{/0°} + 474.79\underline{/-36.9°}(0.02 + j0.03 + j0.031)$$

$$= 302.61\underline{/-3.3°}\text{ V/phase}$$

and

$$V = (\sqrt{3}(302.61) = 524.13\text{ V, L-L}$$

$$R\% = \frac{524.13 - 480}{480}(100) = \underline{9.19\%}$$

3. *With the synchronous motor in the circuit*:

    a. The current drawn by the synchronous condenser is found as follows:

$$Q = \sqrt{3}\ V_{\text{L-L}}I_L \sin \theta = 200\ \text{kVAR}$$

$$P = \sqrt{3}\ V_{\text{L-L}}I_L \cos \theta = 8\ \text{kW}$$

From the above, the current drawn by the synchronous motor is

$$I_{\text{S.M.}} = 240.76\ \underline{/87.7°}\ \text{A}$$

The current in the feeder is the sum of the currents drawn by the induction and synchronous motors. Thus,

$$I_f = 474.79\underline{/-36.9°} + 240.76\underline{/87.7°}$$

$$= 391.97\underline{/-6.5°}\ \text{A}$$

The new power factor is

$$\cos 6.5° = 0.99\ \text{lag}$$

The losses in the feeder are

$$P_{lf} = 3(0.02)(391.97)^2 = \underline{9.22\ \text{kW}}$$

    b. The voltage at the output of the transformer is

$$V_1 = \frac{480}{\sqrt{3}}\ \underline{/0°} + 391.97\underline{/-6.5°}(0.02 + j0.03)$$

$$= 286.15\underline{/-2.2°}\ \text{V/phase}$$

Then

$$V_1 = \sqrt{3}(286.15) = 496.15\ \text{V, L-L}$$

The kVA delivered by the transformer is

$$S = \sqrt{3}(496.15)(391.97) = \underline{336.84\ \text{kVA}}$$

    c. The transformer losses are

$$P_{l_{(t)}} = 3\left(\frac{496.15}{480}\right)^2 + 5\left(\frac{391.97}{601.4}\right)^2 = \underline{5.33\ \text{kW}}$$

d.　The no-load voltage at the substation is

$$V_{nl} = \frac{480\underline{/0°}}{\sqrt{3}} + 391.97\underline{/-6.5°}(0.02 + j0.03 + j0.031)$$

$$= 288.53\underline{/4.55°} \text{ V/phase}$$

Then

$$V_{nl} = \sqrt{3}(288.53) = 499.75 \text{ V, L-L}$$

The voltage regulation is

$$R\% = \frac{499.75 - 480}{480}(100) = \underline{4.1\%}$$

For purposes of comparison, the results are summarized in Table 5-2.*

TABLE 5-2　Summary of results of Example 5-4

Synchronous Motor	Power Factor at the 480 V Terminals	kVA Delivered by the Transformer	Losses in the Transformer in kW	Losses in the Feeder in kW	Current through the Feeder in A	Voltage Regulation in Percent
Disconnected	0.8 lagging	417.8	6.48	13.53	474.79	9.19
Connected	0.99 lagging	336.84	5.33	9.22	391.97	4.1

The advantages of higher-power-factor operation become very evident when we compare the tabulated results. The adverse effects of low-power-factor operation of large industrial customers on the upstream equipment cannot be tolerated.

---

*The student may evaluate some of the financial advantages of the high-power-factor operation by assuming the following:

- Cost of energy: \$1000/kW/year. Cost of transformers: \$50/kVA.
- Cost of a three-conductor cable: \$30/meter per ampere capacity.
- Length of cable: 200 meters.
- Monthly power-factor penalty:

$$\$10 \text{ (actual kW)} \left(\frac{Pf_n}{Pf_a} - 1\right)$$

where
　　Pf_a = actual power factor
　　Pf_n = minimum permissible power factor, usually 0.90

- The cost of a 480 V capacitor bank is about \$40/kVAR.

a. A synchronous motor, when operated at 0.8 Pf leading, has an armature copper loss of 4 kW. *Estimate* the armature copper loss when the motor operates at unity power factor. Assume that the power drawn by the motor remains constant.

b. Derive Eq. (5.37).

Answer (a) 2.56 kW.

The speed of a 480 V, 60 Hz, 10-pole, 0.94 efficient, unity-power-factor synchronous motor is controlled through a frequency converter. The synchronous reactance is 1.04 Ω/phase, and the motor drives a 50 kW constant-power load. When the output voltage of the converter is 432 V, determine:

a. The speed of the motor.

b. The excitation voltage and the torque angle.

Answer **(a)** 648 r/min; **(b)** 256.5 V/phase, $-8.6°$

5.2 Three-Phase Cylindrical Rotor Machines: Generators

5.2.1 Equivalent-Circuit and Phasor Diagrams

This section deals with the principle of operation, governing equations, characteristics, and the measurement of the parameters of three-phase generators. As mentioned earlier, three-phase generators transform mechanical power into electrical power. The mechanical power of the prime mover rotates the shaft of the generator on which the dc field windings are installed. The speed of the prime mover is maintained at a constant level through electronically controlled speed regulators, commonly known as governors. The rotation of the dc flux cuts the windings of the armature, and, because of the induction principle, a three-phase voltage is generated.

Three-phase generators are also called three-phase alternators. Structurally, they are similar to three-phase synchronous motors.

The direction of energy flow is shown in Fig. 5-20(a). The approximate per-phase equivalent circuit as seen from the stator and rotor terminals is shown in Figs. 5-20(b) and (c), respectively. The physical significance and nomenclature of each parameter are identical to those of three-phase synchronous motors.

The governing equations for alternators are

$$V_f \underline{/\delta°} = V_t \underline{/0°} + I_a (R_a + jX_s) \tag{5.50}$$

$$I_f = I_m + I'_a \tag{5.51}$$

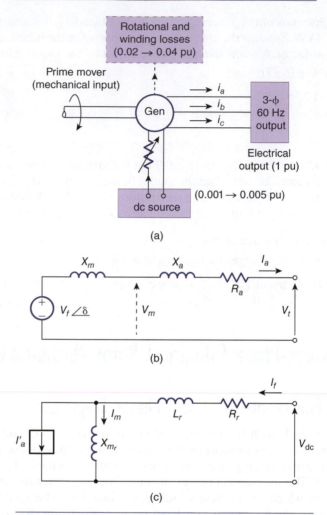

FIG. 5-20 Alternators: **(a)** direction of energy flow
and typical losses for a 400 kVA alternator, **(b)** approximate equivalent circuit as seen from the stator, and
(c) partial equivalent circuit as seen from the rotor.

Note that these equations have a positive sign, whereas those for synchronous
motors have a negative sign.

The torque angle of an alternator, owing to physical considerations, always
leads the terminal voltage.

The Vee curves for alternators are similar to those for synchronous motors,
except that the words "leading" and "lagging" should be switched. For example,
an *underexcited* generator (leading Pf) will correspond to an *overexcited* (leading
Pf) motor.

The power factor of a single alternator does not depend on its field current but on the impedance of the connected load. However, if an alternator operates in parallel with other generators, their individual power factors can be controlled by properly adjusting their corresponding field currents.

The phasor diagrams for unity and lagging power factor are drawn in Figs. 5-21(a) and (b), respectively. The order of construction is identified by the encircled numbers. The phasor diagrams of generators are easier to draw than those of motors because of the positive sign in the governing equations, Eqs. (5.50) and (5.51).

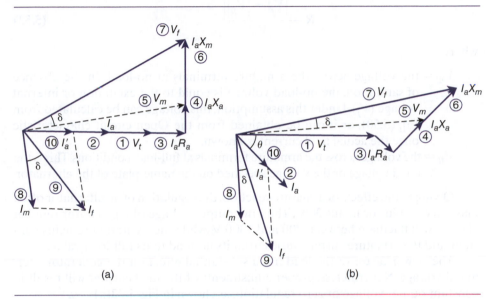

FIG. 5-21 Phasor diagrams for alternators: **(a)** unity power factor, **(b)** lagging power factor.

Notice that at lagging-power-factor operation a large field current is required in order to magnetize the stator and thus to enable the generator to provide rated terminal voltage. In other words, lagging-power-factor currents have a demagnetizing effect on the stator. For this reason, normally a generator *cannot* start when its load is an ac motor of about equal rating. At starting, an ac machine draws a highly inductive current ($\theta \approx 70°$) of a magnitude five to six times larger than the rated value.

Therefore, to select a standby generator, you have to carefully evaluate the nature of the connected load. The normal practice is to consult with the manufacturers regarding the required size and *starting capability* of the unit under consideration. A general rule of thumb is that a generator can start an ac machine provided that the starting kVA of the motor is not more than twice the nominal kVA of the generator. When a properly sized capacitor bank is used, its leading-power-factor currents reduce the magnitude of the downstream inductive currents, thus the generator's starting capability is accordingly increased.

5.2.2 Regulator of Alternators

For practical application, it is important to be able to determine the regulation of a three-phase synchronous generator. Mathematically, the regulation (R), as explained in Chapter 2, is defined by

$$R = \left.\frac{|V_{nl}| - |V_{fl}|}{|V_{fl}|}\right|_{I_f = \text{constant}} \tag{5.52}$$

or

$$R = \left.\frac{|V_{f}| - |V_{fl}|}{|V_{fl}|}\right|_{I_f = \text{constant}} \tag{5.53}$$

where

V_{nl} = the voltage across the armature terminals at no-load. In the absence of saturation, the no-load voltage is equal to the excitation or internal voltage (V_f). Under this assumption, this voltage can be calculated from Eq. (5.50), or it can be obtained from the Open circuit characteristic once the actual field current is known.

V_{fl} = the voltage across the armature terminals at full-load conditions. This is the rated voltage or the voltage specified on the name-plate of the alternator.

Owing to the effects of armature reaction, the regulation of an alternator often may vary by plus or minus 30%. Thus, the output voltage of a generator rated at 1000 V will fluctuate between 700 and 1300 V, while the excitation remains constant and the armature current varies from its no-load to its full-load value.

The power factor of the load has a substantial effect on the generator's terminal voltage. Nevertheless, proper adjustments of the field current will result in constant output voltage or zero regulation, as shown in Fig. 5-22(a).

Figure 5-22(b) shows that, for a particular value of armature current, capacitive loads increase the terminal voltage while inductive loads decrease it. The

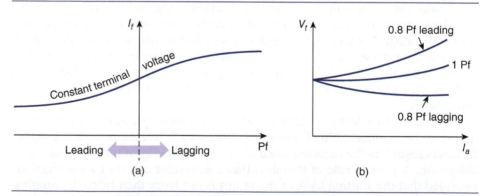

FIG. 5-22 Characteristics of alternators: **(a)** field current versus power factor for constant terminal voltage, **(b)** terminal voltage as a function of armature current for fixed excitation and variable power factor.

reason is that lagging-power-factor armature currents have a demagnetizing effect on the machine, while leading-power-factor currents have a magnetizing effect.

In practice, automatic regulators are used to adjust the field current in such a way as to maintain constant output voltage (zero regulation), independently of load condition.

The regulation of large three-phase alternators cannot be measured directly in the laboratory because it is difficult to find large enough loads to satisfy the required full-load condition. Theoretically, however, a generator's voltage regulation can be determined by the governing equations and from the data on its characteristics. These characteristics are discussed in the following sections.

5.2.3 Characteristics of Alternators

Open-Circuit Characteristic

The open-circuit characteristic (OCC) shows the relationship between the field current and the voltage across the open-circuited terminals of the armature windings. Laboratory data for this characteristic are obtained by driving the machine as a generator at rated synchronous speed (see Fig. 5-23(a)). When the armature and field currents are equal to zero, the power furnished by the prime mover constitutes the mechanical losses of the machine at no-load. A typical OCC is shown in Fig. 5-23(b).

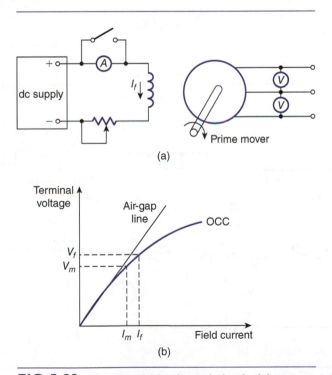

FIG. 5-23 Alternator: **(a)** schematic for the laboratory setup used to obtain data for the magnetization curve, **(b)** typical open-circuit characteristic.

The open-circuit characteristic can be used to obtain the magnetizing component (I_m) and/or the total field current (I_f) once the magnetizing voltage (V_m) and the excitation voltage (V_f) have been calculated from the machine's operating data and equivalent circuit. The open-circuit characteristic is also called the magnetization curve, the no-load characteristic, or the saturation curve.

The line drawn tangentially to the linear portion of the open-circuit characteristic is called the air-gap line. This straight line is the magnetic characteristic of the machine when saturation is neglected.

Short-Circuit Characteristic

The short-circuit characteristic (SCC) shows the relationship between the field current and the armature current when the alternator terminals are shorted. The machine is driven as a generator, and the short-circuit armature current (I_a) is recorded while the field current (I_f) is gradually increased. The power delivered by the prime mover is equal to the rotational losses plus the stator copper losses.

A typical SCC and the laboratory setup that may be used to obtain it are shown in Figs. 5-24(a) and (b), respectively. Since the terminal voltage (V_t) is equal to zero, the excitation voltage, as can be seen from the equivalent circuit of Fig. 5-24(c), is given by

$$V_f = I_a (R_a + jX_s) \tag{5.54}$$

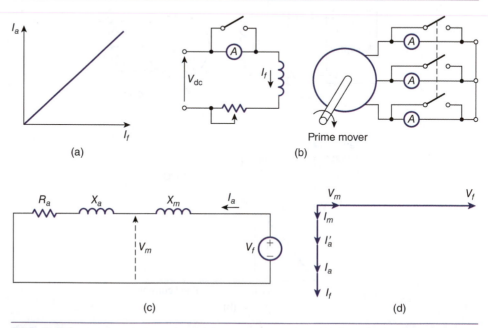

(a)

(b)

(c)

(d)

FIG. 5-24 (a) Typical short-circuit characteristic, (b) schematic for the laboratory setup that may be used to obtain data for the short-circuit characteristic, (c) equivalent circuit, (d) approximate phasor diagram ($R_a = 0$).

Usually, the armature resistance is negligible. Thus,

$$V_f \approx jI_aX_s \qquad (5.55)$$

From the last relationship, it is evident that the armature current lags the excitation voltage by 90°. Rewriting (Eq. 5.51), we have

$$I_f = I_m + I_a' \qquad (5.56)$$

I_m is very small because the magnetizing voltage is negligible. I_f also lags V_f by 90°. Therefore, the field current I_f and the armature reaction in equivalent field amperes (I_a') are approximately in phase. This implies that a linear change in the armature current will be accompanied by a linear change in the field current. The phasor diagram for the short-circuit condition is shown in Fig. 5-24(d).

The short-circuit characteristic is also called the transformation-ratio curve, because its slope gives the approximate value of the effective turns ratio. Thus,

$$\text{slope of SCC} = \frac{I_a}{I_f} = \frac{I_a}{I_m + I_a'} \qquad (5.57)$$

Since I_m is negligible,

$$\text{slope of SCC} \approx \frac{I_a}{I_a'} = N_e \qquad (5.58)$$

(A more accurate value for this parameter can be obtained from the zero-power-factor characteristic that is discussed in the following section).

By reducing the speed from its rated value, the excitation voltage (V_f) will also be reduced. From the generator principle (see Chapter 6, Section 6.1),

$$V_f = K\phi\omega \qquad (5.59)$$

where

K = a constant of the machine.

ϕ = the effective flux.

ω = the angular speed of rotation.

The impedance of the armature circuit, being largely inductive, will also be reduced accordingly.

$$Z_s \approx jX_s \qquad (5.60)$$

and

$$Z_s \approx j\omega L_s \qquad (5.61)$$

where L_s is the synchronous inductance of the machine. Thus, when the machine is short-circuited, the armature current is given by

$$I_a = \frac{V_f}{Z_s} \approx \frac{K\phi\omega}{j\omega L_s} = K_1 I_f \tag{5.62}$$

where K_1 is a constant of proportionality. This expression implies that, for a given field current and a moderate variation in the speed of the prime mover, the armature current is constant. A 25% change in the speed of the prime mover above or below its synchronous value does not appreciably affect the linearity of the short-circuit characteristic.

Zero-Power-Factor Lagging Characteristic

The zero-power-factor (ZPF) characteristic shows the relationship between the terminal voltage and the field current at the constant rated armature current (I_a) and zero-load power factor. Lagging loads with a power factor up to 0.1 have a negligible effect on the actual zero-power-factor curve.

A typical ZPF characteristic and a laboratory setup that may be used to obtain it are shown in Figs. 5-25(a) and (b), respectively. The load to the generator

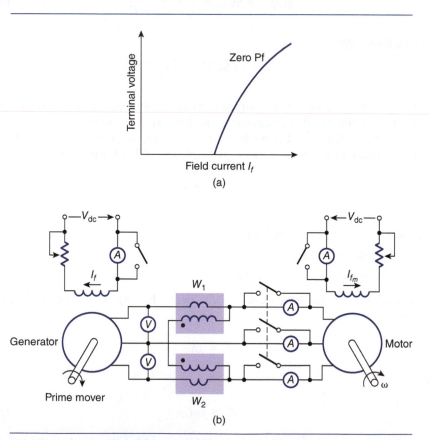

FIG. 5-25 **(a)** Typical lagging zero-power characteristic, **(b)** schematic for a laboratory setup used to obtain data for the zero-power-factor curve.

is a synchronous motor whose power factor and armature current are controlled through its field current. The terminal voltage is controlled by the generator's field current. The output power of the generator must be larger than that of the driven motor because of the motor's losses.

The readings of the instruments on the output terminals of the generator (voltmeters, ammeters, and wattmeters) are also used to calculate the power factor of the load. Instead of the two wattmeters (W_1 and W_2), one may use a polyphase wattmeter. The measurement of three-phase power with two wattmeters is explained in the Appendix (Section A.5).

5.2.4 Measurement of Parameters

Measurement of Synchronous Reactance X_s

The short-circuit characteristic, in conjunction with the magnetization curve, can be used to obtain the unsaturated ($X_{s_{un}}$) and saturated ($X_{s_{sat}}$) values of the synchronous reactance, as follows.

Unsaturated Value

Refer to Fig. 5-26. For any value of field current (I_f), there is a corresponding voltage on the air-gap line (V_{g_1}) and an armature current (I_{a_1}) on the short-circuit characteristic.

The unsaturated value of the synchronous reactance is given by

$$X_{s_{un}} = \frac{\text{voltage obtained from the air-gap line}}{\text{armature current obtained from the SCC}} \bigg|_{\text{at any field excitation}}$$

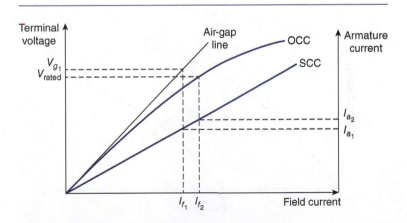

FIG. 5-26 Open- and short-circuit characteristics.

or

$$X_{s_\text{un}} = \frac{V_{g_1}}{I_{a_1}}\Bigg|_{I_{f_1}} \qquad (5.63)$$

Saturated Value

The ratio of the rated voltage (V_1) to the short-circuit current (I_{a_2}) at the same field conditions gives the approximate value of the saturated synchronous reactance (X_{s_sat}).

$$X_{s_\text{sat}} = \frac{V_\text{rated}}{I_{a_2}}\Bigg|_{I_{f_2}} \qquad (5.64)$$

The per-unit value of the saturated synchronous reactance is the reciprocal of the short-circuit ratio (SCR). This parameter is defined as follows:

$$\text{SCR} = \frac{\text{field current obtained from OCC at rated terminal voltage}}{\text{field current obtained from SCC at rated armature current}} \qquad (5.65)$$

The Potier Triangle: Measurement of X_a and I_a'

The zero-power-factor (ZPF) characteristic is used in conjunction with the open-circuit characteristic in order to obtain the Potier triangle. The Potier triangle (Fig. 5-27) is used to determine:

The stator leakage reactance (X_a).

The armature reaction (I_a') in equivalent field amperes.

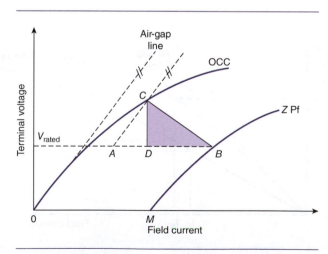

FIG. 5-27 Construction of the Potier triangle.

The following procedure is suggested in establishing the Potier triangle:

1. On the horizontal line at rated voltage, draw AB:

$$AB = OM \qquad (5.66)$$

2. From point A, draw the line parallel to the air-gap line. This intersects the magnetization curve at point C. From this point, draw the perpendicular to AB. This establishes point D.

The base of this triangle gives the armature reaction in equivalent field amperes—that is, the I_a'. Thus,

$$BD = I_a' \qquad (5.67)$$

The effective turns ratio (N_e) can be calculated from

$$N_e = \frac{I_a}{I_a'} \qquad (5.68)$$

The other perpendicular side of the Potier triangle—that is, CD—gives the Potier reactance voltage drop. This is approximately equal to the voltage drop across the leakage reactance. Thus,

$$CD = I_a X_a \qquad (5.69)$$

From this basis, the leakage reactance (X_a) can be calculated. Thus, the leakage reactance and the armature reaction are given, respectively, by the vertical and horizontal sides of the Potier triangle.

The origin of the ZPF characteristic corresponds to the value of field current required to circulate rated armature currents while the terminals of the machine are shorted. This particular point can also be obtained from the short-circuit characteristic.

Some Highlights of the Zero-Power-Factor Characteristic

Up to moderate saturation levels, the ZPF *lagging* characteristic can be obtained from the OCC by shifting the OCC downward by an amount equal to $I_a X_a$, and horizontally to the right by an amount equal to I_a', as shown in Fig. 5-28(a). Similarly, the ZPF *leading* curve can be obtained from the OCC by shifting it horizontally to the left by an amount equal to I_a', and vertically upward by an amount equal to $I_a X_a$, as shown in Fig. 5-28(a). The leakage reactance (X_a) is assumed to remain constant. At high saturation levels, this assumption is not valid.

When the power factor lags, more field current is required to magnetize the machine and to overcome the demagnetizing effects of the armature reaction.

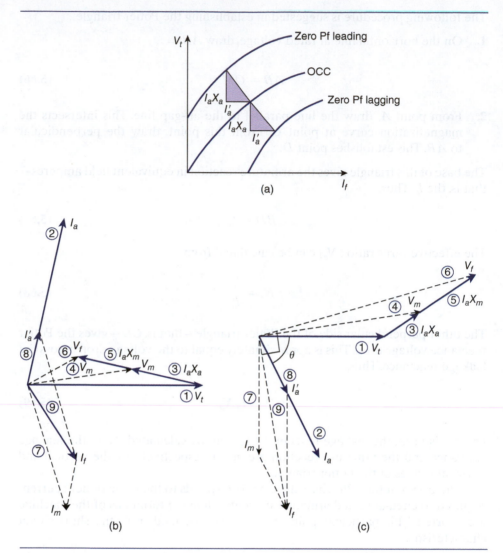

FIG. 5-28 Alternator: **(a)** zero-power-factor and open-circuit characteristics, **(b)** phasor diagram for near zero leading power factor, **(c)** phasor diagram for near zero lagging power factor.

When the power factor leads, less field current is required because the armature reaction aids the field in magnetizing the machine. At the origin of the zero-power-factor leading curve, the actual field current is zero because the armature reaction is equal and opposite to the magnetizing component of the field current.

The phasor diagrams for ZPF leading and lagging loads are shown in Figs. 5-28(b) and (c), respectively.

The three-phase alternator shown in the one-line diagram of Fig. 5-29(a) has a per-phase synchronous impedance of 0.03 + j0.50 per unit. The magnetization characteristic at the operating region is given by

EXAMPLE **5-5**

$$V_f = 70 + 55I_f$$

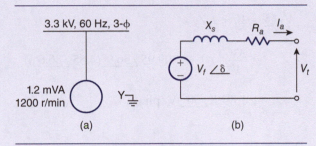

3.3 kV, 60 Hz, 3-ϕ

1.2 mVA
1200 r/min

(a)

(b)

FIG. 5-29

The generator delivers rated current to an infinite bus at 0.8 leading power factor. Determine:

a. The ohmic value of the synchronous impedance.
b. The excitation voltage.
c. The approximate torque angle in electrical and mechanical degrees.
d. The regulation.
e. The field current.

SOLUTION

a. The base parameters are given by the name-plate data of the generator:

$$V_b = 3.3 \text{ kV}, \qquad S_b = 1.2 \text{ MVA}$$

Then, by definition,

$$Z_b = \frac{(3.3)^2}{1.2} = 9.08 \ \Omega/\text{phase}$$

The ohmic value of the synchronous impedance is calculated as follows:

$$Z_{pu} = \frac{Z_{actual}}{Z_b}$$

$$Z_{actual} = 9.08(0.03 + j0.50)$$

$$= 0.27 + j4.54 = 4.55 \underline{/86.6°} \ \Omega/\text{phase}$$

b. The magnitude of the rated current of the generator is

$$I = \frac{S}{\sqrt{3}\,V_{L\text{-}L}} = \frac{1200}{\sqrt{3}\,3.3} = 209.95 \text{ A}$$

The excitation voltage is calculated by applying KVL in Fig. 5-29(b).

$$V_f\underline{/\delta} = V_t + I_a Z_s$$

$$= \frac{3300}{\sqrt{3}}\underline{/0°} + 209.95\,\underline{/36.9°}(4.55\,\underline{/86.6°})$$

$$= 1592.8\,\underline{/30°}\text{ V/phase}$$

and

$$V_f = 2758.8 \text{ V, line-to-line}$$

c. The approximate value of the torque angle in electrical degrees is

$$\delta = \underline{30°}$$

In mechanical degrees, its approximate value is

$$\delta_m = \frac{2\delta}{P} = \frac{2(30°)}{6} = \underline{10°}$$

d. The regulation in percent is

$$R\% = \frac{2758.8 - 3300}{3300}(100) = \underline{-16.4\%}$$

e. The field current is found from the given OC characteristic as follows:

$$I_f = \frac{2758.8 - 70}{55} = \underline{48.89 \text{ A}}$$

EXAMPLE **5-6** Test results for a 3-ϕ, 60 Hz, 480 V, 400 kVA synchronous generator are as follows:

Open-circuit test:

Field current (A)	6	10	14	18	22	26
Armature voltage (V)	225	348	440	500	540	560

Short-circuit test: A field current of 8 A was required to produce the rated armature current.

Zero-power-factor test:

$$I_a = \text{rated}$$

$$V_{\text{L-L}} = 500 \text{ V}$$

$$I_f = 30 \text{ A}$$

Assuming that the stator resistance is negligible, calculate the field current required to produce the rated output voltage and current to a 0.90 Pf lagging load.

SOLUTION

The characteristics of the generator are shown in Fig. 5-30.

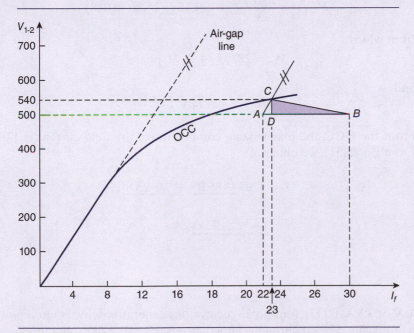

FIG. 5-30

From the diagram:

$$AB = 8 \text{ A}$$

$$AC // \text{air-gap line}$$

From the Potier triangle:

$$CD = I_a X_a = 40 \text{ V, L-L}$$

$$DB = I'_a = 7 \text{ A}$$

The generator's rated current is

$$I_a = \frac{400}{\sqrt{3}(0.480)} = 481.13 \text{ A}$$

In phasor form,

$$I_a = 481.13 \underline{/-25.8°} \text{ A}$$

The leakage reactance is

$$X_a = \frac{40/\sqrt{3}}{481.13} = 0.048 \ \Omega/\text{phase}$$

The magnetization voltage is

$$V_m \underline{/\beta} = \frac{480}{\sqrt{3}} \underline{/0°} + 481.13 \underline{/-25.8°}(j0.048)$$

from which

$$V_m = 287.95 \underline{/4.1°} \text{ V/phase}$$

and

$$V_m = 498.74 \text{ V, L-L}$$

From the OCC, the magnetizing component of the field current is 17.95 A. From Eq. (5.51), we obtain

$$I_f = 17.95 \underline{/-85.9°} + 7 \underline{/-25.8°}$$

$$= \underline{\underline{22.3 \underline{/-70°} \text{ A}}}$$

Exercise 5-6

A 480 V, 50 kVA, 60 Hz, four-pole synchronous generator delivers rated power at 0.8 Pf lagging. The synchronous impedance is $0.2 + j1.4$ ohms per phase. Determine:

a. The excitation voltage.
b. The generator's regulation.

Answer (a) 593.38 V, L-L; (b) 23.62%

5.3 Salient-Pole Synchronous Machines

5.3.1 Introduction

A two-pole salient-rotor synchronous motor is shown in Fig. 5-31. The length of the air gap between the rotor and the stator is not uniform. The axis along which the air gap is minimum is called the *direct axis*, and that along which the air gap is maximum is called the *quadrature axis*.

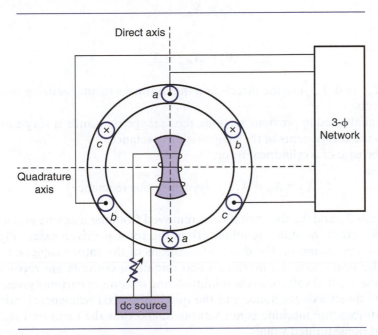

FIG. 5-31 An elementary representation of a two-pole salient-rotor synchronous machine.

Synchronous machines with salient-pole rotors usually have more than two poles and therefore rotate at relatively low synchronous speeds, as required by many industrial drives. They can develop up to 40% of their rated power without any excitation—zero field current—while cylindrical machines cease to function as motors or generators once the dc excitation is removed.

Saliency increases the ability of the machine to oppose the forces that tend to drive a motor out of synchronism. As a result, compared to cylindrical-rotor machines, salient machines are more suitable in applications where the motor might be subjected to occasional sudden torque variations.

In all modern synchronous motors, there is no direct connection between the rotor windings and the field supply voltage source. The field current, as explained in Chapter 7, is obtained by rectifying the voltage induced in another set of rotor windings, commonly known as a rotating armature.

The magnetizing flux through the machine's two distinct air gaps produces two different inductances. The inductance along the direct axis is designated as L_d, and that along the quadrature axis is designated as L_q. Since the inductance is an inverse function of the length of the air gap, the direct-axis (X_{m_d}) component of the magnetizing reactance is larger than its quadrature-axis (X_{m_q}) component. The ratio of these two reactances is about 1.5.

The effective reactance along either the direct (X_d) or the quadrature axis (X_q) is the sum of the armature leakage reactance (X_a) and its corresponding component of the magnetizing reactance. Mathematically,

$$X_d = X_{m_d} + X_a \qquad (5.70)$$

$$X_q = X_{m_q} + X_a \qquad (5.71)$$

where X_{m_d} and X_{m_q} are the direct- and quadrature-axis magnetizing reactances, respectively.

As mentioned in previous sections, the leakage reactance is very small compared to the components of the magnetizing reactance.

In the case of a cylindrical rotor:*

$$X_d = X_q = X_s \qquad \text{(synchronous reactance)} \qquad (5.72)$$

To measure X_d and X_q, the excitation is removed from the machine and a voltage is applied across the stator terminals. Then the rotor is driven externally with a variable-speed dc motor. For different positions of the rotor magnetic axis relative to the stator axis, the maximum and minimum currents are recorded. The ratio of the applied voltage to the minimum and maximum currents gives, respectively, the direct-axis reactance and the quadrature-axis reactance. Under these test conditions, the machine is not saturated, and thus the measured reactances represent nonsaturated values.

Steady-State Analysis

The steady-state performance of salient-pole synchronous machines is satisfactorily predicted and analyzed by using Eq. (5.73). The positive signs are for a generator, and the negative signs are for a motor. This equation is derived from the De Blondel, or two-reaction, theory. This theory takes in the effects of saliency but ignores the effects of magnetic saturation. Magnetic saturation is accounted for by the cylindrical-rotor theory, which ignores saliency. Under normal operating conditions, both theories yield slightly different but generally satisfactory results.

---

*Cylindrical-rotor machines have some saliency, which exists because the slots for the field windings are not uniform around the entire rotor.

According to the De Blondel theory and as shown by Eq. (5.74), the armature current is represented by two components: the direct-axis component (I_d) and the quadrature-axis component (I_q). The relationship between the armature current and its components is not given in phasor form because the direct-axis component, depending on the power factor, may lead or lag the quadrature-axis component for either a motor or a generator. However, the two components are 90° out of phase with each other, and the quadrature-axis component is always in phase with the excitation voltage. All other parameters of Eq. (5.73) are the same as those for cylindrical-rotor synchronous machines.

$$V_f \underline{/\pm\delta°} = V_t \underline{/0°} \pm I_a R_a \pm jI_q X_q \pm jI_d X_d \tag{5.73}$$

$$|I_a| = \sqrt{I_q^2 + I_d^2} \tag{5.74}$$

The development of the per-phase equivalent circuit of a salient-pole machine from which Eq. (5.73) could be derived is rather complex, and it is not presented in this text.

5.3.2 Phasor Diagrams

The steady-state analysis of salient-pole synchronous machines is simplified by using phasor diagrams—that is, the graphical representation of Eqs. (5.73) and (5.74). The following procedure is suggested in drawing the phasor diagrams for a motor.

From Eq. (5.73), write the governing voltage equation:

$$V_f \underline{/-\delta°} = V_t \underline{/0°} - I_a R_a - jI_q X_q - jI_d X_d \tag{5.75}$$

1. Draw to scale the terminal voltage (V_t) and take it as a reference (see Fig. 5-32).
2. Draw the armature current at its given phase angle. In this case, $\theta = 0°$.
3. From the tip of the terminal voltage, draw the phasor $I_a R_a$. This is the voltage drop in the resistance (R_a) and should be in phase with the armature current (I_a). Here, however, $I_a R_a$ must be drawn at 180° to its normal direction because of the negative sign in the voltage equation.
4. Draw to scale a dotted line representing the phasor $I_a X_q$. This is the voltage drop in the reactance (X_q) and should lead the current (I_a) by 90°. Here again $I_a X_q$ must be drawn at 180° to its normal direction because of the minus sign in the voltage equation. The end of this phasor establishes the excitation line (V_f line) and therefore the torque angle. Although it is not in the equation, the term $I_a X_q$ is used to locate the excitation or quadrature-axis line. For this reason, it is drawn with a dotted line. The torque angle can also be obtained mathematically from the following relationship:

$$V_q \underline{/-\delta°} = V_t \underline{/0°} - I_a R_a - jI_a X_q \tag{5.76}$$

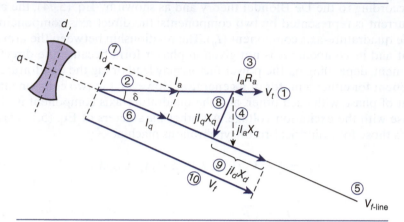

FIG. 5-32 Phasor diagram for a salient synchronous motor operating at unity power factor.

where V_q represents the sum of the three phasors of Eq. (5.76). Its phase angle (δ) establishes the direction of the line on which the excitation voltage must lie.

5. Draw the quadrature or excitation line. This is designated as the V_f line.

6. Resolve the armature current into its rectangular components I_d and I_q. The quadrature-axis component (I_q) is always in phase with the excitation line.

7. The direct-axis current is 90° out of phase with the quadrature axis. In general, I_d may lead or lag I_q, depending on the machine's operating power factor.

8. Draw the phasor I_qX_q. This represents the voltage drop across the reactance (X_q) and should lead the current by 90°. Again, because of the negative sign in the voltage equation, it must be drawn at 180° to its normal direction. The length of I_qX_q does not need to be calculated because it always starts at the tip of I_aR_a and ends perpendicularly on the excitation line.

9. Finally, draw to scale the phasor I_dX_d. As this represents the voltage drop in the reactance (X_d), it should lead the current (I_d) by 90°. But because of the equation's minus sign, it is drawn at 180° to its normal direction.

10. The excitation voltage (V_f) is drawn from the origin to the end of the tip of the phasor I_dX_d.

The procedure for drawing the phasor diagrams for salient alternators is similar. In fact, it is simpler to construct the phasor diagrams for a generator because of the positive sign in the loop Eq. (5.73).

The phasor diagram of a salient synchronous generator operating at unity power factor is shown in Fig. 5-33.

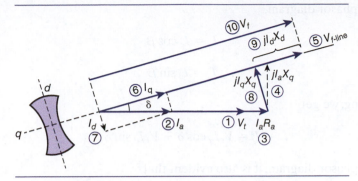

FIG. 5-33 Phasor diagram for salient synchronous generator operating at unity power factor.

5.3.3 Power Developed

To find the expression for the power developed by a salient synchronous motor, consider the phasor diagram shown in Fig. 5-34. Neglect armature resistance and assume a leading power factor. By definition,

$$P = V_t I_a \cos \theta \text{ watts/phase} \tag{5.77}$$

Say,

$$\beta = \theta + \delta \tag{5.78}$$

Then

$$P = V_t I_a \cos (\beta - \delta) \tag{5.79}$$

$$= V_t I_a \cos \beta \cos \delta + V_t I_a \sin \beta \sin \delta \tag{5.80}$$

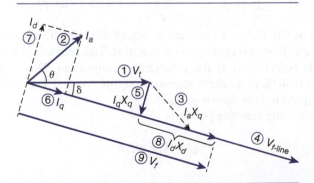

FIG. 5-34 Phasor diagram for a leading power factor synchronous motor.

From the phasor diagram,

$$I_q = I_a \cos \beta \tag{5.81}$$

$$I_d = I_a \sin \beta \tag{5.82}$$

Substituting, we get

$$P = V_t I_q \cos \delta + V_t I_d \sin \delta \tag{5.83}$$

From the phasor diagram, it is also evident that

$$I_d = \frac{V_f - V_t \cos \delta}{X_d} \tag{5.84}$$

and

$$I_q = \frac{V_t \sin \delta}{X_q} \tag{5.85}$$

Substituting Eqs. (5.85) and (5.84) into Eq. (5.83), we obtain

$$P = V_t \frac{V_t \sin \delta}{X_q} \cos \delta + \frac{V_t \sin \delta (V_f - V_t \cos \delta)}{X_d} \tag{5.86}$$

From the above,

$$P = \frac{V_t^2 (X_d - X_q)}{X_d X_q} \sin \delta \cos \delta + \frac{V_f V_t}{X_d} \sin \delta \tag{5.87}$$

using

$$\sin 2\delta = 2 \sin \delta \cos \delta \tag{5.88}$$

The expression in Eq. (5.87) becomes

$$P = \frac{V_f V_t}{X_d} \sin \delta + \frac{V_t^2 (X_d - X_q)}{2 X_d X_q} \sin 2\delta \text{ watts/phase} \tag{5.89}$$

The first term of Eq. (5.89) is identical to Eq. (5.28), which represents the power developed by a cylindrical synchronous machine. The second term represents the saliency and is referred to as the reluctance component. This component may account for up to 40% of a salient machine's output power. The power developed versus the torque angle is shown in Fig. 5-35.

By definition, the reactive power is

$$Q = V_t I_a \sin \theta \tag{5.90}$$

As before,

$$\theta = \beta - \delta \tag{5.91}$$

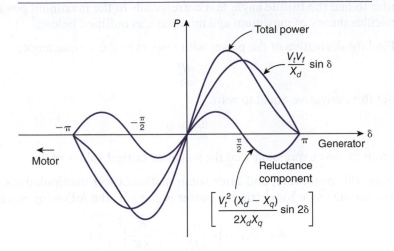

FIG. 5-35 Power versus torque angle.

Thus,

$$Q = V_t I_a \sin (\beta - \delta) \qquad (5.92)$$

From Eqs. (5.84), (5.85), and (5.92), after expansion and simplification, we get

$$Q = \frac{V_t V_f}{X_d} \cos \delta - V_t^2 \left(\frac{\cos^2 \delta}{X_d} + \frac{\sin^2 \delta}{X_q} \right) \text{VAR/phase} \qquad (5.93)$$

When the parameters of the machine are expressed in per unit, Eqs. (5.89) and (5.93) give the total input real and reactive power in perunit.

5.3.4 Torque Angle for Maximum Power

When the motor is supplied from an infinite bus and with a constant field current, then Eq. (5.89) can be written as

$$P = K_1 \sin \delta + K_2 \sin 2\delta \qquad (5.94)$$

where the constants K_1 and K_2 are given, respectively, by

$$K_1 = \frac{V_f V_t}{X_d} \qquad (5.95)$$

$$K_2 = \frac{V_t^2(X_d - X_q)}{2X_d X_q} \qquad (5.96)$$

In order to find the torque angle that corresponds to the maximum power, apply the calculus theory of maximum and minimum, as outlined below:

1. Find the derivative of the power with respect to the torque angle:

$$\frac{dP}{\partial\delta}$$

2. Set this derivative equal to zero:

$$\frac{dP}{d\delta} = 0$$

3. From the last expression, find the so-called critical values of δ.

Applying this procedure, and after some mathematical manipulations, we find that the torque angle for maximum power is given by the following equation:

$$\delta = \arccos\left(-\frac{K_1}{8K_2} \pm \sqrt{\left(\frac{K_1}{8K_2}\right)^2 + \frac{1}{2}}\right) \tag{5.97}$$

The torque angle for maximum power is usually about 70°. The intermediate steps in the derivation of Eqs. (5.93) and (5.97) are left as an exercise (see Exercise 5-7(b)).

5.3.5 Stiffness of Synchronous Machines

The stiffness or toughness of a synchronous machine is a measure of its ability to resist and oppose the forces that tend to pull it out of synchronism. Mathematically, the stiffness is given by the partial derivative of the power developed with respect to the torque angle. In other words, the stiffness of a machine is given by the slope of the power-angle curve at the operating point under consideration. Maximum stiffness is referred to as synchronizing power.

Cylindrical Rotors

From Eq. (5.28), we can easily obtain the stiffness (S_T) of a cylindrical-rotor machine as follows:

$$S_T = \frac{dP}{d\delta} = 3\frac{V_f V_t}{X_s}\cos\delta \quad \text{watts/elect. radian} \tag{5.98}$$

where the factor of 3 accounts for the three phases of the machine.

The stiffness is maximum at the origin of the power-angle curve and minimum at the torque angle that corresponds to zero output power. In other words, at pull-out, the stiffness of the machine is zero.

From Eq. (5.89), it can be shown that the stiffness of a motor with salient rotor is

$$S_T = \frac{dP}{d\delta} = 3\left(\frac{V_f V_t}{X_d}\cos\delta + \frac{V_t^2(X_d - X_q)}{X_d X_q}\cos 2\delta\right) \text{watts/elect. radian} \tag{5.99}$$

Although an increase of the field current does not affect the machine's output power, it does increase its toughness.

EXAMPLE 5-7

The synchronous motor shown in the one-line diagram of Fig. 5-36(a) draws rated current at a power factor of 0.95 leading. Its direct- and quadrature-axis reactances are 0.8 per unit and 0.5 per unit, respectively. Neglecting losses, determine:

1. a.   The approximate torque angle in electrical and mechanical degrees.
    b.   The excitation voltage.
    c.   The maximum torque that the machine can develop when the excitation voltage, as determined in (b), is constant.

2. Repeat part 1 by using the cylindrical rotor theory.

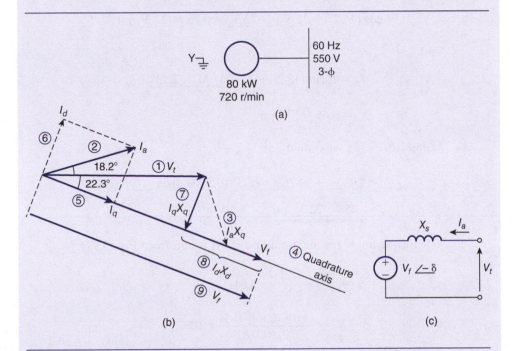

(a)

(b)

(c)

FIG. 5-36

SOLUTION

1. a.   The torque angle is found by using Eq. (5.76).

$$V_q \underline{/-\delta} = V_t - jI_aX_q$$

$$V_q \underline{/-\delta} = 1.0 \underline{/0°} - j1.0 \underline{/18.2°}(0.5) = 1.25 \underline{/-22.3°} \text{ pu}$$

Thus, the torque angle in electrical degrees is

$$\delta = -22.3°$$

In mechanical degrees, it is

$$\delta_m = \frac{2}{10}(22.3) = 4.5°$$

b. The phasor diagram, as shown in Fig. 5-36(b), is drawn by following the procedure outlined in Section 5.3.2. Then, from basic trigonometry,

$$I_q = I_a \cos(\theta + \delta) \angle \delta$$

$$= \cos(18.2° + 22.3°) \angle{-22.3°} = 0.76 \angle{-22.3°} \text{ pu}$$

and

$$I_d = I_a \sin(18.2° + 22.3°)\angle{90° - 22.3°}$$

$$= 0.65 \angle{67.7°} \text{ pu}$$

Using Eq. (5.73), we obtain

$$V_f\angle{-\delta} = 1.0 \angle{0°} - j0.76 \angle{-22.3°} (0.5) - j0.65 \angle{67.7°}(0.8)$$

$$= 1.45 \angle{-22.3°} \text{ pu}$$

c. The torque angle for maximum power is found from Eq. (5.97):

$$K_1 = \frac{1.45(1)}{0.8} = 1.81$$

$$K_2 = \frac{(1.0)^2 (0.8 - 0.5)}{2(0.8)(0.5)} = 0.375$$

$$\delta = \arccos\left(-\frac{1.81}{8(0.375)} \pm \sqrt{\left(\frac{1.81}{8(0.375)}\right)^2 + \frac{1}{2}}\right)$$

from which

$$\delta = 70.9°$$

The maximum torque is

$$T = \frac{\text{maximum power in perunit}}{\text{speed perunit}}$$

$$T = \frac{1.45(1)}{0.8} \sin 70.9° + \frac{(1.0)^2(0.8 - 0.5)}{2(0.8)(0.5)} \sin 141.9°$$

$$= \underline{1.94 \text{ pu}}$$

2. For a cylindrical rotor,

$$X_d = X_q = X_s = 0.8 \text{ pu}$$

a. Using the per-phase equivalent circuit of Fig. 5-36(c), from KVL, we have

$$V_f \underline{/-\delta} = 1.0 \underline{/0°} - j1.0 \underline{/+18.2°} (0.8) = 1.46 \underline{/-31.3°} \text{ pu}$$

Thus,

$$\delta = \underline{-31.3°}$$

and

$$\delta_m = \frac{2}{10}(31.3) = \underline{6.3°}$$

b. The excitation voltage is

$$V_f = \underline{1.46 \text{ pu}}$$

c. For maximum torque,

$$\delta = 90°$$

and

$$T = \left(\frac{1}{1.0}\right)\frac{1.46(1)}{0.8} = \underline{1.83 \text{ pu}}$$

For purposes of comparison, the results are summarized in Table 5-3.

TABLE 5-3 Summary of results of Example 5-7

Type of rotor	Torque Angle in Electrical Degrees	Excitation Voltage in pu	Maximum Torque in pu
Salient	22.3	1.45	1.94
Cylindrical	31.3	1.46	1.83

Exercise

5-7

a. Draw the phasor diagram of a salient synchronous motor operating at unity power factor and show that the power angle is given by

$$\delta = \arctan \frac{I_a X_q}{V_t - I_a R_a}$$

b. Derive Eqs. (5.93) and (5.97).

c. Determine the torque angle for maximum power of a 480 V, 50 kW, 0.9 efficient, unity power factor, salient synchronous motor. The motor's direct- and quadrature-axis reactances are 3 Ω and 2 Ω, respectively.

Answer (c) 71°.

5.4 Conclusion

The operation of synchronous motors is based on the natural tendency of two fields to try to align their magnetic axes. The two fields—the stator and rotor fields—rotate at synchronous speeds.

The rotating stator field is produced by balanced three-phase currents flowing in properly distributed stator coils. The rotor field is produced by the direct current that flows through the synchronously rotating rotor windings.

The speed of a synchronous motor is directly proportional to the frequency of the stator voltage and inversely proportional to the number of poles. For a fixed-frequency source and number of poles, the speed is also fixed.

Because ordinary synchronous motors have these two crucial limitations—they operate at only one speed, and they require a dc voltage for their rotor windings and a balanced three-phase voltage for their stator windings—they have few applications.

At starting, synchronous motors function like squirrel-cage induction machines and thus draw large inrush currents. As a result, the voltage drop in the impedance of the supply network may be considerable. The higher the inrush current, the greater the disturbance on the other electrical apparatus operating within the same distribution network.

High-inrush currents also *reduce* the starting torque to levels that are sometimes too low to meet the requirements of a connected load. In this case, an engineer may select a soft start motor. Such motors have relatively low starting torque and current, and they produce sufficient power to provide the high starting-torque requirements of some loads.

The torque produced by a synchronous motor depends on its terminal voltage, its excitation voltage, and the phase angle between these two voltage phasors (commonly known as the power angle). Under normal operating conditions, the power angle is usually less than 30°.

The stator field may aid or oppose the rotor field. Once a machine is designed, this armature reaction depends on the operating power factor and on the stator current. As a result, a leading-power-factor motor requires more field current than a lagging-power-factor motor. Conversely, in order to produce a given terminal voltage, a synchronous generator requires much more field current when it operates at a lagging power factor than when it operates at a leading power factor. In either case, it is always useful to sketch and inspect a phasor diagram before arriving at any conclusions. As seen from the ac voltage supply, synchronous machines appear as variable capacitive or inductive impedances.

Vee curves give the variation of the power factor, and the armature current drawn by the motor, as a function of field current. In general, synchronous motors should not be designed for, or be operated at, leading power factors because their stator copper losses are thereby increased. It is much more economical to operate at unity power factor and to purchase a capacitor bank to provide the required lagging kVAR power. Reducing the power factor of a synchronous motor has a disadvantage, however: Increasing the power factor from 0.8 leading to unity decreases the breakdown torque of the motor. Depending on load requirements, the motor may therefore stall. In short, each particular situation should be analyzed before any arbitrary adjustments are made.

Synchronous generators are used exclusively to generate three-phase power. Utility companies use large units (up to several hundred MVAs) to generate commercially available three-phase power. Units with lower ratings (up to 1 MVA) are used in various plants to generate the standby power required to operate critical loads when the normal power source is temporarily interrupted. These generators are always driven at constant speeds by so-called prime movers.

Theoretically, depending on the field current and power factor, the terminal voltage of a generator varies over a wide range. In practice, however, owing to closed-loop voltage-regulator systems, this voltage is kept constant at the desired level.

The various circuit parameters of a synchronous machine, when expressed in per unit (as is the practice), are within a closed range and are available through a manufacturer's published data.

Salient-pole machines are stiffer than cylindrical-rotor machines. Being able to tolerate stronger load disturbances, they normally operate at speeds below 1800 r/min. Salient machines can be analyzed using the graphical or mathematical interpretation of the De Blondel equation. Salient machines can develop up to 40% of their rated torque without any excitation, whereas cylindrical-rotor machines cease to operate when the field current is removed.

Synchronous motors are not used (except when they are equipped with flywheels) to drive loads whose torque changes abruptly or is cyclical in nature because they can become unstable. Their stiffness, or capacity to oppose external disturbances, depends on their degree of saliency. Damping windings provide a higher starting torque and a stabilizing force each time the motor is forced to deviate from its synchronous speed.

Fig. 5-37 shows the motor torque and load characteristic versus speed for a 2800 kW synchronous motor.

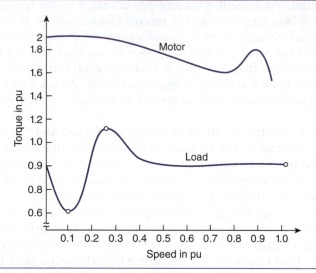

FIG. 5-37 Load and motor torque-speed characteristics. Motor: 2800 kW, 225 r/min, 60 Hz, 4000 V synchronous motor. Load: Ball grinding mill.

5.5 Review of Important Mathematical Relationships

Table 5-5 summarizes the main concepts of this chapter in their mathematical form.

Item	Description and Formula	Remarks
	TABLE 5-5 Review of important mathematical relationships	
	a. General	
1	Strength of the synchronously rotating field	
	$$\mathcal{F}_s = \frac{3}{2}\,\mathcal{F}_1 \cos(\omega t - \beta)$$	Eq. (5.1)
2	Speed of synchronous machines	
	$$n_s = 120\,\frac{f}{P}\ \text{r/min}$$	Eq. (5.2)
	b. Cylindrical Rotor	
3	Synchronous impedance	
	$$Z_s = R_a + jX_s$$	Eq. (5.7)

TABLE 5-5 (Continued)		
Item	Description and Formula	Remarks
4	Relationship between mechanical and electrical radians (power angle) $$\delta_m = \frac{2}{p}\delta$$	Eq. (5.8)
5	Effective turns ratio $$N_e = \frac{I_a}{I_a'}$$	Eq. (5.9)
6	Field current $$I_f = I_m - I_a'$$	Eq. (5.11)
7	Excitation voltage $$V_f \underline{/-\delta} = V_t \underline{/0°} - I_a\,(R_a + jX_s)$$	Eq. (5.12)
8	Power developed ($R_a = 0$) $$P = \frac{V_t V_f}{X} \sin \delta \ \text{watts/phase}$$	Eq. (5.28)
9	Reactive power ($R_a = 0$) $$Q = \frac{V_t}{X}(V_t - V_f \cos \delta)$$	Eq. (5.36)

c. Three-phase Generators

10	Excitation voltage $$V_f \underline{/\delta°} = V_t \underline{/0°} + I_a(R_a + jX_s)$$	Eq. (5.50)
11	Field current $$I_f = I_m + I_a'$$	Eq. (5.51)
12	Effective turns ratio $$N_e = \text{slope of SCC}$$ $$= \frac{I_a}{I_a'}$$	Eq. (5.58)
13	Saturated value of synchronous reactance $$X_{s_{sat}} = \frac{V_{rated}}{I_{a_2}}\bigg|_{I_{f2}}$$	Eq. (5.64)

(Continues)

TABLE 5-5 *(Continued)*

Item	Description	Remarks
	d. Salient-Pole Machines	
14	Excitation voltage	
	$$V_f \underline{/\pm\delta^\circ} = V_t \underline{/0^\circ} \pm I_a R_a \pm j I_q X_q \pm j I_d X_d$$	Eq. (5.73)
	(positive sign for generator; negative sign for motor)	
15	Power developed	
	$$P = \frac{V_f V_t}{X_d} \sin \delta + \frac{V_t^2 (X_d - X_q) \sin 2\delta}{2 X_d X_q} \text{ watts/phase}$$	Eq. (5.89)

5.6 Manufacturer's Data

Tables 5-6, 5-7, and 5-8 present typical machine data, as furnished by manufacturers.

TABLE 5-6 Estimated characteristics for a three-phase, 514 r/min, 60 Hz, 4000 V, 3300 kW salient-pole synchronous motor—soft start—brushless exciter

		Power Factor Leading	
Item	Parameter	0.90	0.80
1	Inrush current in percent	370	340
2	Torque in percent:		
	Starting	50	50
	Pull-in	65	62
	Pull-out	175	195
3	Efficiency in percent:		
	Full-load	95.2	94.8
	$\frac{1}{2}$ full-load	94.2	93.8
4	Reactances in per unit:		
	X_d	2.2	2.46
	X_d'	0.45	0.55
	X_d''	0.28	0.32
5	Field excitation in kW	40	40

Based on data from General Electric Canada, Inc.

TABLE 5-7   Data for a three-phase, 225 r/min, 4000 V, 0.80 power factor leading. 2800 kW synchronous motor—hard start—brushless exciter

Item	Parameter	Value	Remarks
1	Inrush current in percent	600	The starting characteristics of the motor and load (ball grinding mill) are shown in Fig. 5-37.
2	Torque in per unit:		
	Starting	2.2	
	Pull-in	1.4	
	Maximum	275	
3	Reactances in per unit:		
	X_d	1.1	
	X_d'	0.29	
	X_d'' (unsaturated value)	0.126	
4	Efficiency at full-load	0.95	
5	Exciter's amps at rated load	13.9	
6	Field resistance at 25°C in ohms	6.32	
7	Allowable stall time in seconds	5	
8	Accelerating time in seconds	4.5	

Based on data from General Electric Canada, Inc.

TABLE 5-8   Data for high- and low-starting-torque synchronous motors: 514 r/min, 60 Hz, 0.8 leading Pf, 4500 kW

Item	Description	Type of Starting Torque	
		High	Low
1	Per-unit cost[†]	1.0	1.15
2	Inrush current in per unit	6.5	3.75
3	Power factor at starting in percent	28	12
4	Efficiency at full-load in percent	96	96.5
5	Torques in per unit:		
	Starting	1.8	0.4
	Pull-in	1.4	0.7
	Pull-out	2	2
6	Torque angle at nominal operating conditions	28°	27°
7	Relative diameters	1	1.2

†*In 2010, the per-unit cost was about $600,000.*
Based on data from General Electric Canada, Inc.

5.7 Review Questions

1. What are the advantages and disadvantages of synchronous motors?

2. Explain the magnetizing and demagnetizing effects of the armature current.

3. Draw and explain the Vee curves for a synchronous machine.

4. When starting a synchronous motor, a small resistance is placed in the rotor circuit; otherwise, the open-circuited windings may be damaged. Explain the reasons for this.

5. What conditions are necessary to produce a synchronously rotating magnetic field? How does this field compare with that produced by a permanent magnet rotating at the same speed?

6. The excitation voltage in a synchronous machine can be either larger or smaller than the terminal voltage, depending on the power factor. What are the reasons for this?

7. What is torque angle, and on what external factors does it depend? How could it be observed in a laboratory?

8. What does the Potier triangle give? How would you measure it?

9. Define voltage regulation, and explain its significance and effects on the operation of electrical equipment.

10. What are the advantages and limitations of synchronous machines with salient rotors?

11. How does the impedance of a transmission line affect the power that can be transmitted between a generator and an infinite bus?

5.8 Problems

5-1 A three-phase, 60 Hz, 10-pole, 480 V, 100 kW synchronous motor has an overall efficiency of 0.92 and draws rated current at unity power factor. Its per-phase synchronous impedance is $0.10 + j1.1$ ohms. Determine:

a. The excitation voltage.

b. The approximate value of the torque angle, in mechanical degrees.

c. The stator electrical losses and the rotational losses.

5-2 A 300 kW, 4.16 kV, 600 r/min, 60 Hz, 3-ϕ synchronous motor has a synchronous impedance of $1 + j5$ ohms/phase. The motor delivers rated power and draws nominal current at a power factor of 0.9 leading. Its rotational losses are 12 kW. Determine:

a. The armature current.

b. The excitation voltage and the torque angle.

c. The per-unit value of the synchronous impedance.

5-3 A 230 V, three-phase, 60 Hz, eight-pole, 37.3 kW star-connected synchronous motor has a synchronous reactance of 0.60 ohms per phase. The rotational losses are constant at 1800 watts, and the stator resistance is negligible. Data for the open-circuit characteristic, at rated speed, is as follows:

Line-to-line voltage (V)	138	228	292	332	347
Field current (A)	2.0	4	6	8	10

Determine:

a. The field current when the motor draws rated current at a power factor of 0.80 leading.

b. The armature current when the load is halved and the excitation remains the same as in (a).

5-4 A 300 kW, four-pole, 2.3 kV, 60 Hz, 3-ϕ synchronous motor has a synchronous impedance of $0.3 + j4.5$ ohms/phase. Determine:

a. The torque angle at which the motor will draw maximum power from the infinite bus.

b. The torque angle at which the motor will draw maximum reactive power from the infinite bus.

5-5 A 750 kW, 1200 r/min, 2.2 kV, 60 Hz, 3-ϕ, 0.965 efficient synchronous motor has a per-phase synchronous reactance of 3.0 ohms and delivers power to a constant-torque mechanical load. The open-circuit characteristic at the operating region is given by

$$V_f = 254.37 I_f$$

When the motor draws rated current at unity power factor, determine:

a. **1.** The excitation voltage.

2. The torque angle.

3. The field current.

4. The magnitude of the armature current.

5. The real, reactive, and complex power drawn by the motor.

b. If the excitation is changed by ±25% from its nominal value, while the load remains constant, repeat (a).

5-6 A 300 kW, 2-pole, 1.2 kV, 60 Hz, 3-ϕ synchronous motor has a leakage impedance of $0.1 + j1.0$ ohms/phase. The effective turns ratio per pole is 20:1. The open-circuit characteristic is given by

$$V_f = 30 + 55 I_f$$

The motor receives electrical power at unity power factor. Its rotational losses are 17.05 kW. Determine:

a. The magnetization voltage.

b. The magnetizing component of the field current.

c. The armature reaction in equivalent field amperes.

d. The field current and the exact torque angle, in mechanical degrees.

e. The voltage induced in the armature windings due to the field current.

5-7 A 500 kVA, 1800 r/min, 4.16 kV, 60 Hz, 3-ϕ, 0.94 efficient synchronous machine supplies rated current at 0.80 power factor lagging. Its synchronous impedance per phase is $1 + j12$ ohms. Determine:

a. The number of poles.

b. The per-unit value of the synchronous impedance.

c. The stator copper loss.

d. The electromechanical torque developed.

5-8 **a.** A 1.2 MVA, 1800 r/min, 6.6 kV, 60 Hz, 3-ϕ generator delivers rated current to an infinite bus at unity power factor. The synchronous reactance is 1.0 per unit, based on its own rating. The open-circuit characteristic in the operating region is given by $V_f = 100 + 65 I_f$. Determine:

1. The excitation voltage.

2. The torque angle.

3. The armature current.

4. The field current.

5. The complex power delivered to the load.

b. Repeat part (a) if the field current is kept constant while the power supplied by the prime mover is changed by ±25%.

5-9 A 750 kW, 1800 r/min, 6.3 kV, 60 Hz, 3-ϕ, 0.94 efficient synchronous motor when tested gave the following results:

Open-circuit test at rated speed:

Line voltage in volts	3500	5500	6500	7500	8000	8500
Field current in amperes	6	9.6	11.8	15	17.2	22

Short-circuit test at rated speed:

Line current	105 A
Field current	9 A

Zero-power-factor test:

Line current	105 A
Field current	20 A
Line voltage	5900 V

The motor delivers rated power and operates at a power factor of 0.80 leading. The resistance of the armature is 1.35 Ω/phase. Determine:

a. The Potier triangle.

b. The effective turns ratio.

c. The field current.

5-10 A 373 kW, 480 V, 0.8 Pf leading salient-pole synchronous motor has a direct-axis synchronous reactance of 1.0 per unit and a quadrature-axis synchronous reactance of 0.65 per unit. The rotational and stator copper losses are negligible. The field current is adjusted so that the motor draws minimum armature current while delivering rated power to a mechanical load. Under this operating condition, determine:

a. The power factor.

b. The power angle.

c. The excitation voltage.

d. The power developed. What percent of the total power is reluctance power?

5-11 A 5 MVA, 600 r/min, 6.6 kV, 60 Hz, 3-ϕ synchronous alternator delivers rated current at a power factor of 0.95 leading. Its direct-axis reactance and quadrature-axis reactance are 8.0 and 6.0 ohms, respectively. Determine:

a. The torque angle at full-load.

b. The regulation.

c. The reluctance torque in percent of the rated torque.

5.12 Explain the following:*

a. Increasing the load to a power generating system, the frequency of the distribution system is momentarily decreased, while on decreasing the load, the frequency is increased.

(This frequency variation is used in the momentary control of the power output in a multi-machine generating station when the power of any of the supplied loads is suddenly changed).

b. Generating systems that include rotating inertia are less sensitive to electrical disturbances than those which do not include it (batteries, photo voltaics, etc.).

c. Utilities without downstream reactive power cannot push real power towards consumers.

* Extensive articles on wind power generation and photovoltaics one may find in "Transmission and Distribution World," July 2013.

6

DC Machines

6.0   Introduction

6.1   Steady-State Analysis

6.2   Modern Methods of Speed Control

6.3   Conclusion

6.4   Review Questions

6.5   Problems

What You
Will Learn in
This Chapter

A Theoretical Highlights

1 The principles of motor and generator operation

2 Flux-density distribution and armature reaction

3 Types of dc machines and their characteristics

4 Control of starting current

5 Compound generators

6 Modern methods of speed control

B Outstanding Practical Characteristics

1 Principles of operation and motor-generator characteristics

2 Effects of armature reaction on the machines' performances

3 Control of the torque-speed characteristics of series and separately
   excited dc motors

4 How to minimize the motor's high starting current

5 Thyristor speed control of dc motors

6 Power factor of thyristor-controlled dc motor

7 Typical manufacturers' machine characteristics

C Additional Students' Aid on the Web

1 Transient analysis

2 Time-constant considerations

3 Block diagrams and transfer functions

4 Transformation of mechanical parameters

5 General machine equations

6 Examples, exercises, and problems

6.0 Introduction

Direct-current motors are widely used in hoists, steel mills, cement plants, the pulp and paper industry, and mining operations. They are versatile and can be adjusted to meet tough job requirements, such as high starting torque, constant-power or constant-torque loads, quick acceleration and deceleration, and speed variations of up to $\pm$ 400% of base value.

Although their torque-speed characteristics are easily adjustable, dc machines are not used for ordinary industrial applications because of the limitations, cost, and unavailability of dc voltage sources.

This chapter deals with the dc machine's principles of operation, steady-state operation, and modern techniques of speed control.

6.1 Steady-State Analysis

6.1.1 General

Like all the three-phase ac machines, direct-current machines have a stator and a rotor winding. The stator winding is referred to as the field winding, and the rotor winding is known as the armature winding. Manufacturer's pictures of a dc motor and its stator and rotor windings are shown in Figs. 6-1, 6-2, and 6-3.

FIG. 6-1 A 375 kW dc motor. *Courtesy of General Electric*

FIG. 6-2 Stator of a 375 kW dc motor. *Courtesy of General Electric*

FIG. 6-3 Rotor of a 1200 kW, 600 V, 900 r/min dc motor. *Courtesy of General Electric*

Field windings are wound around the poles of the stator and are supplied with dc current, which produces the main magnetic field of the machine. Depending on the machine type and rating, the stator structure may incorporate additional windings, such as the compensating, the commutating, and the auxiliary field windings. The functions and the relative physical location of these windings are described in Sections 6.1.5 and 6.1.7.

Armature windings are placed in the rotor slots, which are uniformly distributed around the rotor's periphery. The end connections of the armature windings terminate in the commutator segments (longitudinal bars of copper), which ride on the stationary brushes. The commutator segments are insulated from each other. In conjunction with the brushes, they rectify the ac current of the armature windings. The armature current of a dc motor is conducted from the external machine terminals to the brushes, then to the commutator segments, and from there to the armature conductors.

Depending on how the field winding is excited relative to the armature coil, dc machines are generally classified as shunt, series, or separately excited. As shown in Fig. 6-4, the field winding in a shunt machine is connected in parallel to the armature.

In series dc machines, the field winding is connected in series with the armature winding (see Fig. 6-14(a)). Depending on the location and relative magnetic orientation of the auxiliary field winding, dc shunt machines are further subdivided into other categories (see Section 6.1.7).

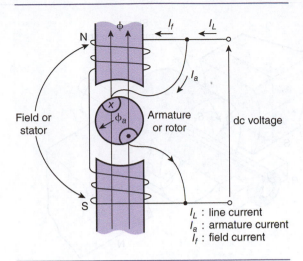

FIG. 6-4 An elementary representation of a two-pole dc shunt motor. (The effect of the armature flux on the field flux is neglected.)

Regardless of the winding connection, as shown in Fig. 6-4, the current through the field windings (I_f) produces the field flux (ϕ_f), and the current through the armature windings (I_a) produces the armature flux (ϕ_a). The field flux combines with the armature flux to produce the machine's net effective flux (ϕ) per pole. The natural tendency of the stator and rotor fields to align their magnetic axes produces torque.

6.1.2 Principles of Operation

The equation that describes the generator-motor operation can be derived from the concept of the speed voltage (Eq. (1.139)) and Ampere's law (Eq. (1.194)), or from use of the concepts associated with the energy stored in a coil and the definition of flux linkages. This section uses the last-named method because it gives a better understanding of the fundamentals.

Motor Principles

The operation of a dc motor is based on the tendency of a current-carrying coil, when placed in an external magnetic field, to rotate in such a way that the field it creates is parallel to, and in the same direction as, the external field. The general expression for the torque developed by a dc motor is derived as follows.

For the sake of simplicity, the armature winding shown in Fig. 6-5 has only a single-loop coil. The instantaneous value of the torque developed (T_i) is

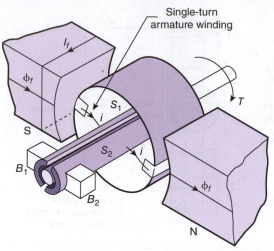

Single-turn armature winding

S_1, S_2 : commutator segments
B_1, B_2 : stationary brushes
N, S : north and south magnetic poles

FIG. 6-5 An elementary representation of a dc machine.

determined by the partial derivative of the energy stored (W_f) in the armature winding, with respect to its angle of rotation (θ). Mathematically expressed,

$$T_i = \frac{\partial}{\partial \theta} W_f \qquad (6.1)$$

The incremental energy stored is given by

$$dW_f = v_g i \, dt \qquad (6.2)$$

where v_g is the voltage induced in a single turn of the armature coil and i is its current. From the above, we obtain

$$T_i = \frac{v_g i \, dt}{\partial \theta} \qquad (6.3)$$

If we substitute for the voltage induced its equivalent expression in terms of the flux linkage (λ), we get

$$T_i = \frac{1}{\partial \theta} \left(i \frac{d\lambda}{dt} \, dt \right) \qquad (6.4)$$

$$T_i = i \frac{d\lambda}{\partial \theta} \qquad (6.5)$$

As the rotor turns dx of a revolution, the incremental change in the angle of rotation is

$$d\theta = 2\pi(dx) \text{ radians} \qquad (6.6)$$

Each side of the conductor cuts only once through the regions of the north and south magnetic field. That is, the incremental change of the flux linkage for each side of the single-turn armature winding is

$$d\lambda = 2\phi(dx) \qquad (6.7)$$

where dx, as before, is the incremental change in the rotation of the rotor. The change in the flux linkage for the entire length of the winding is

$$d\lambda = 4\phi(dx) \text{ for a two-pole machine} \qquad (6.8)$$
$$= 2p\phi(dx) \text{ for a } p\text{-pole machine} \qquad (6.9)$$

The current through the single winding is determined by the number of parallel paths of the armature winding and is related to the external armature current (I_a) by

$$i = \frac{I_a}{\beta} \qquad (6.10)$$

where

β = the number of parallel paths of the armature winding

β = number of machine poles, for *lap-type** armature winding

$\beta = 2$, for *wave-type*[†] armature winding.

From Eqs. (6.5), (6.6), (6.9), and (6.10), we see that the expression for the torque developed by a single conductor is

$$T_i = \frac{p}{\beta\pi} \phi I_a \text{ N} \cdot \text{m/turn} \qquad (6.11)$$

For N turns of the armature winding, the total torque developed (T) by the motor is

$$T = \frac{Np}{\beta\pi} \phi I_a \text{ N} \cdot \text{m} \qquad (6.12)$$

or

$$T = K\phi I_a \text{ N} \cdot \text{m} \qquad (6.13)$$

Lap-type winding: The starting and terminating ends of adjacent coils overlap each other's winding and are connected to adjacent commutator segments. The number of parallel paths is equal to the number of poles. This type of winding is used for low-voltage, high-current applications.

[†]*Wave-type winding*: The coils are connected in two parallel paths regardless of the number of poles. Most dc machines with ratings less than 75 kW are constructed using this type of winding.

where the constant of proportionality K is given by

$$K = \frac{Np}{\beta\pi} \tag{6.14}$$

Thus, the torque developed by a dc motor is directly proportional to the machine's effective field, to the armature current, to the number of poles, and to the number of armature conductors.

Generator Principles

The operation of generators is based on the induction principle, whereby voltage is induced in a conductor that is rotated through an external magnetic field. The polarity of the induced voltage, as per Lenz's law, is in such a direction as to produce a current whose mmf opposes the external magnetic field. The conductor under consideration is the coil of the rotor, the external magnetic field is provided by the stator coil, and the conductor's rotation is derived from a prime mover such as an induction motor, a steam or hydroturbine, and so forth.

The derivation of the general expression for the voltage generated is as follows. The voltage induced in a single-turn armature winding (Fig. 6-5) is given by the time rate of change of the armature coil's flux linkage. Considering only magnitudes, we have

$$v_g = \frac{d\lambda}{dt} \tag{6.15}$$

where v_g is the voltage generated and λ is the armature coil's flux linkage.

The change in the flux linkage of the armature winding that rotates through dx of a revolution is, as before,

$$d\lambda = 4\phi(dx) \quad \text{for a two-pole machine} \tag{6.16}$$
$$= 2p\phi(dx) \quad \text{for a } p\text{-pole machine} \tag{6.17}$$

The increment of time that the winding requires to travel through dx of a revolution is

$$dt = \frac{2\pi}{\omega}(dx) \text{ seconds} \tag{6.18}$$

where ω is the angular speed of the armature in radians per second. From Eqs. (6.15), (6.17), and (6.18), we obtain

$$v_g = \frac{p}{\pi}\phi\omega \text{ V/turn} \tag{6.19}$$

The total voltage generated in an armature winding depends on the number of coil turns connected in series. The number of coil turns connected in series is equal to the number of conductors divided by the number of parallel paths. That is,

$$\text{number of series conductors} = \frac{N}{\beta} \qquad (6.20)$$

From Eqs. (6.19) and (6.20), the total voltage generated (V_g) in the armature conductors is

$$V_g = \frac{Np}{\beta\pi}\,\Phi\omega \qquad (6.21)$$

or

$$V_g = K\phi\omega \qquad (6.22)$$

where the constant of proportionality K is the same as that given for the torque equation (see Eq. (6.14)).

The voltage generated in a single conductor, the machine's total voltage, and its rectification are further discussed in Section 6.1.6 and are shown in Fig. 6-10.

The voltage generated in the armature conductors is also referred to as the armature voltage V_a, the countervoltage V_c, the back emf V_b, or simply the speed voltage. You must differentiate between the voltage generated in a machine and its terminal voltage. In this text, the voltage generated will be designated as V_g and the terminal voltage as V_t.

From the preceding discussion it is evident that the principles of dc motor and generator operation are the same as those of polyphase ac machines. The essential difference between a motor and a generator is the direction of energy flow. A motor transforms electrical energy into a mechanical form, whereas a generator transforms mechanical energy into an electrical form.

6.1.3 Power Considerations

The electromagnetic power developed (P_d) by a machine is equal to the product of the electromagnetic torque developed (T) times the speed (ω) of its rotor. In mathematical form,

$$P_d = T\omega \qquad (6.23)$$

The torque is expressed in N · m, the speed in rad/s, and the power in watts. Applying the power-balance concept for a motor, we obtain

$$\left(\begin{array}{c}\text{input}\\ \text{electrical power}\end{array}\right) = \left(\begin{array}{c}\text{sum of various}\\ \text{winding copper losses}\end{array}\right) + \left(\begin{array}{c}\text{power}\\ \text{developed}\end{array}\right)$$

Using mathematical symbols,

$$V_t I_L = I_f^2 R_f + I_a^2 R_a + P_d \qquad (6.24)$$

where

V_t and I_L = the motor's supply (terminal) voltage and line current, respectively

$I_f^2 R_f$ = the copper losses of the field winding

$I_a^2 R_a$ = the copper losses of the armature winding

The power developed provides the output power (P_{out}) plus the rotational losses ($P_{r.l.}$). Mathematically expressed,

$$P_d = P_{r.l.} + P_{out} \qquad (6.25)$$

To a certain extent, rotational losses depend on the speed of the motor, but throughout this chapter, they will be assumed to remain constant. At nominal operating conditions, the output power of the motor is referred to as the full-load power, rated power, or nameplate power.

The various losses for a 100 kW dc machine and the direction of energy flow are shown in Fig. 6-6.

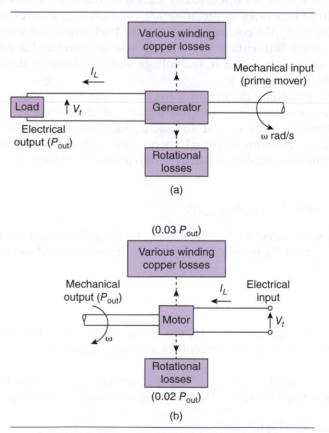

(a)

(b)

FIG. 6-6 DC machines—direction of power flow and typical losses for a 100 kW machine: **(a)** generator, **(b)** motor.

A 10 kW, four-pole dc machine is magnetized at a flux level of 0.08 Wb and has a lap-type armature winding with 24 turns. When the motor delivers rated power at 1200 r/min, determine:

EXAMPLE 6-1

a. The voltage induced in the armature windings.
b. The armature current.

SOLUTION

a. The speed of the motor is

$$\omega = 1200 \frac{2\pi}{60} = 125.7 \text{ rad/s}$$

For a lap-type winding, the number of parallel paths is equal to the number of poles. Thus,

$$\beta = 4$$

Substituting the given data into Eq. (6.21), we obtain

$$V_g = \frac{24(4)}{4\pi} (0.08)(125.7) = \underline{76.8 \text{ V}}$$

b. From Eq. (6.23), the torque is

$$T = \frac{\text{power}}{\text{speed}} = \frac{10,000}{125.7} = 79.6 \text{ N} \cdot \text{m}$$

From Eq. (6.12),

$$I_a = \frac{79.6}{24 \times 4(0.08)} (4\pi) = \underline{130.21 \text{ A}}$$

A 5 kW, two-pole dc motor has 200 W of rotational losses and draws an armature current of 30 A when it delivers rated power at 900 r/min. Determine:

Exercise

6-1

a. The voltage generated in the armature windings.
b. The torque developed and the output torque of the motor.

Answer   (a) 173.3 V;   (b) 55.2 N · m, 53.1 N · m.

A two-winding transformer has a primary and a secondary winding and two corresponding mmf's. The interaction of these two fields does not produce any torque. What are the reasons for this?

6.1.4 Voltage and Torque Relationships as Functions of Mutual Inductance

This section derives general expressions for the voltage generated and the torque developed in a dc machine as functions of the mutual inductance between the stator and rotor windings.

The voltage induced in the armature of a dc machine with constant field current is given by

$$v_g = -I_f \frac{d}{dt} L_{af(\theta)} \tag{6.26}$$

where $L_{af(\theta)}$ is the mutual inductance between the armature and field windings as a function of the angle of rotation. The mutual inductance is nonsinusoidal in waveform. It has a fundamental term and a higher order harmonics. The negative sign gives the polarity of the induced voltage as per Lenz's law.

Rewriting the relationship of mechanical (θ_m) and electrical (θ) radians (see Eq. (3.15)), we have

$$\theta = \frac{p}{2} \theta_m \tag{6.27}$$

Differentiating with respect to time, we get

$$\frac{d\theta}{dt} = \frac{p}{2} \frac{d\theta_m}{dt} \tag{6.28}$$

From the above,

$$\frac{d\theta}{dt} = \frac{p}{2} \omega \tag{6.29}$$

where p is the number of poles and ω is the speed of the motor in rad/s.

Equation (6.26) can be rewritten as:

$$v_g = -I_f \frac{d\theta}{dt} \frac{dL_{af(\theta)}}{d\theta} \tag{6.30}$$

From the above and Eq. (6.29), we obtain

$$v_g = -I_f \frac{p}{2} \omega \frac{dL_{af(\theta)}}{d\theta} \tag{6.31}$$

Assuming that the mutual inductance has only a fundamental component, the maximum value of the induced voltage is

$$V_g = \frac{p}{2} \omega I_f L_{\text{afm}} \tag{6.32}$$

where L_{afm} is the maximum value of the fundamental component of the mutual inductance.

The equation of the torque is derived as follows:

$$T = \frac{\text{power}}{\text{speed}}$$

$$= \frac{(\text{voltage induced in the armature})(\text{armature current})}{\text{speed}}$$

Thus,

$$T = \frac{V_g I_a}{\omega} \tag{6.33}$$

From the above and from Eq. (6.32), we obtain

$$T = \frac{p}{2} I_f L_{\text{afm}} I_a \tag{6.34}$$

Comparing Eqs. (6.13) and (6.34) shows that the constant of proportionality of a given dc machine may also be expressed in terms of the mutual inductance between the field and armature windings. That is,

$$K = \frac{p}{2\phi} I_f L_{\text{afm}} \tag{6.35}$$

The maximum value of mutual inductance between the rotor and stator windings can be measured by one of the standard methods (see Chapter 1, Example 1-30), or it can be evaluated indirectly from Eq. (6.32).

The voltage induced in the field windings due to armature current is negligible because the structure of the machine drastically reduces the magnetic coupling of the stator windings to the flux of the armature current.

Exercise

6-3

A 5 kW, 1200 r/min, four-pole dc machine has an armature current of 20 A and a field current of 2 A. Determine:

a) The amplitude of the fundamental component of the mutual inductance between the stator and rotor windings.
b) The voltage generated.

Answer   (a) 0.497 H;   (b) 250 V

6.1.5 Magnetic System, Flux Distribution, and Armature Reaction

This section discusses the dc machine's magnetic system, flux distribution, and the interaction of the stator and rotor fields, commonly known as armature reaction.

Magnetic System and Flux Distribution

Figure 6-7 shows an elementary magnetic system for a two-pole dc machine. The flux produced by the mmf of the field winding completes its loop by passing through the air gaps, stator, and rotor structures.

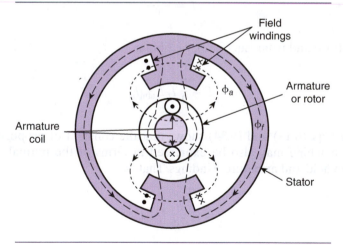

FIG. 6-7 An elementary magnetic system for a two-pole dc machine.

Applying Ohm's law for magnetic circuits, we have

$$\phi_f = \frac{N_f}{\mathscr{R}_f} I_f \qquad (6.36)$$

where the subscript f indicates field parameters. The reluctance ($\mathcal{R}_f$) represents the opposition presented to the field flux by the intrinsic properties of the magnetic circuit components. This reluctance is determined by the sum of the magnetic resistances of the rotor, stator, and air-gap paths, through which the field flux completes its loop.

In general,

$$
\mathcal{R} = \frac{l}{\mu A}
$$

$$
= \frac{\text{length of closed magnetic path}}{(\text{permeability}) \times (\text{cross-section area perpendicular to the flux})} \qquad (6.37)
$$

Taking into consideration the three distinct parts of the field's magnetic circuit, we have

$$
\mathcal{R}_f = \frac{l_r}{\mu_r A_r} + \frac{l_s}{\mu_s A_s} + \frac{l_g}{\mu_g A_g} \qquad (6.38)
$$

where the subscripts r, s, and g represent, respectively, the usual parameters of rotor, stator, and air gaps through which the field flux closes its magnetic circuit.

For operations along the linear section of the magnetic characteristic, the permeability is constant. Thus from Eqs. (6.36) and (6.38), we obtain

$$
\phi_f = K_f I_f \qquad (6.39)
$$

where K_f is a constant of proportionality that is proportional to the number of field winding turns and inversely proportional to the reluctance of the field's magnetic circuit.

Similarly, the flux of the armature coils (ϕ_a) is determined by

$$
\phi_a = K_a I_a \qquad (6.40)
$$

where I_a is the armature current and K_a is a constant of proportionality which, as before, depends on the reluctance of the armature's magnetic circuit and on the number of turns of the armature winding.

Equations (6.39) and (6.40) are *not* applicable for machine operations along the nonlinear part of the magnetization characteristic.

Flux-Density Distribution

Figure 6-8 shows a simplified version of the flux-density distribution of a two-pole dc machine. The armature (B_a) and field (B_f) flux densities combine to produce the resultant (B_r) or effective flux density.

The field flux density under the north magnetic pole is designated as positive, while that under the south magnetic pole is negative. The relative polarity becomes obvious when one considers that in an N–S magnetic system, the lines of force leave the north pole and enter the south pole.

The distribution of the armature flux density is of lower amplitude because of the larger reluctance path through which the armature flux completes its magnetic

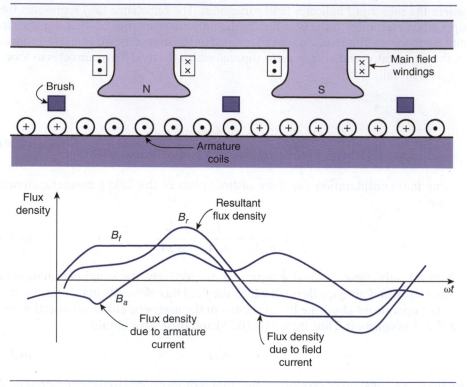

FIG. 6-8 Simplified view of a two-pole machine and the corresponding flux-density distributions. (Interpoles are not shown.)

circuit. Notice also that for one complete revolution of the rotor, each side of the armature coil goes through a complete magnetic circuit. As a result, the armature coil will appear to travel two cycles for each magnetic field cycle.

The flux-density distribution is zero under the stationary brushes because the stationary brushes are physically located at a point between the north and south magnetic poles where the field is zero.

The Concept of Armature Reaction

The flux of the armature current effectively *opposes* the flux of the field current and *distorts* the waveform of the field's flux-density space distribution. These two adverse effects of the armature flux are referred to as armature reaction. The qualitative analysis of the armature reaction is described next, while its quantitative analysis is given in Section 6.1.9.

Opposition to the Field

Depending on the rotor position, the flux of the armature *aids* the flux of the field on one side of the pole and *opposes* it on the other side. Aiding the field in one pole partly saturates the magnetic material, and thus the effective field is reduced.

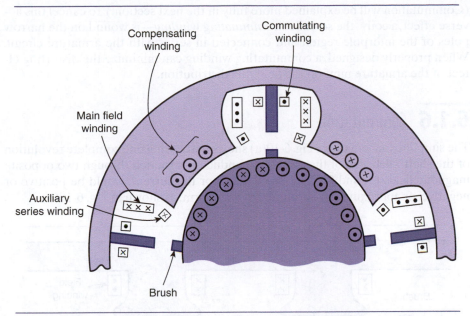

FIG. 6-9 DC machines. This figure illustrates the physical location and relative polarity of compensating and commutating windings.

The overall effect of the armature mmf is, then, to distort and/or to oppose the flux of the field. This opposition, or armature reaction (AR), is usually less than about 5% of the field's flux at no-load.

With constant field current and rated armature current, an armature reaction of 4% means that

$$\phi_2 = 0.96\phi_1$$

where ϕ_1 is the machine's flux at negligible armature current ($I_{a_1} = 0$) and ϕ_2 is its effective flux at rated armature current (I_{a_2}). In other words, at full-load, the effective magnetic field is 96% of the field at no-load.

The cross-magnetizing effect of the armature current may be minimized by placing, in longitudinal slots in the pole faces, a *compensating winding* (see Fig. 6-9). This winding is connected in series with the armature and hence carries the same current, but in the direction opposite to that of the adjacent armature conductors.

Distortion of the Field's Flux Distribution

As shown in Fig. 6-8, the flux of the armature coil distorts the field in the inter-pole region, particularly where the commutator segment rides under the positive or the negative polarity brush. This results in a nonlinear variation of the armature current, which can lead to severe sparking. In fact, one of the essential limitations of dc machines is the arcing and possible flashover that may accompany the reversal of the armature current, commonly known as commutation.

(Commutation will be explained more fully in the next section.) To cancel this adverse effect, a coil—the so-called *commutating winding*—is wound on the narrow poles of the interpole region and connected in series with the armature circuit. When properly designed, a commutating winding can minimize the distorting effect of the armature mmf on the field flux distribution.

6.1.6 Commutation

The simplified schematic of Fig. 6-10(a) shows that during one complete revolution of the shaft, each side of the armature conductor is rotated through two opposite magnetic fields (N and S). Thus the voltage, or the current, would be positive or negative, or approximately sinusoidal in waveform, as shown in Fig. 6-10(b).

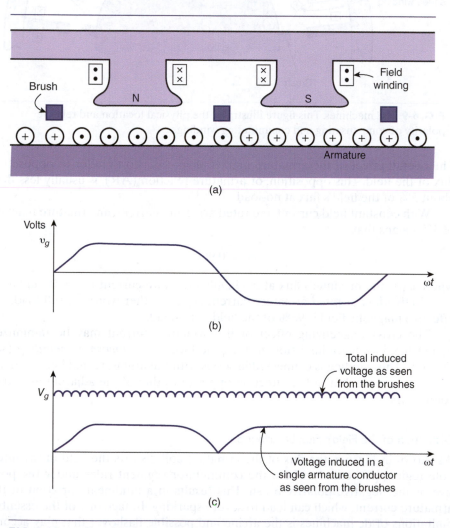

(a)

(b)

(c)

FIG. 6-10 **(a)** Simplified view of a two-pole machine; **(b)** voltage induced in a single armature conductor; **(c)** effect of the brushes on the induced voltage.

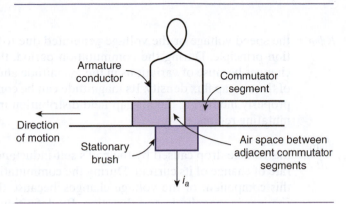

FIG. 6-11 Commutation, showing the location of brush and commutator segments when an armature conductor is shorted.

Properly reversing the end connections of the conductor causes a unidirectional or dc current to flow in the machine's external circuit. The current is reversed through the brushes, whose polarity is fixed by their physical location relative to the north and south poles of the field winding. The reversal of the armature current during half the period of its cycle is called *commutation*.

As the location of the brushes indicates, the commutation starts when the conductors are at the region of the weakest magnetic field (see Fig. 6-10(c)). Ideal or sparkles commutation occurs when the armature current is made to vary linearly during the commutation period. This occurs when the resistance of the shorting brush(es) is the predominant characteristic of the shorted armature conductor.

The instant the conductor leaves the north pole of the magnetic field and is about to enter the south pole, the brush, as shown in Fig. 6-11, short-circuits the conductor. As a result, sparking may take place. The degree of the arcing will depend on the magnitude of the armature current, on the voltage induced in the shorted conductor, and on the resistance of the short-circuited path. Arcing is also increased by the distortion of the air-gap field that results from the armature reaction. The distortion of the air-gap field, as already discussed, can be completely neutralized by properly designing the commutating winding.

Excessive sparking may result in the breakdown of the air space between the commutator segments, and thus a flashover may engulf the entire periphery of the commutator. The flashover, as seen from the terminals of the motor, constitutes a short circuit to the external voltage source.

Mathematical Considerations

If one neglects the coil resistance, the voltage induced in an armature coil (V_{coil}) depends on its self-inductance, its mutual inductance, and the speed of the rotor. Mathematically,

$$V_{coil} = K\phi\omega - \frac{d\lambda_a}{dt} - \frac{d}{dt}(i(t)L(t)) \qquad \textbf{(6.41)}$$

where:

$K\phi\omega$ = the speed voltage or the voltage generated due to the induction principle. During the commutation period, this voltage changes because of variations in the magnitude and direction of the air-gap flux density. Its magnitude can be controlled by properly modifying the air-gap field distribution in the commutating region.

$\dfrac{d}{dt}(\lambda_a) = L_a \dfrac{di_{a(t)}}{dt}$ = the voltage drop caused by the coil's self-inductance and the rate of change of its current. During the commutation period, this component of the voltage changes, because the current changes in magnitude and direction. By definition, this voltage is also given by the rate of change of the flux linkage of the coil undergoing commutation.

$\dfrac{d}{dt}(i(t)L(t))$ = the voltage drop due to the conductor's mutual $(L(t))$ inductance with respect to the adjacent conductors that carry an instantaneous current of $i(t)$ amperes. Here, only one term is considered, but it is obvious that almost all coils are mutually coupled to the conductor that is undergoing commutation.

Before commutation takes place, these voltages are balanced by the external voltage. During commutation, however, the coil is short-circuited, and the sum of these voltages must be equal to zero. The generation or development of sparks results from nature's attempt to set up a voltage that satisfies KVL.

6.1.7 Equivalent Circuits and External Machine Characteristics

The equivalent circuit of a dc machine explicitly shows how the field winding is excited in relation to the armature winding. As such, an equivalent circuit reveals the type of machine (series, shunt, separately excited, etc.) it represents. An equivalent circuit may include the actual values of the winding impedances, so that it can be used for transient and steady-state analyses. In this section, however, only steady-state analyses are discussed. Consequently, the equivalent circuits include only the resistance of the armature (R_a) and field windings (R_f).

The resistance of the armature winding is relatively small, normally a fraction of an ohm. It includes the resistances of the armature conductors and the resistances of the compensating and commutating windings. The effects of the contact between the brush and the commutator segments is represented by a constant voltage drop (about 1 V).

The resistance of the field winding, when connected in series with the armature winding, is very small. When connected in parallel to the supply voltage, however, it is several hundred times larger than the resistance of the armature windings.

The external characteristics of dc machines can easily be derived from the basic motor-generator relationships and from the equations that can be obtained from the machine's equivalent circuit.

Shunt dc Motors

The equivalent circuit of a dc shunt motor is shown in Fig. 6-12(a). The voltage V_t represents the motor's dc supply or terminal voltage, and I_a represents the current of the armature windings. The armature current is determined by the supply voltage, the driven load, and the field current. The resistance of the field winding ($R_f > 100\ \Omega$) is much greater than the resistance of the armature coil ($R_a < 0.2\ \Omega$), and under normal operating conditions, the field current is much smaller than the armature current.

Ordinarily, a dc shunt motor is equipped with a "starting box" that houses an external resistor whose value can be decreased in several steps. The purpose of the external resistor is to limit the armature current during starting (see Section 6.1.8).

When stopping a dc shunt motor, the armature circuit should be opened either before or at the same time as the field circuit. If the field current is interrupted while the armature circuit is energized, severe sparking on the commutator will result, and the motor's speed will increase to dangerously high levels.

For purposes of speed control, the motor may be equipped with variable external resistors in the armature (R_{a_x}) and field (R_{f_x}) circuits. The motor's external characteristics are derived as follows.

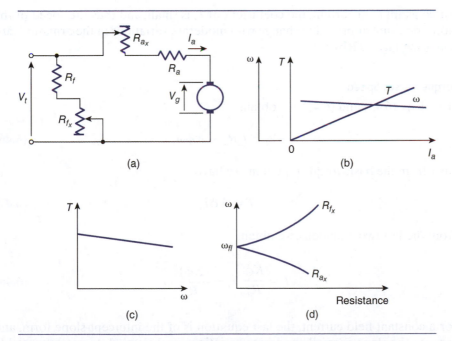

(a)   (b)

(c)   (d)

FIG. 6-12 Shunt dc motors: **(a)** equivalent circuit including the rheostat's resistances, **(b)** torque and speed versus armature current, **(c)** torque versus speed, **(d)** speed versus rheostat resistances.

Torque versus Armature Current

The basic motor-torque equation for a constant field current and negligible armature reaction becomes

$$T = K_1 I_a \tag{6.42}$$

That is, the torque is directly proportional to the armature current. The torque-armature current characteristic is shown in Fig. 6-12(b).

Speed versus Armature Current

From the basic generator equation and KVL in the equivalent circuit ($R_{a_x} = 0$), we have

$$V_g = K\phi\omega \tag{6.43}$$

$$V_t = V_g + I_a R_a \tag{6.44}$$

From the above,

$$\omega = \frac{V_t}{K\phi} - \frac{R_a}{K\phi} I_a \tag{6.45}$$

For constant field current, the coefficient of I_a is small, and thus the speed of the motor does not appreciably change with moderate variations in the armature current. (See Fig. 6-12(b).)

Torque versus Speed

From Eqs. (6.43) and (6.44), we obtain

$$V_t = I_a R_a + K\phi\omega \tag{6.46}$$

Also, from the basic torque equation, we have

$$T = K\phi I_a \tag{6.47}$$

From the last two equations, we obtain

$$T = \frac{K\phi}{R_a} V_t - \frac{(K\phi)^2}{R_a}\omega \tag{6.48}$$

For a constant field current, the last equation is of the intercept-slope form, and because the slope is small, moderate variations in the speed do not appreciably change the torque. The torque-speed characteristic of the shunt motor is shown in Fig. 6-12(c).

Speed versus External Resistors

The speed of the shunt motor can be controlled through the external armature and field resistors.

Effects of Armature Rheostat (R_{ax})

Applying KVL in the equivalent circuit, we obtain

$$V_t = V_g + I_a(R_a + R_{ax}) \qquad (6.49)$$

From Eqs. (6.43) and (6.49), we get

$$\omega = \frac{V_t - I_a R_a}{K\phi} - \frac{I_a}{K\phi} R_{ax} \qquad (6.50)$$

For constant field and armature currents, the larger the external armature resistance, the smaller will be the motor's speed.

Equation (6.50) makes it evident that the external armature resistance can control the speed of the motor from zero up to its full-load value. This method of speed control is accompanied by high copper losses ($I_a^2 R_{ax}$). For this reason, almost all modern speed-control apparatuses use solid-state devices to reduce the terminal voltage and thus to control the speed of the motor. The speed of a shunt motor as a function of the external armature resistance is shown in Fig. 6-12(d).

Effects of the Field Rheostat (R_{fx})

From the speed-voltage equation and Ohm's law, across the field winding we have

$$V_g = K_1 I_f \omega \qquad (6.51)$$

$$I_f = \frac{V_t}{R_f + R_{fx}} \qquad (6.52)$$

From Eqs. (6.44), (6.51), and (6.52), we obtain

$$\omega = \frac{(V_t - I_a R_a)(R_f + R_{fx})}{K_1 V_t} \qquad (6.53)$$

Since $V_t \gg I_a R_a$, the last expression becomes

$$\omega \approx \frac{R_f + R_{fx}}{K_1} \qquad (6.54)$$

Equation (6.54) shows that by increasing the external field resistance, the speed of the motor is also increased. The speed of the motor as a function of the field rheostat resistor is shown in Fig. 6-12(d).

The maximum attainable speed is limited by sparking in the brushes or by mechanical considerations of the centrifugal forces.

EXAMPLE **6-2**

A 10 kW, 220 V shunt dc motor has field and armature resistances of 110 ohms and 0.20 ohms, respectively. At no-load, the motor runs at 1200 r/min and has an armature current of 5.0 A. Determine the speed and electromagnetic torque when the motor draws 50 A from the supply. Assume an armature reaction that is:

a. Negligible
b. 4%.

SOLUTION

a. The parameters marked with the subscript 1 will correspond to the no-load condition, and those marked with subscript 2 will correspond to the load condition of 50 A. Applying KVL to Fig. 6-13, we obtain

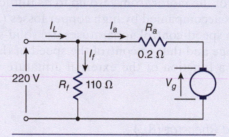

FIG. 6-13

$$V_{g_1} = V_t - I_{a_1} R_a$$
$$= 220 - 0.2(5) = 219 \text{ V}$$

When the motor draws 50 A, the armature current will be

$$I_{a_2} = I_L - I_f = 50 - \frac{220}{110} = 48 \text{ A}$$

and

$$V_{g_2} = 220 - 0.2(48) = 210.4 \text{ V}$$

For negligible armature reaction and constant field current,

$$\phi_1 = \phi_2$$

From the generator principle, we have

$$\frac{V_{g_2}}{V_{g_1}} = \frac{n_2}{n_1}$$

From the above,

$$n_2 = 1200 \frac{210.4}{219} = \underline{1152.88 \text{ r/min}}$$

Besides giving the rotational losses, the no-load data can also be used to establish the motor constant, as follows:

$$V_{g_1} = K\phi\omega_1$$

For constant field,

$$V_{g_1} = K_1\omega_1$$

Thus,

$$K_1 = \frac{219}{1200 \frac{2\pi}{60}} = 1.74 \text{ V/rad/s}$$

The electromagnetic torque developed is

$$T = K\phi I_a$$
$$= K_1 I_a$$
$$= 1.74 \times 48 = \underline{83.65 \text{ N} \cdot \text{m}}$$

Alternatively,

$$T = \frac{V_{g_2} I_a}{\omega} = \frac{210.4\,(40)}{1152.88 \frac{2\pi}{60}} = \underline{83.65 \text{ N} \cdot \text{m}}$$

b. For an armature reaction of 4%, we have

$$\phi_2 = 0.96\phi_1$$

From the generator principle, we obtain

$$n_2 = \frac{V_{g_2}\phi_1}{V_{g_1}\phi_2} n_1$$

Substituting the known parameters, we obtain

$$n_2 = 1200 \frac{210.4}{219} \left(\frac{1}{0.96}\right) = \underline{1200.91 \text{ r/min}}$$

The torque is found as follows:

$$T = K\phi_2 I_{a_2}$$
$$= 0.96(K\phi_1 I_{a_2}) = 0.96(1.74 \times 48)$$
$$= \underline{80.31 \text{ N} \cdot \text{m}}$$

For purposes of comparison, the results are summarized in Table 6-1. Clearly, the armature reaction increases the speed and reduces the torque of the motor by a percentage about equal to that of the armature reaction.

TABLE 6-1 Summary of results of Example 6-2

Assumed Armature Reaction	Armature Current	Speed r/min	Torque Developed (N · m)
0	48	1152.88	83.65
4%	48	1200.91	80.31

Series dc Motors

In a series dc motor, the field winding is connected in series with the armature winding (see Fig. 6-14(a)). For speed control, a variable resistor (R_{f_x}) is often connected in parallel to the field winding. The motor's external characteristics are derived as follows.

Torque versus Armature Current

In a series dc motor *without* an external field resistor, the field current is equal to the armature current. Hence, for operations along the linear portion of the magnetic characteristic, and in the absence of armature reaction, the machine's flux is directly proportional to the armature current. Rewriting the basic torque equations, we have

$$T = K\phi I_a \tag{6.55}$$

or

$$T = K_1 I_a^2 \tag{6.56}$$

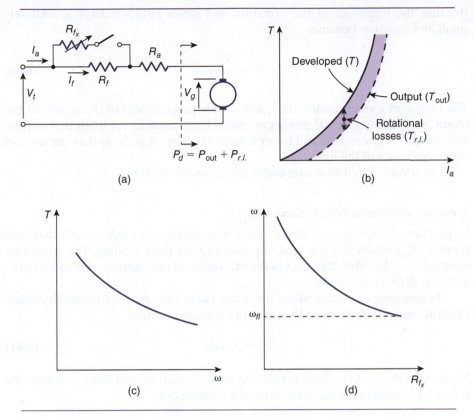

FIG. 6-14 Series dc motors: (a) equivalent circuit, (b) torque versus armature current, (c) torque versus speed, (d) speed versus external field resistance.

In other words, the torque developed by a series dc motor is directly proportional to the square of the armature current. The torque-versus-armature-current characteristic is shown in Fig. 6-14(b). The actual curve, however, is displaced downward because of the machine's rotational losses.

Torque versus Speed

Applying KVL in the circuit of Fig. 6-14(a), we obtain

$$V_t = V_g + I_a(R_a + R_f) \tag{6.57}$$

Substituting for voltage generated its equivalent equation, we obtain

$$V_t = K_1 I_a \omega + I_a(R_a + R_f) \tag{6.58}$$

From Eqs. (6.56) and (6.58), we find

$$T = K_1 \left(\frac{V_t}{K_1\omega + R_a + R_f} \right)^2 \tag{6.59}$$

Because the resistance of the armature and series field windings is relatively small, the equation becomes

$$T \approx \frac{V_t^2}{K_1 \omega^2} \qquad (6.60)$$

The torque of a series motor, then, is inversely proportional to the square of the speed. As a result, series dc motors are suitable for mechanical loads that require high torque at low speeds and lower torque at higher speeds, so they are used in trains, trolleys, and the like.

The torque-speed characteristic is shown in Fig. 6-14(c).

Speed versus External Field Resistance (R_{f_x})

In practice, the speed of a series motor is controlled through an external field resistor (R_f), which is connected in parallel to the field winding. The governing equation that describes the effect of the external field resistance on the speed of the motor is derived as follows.

By assuming operation along the linear part of the magnetization characteristic, the basic expression for the armature voltage becomes

$$V_g = K_1 I_f \omega \qquad (6.61)$$

Neglecting the voltage drop across the small armature and field resistors, the voltage generated is equal to the terminal voltage; thus,

$$V_t \approx K_1 I_f \omega \qquad (6.62)$$

Considering the field resistors, from the current-divider concept, we obtain

$$I_f = I_a \frac{R_{f_x}}{R_f + R_{f_x}} \qquad (6.63)$$

From Eqs. (6.62) and (6.63), we get

$$\omega = \frac{V_t}{K_1 I_a R_{f_x}} (R_f + R_{f_x}) \qquad (6.64)$$

or

$$\omega = \frac{V_t}{K_1 I_a} \left(1 + \frac{R_f}{R_{f_x}} \right) \qquad (6.65)$$

For a given armature current, then, the larger the external field resistance is, the smaller the speed of the motor will be. The corresponding motor characteristic is shown in Fig. 6-14(d).

EXAMPLE 6-3

A 300 V, 60 A, 1200 r/min series dc motor has an armature and a series field resistance, each of which is 0.20 ohm. When a 0.10 ohm resistance is connected in parallel to the field winding, the motor's torque is doubled. Assuming negligible armature reaction and rotational losses, determine the motor's:

a.   Armature current.
b.   Speed.
c.   Efficiency.

SOLUTION

The given nameplate data establishes the constant of the motor. Using the equivalent circuit of Fig. 6-15(a), at 1200 r/min, we have

$$V_{g_1} = V_t - I_{a_1}R_{t_1}$$
$$= 300 - 60(0.2 + 0.2) = 276 \text{ V}$$

As before, subscript 1 designates the motor's parameters at rated conditions, and subscript 2 the parameters that correspond to the new motor's operating condition.

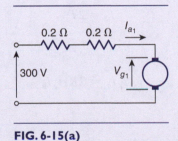

FIG. 6-15(a)

From the generated voltage equation, we also have

$$V_{g_1} = K_1 I_{f_1} \omega$$

Solving for the constant of proportionality, we obtain

$$K_1 = \frac{276}{(60)1200\left(\dfrac{2\pi}{60}\right)}$$
$$= 0.0366 \text{ V/rad/s/field A}$$

a. The current through the 0.10 ohm resistor that is connected in parallel with the field winding (see Fig. 6-15(b)) does *not* contribute to the effective flux within the machine because this resistance is physically connected outside the machine's structure.

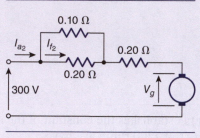

FIG. 6-15(b)

Using the current-divider concept, we obtain

$$I_{f_2} = I_{a_2} \left(\frac{0.1}{0.1 + 0.2} \right) = \frac{I_{a_2}}{3}$$

From the statement of the problem,

$$T_2 = 2T_1$$

or

$$K_1 I_{f_2} I_{a_2} = 2K_1 I_{f_1} I_{a_1}$$

From the above,

$$I_{a_2}^2 \left(\frac{1}{3} \right) = 2I_{a_1}^2$$

Substituting and solving for the unknown, we obtain

$$I_{a_2} = 60\sqrt{6} = \underline{146.97 \text{ A}}$$

b. The speed of the motor is determined as follows. From KVL, we have

$$V_{g_2} = 300 - 146.97(0.2 + 0.10 \mathbin{//} 0.20)$$
$$= 260.81 \text{ V}$$

and

$$\omega_2 = \frac{V_{g_2}}{K_1 I_{f_2}}$$

$$= \frac{260.81}{0.0366 \left(\dfrac{146.97}{3}\right)}$$

$$= 145.46 \text{ rad/s}$$

and

$$n_2 = \underline{1389 \text{ r/min}}$$

c. The efficiency of the motor is

$$\eta = \frac{P_{\text{out}}}{P_{\text{in}}} = 1 - \frac{P_{\text{loss}}}{P_{\text{in}}}$$

$$= 1 - \frac{I_a^2(R_a + R_f)}{V_t I_a}$$

$$= 1 - \frac{(146.97)^2(0.2 + 0.10 \,/\!/\, 0.20)}{300\,(146.97)} = 0.87$$

Alternatively,

$$\eta = \frac{V_{g_2} I_{a_2}}{V_t I_{a_2}} = \frac{260.81}{300}$$

$$= \underline{0.87}$$

For purposes of comparison, the results are summarized in Table 6-2.

TABLE 6-2	Summary of results of Example 6-3		
Load Torque	External Field Resistor (ohms)	Speed (r/min)	Line Current (A)
T_1	∞	1200	60
$2T_1$	0.1	1389	146.97

Exercise

6-4

For a series dc motor, show that the speed as a function of armature current is given by

$$\omega = \frac{V_t}{K_1 I_a} - \frac{R_a + R_f}{K_1}$$

where K_1 is the motor's constant in V/A/rad/s.

Separately Excited dc Motors

In a separately excited dc motor, the field coil is supplied from a different voltage source than that of the armature coil, as shown in Fig. 6-16(a). The field circuit normally incorporates a rheostat through which the field current, and thus the motor's characteristics, can be externally controlled. This motor is suitable primarily for two types of loads: those that require constant torque for speed variations up to full-load speed and those whose power requirements are constant for speed variations above nominal speed.

The power-versus-speed and torque-versus-speed characteristics of such loads are shown in Fig. 6-16(b) and 6-16(c), respectively. These diagrams also identify the method of a motor's speed control. The terminal voltage (V_t) control is used for loads that require constant torque for speed variations *up to* full-load, while the field current (I_f) control is used for constant power requirements for speed variations *above* the full-load speed. Both methods of speed control aim to

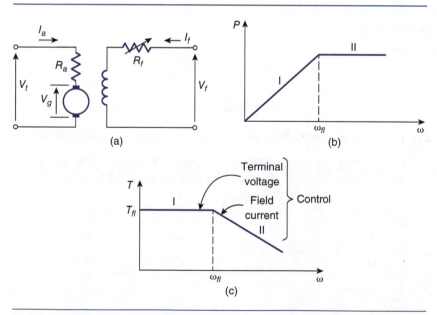

FIG. 6-16 Separately excited dc motor: **(a)** equivalent circuit, **(b)** power-speed characteristic of a load, **(c)** speed-torque characteristic of the load in (b).

supply the requirements of the mechanical load while not exceeding the nominal capability or rating of the machine.

Each mode of speed control is justified by considering the following basic torque and armature–current relationships:

$$T = K\phi I_a \tag{6.66}$$

$$I_a = \frac{V_t - K\phi\omega}{R_a} \tag{6.67}$$

Terminal Voltage Control

From Eq. (6.66) for constant torque and armature current, the field current must also be constant. Then the armature current, as can be seen from Eq. (6.67), will be kept at its rated value, provided that the terminal voltage increases according to the increase in the speed.

Field Control

For loads that require constant power at speeds higher than rated value, the motor's terminal voltage is kept constant, and the increase of the speed is achieved by reducing the field current. The reduction of the field current for constant armature current results in a reduction of the torque and a corresponding increase in the speed. In practice, the speed control through the field current or through the terminal voltage is accomplished by using the solid-state speed-control apparatus (see Section 6.2).

EXAMPLE 6-4

A 220 V, 4 kW, 22 A, 1260 r/min, separately excited dc motor is driving a fan whose torque is proportional to the square of the motor's speed. The resistance of the armature winding is 0.50 ohm. Neglecting armature reaction and assuming that the excitation remains constant:

a. Determine the terminal voltage required to lower the speed of the fan by 100 r/min.
b. Sketch the torque-speed characteristics of the motor and the fan. Identify the operating points.

SOLUTION

a. The generated voltage at 1260 r/min is found by applying KVL in the armature circuit of Fig. 6-17(a).

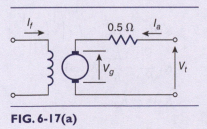

FIG. 6-17(a)

$$V_{g_1} = V_t - I_a R_a$$

$$= 220 - 22(0.5) = 209 \text{ V}$$

From the generator principle, for constant field current we have

$$K_1 = \frac{V_{g_1}}{\omega} = \frac{209}{1260 \dfrac{2\pi}{60}}$$

$$= 1.58 \text{ V/rad/s}$$

From the motor principle and the given torque-speed relationship, we obtain

$$\frac{T_1}{T_2} = \frac{I_{a_1}}{I_{a_2}} = \left(\frac{\omega_1}{\omega_2}\right)^2$$

Solving for I_{a_2}, and substituting the known parameters, we obtain

$$I_{a_2} = 22 \left(\frac{1160}{1260}\right)^2 = 18.65 \text{ A}$$

The new generated voltage is

$$V_{g_2} = K_1 \omega = 1.58 \left(\frac{1160}{60}\right) 2\pi = 192.41 \text{ V}$$

From KVL, the required terminal voltage is

$$V_t = 18.65(0.5) + 192.41 = 201.74 \text{ V}$$

b. The torque-speed characteristic of the fan is as shown in Fig. 6-17(b). The torque of the motor at a given speed must be equal to the torque requirement of the fan at the same speed.

   The general expression for the torque-speed characteristic of the motor as a function of the terminal voltage is derived as follows:

$$T = K\phi I_a = K_1 I_a = K_1 \left(\frac{V_t - V_g}{R_a}\right)$$

$$= K_1 \frac{V_t}{R_a} - \frac{K_1}{R_a}(K\phi\omega)$$

$$= K_2 V_t - K_3 \omega$$

The constants of proportionality, K_2 and K_3, have been introduced in order to shorten the algebra. The last expression is sketched in Fig. 6-17(b).

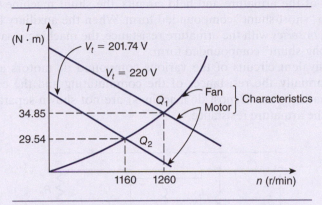

FIG. 6-17(b)

The torque at the given speeds is

$$T_1 = 1.58(22) = 34.85 \text{ N} \cdot \text{m}$$

$$T_2 = 1.58(18.65) = 29.54 \text{ N} \cdot \text{m}$$

The points Q_1 and Q_2, located at the intersection of the motor-load characteristics, represent the operating points.

Redo Example 6-4 by assuming that the load's speed requirement is 1360 r/min instead of 1160 r/min. Will the motor overheat?

E x e r c i s e

6-5

Answer   238.41 V

Compound dc Machines

Shunt dc machines may be equipped with an *auxiliary field* winding. The function of this winding is to modify the machine's characteristics to make it more suitable for a particular load requirement.

The auxiliary and shunt field windings are wound around the main field poles. The auxiliary winding has very few turns. It carries either line or armature current, depending on its connection.

When equipped with an auxiliary field winding, shunt dc machines are called *compound* dc machines. Compound machines are further designated as cumulatively compounded or differentially compounded, depending on whether the auxiliary winding aids or opposes the main field winding.

When the auxiliary winding is connected *between* the terminal voltage and the junction of the armature and field circuits, the shunt machine is said to be connected in "short-shunt" compounded form. When the auxiliary field winding is connected *in series* with the armature resistance, the machine is said to be connected in "long-shunt" compounded form.

The equivalent circuits of the various compound dc motors are shown in Fig. 6-18. Normally, the resistances of the commutating and the compensating windings (that are part of most dc machines) are not shown separately but are included in the armature resistance.

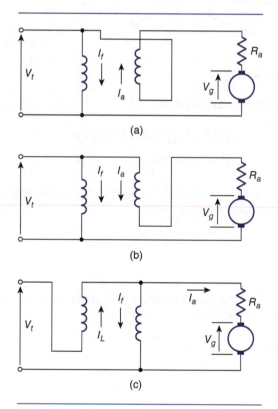

FIG. 6-18 Equivalent circuits of compound dc motors: **(a)** long-shunt, differentially compounded, **(b)** long-shunt, cumulatively compounded, **(c)** short-shunt, differentially compounded.

As shown in Fig. 6-19, the torque-speed characteristic of a compound dc motor is essentially between that of the shunt and the series motor. Actually, a variation of the motor characteristic can be obtained by properly sizing and connecting the auxiliary winding.

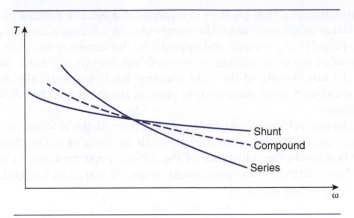

FIG. 6-19 Torque-speed characteristics.

Magnetic Circuit

Figure 6-20 shows the equivalent magnetic circuit per pole of a cumulatively compounded, long-shunt motor. The effective magnetic flux within the machine is essentially produced by the mmf $(N_f I_f)$ of the main field.

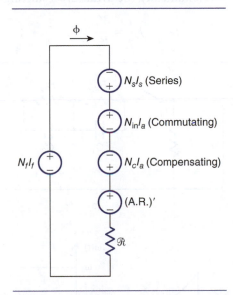

FIG. 6-20 Magnetic equivalent circuit of a long-shunt, cumulatively compounded motor.

The mmf contribution of the *auxiliary series* winding $(N_s I_s)$ is shown as aiding the mmf of the main winding. However, the auxiliary series winding should be of opposite polarity if the machine is differentially compounded. In the long-shunt connection, the current through the auxiliary winding is equal to the armature current. In the short-shunt connection, the current through this winding is equal to the line current.

The armature reaction (A.R.)', in equivalent magnetic volts, is of a polarity opposite to that of the main field. The compensating winding is designed in such a way that its mmf ($N_c I_a$) is equal and opposite to the armature reaction.

The mmf of the *commutating* or *interpole* winding ($N_{in} I_a$) has a qualititative effect on the flux density of the field winding, for it is specifically designed to minimize the distortion of the armature mmf on the main field flux density at the interpole region.

The effective reluctance $\mathcal{R}$ represents the sum of the reluctances of the air gaps, of the stator path, and of the path through the rotor of the machine. At relatively low flux levels, the reluctance of the air gap predominates. At higher flux densities, the reluctance of the constituent magnetic materials, because of saturation, is the controlling factor.

6.1.8 Starting

The equivalent circuit of a shunt dc motor is shown in Fig. 6-21(a). At the instant of starting, the rotor in all dc motors is not revolving, and thus the voltage generated is zero. As a result, the terminal voltage is applied across the relatively small armature resistance. Consequently, a very large current flows through the conductors of the armature. Mathematically, the armature's starting current ($I_{a_{st}}$) is

$$I_{a_{st}} = \frac{V_t}{R_a} \tag{6.68}$$

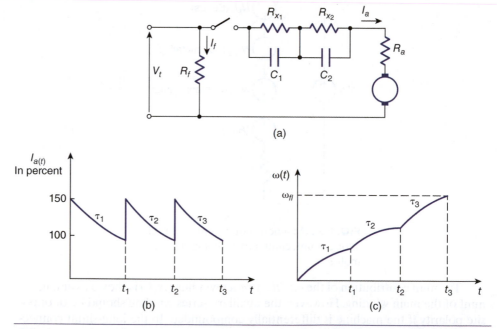

(a)

(b)

(c)

FIG. 6-21 Starting a dc shunt motor through step-by-step reduction of the external armature resistors: **(a)** equivalent circuit, **(b)** armature current versus time, **(c)** speed versus time.

High starting current is undesirable because it adversely affects the dc power supply, the commutation process, and the rate of heat loss within the machine.

In order to limit excessively large starting currents, dc shunt motors—like all other types of dc machines—are provided with external resistors that are connected in series with the armature circuit (see Fig. 6-21(a)). During starting, these resistors are shorted,* in steps, by the contactors (C_1 and C_2) until the armature current reaches its rated value. At the same time, the speed of the motor increases from zero up to its nominal value.

The variations in armature current and speed, as functions of time during the step-by-step reduction in the external resistors for a particular motor, are shown in Figs. 6-21(b) and (c), respectively.

At starting, both of the contactors (see Fig. 6-21(a)) are open. Therefore, the armature current is limited to

$$I_{a_{st}} = \frac{V_t}{R_a + R_{x_1} + R_{x_2}} \qquad (6.69)$$

As soon as the motor starts to rotate, the armature voltage increases. Consequently, the armature current is gradually reduced. From KVL, we have

$$I_a = \frac{V_t - K_1\omega}{R} \qquad (6.70)$$

where K_1 is the constant of the machine, R is the total resistance of the armature circuit, and ω is the value of the motor's speed at the instant under consideration.

When the armature current reaches its rated value, the resistance R_{x_1} is shorted through the contactor C_1. At this instant, the speed of the motor cannot change instantaneously. Thus, the armature current must instantaneously increase in order to satisfy KVL. In Fig. 6-21(a), the resistors have been selected in such a way as to limit the armature current to 150% of its rated value.

For this variation of the armature current immediately before and after the removal of the first resistor (R_{x_1}), we have

$$I_{a_r} = \frac{V_t - K_1\omega_1}{R_a + R_{x_1} + R_{x_2}} \qquad (6.71)$$

and

$$1.5I_{a_r} = \frac{V_t - K_1\omega_1}{R_a + R_{x_2}} \qquad (6.72)$$

where I_{a_r} is the rated current of the armature windings and ω_1 is the speed of the motor at the instant when the resistor R_{x_1} is shorted. When the second external

---

*The operation of the contactors is explained in Chapter 7.

resistor is shorted, the above equations can be rewritten with a new value for the speed of the motor. The resulting equations will yield the required values of the external resistors, and the speed of the motor, at the instant a resistor is removed.

At any instant after starting, the motor's armature current and speed, as functions of time, can be found by solving the governing differential equations. The solution of these equations gives the transient response of the dc machines, which is discussed in the Web section, Chapter 6W.

The time constants that control the decay of the armature current and the build-up of the speed (see Fig. 6-21(b)) are determined by the parameters of the motor plus load. Each time a resistor is removed, the corresponding time constant will, of course, increase.

6.1.9 Open-Circuit Characteristics and DC Generators

This section discusses the open-circuit characteristics of dc machines, their effective field current, and the various types of dc generators.

Open-Circuit Characteristic

The magnetization, or open-circuit characteristic (OCC), of a machine is obtained by driving the machine as a generator at constant speed. The terminal voltage, which is equal to the voltage generated (V_g), is recorded for various values of field current (I_f).

Figure 6-22(a) shows a laboratory set-up that may be used to obtain data to plot the open-circuit characteristic of a separately excited dc machine. The mechanical power input to the generator is equal to the sum of its mechanical losses and iron losses. The OCC of the machine is shown in Fig. 6-22(b).

At low flux densities, the magnetization characteristic (MC) is linear, owing to the constant permeability of the air gap. At higher flux densities, the MC becomes nonlinear, owing to the variations in the permeability of the magnetic material.

Alternatively, at relatively low flux densities, the reluctance of the air gap predominates, and all input mmf is used to overcome the opposition to the flux that is presented by the air gap. As the flux density or the level of magnetization increases, the reluctance of the magnetic material becomes the controlling parameter of the magnetic circuit.

The machine's equivalent magnetic circuit per pole is shown in Fig. 6-22(c). The armature reaction (A.R.)$'$ is expressed in equivalent field ampere-turns.

The basic relationships of Chapter 1 show that the generated voltage (V_g) is proportional to the magnetic flux density (B), and also that the magnetic field intensity (H) is proportional to the field current (I_f). That is, the ordinate and the abscissa of the MC correspond to the magnetic field's flux density and flux intensity, respectively.

The flux of a coil is proportional to the coil's current and to the permeability of the magnetic material. That is,

$$\phi = K\mu I \qquad \text{(6.73)}$$

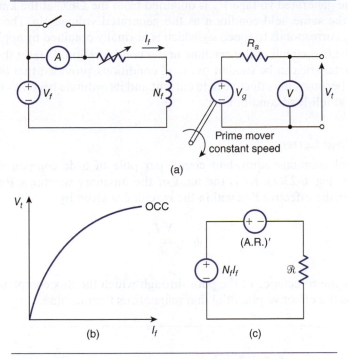

FIG. 6-22 Separately excited dc machine: **(a)** schematic for the laboratory setup used to obtain data for the magnetization curve, **(b)** typical open-circuit characteristic, **(c)** equivalent magnetic circuit per pole. (At no-load, the mmf of the armature reaction is zero.)

where K is a constant that is dependent on the physical dimensions of the magnetic material and on the winding's number of turns. As shown in Fig. 1-61, the permeability depends on the operating point or the level of magnetization.

From the generator principle, for two different speeds of operation ω_1 and ω_2, we have

$$\frac{V_{g_1}}{V_{g_2}} = \frac{\phi_1 \omega_1}{\phi_2 \omega_2} \tag{6.74}$$

When the effective flux ϕ_1 and ϕ_2 are the same, we obtain

$$\frac{V_{g_1}}{V_{g_2}} = \frac{\omega_1}{\omega_2} \tag{6.75}$$

In Eq. (6.75), the generated voltages V_{g_1} and V_{g_2} must be calculated at the same effective field conditions. The magnetization curve can be used in conjunction with Eq. (6.75) to find the speed of the machine for a particular operating condition.

Usually, the generated voltage V_{g_1} is obtained from the OCC at the known speed ω_1 but at the same field condition as the generated voltage V_{g_2}. The generated voltage V_{g_2} corresponds to speed ω_2, which is normally obtained by applying KVL in the equivalent circuit of the machine under consideration. Clearly, then, the no-load characteristic can be used at full-load conditions, provided that the abscissa of the MC becomes the effective field current and its ordinate becomes the voltage generated at full-load conditions.

Effective Field Current

The general magnetic equivalent circuit per pole of a dc compound machine is shown in Fig. 6-23(a). NI is the mmf of the auxiliary windings. Referring to Fig. 6-23(b), the effective flux within the machine is given by

$$\phi = \frac{N_f I_{ef}}{\mathcal{R}} \tag{6.76}$$

where $\mathcal{R}$ is the reluctance of the path through which the flux completes its loop, and $N_f I_{ef}$ is the effective potential that magnetizes the machine.

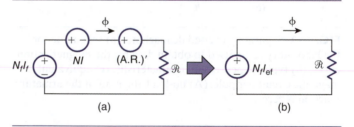

(a)   (b)

FIG. 6-23 Compound dc machine: **(a)** equivalent magnetic circuit per pole, **(b)** simplified equivalent circuit.

The effective mmf is equal to the mmf contribution of the main and auxiliary windings minus the opposition of the armature reaction. Mathematically,

$$N_f I_{ef} = N_f I_{f_a} \pm N_s I_s - (A.R.)' \tag{6.77}$$

Dividing both sides by the turns of the main field winding N_f, we obtain

$$I_{ef} = I_{f_a} \pm \frac{N_s}{N_f} I_s - A.R. \tag{6.78}$$

where

$I_{f_{ef}}$ = the effective field current of the machine. Although this current cannot be seen in the machine's electrical equivalent circuit, it constitutes the abscissa of its OCC under load conditions.

I_{f_a} = the actual current through the main field winding of the machine. This current can easily be calculated from the given data and from use of the equivalent circuit of the machine under consideration.

N_sI_s = the magnetomotive force of the auxiliary field winding. This winding is a feature of a compounded machine, and its flux may aid (cumulative connection) or oppose (differential connection) the flux of the main winding.

(A.R.)' = the armature reaction in equivalent field ampere-turns.

A.R. = the armature reaction in equivalent field amperes.

The cross-magnetizing mmf of the armature reaction is directly proportional to the armature current, but the resulting reduction in the effective, *voltage-generating flux* is approximately proportional to the *square of the armature current*.

The effects of the armature reaction on the OCC and on the full-load characteristics of a separately excited generator are shown in Fig. 6-24. In Fig. 6-24(b), the A.R. is expressed as a percentage of the flux at no-load. The letters A and B are used to identify the no-load and full-load conditions, respectively. In Fig. 6-24(c), the effects of the armature reaction are expressed in volts. This is seldom used in the analysis of problems because full-load terminal characteristics are unavailable. In Fig. 6-24(d),

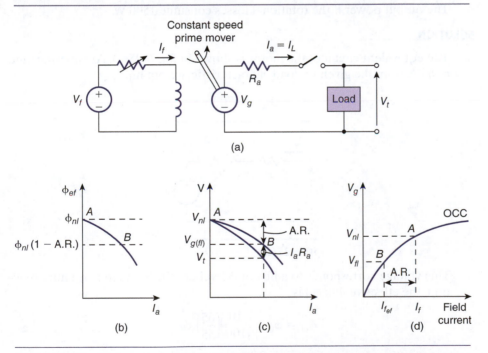

(a)

(b)   (c)   (d)

FIG. 6-24 Separately excited generator—an illustration of the alternative units of armature reaction: **(a)** equivalent circuit, **(b)** effective flux versus load current (armature-reaction effects in percent of the flux at no-load, **(c)** generated and terminal voltages as functions of load current (armature-reaction effects in volts), **(d)** open-circuit characteristic (armature-reaction effects in terms of equivalent field amperes).

the armature reaction is expressed in equivalent field amperes, which is the most popular unit of measure of armature reaction. The open-circuit characteristic is used at no-load and at full-load conditions.

The effects, then, of the armature reaction can be expressed in terms of no-load flux, generated voltage, and/or field current.

EXAMPLE **6-5** A dc series motor is operated from a 450 V supply. The total armature and series field resistance is 0.40 ohm. Its magnetization characteristic, taken at 800 r/min, is

Volts	360	412	460	470
Amperes	35	45	55	65

The demagnetizing effect of the armature reaction, which may be assumed to be proportional to the square of the armature current, is such that for an armature current of 55 A, the flux per pole is reduced by 10%. When the motor draws 65 A from the 450 V supply, determine:

a. The speed.
b. The electromagnetic torque developed.
c. The output power if the rotational losses consume 1500 W.

SOLUTION

a. The equivalent circuit of the motor is shown in Fig. 6-25(a). At 800 r/min and at 65 A from the given no-load characteristic, we obtain

$$V_{g_1} = 470 \text{ V}$$

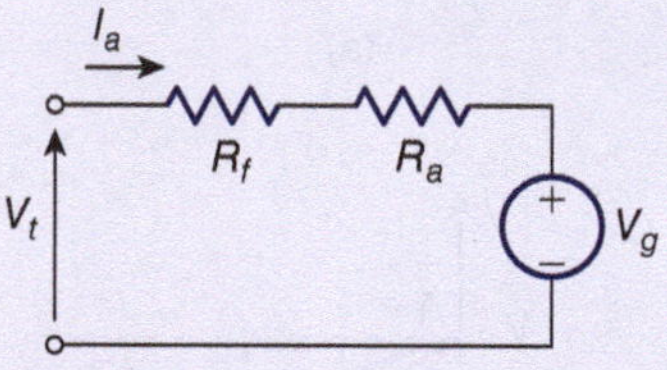

FIG 6-25(a)

This voltage corresponds to a flux of ϕ_0 webers. Owing to the armature reaction, the effective flux ϕ_{ef} is

$$\phi_{ef} = \phi_0 - \left(\frac{10}{100}\right)\left(\frac{65}{55}\right)^2 \phi_0$$

$$= 0.86\phi_0$$

From the above, the actual voltage generated at full-load conditions, but at 800 r/min, is

$$V_{g_1} = 470(0.86) = 404.36 \text{ V}$$

The generated armature voltage at the unknown speed, but at the same flux level as V_{g_1}, is

$$V_{g_1} = V_t - I_a R$$
$$= 450 - 65(0.4)$$
$$= 424 \text{ V}$$

Substituting the above data into Eq. (6.75), we obtain

$$n_2 = 424\left(\frac{800}{404.36}\right) = \underline{838.87 \text{ r/min}}$$

b. The torque developed is

$$T = \frac{V_{g_2} I_a}{\omega_2} = \frac{424(65)}{838.87\left(\dfrac{2\pi}{60}\right)} = \underline{313.73 \text{ N} \cdot \text{m}}$$

c. From the power-balance equation, we have

$$P_d = P_{r_1} + P_0$$

from which

$$P_0 = 424 \times 65 - 1500 = \underline{26.06 \text{ kW}}$$

An Alternative—But Incorrect—Solution of Part (a)

The armature voltage at the unknown speed is, as before,

$$V_{g_2} = 424 \text{ V}$$

The effective value of the field current is

$$I_{\text{ef}} = (\text{actual field current}) - (\text{armature reaction})$$

At 65 A,

$$I_{\text{ef}} = 65 - 65\left(\frac{10}{100}\right)\left(\frac{65}{55}\right)^2$$
$$= 59.92 \text{ A}$$

Projecting this current on the OCC (Fig. 6-25(b)), we obtain

$$V_{g_1} = 466 \text{ V}$$

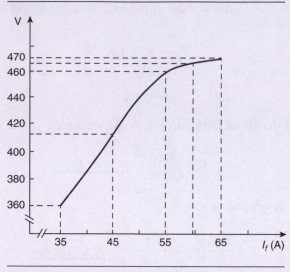

FIG 6-25(b)

and

$$n_2 = 424\left(\frac{800}{466}\right)$$
$$= 727.90 \text{ r/min}$$

which is lower than the speed already calculated by an amount that is equal to

$$\frac{838.87 - 727.9}{727.9}(100) = 15.25\%$$

In other words, it is a mistake to assume that the effective flux is dependent only on the field current. The change of permeability, and its effects on the voltage generated, cannot be assumed to be negligible.

Exercise

6-6

Given the following data for a dc motor that operates at saturation:

Condition	Generated Voltage (V)	Effective Field (A)	Speed (r/min)
1	213	1.4	900
2	208	1.2	?

a. Explain why the tabulated data, when substituted into Eq. (6.75), will not give the speed of the motor at operating condition 2.

b. What additional information is required to calculate the unknown speed?

Shunt Generators

Figures 6-26(a) and (b) show, respectively, the equivalent circuit and the OCC of a shunt-type generator. The shunt generator does not require any excitation for starting or operating. It is self-excited. This advantage results from the machine's residual magnetism—a characteristic of the magnetic material—which provides the required field. As soon as the prime mover starts to turn the rotor, a voltage will be generated in the armature windings because they are rotated through the field of the stator's residual magnetism.

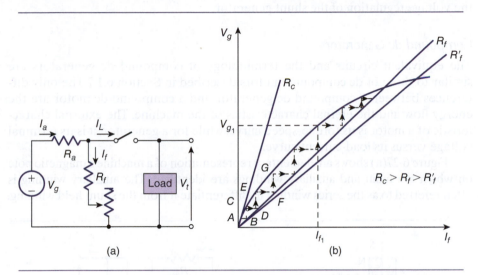

(a)   (b)

FIG. 6-26 Shunt generator: **(a)** equivalent circuit, **(b)** magnetization characteristic illustrating the build-up voltage for the no-load conditions.

Refer to Fig. 6-26(b). When the armature starts to rotate, a small voltage will be generated, as indicated by point A. This voltage, impressed across the field circuit (point B), will cause a current to flow. If the polarity of the connection is correct, the mmf of this current will reinforce that of the residual magnetism; the strengthened magnetic field will then generate the voltage corresponding to point C, which in turn will increase the field current to the value corresponding to point D. The build-up process (*positive feedback*) will continue until the saturation curve intersects the field-resistance line. The slope of this line is equal to the total resistance of the field winding (resistance of the winding plus the portion of the external adjustable resistor that may be incorporated). The intersection of the magnetization curve and the field resistance line furnishes the operating point at no-load.

For a particular field current I_{f_1}, the generated voltage and the terminal voltage are given respectively by

$$V_{g_1} = I_{f_1} R_f + I_{a_1} R_a \qquad \textbf{(6.79)}$$

$$V_{t_1} = I_{f_1} R_f \qquad \textbf{(6.80)}$$

From the above it is evident that the vertical lines drawn from the OCC to the field-resistance line represent the voltage drop across the armature resistance. The longest of these lines will give—indirectly—the maximum permissible value of armature current at rated speed.

By decreasing the field resistance (R_f') below its rated value, the no-load voltage will increase slightly. Conversely, by increasing the field resistance, the no-load voltage will decrease. The value of the field resistance, at which the build-up of the generated voltage cannot take place, is designated as the critical field resistance (R_c).

In general, the field resistance controls the starting, the terminal voltage, and the voltage regulation of the shunt generator.

Compound dc Generators

The equivalent circuits and the terminology of compound dc generators are similar to those of dc compound motors described in Section 6.1.7. The only differences between a compound dc generator and a compound dc motor are the energy flow and the eternal characteristics of the machine. The external characteristic of a motor is its torque-speed curve, while for a generator, it is its terminal voltage versus its load-current curve.

Figure 6-27(a) shows an elementary representation of a machine's magnetic pole on which the main and auxiliary windings are identified. The auxiliary winding is often referred to as the series winding to differentiate it from the shunt field winding.

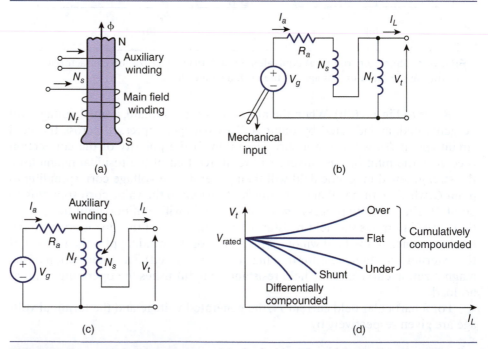

FIG. 6-27 Compounded dc generators: **(a)** physical location of the main and auxiliary field windings, **(b)** long-shunt, **(c)** short-shunt, **(d)** characteristics of differentially and cumulatively compounded generators.

The equivalent circuits of a long-shunt and a short-shunt compounded genera-tor are shown in Figs. 6-27(b) and (c), respectively. The external characteristics of the compounded generators are depicted in Fig. 6-27(d). The cumulatively compounded dc generators are further classified as overcompounded, flat-compounded, or undercompounded, depending on whether the output voltage increases, remains constant, or decreases as the load current increases.

Figure 6-28(a) shows the equivalent circuit of a short-shunt differentially compounded dc generator. Figure 6-28(b) shows its OCC and the effects of both the auxiliary winding and the armature reaction on the magnetization of the machine.

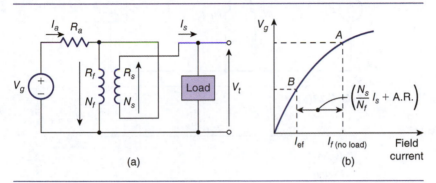

FIG. 6-28 Short-shunt differentially compounded generator: **(a)** equivalent circuit, **(b)** open-circuit characteristic.

A long-shunt, cumulatively compounded dc generator has 1800 shunt-field turns per pole and 12 series-field turns per pole. The armature resistance is 0.12 ohm, the series-field resistance is 0.04 ohm, and the shunt-field resistance is 210 ohms. Data for the open-circuit characteristic at 1700 r/min is as follows:

EXAMPLE **6-6**

Volts	110	200	280	300
Amperes	0.5	1.1	1.9	2.35

The generator supplies a load of 16 kW at 215 V when it is driven at 1700 r/min. Determine the speed at which the generator must be driven to supply a load of 23 kW at 270 V. Assume armature reaction is proportional to the armature current squared.

SOLUTION

The equivalent circuit of the long-shunt, cumulatively compounded dc generator is shown in Fig. 6-29(a). The series-field winding, as the name of the generator implies, aids the shunt-field winding. It is for this reason that the field currents flow in the same direction in the equivalent circuit. The OC characteristic is shown in Fig. 6-29(b). This characteristic can be used to obtain the no-load and full-load machine voltages and field currents only at 1700 r/min.

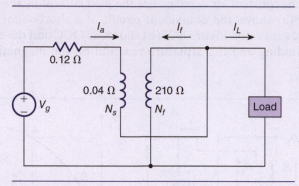

FIG. 6-29(a)

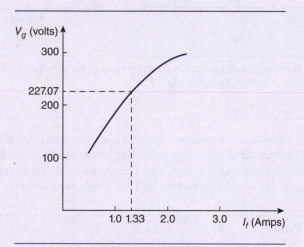

FIG. 6-29(b)

Load Condition: 16 kW at 215 V

The given data will be used to calculate the armature reaction as follows. Referring to Fig. 6-29(a), we have

$$\text{load current} = I_L = \frac{P}{V} = \frac{16,000}{215} = 74.42 \text{ A}$$

The current through the main field winding is

$$I_f = \frac{215}{210} = 1.02 \text{ A}$$

The armature current is

$$I_a = I_L + I_f$$
$$= 74.42 + 1.02$$
$$= 75.44 \text{ A}$$

The generated voltage is

$$V_g = V_t + I_a(R_a + R_s) = 215 + 75.44\,(0.12 + 0.04)$$
$$= 227.07 \text{ V}$$

Projecting this voltage on the given OCC, we obtain the effective field current:

$$I_{ef} = 1.33 \text{ A}$$

The armature reaction—at an armature current of 75.44 A—is found by using Eq. (6.78). Substituting the known data, we obtain

$$1.33 = 1.02 + \frac{12}{1800}\,(75.44) - \text{A.R.}$$

from which

$$\text{A.R.} = 0.197 \text{ equivalent field amperes}$$

Load Condition: 23 kW at 270 V

The load current at the unknown speed is

$$I_L = \frac{23{,}000}{270} = 85.18 \text{ A}$$

The current through the shunt field is

$$I_{f_a} = \frac{270}{210} = 1.28 \text{ A}$$

The armature current is

$$I_a = 85.18 + 1.28 = 86.47 \text{ A}$$

From the statement of the problem, the new armature reaction is

$$A.R. = 0.197 \left(\frac{86.47}{75.44} \right)^2$$

$$= 0.258 \text{ equivalent field amperes}$$

Using Eq. (6.78), we find that the new effective field current is

$$I_{ef} = 1.28 + \frac{12}{1800}(86.47) - 0.258$$

$$= 1.60 \text{ A}$$

Projecting this current on the given OCC gives the armature voltage at 1700 r/min. From the curve,

$$V_{g_1} = 255 \text{ V}$$

At the same field condition, but at the unknown speed, the generated voltage is obtained from the equivalent circuit as follows:

$$V_{g_2} = 270 + 86.47(0.12 + 0.04) = 283.84 \text{ V}$$

The required speed is found by using Eq. (6.75). Substituting, we obtain

$$\frac{283.84}{255} = \frac{n_2}{1700}$$

from which

$$n_2 = \underline{1892.24 \text{ r/min}}$$

In other words, the prime mover must rotate at 1892.24 r/min to meet the new load requirements.

For comparison purposes, the results are summarized in Table 6-3.

TABLE 6-3 Summary of results of Example 6-6

Condition	Output Power (kW)	Voltage (V)	Speed (r/min)	Shunt Field (A)	Effective Field (A)	Armature Reaction (A)
1	16	215	1700	1.02	1.33	0.197
2	23	270	1892.24	1.28	1.6	0.258

6.2 Modern Methods of Speed Control

6.2.1 Rectifiers

The most efficient and accurate method of controlling the starting current and the torque-speed characteristics of a dc motor is by using a variable armature and field voltage. The variable dc voltage is obtained through controlled rectifiers. Controlled rectifiers are used, as shown in Fig. 6-30(a), to convert single or poly-phase ac voltage sources to dc-controlled voltage supplies.

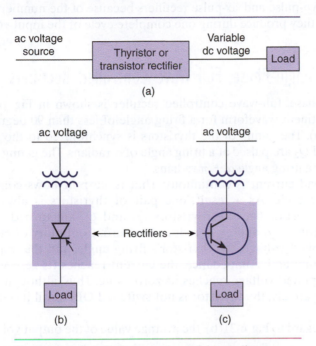

Fig. 6-30 AC-to-DC conversion: **(a)** block diagram representation, **(b)** one-line diagram representation, using thyristors, **(c)** one-line diagram representation, using transistors.

The controlling device for dc machines with low ratings is usually the transistor or the *IGBT* (Fig. 6-30(c)), while for higher machine ratings, it is the thyristor* (Fig. 6-30(b)). Chapter 3 discusses use of the transistor as a control switch. In this section, only the thyristor-rectifying circuits will be analyzed. The dc voltage that these circuits produce at their output terminals can be controlled smoothly by properly varying the thyristor's firing angle.

Controlled rectifiers, however, distort the power system's voltage waveforms, generate harmonics, and always result in an overall power factor of

---

*The basic operation of transistors and thyristors is discussed in the Appendix of the online Section DW.4.

negative nature. The harmonics may be reduced or eliminated completely by using suitable filters, and the power factor can be improved by adding equipment that generates lagging (leading-power-factor) kVAR. Of course, this significantly increases the overall cost of the rectifying unit and decreases its reliability.

In general, a controlled rectifier system consists of an isolating or a voltage step-down power transformer, a thyristor bridge network, and various other control and protection apparatuses.

In the next section the full-wave, single-phase rectifiers and the three-phase controlled rectifiers are discussed briefly. These rectifiers are also known, respectively, as two-pulse and six-pulse rectifiers because of the number of output voltage pulses they produce during one complete cycle of the input voltage.

6.2.2 Single-Phase, Full-Wave Controlled Rectifiers

A single-phase, full-wave controlled rectifier is shown in Fig. 6-31(a). The resulting pertinent waveform for a firing angle of less than 90 degrees is shown in Fig. 6-31(b). The gating of the thyristors is synchronized to the supply voltage, and Q_1 and Q_2 are pulsed at a firing angle of α radians. The gating of Q_3 and Q_4 is initiated at a firing angle $\alpha + \pi$ radians.

The load current is continuous: That is, current flows continuously over a complete cycle. As a result, one pair of thyristors is always conducting. Continuity results because thyristors Q_3 and Q_4 are turned ON before the current through Q_1 and Q_2 is reduced to zero. Continuity of current depends on the applied voltage, the thyristor's firing angle, and the impedance of the load. In an inductive impedance, the current reaches its zero value sometime after the applied voltage reaches its zero value. Thus, although the voltage increases negatively, the thyristor is not switched OFF until its current becomes zero.

With regard to Fig. 6-31(b), the average value of the output voltage is obtained as follows:

$$V_{av_o} = 2\left(\frac{1}{2\pi}\right)\int_{\alpha}^{\pi+\alpha} V_m \sin \omega t \, d\omega t \tag{6.82}$$

from which

$$V_{av_o} = 2\frac{V_m}{\pi} \cos \alpha \tag{6.83}$$

As can be seen from Eq. (6.83), the average value of the dc voltage can be varied smoothly by changing the firing angle of the thyristors accordingly. The average value of the output voltage, as a function of the angle of retard, is sketched in Fig. 6-32. At $\alpha = \pi/2$ radians, the dc value of the output voltage is equal to zero, and a further increase in the firing angle will result in negative output dc voltage.

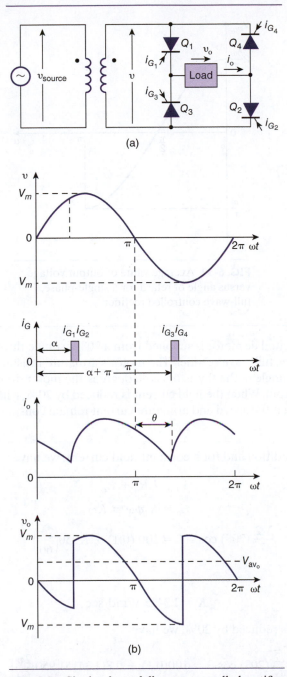

FIG. 6-31 Single-phase, full-wave controlled rectifier continuous load current: **(a)** circuit, **(b)** pertinent waveforms.

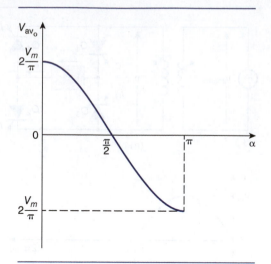

FIG. 6-32 Average value of output voltage versus angle of retard for a single-phase, full-wave controlled rectifier.

EXAMPLE **6-7**

A separately excited dc motor is supplied from a 208 V source through a thyristorized, single-phase, full-wave rectifier. The motor's armature resistance is 0.10 ohm. When the firing angle of the thyristors is 40 degrees, the motor draws 100 A and operates at 950 rpm. When the field current is reduced by 20%, what should be the firing angle so that the speed and armature current remain constant?

Solution

At the given condition and for a constant field current, we have

$$V_t = I_a R_a + E_a$$

$$E_a = K_1 \phi \omega = K \omega$$

$$\frac{2\sqrt{2}}{\pi}(208)\cos 40° = 100\,(0.1) + K950\left(\frac{2\pi}{60}\right)$$

from which

$$K = 1.3415 \text{ V/rad/sec}$$

When the field is reduced by 20%, we have

$$\frac{2\sqrt{2}}{\pi}(208)\cos \alpha = 100(0.1) + 0.8(1.3415)\left(950\,\frac{2\pi}{60}\right)$$

from which

$$\alpha = \underline{51.4 \text{ degrees}}$$

Exercise

6-1

A separately excited dc motor is supplied from a 208 V source through a thyristorized, single-phase, full-wave rectifier. The motor's armature resistance is 0.10 ohm. When the firing angle of the thyristors is 40 degrees, the motor draws 100 A and operates at 950 rpm. When the firing angle is reduced by 20%, by how much should the field current be increased so that the speed of the motor and its armature current will remain constant?

Answer   1.115

6.2.3 Three-Phase, Full-Wave Controlled Rectifiers

A three-phase, full-wave controlled rectifier is shown in Fig. 6-33(a). This circuit is commonly known as a 3-ϕ bridge rectifier, a Graetz rectifier, or simply a six-pulse rectifier. Each of the output voltage pulses has a fundamental frequency, which is six times that of the supply voltage. The impedance Z_n of the R-L load to the nth harmonic current is

$$Z_n = R + jn\omega L \qquad (6.84)$$

where $n = 6, 12$, and so on, and ω is the angular rotation of the source voltage. Owing to increased impedances, the various harmonic components of the load current are much smaller than the corresponding ones produced by a two-pulse controlled rectifier.

Analysis of a six-pulse rectifier is simplified by considering the effects of the positive and negative line voltages, as shown in Fig. 6-33(b). During one complete cycle, the controlling range of each of these voltages is given by the interval through which the particular voltage has, relatively, the maximum positive value. For example, as can be seen in Fig. 6-33(b), the controlling range of the voltage v_{ab}

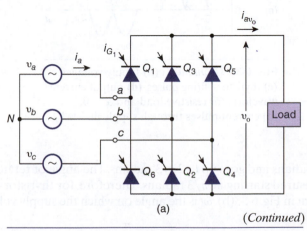

(a)

(*Continued*)

FIG. 6-33 Three-phase, full-wave rectifier: **(a)** circuit.

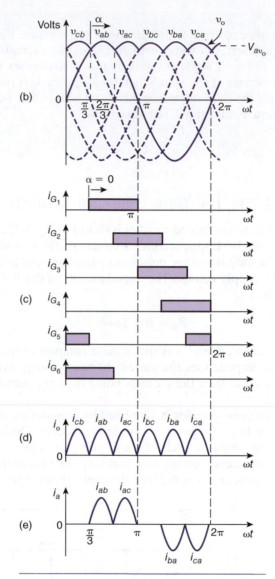

FIG. 6-33 (*Continued*) **(b)** supply voltages,
(c) thyristors' firing pulses, **(d)** output current
waveform for resistive load and $\alpha = 0$,
(e) current pulses through supply line "a."

starts at $\pi/3$ radians and goes up to $2\pi/3$ radians. The angle of retard for thyristor
Q_1 is thus measured starting at $\pi/3$ radians. Therefore, for thyristor Q_1, Eq. (6.85)
where, as shown in Fig. 6-33(b), ωt is the angle on which the supply voltages depend.

$$\alpha = \omega t - \frac{\pi}{3} \tag{6.85}$$

The "turning ON" of the other thyristors is synchronized to the supply frequency, and their firing angles are spaced $\pi/3$ radians apart. The firing pulse of each thyristor is shown in Fig. 6-33(c). The end of conduction, as in all other circuits using thyristors, depends on the applied voltage and the load impedance. For rectifier operation, the angle of retard must be restricted to less than $\pi/2$ radians.

Figure 6-33(d) shows the output current pulses for a firing angle of zero degrees. A firing angle for a particular thyristor indicates the time when the thyristor is gated relative to its corresponding line-to-line voltage.

Figure 6-33(e) shows the output current pulses conducted through supply line "a." Then in the six-pulse circuit, each supply line carries four of the six output current pulses in an alternating symmetry and thus has no dc component. This permits the use of a 3-ϕ transformer to supply the voltage required by the bridge. Furthermore, the line-current waveform, as can be shown by Fourier series analysis, contains a fundamental component and various harmonics. The harmonics transport no real power to the load and distort the upstream current waveforms. This may lead to original text resonance problems and communication interference.

The type and duration of the output current pulses depend on the firing angle and the parameters of the load. The limits of integration for calculating the various parameters depend on whether the output current is continuous or discontinuous.

For continuous load current, the dc value (V_{av_o}) of the output voltage is

$$V_{av_o} = \frac{6}{2\pi} \int_{(\pi/3)+\alpha}^{(2\pi/3)+\alpha} V_m \sin \omega t \, (d\omega t) = 3 \frac{V_m}{\pi} \cos \alpha \qquad (6.86)$$

When the load to a six-pulse rectifier is a dc machine, Eq. (6.86) gives the motor's terminal voltage. Neglecting the resistance of the armature winding, from the generator principle and from KVL, we have

$$V_g \approx 3 \frac{V_m}{\pi} \cos \alpha = K\phi\omega \qquad (6.87)$$

Thus, the speed of the motor, as a function of the firing angle, is

$$\omega \approx \frac{3V_m}{K\phi\pi} \cos \alpha \qquad (6.88)$$

That is, for constant field current, the speed of the motor is directly proportional to the cosine of the firing angle. Therefore, it is evident that the thyristor's firing angle controls the motor's starting current and speed of operation.

Refer to Fig. 6-33(e). By assuming the load current is constant at I_{av_o}, we find the rms value of the supply line currents as follows:

$$I_{rms}^2 = \frac{2}{2\pi} \int_{\pi/3}^{\pi} I_{av_o}^2 \, d\omega t \qquad (6.89)$$

$$= \frac{2}{3} I_{av_o}^2 \qquad (6.90)$$

From Eq. (6.90),

$$I_{rms} = \sqrt{\frac{2}{3}} I_{av_o} \tag{6.91}$$

Equation (6.91) gives the rms value of the line current, not of its fundamental component.

The apparent power of the supply source is

$$|S| = 3V_{an}I_a = 3\frac{V_{L\text{-}L}}{\sqrt{3}} \sqrt{\frac{2}{3}} I_{av_o} = \sqrt{2} \, V_{L\text{-}L} \, I_{av_o} \tag{6.92}$$

where $V_{L\text{-}L}$ is the rms value of the line-to-line supply voltage. Equation (6.92) can be used to size the isolating transformer, which supplies power to the rectifying bridge.

The apparent or total power factor ($\cos \theta_T$) at the input terminals of the bridge is found as follows:

$$\text{power} = \sqrt{3} \, V_{L\text{-}L} \, I_L \cos \theta_T \tag{6.93}$$

from which

$$\cos \theta_T = \frac{V_{av_o} I_{av_o}}{\sqrt{3} V_{L\text{-}L} I_L} \tag{6.94}$$

By substituting for V_{av_o} and I_L their equivalent expressions (given, respectively, by Eqs. (6.86) and (6.91)), we obtain

$$\cos \theta_T = \frac{3\frac{V_m}{\pi} \cos \alpha I_{av_o}}{\sqrt{3} \frac{V_m}{\sqrt{2}} \sqrt{\frac{2}{3}} I_{av_o}} \tag{6.95}$$

From the above,

$$\cos \theta_T = \frac{3}{\pi} \cos \alpha \tag{6.96}$$

The line current to the input terminals of the bridge has a fundamental component and various harmonics. Since the harmonics transport no average power to the load, the power measured with a nondigital power meter is

$$P = \sqrt{3} V_{L\text{-}L} I_L \cos \theta \tag{6.97}$$

where I_L is the rms value of the fundamental component of the line current and $\cos \theta$ is the apparent power factor of the circuit. From Eqs. (6.93) and (6.97), we obtain

$$\cos \theta_T = g \cos \theta \qquad\qquad (6.98)$$

where g is the ratio of the rms value of the fundamental component of the line current divided by the rms value of the current waveform.

EXAMPLE **6-8**

The Δ-Y transformer shown in Fig. 6-34(a) supplies power to a six-pulse bridge rectifier. The line-to-line voltages across the secondary lines of the transformer are as shown in Fig. 6-34(b). The load current is constant and the transformer is ideal, with the turns ratio equal to unity. For a firing angle of zero degrees, draw the waveforms of the current through:

a.   The load.

b.   The transformer windings.

c.   The primary supply line $\left(i_{L_1}\right)$.

SOLUTION

a.   The waveform of the load current I_{av_o} is shown in Fig. 6-34(c). This current is continuous and of constant amplitude because of the load's inductance.

b.   Each of the secondary transformer lines conducts during only two-thirds of the applied voltage's cycle. Thus, for continuous load current, each secondary transformer line conducts for $\frac{2}{3}\pi$ radians in the positive direction and $\frac{2}{3}\pi$

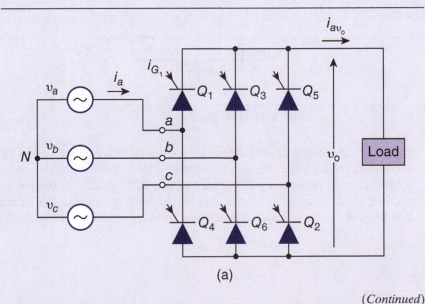

(a)

(Continued)

FIG. 6-34 Six-pulse rectifier. Waveforms of the currents through the load, the secondary windings, and a primary line of a Δ-Y transformer.

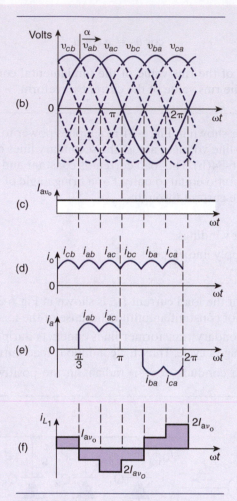

FIG. 6-34 (*Continued*)

radians in the negative direction. The particular conducting period of each secondary transformer line can be obtained by inspection of the given line-to-line voltage waveforms. For example, line "a" conducts in the positive direction when voltages v_{ab} and v_{ac} are the highest voltages in the circuit, and conducts in the negative direction when voltages v_{ba} and v_{ca} have the highest potential in the circuit. Therefore, line "a," as shown in Fig. 6-34(d), conducts only during the time intervals between 60 to 180 degrees and 240 to 360 degrees. The currents in the other secondary lines (i_b and i_c) are drawn as shown in Fig. 6-34(d).

The transformer's primary winding currents are obtained from the secondary line currents by using KVL in magnetic circuits. That is,

$$N_A i_{AC} = N_a i_a$$

Since the numbers of primary and of secondary winding turns are equal, the magnitude of the winding currents must also be equal. However, these currents must be in opposite directions, as per Lenz's law.

In a similar manner, the currents through the other primary transformer windings can easily be determined. The resulting waveforms are shown in Fig. 6-34(e).

c. The primary line current is found by applying KCL in the delta-connected winding. Considering line "A," we have

$$i_{L_1} = i_{AC} - i_{BA}$$

From the above, the current through line "A" is obtained graphically by superimposing the negative of i_{BA} on the winding current i_{AC} (See Fig. 6-34(f).)

A 20 kW, 220 V, 1200 r/min, 0.94 efficient dc motor is supplied through a six-pulse rectifier with a constant current. The line-to-line voltage across the secondary terminals of the transformer is 208 V. Determine:

EXAMPLE 6-9

a. The nominal firing angle of the thyristors.
b. The apparent power factor of the load, as seen from the secondary terminals of the rectifying transformer.
c. The line current toward the rectifying bridge.
d. The kVA rating of the rectifying transformer.

SOLUTION

a. The thyristor's firing angle is found from Eq. (6.86). Solving for α and substituting the given values, we obtain

$$\alpha = \arccos \frac{(220)\pi}{3\sqrt{2}\ 208}$$

$$= \underline{38.45°}$$

b.   From Eqs. (6.96), we have

$$\cos \theta_T = \frac{3}{\pi} \cos \alpha$$

$$= \frac{3}{\pi} \cos 38.45$$

$$= 0.75 \text{ lagging}$$

c.   The average value of the load current is

$$I_{av_o} = \frac{P}{V_{av_o}} = \frac{20}{0.94(0.220)}$$

$$= 96.71 \text{ A}$$

Thus, from Eq. (6.91), the rms value of the line current is

$$I_{av_o} = \sqrt{\frac{2}{3}} \, 96.71$$

$$= 78.96 \text{ A}$$

d.   The rms value of the load current is equal to its dc value. Thus, from Eq. (6.92), the kVA rating of the transformer is

$$|S| = \sqrt{2} \,(0.208)(96.71) = \underline{28.45 \text{ kVA}}$$

For practical purposes, a 30 kVA, commercially available transformer should be used.

Exercise

6-8

a.   Draw the output voltage and current waveforms of a single-phase, full-wave rectifier. Assume that the load current is continuous and $\alpha = 30°$.

b.   A six-pulse, three-phase thyristorized rectifier supplies 100 A dc to a load. Determine the average and rms values of a thyristor's current.

Answer   33.33 A, 57.74 A

Exercise

6-9

A separately excited, 100 kW, 480 V, 0.9 efficient dc motor has an armature resistance of 0.10 Ω. The motor receives its dc supply through a six-pulse rectifier as shown in Fig. 6-35. Determine:

a.   The motor's current under nominal operating conditions.

b. The thyristor's firing angle under nominal operating conditions and at starting if the motor's starting current is limited to 800 A.

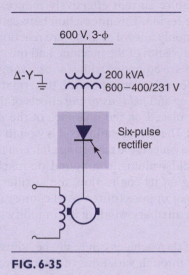

FIG. 6-35

Answer (a) 231.48 A; (b) 27.3°, 81.5°

6.3 Conclusion

The operation of a dc motor is based on the tendency of a current-carrying conductor—when placed in an external magnetic field—to rotate so that the field it creates is parallel to, and in the same direction as, the external field.

The operation of dc generators is based on the induction principle, according to which a voltage is induced in conductors that are rotated through an external magnetic field.

In their basic form, dc machines have a field coil and an armature winding. The field coil is wound around the machine's stationary magnetic poles, while the armature winding is wound on its rotating shaft. The armature current is conducted from the external motor terminals to the stationary brushes, then to the rotating segments of the commutator, and from there to the armature conductors.

The rotated armature conductors successively cut the fields of the north and south magnetic poles. As a result, the generated armature current is of alternating waveform. However, as seen from the machine's external terminals, the armature current is of dc waveform because it is rectified by the commutator-brush action. The process by which the armature current reverses direction during half the period of its cycle is called commutation.

During part of the rectification process, each armature conductor is shorted through the brushes. As a result, sparking may take place. Sparking and flashover used to be the main limitations of dc machines. However, in recent years, the design of the machines has been improved, and the commutation process is now sparkless. Nevertheless, the brushes require periodic replacement. In general, this process increases the maintenance cost of dc machines.

The flux of the armature current effectively reduces the magnetic field and distorts it in the interpole region. This interaction between the flux of the field and armature current is commonly known as armature reaction. Armature reaction adversely affects the characteristics of the machine and the commutation process.

All large dc machines are equipped with *compensating* and *commutating windings*. These windings are connected in series with the armature coil. Their function is to cancel the *demagnetizing* and *field-distorting* effects of the armature current. The compensating winding is placed on the pole face of the machine and cancels the reduction of the field flux. The commutating coil is wound around the narrow poles in the interpole region and cancels the distorting effect on the field-flux distribution.

Besides the main field winding, compound dc machines have an auxiliary field winding. Depending on its connection, the auxiliary field winding may aid (cumulative connection) or oppose (differential connection) the flux of the main field. The function of the auxiliary winding is to modify the external characteristics of the machine.

The characteristics of dc machines can easily be adjusted through armature or field-variable resistors or through variations of their dc voltage supply. The use of potentiometers results in higher energy losses. The use of variable dc voltage sources results in lower energy losses and in smoother torque-speed characteristics.

All modern dc drives are equipped with controlled rectifiers through which the motor characteristics are easily controlled. The six-pulse bridge rectifier is the most popular compromise between cost and desired characteristics. In general, controlled rectifiers are reliable and highly efficient, generate harmonics toward their ac voltage supply and a ripple voltage toward their dc output lines, and require kVAR compensation.

Owing to their efficiency, variable-speed dc drives have gradually displaced the more inefficient drives, such as eddy-current variable-speed drives, fluid couplings, the classical generator-motor set, and induction motors that drive fans whose flow is controlled through dampers.

Table 6-4 summarizes the essential relationships of dc machines and their control apparatus. Table 6-5 gives manufacturer's data for a 300 kW dc motor. Table 6-6 lists typical component losses of dc drives.

TABLE 6-4   Summary of important equations

Item	Description	Remarks
1	The motor principle	
	$$T = K\phi I_a$$	Eq. (6.13)
2		
	$$K = \frac{Np}{\beta\pi}$$	Eq. (6.14)

(Continued)

TABLE 6-4   (Continued)

Item	Description	Remarks
3	The generator principle $$V_g = K\phi\omega$$	Eq. (6.22)
4	Mutual coupling $$V_g = \frac{p}{2}\,\omega I_f L_{\text{afm}}$$	Eq. (6.32)
5	$$T = \frac{p}{2}\,I_f L_{\text{afm}} I_a$$	Eq. (6.34)
6	Effective field $$I_{ef} = I_{f_a} \pm \frac{N_s}{N_f} I_s - \text{A.R.}$$	Eq. (6.78)
	Two-pulse rectifier	
7	$$V_{\text{av}_o} = 2\,\frac{V_m}{\pi}\cos\alpha$$	Eq. (6.83)
	Six-pulse rectifier	
8	$$V_{\text{av}} = 3\,\frac{V_m}{\pi}\cos\alpha$$	Eq. (6.86)
9	$$I_{\text{rms}} = \sqrt{\frac{2}{3}}\,I_{\text{av}_o}$$	Eq. (6.91)
10	$$|S| = \sqrt{2}\,V_{L\text{-}L} I_{\text{av}_o}$$	Eq. (6.92)
11	$$\cos\theta_T = \frac{3}{\pi}\cos\alpha$$	Eq. (6.96)
12	$$\cos\theta_T = g\cos\theta$$	Eq. (6.98)

TABLE 6-5   Data for a 300 kW, 1150 r/min, 0.95 efficient, 1052 A, 300 V, four-pole, separately excited dc motor.

a. Resistances

Winding(s)	Resistance in ohms at 25°C
Armature	0.005
Commutating and Compensating	0.0045
Separate Field	2.07

(*Continued*)

b. Load characteristic

Load in Percent	Armature Voltage (V)	Armature Current (A)	Field Voltage (V)	Field Current (A)	Speed (r/min)	Commutation
0	300	0	55	22.9	1175	Sparkless
25	300	270	55	22.9	1165	Sparkless
50	300	540	55	22.9	1160	Sparkless
75	300	800	55	22.9	1150	Sparkless
100	300	1052	55	22.9	1150	Sparkless
125	300	1350	55	22.9	1150	Sparkless
150	300	1600	55	22.9	1150	Sparkless

c. Temperature rise test

Time in Hours	Load	Air supply: 60 m³/min Armature Voltage	Armature Current	Field Voltage	Field Current	Speed (r/min)
2	100%	300 V	1075 A	55 V	22.9 A	1150

	Armature Coil	Core	Commut.	Frame	Field Windings Main	Commut.	Comp.	Bearings Commutator Side	Bearings Opposite Side
Temp., °C	48	45	55	18	35	33	32	20	15

d. Measurement of air gap

Main pole: 3.5 mm

Interpole: 7.5 mm

Based on data from Siemens Electric Limited

TABLE 6-6   Typical efficiency of dc drives at rated speed

Rating kW	Efficiency Transformer	Control Circuit	Motor	Total
15		0.98	0.92	0.90
30		0.99	0.88	0.87
150	0.97	0.99	0.89	0.85
225	0.98	0.99	0.90	0.87
750	0.99	0.99	0.94	0.92
1500	0.99	0.99	0.94	0.92

Based on data from Siemens Electric Limited

6.4 Review Questions

1. Describe armature voltage (V_a). Is this voltage ac or dc in waveform? Why is armature voltage also called voltage generated (V_g), back emf (V_b), countervoltage (V_c), or speed voltage?

2. Differentiate between speed voltage and transformer voltage.

3. What is armature reaction, and what are its *two adverse effects* on the operation of dc machines?

4. Under what condition is the armature reaction proportional to the armature current squared?

5. What controls the shape of the magnetization characteristic?

6. What limits the operation of dc machines?

7. What are the *three adverse effects* of high starting current on dc machines?

8. Are the rotating parts of a dc machine the brushes or the commutator segments?

9. At what part of the flux-density distribution is the armature conductor located at the instant when commutation takes place? What is the reason for this?

10. How would you measure the mutual inductance between the armature and field windings of a separately excited motor?

11. Where are the series, the commutating, and the compensating windings located physically? What are their functions?

12. Draw the torque-versus-speed characteristics for series, shunt, and compound motors.

13. Draw the equivalent circuits of long-shunt and short-shunt dc compound machines. Consider differential and cumulative-type connections.

14. Why are dc shunt motors normally equipped with a variable external resistor that is connected in series with the armature?

15. What is the power factor of a six-pulse thyristorized rectifier, as seen from its ac input terminals?

16. What is the average value of the dc voltage for a six-pulse thyristorized rectifier?

6.5 Problems

6-1 At 1800 r/min, a separately excited dc generator with constant excitation develops an open-circuit terminal voltage of 400 V. Determine the electromagnetic torque when the machine operates as a motor and draws an armature current of 60 A. Assume an armature reaction of 5%.

6-2 A 10 kW, 230 V, shunt dc motor has an efficiency of 88% at full-load. The resistances of the armature and shunt field are 0.12 and 230 ohms, respectively. At

no-load, the motor rotates at 800 r/min and draws 4 A from the 230 V source.

a. At full-load conditions, determine the armature current, the electromagnetic torque, and the speed of the motor (neglect armature reaction).

b. Repeat (a), assuming an armature reaction of 5%.

6-3 A 200 V, 1200 r/min, 30 A, dc shunt motor drives a ventilating fan whose torque requirement is proportional to the speed squared. The resistances of shunt-field and armature windings are 100 and 0.15 ohms, respectively. Neglecting rotational losses and armature reaction, determine:

a. The starting current of the motor.

b. The resistance that must be inserted in series with the armature to produce a motor speed of 1000 r/min.

c. The resistance of the field rheostat that must be connected in series with the field winding to yield a speed of 2000 r/min.

6-4 The generator-motor system shown in Fig. P6-4 is used as a variable-speed drive. In actual setups, the speed of the motor is controlled by varying the field of the generator. For this problem, however, assume that the fields of the motor and generator remain constant. The system's nameplate data and armature reaction are as shown on the diagram. The generator's prime mover is a 3-ϕ induction motor whose slip at no-load is negligible and at full-load is 4%. Estimate:

a. The voltage regulation of the generator.

b. The speed regulation of the motor.

6-5 The open-circuit characteristic of a 220 V dc series motor at 1800 r/min is approximated by

$$V_g = \frac{260 \, I_a}{25 + I_a}$$

The motor draws 90 A when it runs loaded at 1800 r/min. If the load conditions change and the motor draws 70 A, determine:

a. The speed.

b. The electromagnetic torque developed.

c. The useful output power if the mechanical losses within the motor consume a torque of 2 N · m.

d. The mathematical relationship that describes the open-circuit characteristic for the condition in (a).

6-6 A short-shunt compound generator has its auxiliary field winding connected in such a way that its magnetic field opposes that of the main field winding. The ratio of the effective number of turns per pole of the main field to auxiliary field winding is 175:1. The armature resistance is 0.10 ohm, the series-field resistance is 0.05 ohm, and the shunt-field resistance is 125 ohms. Data for the open-circuit characteristic at 1800 r/min are as follows:

Armature voltage (V):	50	100	200	260	300	318
Field current (A):	0.21	0.41	1.0	1.51	2.0	2.7

When driven at 1800 r/min, the generator delivers 60 A to a load at 250 V. Determine the speed at which the generator must be driven in order to deliver 75 A at

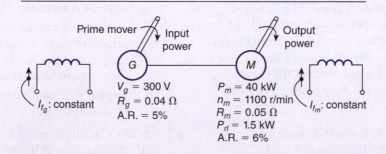

FIG. P6-4

300 V. Assume that armature reaction is proportional to the armature current.

6-7 A 600 V, 7.46 kW, 1200 r/min series dc motor drives a constant torque load of 50 N · m. The series and armature resistances are 0.10 and 0.20 ohm, respectively. In order to obtain variable speed, a resistance R_x is connected in parallel with the field winding. Neglecting armature reaction and rotational losses,

a. Show that the armature current as a function of R_x is given by

$$I_a = 11.48 \sqrt{\frac{R_x + 0.1}{R_x}}$$

b. Show that the speed as a function of R_x is given by

$$I_a = 138.47 \sqrt{\frac{R_x + 0.1}{R_x}} - \frac{0.53}{R_x}(1.5R_x + 0.1)$$

c. Sketch the expressions found in (a) and (b).

6-8 In the Ward-Leonard speed control system shown in Fig. P6-8, the armature of the separately excited dc generator is connected directly to that of a separately excited dc motor. The generator is driven at constant speed, and its open-circuit characteristic at the operating region is described by

$$V_g = 5 + 10I_{fg}$$

The motor has an efficiency of 95%. When the voltage at the terminals of the generator is 320 V, the motor runs at 1200 r/min and delivers 50 kW to the load.

a. For a constant torque-load requirement of 200 N · m, show that the field current of the generator is related to the speed of the motor by

$$I_{fg} = 0.234\omega + 0.78$$

b. For a constant load-power (50 kW) requirement, show that the field current of the generator is related to the speed of the motor by

$$I_{fg} = 0.234\omega + 1.97$$

c. Sketch the expressions derived in (a) and (b).

6-9 A single-phase, full-wave rectifier operating at a firing angle of α degrees supplies power to a resistive load. Show that the apparent power factor of the load, as seen from the transformer's secondary terminals, is given by

$$\cos\theta_T = \sqrt{\frac{\pi - \alpha + \dfrac{\sin 2\alpha}{2}}{\pi}}$$

6-10 A separately excited dc motor is supplied from a 208 V source through a thyristorized, single-phase, full-wave

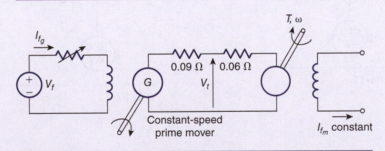

FIG. P6-8

rectifier. The armature resistance is 0.10 ohm and at nominal operating conditions,

$$n = 800 \text{ rpm}, \qquad I_a = 80 \text{ A}$$

and the thyristor's firing angle is 40 degrees.

By assuming that the field current is constant, sketch the motor's speed as a function of the firing angle when:

a. Constant torque is developed.

b. Constant power is developed.

6-11 A 208 V, separately excited dc motor is controlled through a single-phase, full-wave thyristorized bridge. The armature resistance is 0.1 ohm, and when the speed is 950 rpm, the thyristor's firing angle is 40 degrees and the motor draws 100 A. Sketch the field current as a function of the firing angle while the motor's power is constant.

6-12 As shown in Fig. P6-12, a six-pulse rectifier supplies a 200 kW, 600 V, 0.93 efficient

dc motor. The transformer is fed from a 4160 V substation. At nominal operating conditions, the thyristor's firing angle is 36 degrees. Determine:

a. The transformer's turns ratio.

b. The kVA rating of the transformer.

c. The apparent power factor, as seen from the terminals of the substation.

Assume that the transformer is 96% efficient.

6-13 Two dc drives, each 1000 kW, 800 V, and 0.95 efficient, are connected, as shown in Fig. P6-13, to a 4.16 kV substation. Assuming that the motors draw rated current, determine:

a. The thyristor's firing angle.

b. The magnitude of the total current through the 4.16 kV feeder.

c. The apparent power factor, as seen from the secondary lines of one of the transformers.

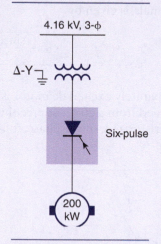

FIG. P6-12

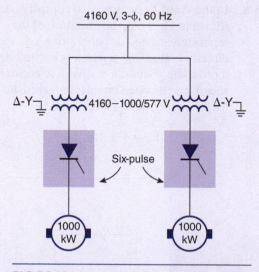

FIG P6-13

7

Control Schematics

7.0   Introduction

7.1   Basic Devices and Symbols

7.2   The Concept of Protection

7.3   Actual Control Schematics

7.4   Conclusion

7.5   Review Questions

7.6   Problems

497

What You Will Learn in This Chapter

A **Theoretical Highlights**
1 The basic electromagnetic relays and contactors
2 Thermal overload relays
3 Protection of motors
4 Control schematics of 3-ϕ motors
5 Control schematics of brushless synchronous motors

B **Outstanding Practical Highlights**
1 Basic control schematics
2 Overload and overcurrent protection of motors
3 Electromagnetic timers
4 Sensors
5 Electronic control of brushless synchronous motors

C **Additional Students' Aid on the Web**
1 Manual resets
2 Control schematics of MV induction motors
3 Control schematics of dc machines
4 Logic circuits
5 Programmable logic controls

7.0 Introduction

The control schematic of a machine is an electrical drawing that uses conventional symbols to reveal and demonstrate the machine's starting, control, protection, deenergization, and safety features. Hence, the control schematic constitutes the interface between the engineer and the machine. As such, an understanding of control schematics bridges the domains of theoretical and practical work in the field of electric machines.

Control schematics may be designed by using electromechanical components or solid-state devices. The trend, of course, is toward solid-state devices.

The design of a *good* control schematic requires basic knowledge of the control components and the principles of operation of the machine under consideration. The design of the *best* control schematic also requires some field experience and up-to-date knowledge of the cost, availability, and limitations of the control components.

The functions and components of control schematics are represented on an engineering drawing by standard conventional symbols. Field specialists, however, do the actual hook-up of these devices using "wiring diagrams." A machine's *wiring diagram* shows the actual or physical interconnections between the various components. It does not identify the functional characteristics of each component.

This chapter explains the operation of the basic devices. Subsequent sections discuss the actual control schematics of three-phase induction motors and synchronous motors. The Web Section Chapter 7W describes the control schematics of 1-ϕ motors, MV induction machines, dc motors, and Programmable Logic Controllers (PLC). The latter are used extensively because of their simplicity, flexibility, and low cost.

7.1 Basic Devices and Symbols

This section discusses relays, electromagnetic starters, pushbuttons, indicating lights, temperature sensors, and thermostats. The principle of operation and the symbol representation of these devices provide the basis for understanding and designing electrical control schematics.

7.1.1 Electromagnetic Relays

As shown in Fig. 7-1(a), the basic electromagnetic relay has a coil and a contact whose status changes when the relay coil is energized. The coil is wound around a stationary core, while the moving part of the contact is attached on a hinged iron bar called the armature. When the relay is energized, the armature is attracted to the electromagnet, and thus the contact of the relay changes status (Fig. 7-1(b)).

Physically, the coil and the contact are within the same enclosure. Electrically, however, they are generally connected to different circuits. As shown in Fig. 7-1(c), the input terminals 1 and 2 of the relay receive the excitation signal, while the output terminals 3 and 4 are the external connections of the relay's contact.

Depending on the design, to achieve energization, the coil of a given relay may require a certain voltage (voltage coil) or current (current coil), or a combination of voltage and current. The energizing signal, where it is a voltage or a current, could be produced by either normal or abnormal machine operations, or by a sequence of events that accompany the operation of an electrical apparatus.

The energizing signal may be dc or ac; each has its advantages and disadvantages. Relays with ac coils cost less than those with dc coils but react slowly, consume more energy, and may malfunction owing to occasional transient fluctuations of the supply voltages. Although relays with dc coils are more expensive, they respond more quickly and, within a certain voltage range, are independent of the ac voltage fluctuations.

In a control schematic, the coil of a relay is represented by a circle in which is inscribed an arbitrary letter, as shown in Fig. 7-1(d). This letter is also used to designate the relay's contact. The letter X appears in this illustration, but very often the letter R is employed instead.

The contact(s) of the relay may be used to switch on an indicating light, to sound an alarm, to start a machine, or to remove a machine from its voltage

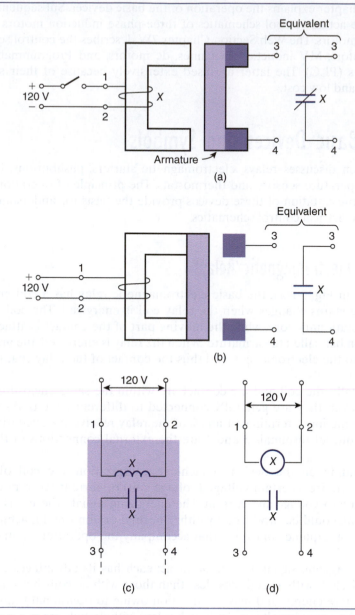

FIG. 7-1 Representation of an electromagnetic relay: **(a)** deenergized, NC contact, **(b)** energized, contact opened, **(c)** wiring diagram of (b), **(d)** control schematic of (b). The armature hinge is not shown.

source. The contacts may be of the momentary or of the time-delay type, or a combination of both. When a relay's coil is energized, the momentary type of contacts change status instantaneously, while time-delay relays change the status of their contacts with some specific time delay. In a relay with a "time delay on energization," the beginning of the time delay starts immediately after its coil is energized. In a relay with a "time delay on deenergization," the time delay begins immediately after its coil is deenergized.

As shown in the control schematics, the status of the contacts of the various relays corresponds, by convention, to the deenergized status of the relays. The symbols and the general form of a device's contacts are discussed in greater detail in Section 7.1.4.

Electromechanical relays can be divided into two categories: (1) relays for general purposes, and (2) relays for the protection of equipment. General-purpose relays are simple and economical and are used mainly for switching and annunciating functions. A simple general-purpose 120 V relay is shown in Fig. 7-1. The voltage rating of its coil is 120 V. However, the voltage and current ratings of its contact must be suitable for the circuit in which it will be connected.

Relays for the protection of equipment can have their current coils connected in series with the protected load, or the input signal to their coils can be obtained from instrument transformers that monitor the electrical condition of the load. Relays whose current coils are connected in series with the load are called *thermal overload relays*. Relays that receive their coil excitation through current or voltage transformers are generally known as *protective relays*. In either case, the function of the relay is to protect the equipment, and therefore the assigned terminology is not functionally very meaningful. Because of their industrial importance, both of these types of relays are discussed extensively in the sections that follow.

As shown in Fig. 7-2, a single-phase electromagnetic starter—often referred to as a magnetic contactor—incorporates an electromagnetic coil (C), two main contacts (C_1), and an auxiliary contact (C_2). The coil and its contacts are usually identified by the same letter. This simplifies the understanding of multimachine control schematics. For the sake of clarity, however, in the introductory sections of this chapter the subscripts 1 and 2 will be used to differentiate between the main and auxiliary contacts.

Magnetic contactors are similar to relays. Their only electrical difference is that the main contacts of magnetic starters have higher voltage and current capacities than the contacts of the relays. The function of the magnetic contactor's assembly is to energize or deenergize a motor or any other electrical apparatus. When a suitable voltage is supplied to the coil (see Fig. 7-2(a), terminals 1 and 2) of the magnetic contactor, an electromagnetic force is developed that closes the normally open (NO) contacts C_1 and C_2. Closing the contacts C_1 connects the load to its voltage source.

For its energization, the coil of the starter (depending on its manufacturer) requires a minimum of about 80% of its rated voltage. Thus, the dipping of the supply voltage that usually accompanies the starting of ac machines may deenergize the electromagnetic starter.

The main contacts C_1 are connected in series with the controlled electrical load, so they must be capable of safely carrying and withstanding the currents that are produced by starting, operating, and deenergizing the load. The current capacities of the main contacts, on starting and on deenergizing the load, are referred to as the "making" and "breaking" capacity of the contacts, respectively. Typical ratings for three-phase electromagnetic starters are shown in Table 7-1.

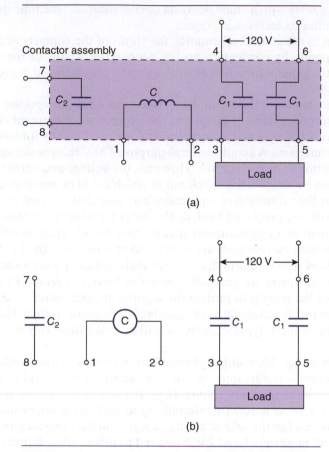

FIG. 7-2 Electromagnetic starter or contactor: **(a)** wiring diagram, **(b)** representation in a control schematic.

TABLE 7-1 Typical ratings of electromagnetic starters

	Electromagnetic Starter					
	Characteristics of the 120 V Coil			Capacity of Main Contacts (A)		
Motor Rating: 480 V, 60 Hz, 3-φ	Nominal Power (W)	Inrush Current (A)	Continuous Current (A)	Make	Carry	Break
1 kW	10	0.42	0.11	100	5	90
10 kW	13	0.65	0.12	400	50	350
100 kW	35	2.5	0.40	1400	200	1150

Based on data from Siemens Electric Limited

As discussed in Section 7.1.6, the auxiliary contact C_2 is normally used to "seal in" the start pushbutton. At an extra cost, the starter may be equipped with additional auxiliary contacts of either the momentary or time-delay type. The three-phase electromagnetic starter has three main contacts, one for each of the three power supply lines, as shown later in this chapter in Fig. 7-17.

7.1.2 Thermal Overload Relays

Thermal overload relays are temperature-sensitive devices whose contacts close or open depending on the degree of heat produced by the current through the relay's coil. The coil of the relay, often referred to as its "heating element," is connected *in series* with the protected load. Since there is a definite time interval between the current flow and the heat produced by the heating element, thermal overload relays are used not for overcurrent* protection but for *overload* protection.

The heat generated on overload is proportional to the current squared. This heat is used either to melt an alloy, thereby permitting a rachet wheel to turn and open a control contact, or to heat a bimetallic strip, causing it to bend and open a control contact.

As seen in Fig. 7-3(a), the thermal overload relay has, as a standard, a normally closed (NC) contact that is *manually* reset after current interruption. The more expensive thermal overload relays have an automatically reset NC contact and may also be equipped with a normally open (NO) contact for alarm or

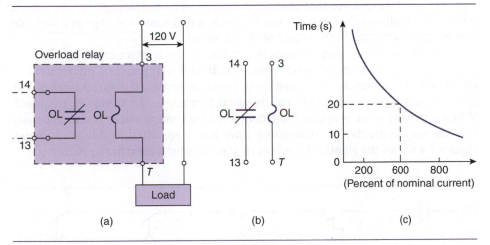

FIG. 7-3 Overload relay: **(a)** wiring diagram, **(b)** representation in a control schematic, **(c)** time-current characteristic.

---

* An overcurrent or short-circuit current is more than six times the machine's nominal current. Overload is the current that is from one to six times the machine's rated current. Etymologically speaking, no clear-cut division exists between an overload and a short-circuit current. Generally, an overload current flows along the same path as the machine's full-load current, while an overcurrent flows partially along a different path.

indication purposes. The control schematic representation of a thermal overload relay is shown in Fig. 7-3(b).

The current level that can energize a thermal overload relay is externally adjustable by about ±25% of its base value. The typical time-current characteristic of a thermal overload relay is shown in Fig. 7-3(c). The thermal overload relays will trip in about 20 s when their coil's current is six times its rated value. When the load's current is 1.25 pu, the thermal overload relay will trip in about 90 s. In some thermal overloads, the temperature of the heating element at the instant of tripping is about 140°C.

Prolonged overloads deteriorate the winding's insulation, while uncontrolled overcurrents can damage all current-carrying and noncurrent-carrying parts of the machine.

The thermal overload protection of motors with ratings smaller than 4 kW is normally an internal thermal protector that breaks the motor's circuit at a certain pre-determined current level. After current interruption in such cases, one has to wait until the motor cools off before the thermal protector closes the opened circuit.

7.1.3 Electrical Contacts

An electrical contact makes or breaks a circuit and therefore is used to energize or deenergize a load that is connected in series to it. A contact is either normally closed (NC) or normally open (NO). Generally speaking, a contact is the output of a device that senses a change in a variable and, at a predetermined level of the variable, produces a force that changes the status of its contacts. Relays, magnetic contactors, limit switches, and sensors are devices that sense a change in a variable and produce a change in the status of their contacts.

The driving force that changes the status of the contact could be manual, electromechanical, thermal, optic, or the like. In fact, many natural phenomena— changes in liquid levels (float switches), pressure levels (pressure switches), or speed levels (speed switches)—can be used to change the status of a contact. When a driving force is applied, the contacts may change status instantaneously (momentary contacts) or after some time has elapsed (time-delay contacts). Figure 7-4 shows the standard symbols of some contacts used in the industry.

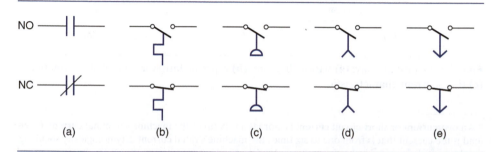

FIG. 7-4 Symbols of contacts: **(a)** instantaneous (relay or magnetic contactor), **(b)** temperature, **(c)** pressure, **(d)** time delay on energization, **(e)** time delay on deenergization.

The contacts of various devices have different configurations and are commonly known as contact forms. The single-pole single-throw (SPST) contact form (see Table 7-2) has only one contact, while the single-pole double-throw (SPDT) configuration has two or more contacts. The pole of the device is indicated with a bar terminating at a square. The pole's force, and thus its making or breaking capacity, is determined by a physical force on which the design of the device is based. A device with a double-pole, double-throw (DPDT) configuration has two poles and four contacts. Usually, two of the contacts are NO and the other two NC. Devices with more than four contacts are more expensive and hence find limited use. In selecting a device with the SPDT contact form, it is important to understand the effects of the status of the contacts on the connected circuits during the transition period.

TABLE 7-2 Contact forms

Design.	Sequence of Operation, Following the Application of the Force	Actual Configuration	Symbol
SPST-NO	Make (1)		
SPST-NC	Break (1)		
SPDT	Break (1) and then make (2)		
SPDT	Make (1) and then break (2)		

Since the function of a contact is to connect or disconnect a load from its supply voltage, its current and voltage specifications must equal or exceed those of the controlled load.

7.1.4 Indicating Lights

Indicating lights are usually of the incandescent type. As the name implies, they reveal the status (ON or OFF) of an electrical apparatus. In more sophisticated systems, the indicating lights may annunciate the cause of current interruption, such as line-to-ground fault, overload, and overcurrent condition. In solid-state control systems, the incandescent lights are, of course, replaced with light-emitting diodes (LED). The customer usually selects the color of the lights for a particular indicating function.

In selecting indicating lights, it is of paramount importance to know their voltage and resistance rating because these parameters establish the minimum current rating of the contact that controls the switching of the lights.

A motor's fuses, circuit breakers, magnetic starters, and relays are all housed within a metallic cabinet—located on what is commonly known as a motor control center—on the front panel where the indicating lights and the start-stop pushbuttons are located.

7.1.5 Start-Stop Pushbuttons

Start-stop pushbuttons can momentarily make or break an electric circuit, which in turn energizes or deenergizes an electrical load, normally through an electromagnetic starter. In other words, ordinary start-stop pushbuttons are electrical contacts that change status manually. Their symbol is shown in Fig. 7-5(a). Standard pushbuttons are spring operated. As soon as the depressing force is released, the pushbutton goes back to its original state. For this reason, the "start" pushbutton must be connected in parallel to a NO contact (referred to as the "holding" or "maintaining" or "seal-in" contact), which becomes closed as a result of its coil's energization.

Since a "start" or "stop" pushbutton is used to make or break a circuit, it must be able to withstand the currents that flow through its circuit. These currents (make, break, and continuous) are determined by the load of the circuit, which might be light or heavy. Therefore pushbuttons are classified as standard or heavy duty. Their characteristics are shown in Table 7-3.

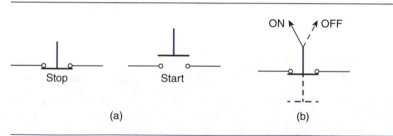

(a)   (b)

FIG. 7-5 Manually operated devices: **(a)** pushbuttons, **(b)** selector switch.

TABLE 7-3	Pushbutton characteristics		
Type of Pushbutton	Current Capacity		
	Continuous	Make	Break
Standard	15	30	20
Heavy duty	20	40	30

In an automated plant, start-stop pushbuttons are replaced by equivalent contacts (using solid-state devices) that are controlled by a signal originating in the plant's main or auxiliary computer.

A manually operated device that *maintains* the circuit—as opposed to the pushbuttons, whose action momentarily makes or breaks the circuit—is the selector switch. The symbol for a selector switch is shown in Fig. 7-5(b).

7.1.6 Industrial Timers

An industrial timer, following the energization of its coil, changes the status of its contacts cyclically. Usually, the time period during which the contacts could remain closed or open is externally adjustable, and the cycling period can vary from some milliseconds to several hundred hours. The number of the controlled contacts and the cycling period depend on the design of the timer. Figure 7-6(a) shows a timer's schematic. Figure 7-6(b) represents the status of its SPDT output contacts as a function of time.

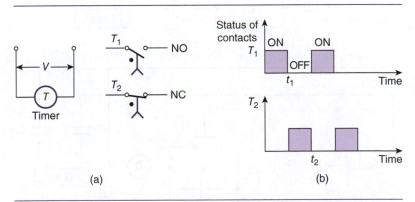

FIG. 7-6 Industrial timers: **(a)** symbol, **(b)** status of its output contacts as a function of time.

Timers can be pneumatic, solid-state, or motor-driven mechanisms. Solid-state timers are normally incorporated within so-called programmable controllers (PCs), which can be programmed to give the desired cycling period to any number of contacts.

EXAMPLE **7-1**

Referring to Fig. 7-7(a):

a. Explain its operation.
b. Modify the control circuit so that the load could be energized or deenergized remotely.

SOLUTION

a. When the start pushbutton is depressed, the coil of the electromagnetic starter C is energized, and as a result its NO contacts close. The closing of the auxiliary contact C_2 maintains the 120 V supply to the coil, while the closing of the main contacts C_1 connects the load to its source voltage. When the stop pushbutton is depressed, the coil of the electromagnetic starter is deenergized. Its contacts open, and thus the load is removed from the voltage source.

b. As shown in Fig. 7-7(b), the remote stop pushbutton is connected in series with the local stop pushbutton, while the remote start pushbutton is connected in parallel to the local one. In either case, the distance of the remote control station is limited by the impedance of the cable used. The higher the cable's impedance, the lower the voltage across the coil of the starter will be. This may adversely affect its operation.

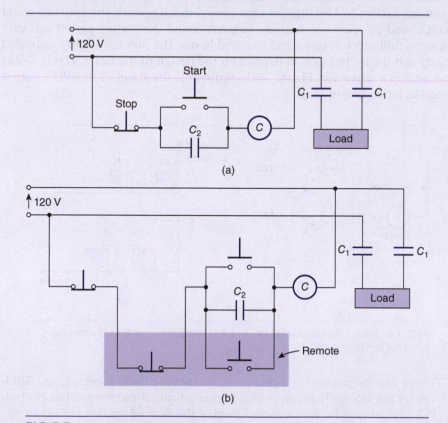

(a)

(b)

FIG. 7-7

Exercise
7-1

For ideal indoor comfort, such as can be found outdoors in the early mornings of some spring days, the following parameters should be at their optimum level:

1. Temperature (20°C)
2. Relative humidity (50%)
3. Rate of fresh air flow (0.007 m^3/s/person)
4. Purity of air ($CO_2 < 600$ ppm).

When any of these conditions are not met, a switch is activated. You should design a 120 V control schematic with indicating lights (red for undesirable condition, green for acceptable condition) that will identify the undesirable status of any of these parameters.

Exercise
7-2

A liquid storage tank is equipped with one inlet and two outlet pipes. The flow through the pipes is controlled by 120 V solenoid valves. The inlet valve is closed by the tank's high-level limit switch, while the outlet valves are operated by start-stop pushbuttons. The status of each valve is annunciated through standard green and red lights. You should design the control schematic for the operation of the valves and of the lights.

7.1.7 Temperature Sensors

A temperature sensor is a device that produces a signal at its output terminals that has some specific relationship to the sensor's ambient temperature. The output signal may be a variable resistance or simply an electrical contact (thermostat) that closes or opens, depending on the sensor's ambient temperature and set point.

Temperature sensors of the variable-resistance type include thermistors and resistance temperature detectors (RTDs). These sensors have a specific resistance change in a given temperature range. Because thermistors and RTDs are the most widely used temperature sensors, a brief description of each will be useful.

Thermistors

Thermistors are nonlinear temperature devices whose resistance decreases exponenttially with an increase in temperature. Thermistors are made from semiconductor materials that are doped with ceramics and metallic oxides. Their temperature characteristics are determined by their special atomic structures. In these materials, the atom's valance electrons form covalent bonds with the electrons of adjacent atoms. When the temperature increases, the thermal vibration of the electrons breaks some of these bonds, and thus some electrons are released.

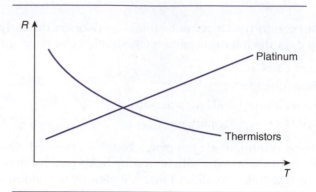

FIG. 7-8 Resistance-versus-temperature characteristics of platinum RTDs and thermistors.

As a result, this changes the resistivity of the material and, consequently, its resistance. In the temperature range 0–100°C, a thermistor's resistance as a function of temperature is described by

$$R = R_0 e^{-K_1 [(1/T_0)-(1/T)]} \tag{7.1}$$

where R and R_0 represent, respectively, the thermistor's resistances at temperature T and T_0. The temperatures are expressed in degrees Kelvin (°K = 273 + °C). K_1 is a constant of the material and has a value that is in the 350° to 400°K range. The thermistor's temperature response has a time constant of the order of a few milliseconds. The resistance-temperature characteristic of thermistors is shown in Fig. 7-8.

RTDS

The resistance of all metals changes as a result of changes in temperature, and thus many materials can be used as temperature sensors. Platinum has the most desirable characteristics (linear coefficient, high resistivity, acceptable thermal emf, resistance stability, broad temperature range, relatively low cost). For these reasons, it is used extensively in devices that monitor and control the temperature of electric machines.

Almost all electric machines rated above 200 kW are equipped with two RTDs per phase. RTDs are permanently connected in the inaccessible parts of a machine's winding and furnish a signal to a monitor in a control cabinet. The monitor is designed to give a temperature indication and to provide an alarm and a trip signal when the temperature of the machine's winding exceeds a predetermined level. The resistance-temperature characteristic of platinum is shown in Fig. 7-8.

Thermostats

Thermostats indicate and control ambient temperature. They constitute the main component of heating and cooling systems.

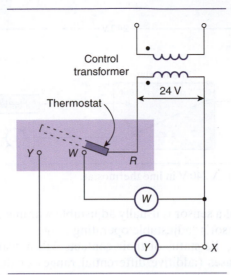

FIG. 7-9 Representation of a thermostat.

In its basic form, a thermostat is a temperature sensor. It alters the status of its contacts each time the ambient temperature is above or below the thermostat's set point. Figure 7-9 shows the standard representation of a thermostat. A temperature-sensitive mercury column* within the thermostat—represented by dotted lines—will connect, depending on the set point and ambient temperature, the 24 V source to terminals X–W, in the case of heating demand, or to terminals X–Y in the case of cooling demand. In other words, depending on the ambient temperature, a contact is made between terminals R and W or between terminals R and Y. The temperatures at which the thermostat will make these contacts are externally adjustable.

The terminals W or Y will energize their corresponding auxiliary relays, which in turn will control the operation of all those apparatuses that make up the heating or cooling system.

All modern thermostats make use of a microprocessor, which can be programmed to alter the status of a thermostat's contacts in response not only to changes in temperature but also to changes in time. In this way, optimum efficiency of the heating and cooling system is achieved, which results in energy savings.

Operating Range and Differential Range

The set point of a sensor is the level at which the status of its contacts changes as a result of a change in the monitored variable. For example, when a temperature sensor with a NO contact has a set point at 50°C, it means that its contact is open for all temperatures below 50°C and becomes closed for all temperatures above

---

* Not all thermostats operate on the effects of temperature on a column of mercury.

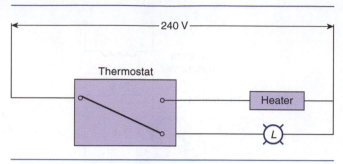

FIG. 7-10 A 240 V in line thermostat.

50°C. The set point of a sensor is usually adjustable within a given range, which is referred to as the sensor's "adjustable operating range."

In some sensors, the status of their contacts will not alter when the monitored variable increases (additive differential range) or decreases (subtractive differential range) about the set point within a given range of the controlled variable change. For example, a temperature sensor with a NC contact, a set point of 30°C, and an additive differential range of 10°C means that on rising temperature, its contact becomes closed for all temperatures less than 40°C. At temperatures, above 40°C, the contact opens and remains open until the temperature decreases to 30°C. At this temperature, the contact closes and remains closed until the temperature is again increased to 40°C. The differential range characteristic of sensors stabilizes the operation of the controlled equipment, for it prevents them from continuous switching ON and OFF around the set point of the controlled variable.

Fig. 7-10 shows what is called the "in line thermostat." It is connected in series with load and can operate at 120 V or 240 V. As a result the intermediary relays are not required.

7.2 The Concept of Protection

7.2.1 General

Aside from the start-stop control components, the control schematics include additional devices that provide equipment protection and help to ensure the safety of personnel. Equipment protection involves the design, selection, and setting of those devices that detect an abnormal operating condition and, as a result, disconnect the faulted network or apparatus from the supply voltage.

An abnormal operating condition produces any one or a combination of the following: overvoltages, overtemperatures, overloads, and overcurrents. An abnormal operating condition may arise from within the apparatus itself or may be caused by external factors, such as high transient voltages, unbalanced voltage supply, and high environmental temperatures.

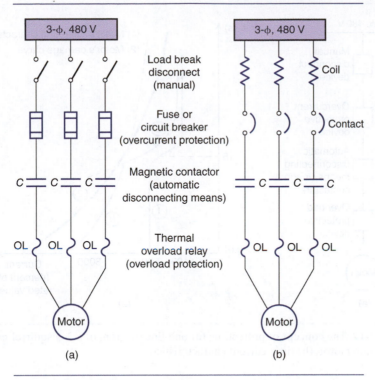

FIG. 7-11 Three-line representation of a 3-ϕ squirrel-cage induction motor: **(a)** fuses for overcurrent protection, **(b)** circuit breaker for overcurrent protection.

Figure 7-11(a) shows the three-wire representation of a standard squirrel-cage, three-phase induction motor. The load-break disconnect and the contacts of the magnetic starter provide the motor's direct (manual) and indirect disconnecting means, respectively. The fuses and the overload devices provide, respectively, the motor's overcurrent and overload protection. Figure 7-11(b) shows the three-wire representation of a squirrel-cage, three-phase induction motor that has a circuit breaker for overcurrent protection. These devices are discussed in some detail in the next section.

An overview of the concept of protection is also given in Fig. 7-12. Figure 7-12(a) shows the one-line diagram of a 3-ϕ, squirrel-cage induction motor. The figure identifies the functional characteristics of the various control and protection devices. Figure 7-12(b) presents the general time-current characteristics of the protective devices and the motor. The dotted line extensions of the characteristics of overload (curve d) and overcurrent (curve b) devices signify those parts of the curves that do not provide any effective protection.

A motor's operating and damage characteristics can be calculated from basic data or obtained from the motor's manufacturer. The characteristics of the protective devices can be found in standard manufacturer's catalogues.

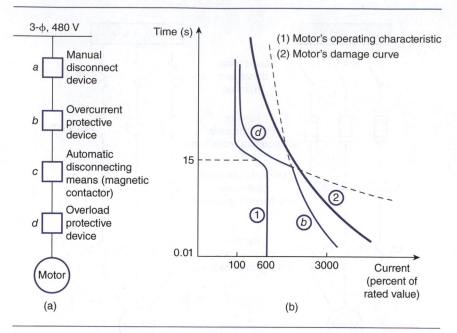

FIG. 7-12 The concept of protection: **(a)** one-line diagram of a 3-ϕ squirrel-cage induction motor, **(b)** time-current characteristics.

The first and most important step in motor protection studies is to obtain and sketch the motor's operating and damage characteristics on log-log paper. The protective devices must then be selected to ensure that their effective time-current characteristics lie between the motor's operating and damage characteristics. Protection devices must interrupt any undesirable current before the motor's damage condition is reached.

7.2.2 Protective Devices

The equipment's protective devices include fuses, relays, and circuit breakers. Fuses and most low-voltage circuit breakers have fixed time-current characteristics. These characteristics indicate how long it will take a particular device to interrupt a given current. For purposes of illustration, consider the 15 A devices shown in Fig. 7-13. A current of 60 A will be interrupted by a *time-delay* fuse in 18 s, while the same current will be interrupted by the circuit breaker in 4 s. A fast-acting fuse will interrupt the same 60 A current in 0.6 s.

Protective devices must be capable of removing—disconnecting—the affected circuit from the supply when overcurrents flow. This feature is known as "interrupting capability." These devices also should be able to withstand the forces that will be produced when they close the circuit while a downstream fault condition exists. This capability is referred to as "making capacity" or "momentary capability."

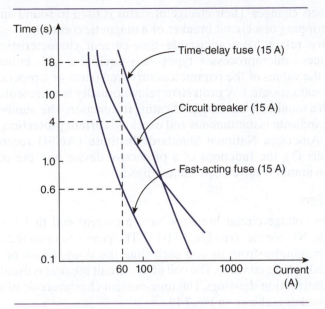

FIG. 7-13 Typical time-current characteristics of 15 A fuses and of a 15 A circuit breaker.

Disconnects

Disconnects, or switches (shown in Fig. 7-11(a)), are normally operated manually and cannot interrupt the current on their own. There are two main types of interrupters: load-break switches and no-load disconnects. Load-break switches are capable of interrupting nominal load current, whereas no-load disconnects can open up the circuit only if the main downstream load is disconnected. All LV switches are of the load-break type, while some of the MV switches are no-load interrupters. In the second category belong the primary switches of some transformers that can interrupt only the no-load current of the transformer.

Fuses

The function of a properly selected fuse is to interrupt the current when it reaches an undesirable level. Low-voltage and low-current (below 60 A rating) fuses will interrupt 135% of their nominal current rating in 60 s. This is referred to as their fusing factor. A 50 A fuse, for example, will open up the circuit to which it is connected in 60 s when, during this time interval, the current through it is 67.5 A.

Considering the function they perform, fuses are quite economical. They are commercially available with different time-current characteristics because various types of equipment have different time-current start and damage characteristic curves.

Protective Relays

A protective relay is a device that detects and identifies a piece of equipment's abnormal operating condition. Upon detection and identification, the status of its output contacts changes. Their change of status is used to sound an alarm, or to initiate the tripping of a circuit breaker or a magnetic contactor.

Protective relays have adjustable time-current characteristics. The more expensive ones—microprocessor types—are equipped with indicating meters that display the values of the parameters surveyed (such as temperature, current, unbalanced voltages, etc.). A protective relay normally is represented by a circle within which a number is inscribed indicating its function. The numbers 50 and 51, for example, indicate instantaneous and overload current protection, respectively. As per the American National Standards Institute (ANSI) recommendations (see Appendix G), the functions of a protective device and the corresponding identification numbers have been standardized.

Circuit Breakers

Ordinary low-voltage circuit breakers have a current coil that is connected in series with the NC contact (see Fig. 7-11(b)). The power to open the circuit breaker's contact originates from its coil each time the downstream or load current exceeds a predetermined value. The coil of the circuit breaker is usually not shown in one-line distribution drawings. The time-current characteristic of a 15 A, 480 V, 3-ϕ circuit breaker is shown in Fig. 7-13.

These low-voltage circuit breakers are commercially referred to as molded-case circuit breakers because they are constructed from some kind of molded material.

The more expensive circuit breakers are tripped through relays and consequently can perform additional protective functions, such as line-to-ground or overcurrent protection. As illustrated in Fig. 7-14, when the current drawn by the

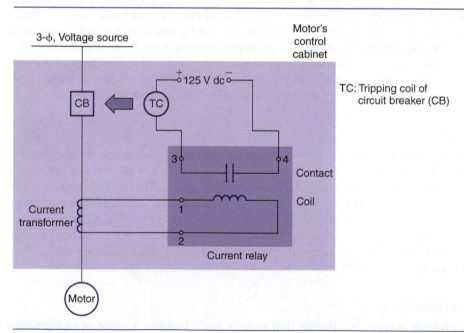

FIG. 7-14 Elementary representation of a relay and its tripping circuit.

motor increases to an undesirable level, the coil of the current relay is energized and its output contact closes. This action in turn energizes the tripping coil (TC) of the circuit breaker. As a result, the motor is disconnected from its voltage supply.

The interruption of the current is within a chamber that is filled with a special oil, SF_6 gas, air, or vacuum. The commercial name of the circuit breaker reveals the quenching medium. For example, an air-circuit breaker indicates that the current interruption is in an air chamber.

Compare and contrast a contactor, a LV molded-case type of circuit breaker, and a thermal overload relay.

EXAMPLE **7-2**

SOLUTION

A contactor provides a means of disconnecting but no protection, whereas a circuit breaker provides both a means of disconnecting and overcurrent protection. An overload relay only gives thermal overload protection. Through the contactor it can switch OFF the equipment it protects.

Because the coil of the contactor is a voltage coil, it is connected in parallel to its supply voltage. The coils of the molded case circuit breakers and of the thermal overload relays, being current coils, are connected in series with the protected load.

For purposes of comparison, Table 7-4 summarizes the main characteristics of each device.

TABLE 7-4 Comparison of protective devices					
		Function		Type of Coil	
			Protection		
Device	Disconnecting Means	Overload	Overcurrent	Voltage	Current
Contactor	Yes	No	No	Yes	
LV circuit breaker	Yes	Yes, but often not effective	Yes		Yes
Thermal overload relay	Yes (indirectly)	Yes	Yes, but not effective		Yes

EXAMPLE **7-3** Figures 7-15(a) and (b) show, respectively, a one-line diagram of a motor and the time-current characteristics of the motor and its protective devices.

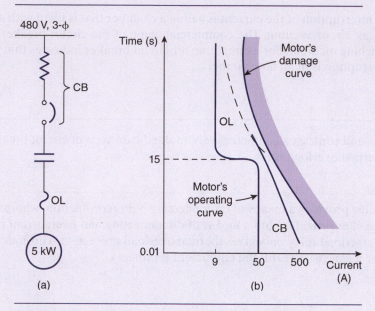

FIG. 7-15 Motor protection: **(a)** one-line diagram, **(b)** time-current characteristics of motor and protective devices.

a. Is the motor properly protected?

b. According to 1969 statistics, 19% of all fires in the United States were caused by motors. What were some of the reasons for the motors' damage and the subsequent fires?

SOLUTION

a. The motor is properly protected because the time-current characteristics of the protective devices indicate that for any overload or overcurrent condition the motor will be switched OFF before its damage condition is reached. The given characteristics of the overload (OL) and circuit breaker (CB) are not extended over their entire range because this would have made functional representation of the curves confusing. After all, the protective curves to the right of the motor's damage curve serve no purpose.

b. The damage to the motors and the subsequent fires could have been caused by any one, or a combination, of the following conditions:

1. The number of motor starts in a given time interval was higher than the maximum permissible.

2. The protective devices malfunctioned owing to aging, or poor maintenance, or some kind of manufacturing defect.

3. Poor motor maintenance resulted in an accumulation of dust that blocked the ventilation of the motor's windings. This is equivalent to shifting the damage curve of the motor to the left of its original position.
4. The overload relay's ambient temperature was much lower than that of the motor.
5. The motor's ventilation system failed.

Figure 7-16 shows a typical one-line diagram of a building's exterior lighting control. The photocell (PC) is normally an open contact that becomes closed at low illumination levels. The timer (T) requires 120 V for its operation and according to its setting can override the function of the photocell. The contactor (C), when energized, switches ON the exterior lights. Draw the corresponding control schematic.

Exercise

7-3

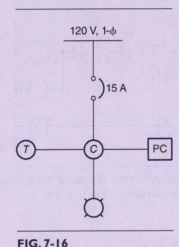

120 V, 1-ϕ

15 A

FIG. 7-16

7.3 Actual Control Schematics

This section explains how the basic control devices discussed earlier are used to construct the control schematics of three-phase motors and synchronous motors. The control schematics of single phase motors and dc machines are described in the Web section, Chapter 7W.

7.3.1 Control Schematics of Three-Phase Induction Motors

The following sections discuss the control schematics of LV squirrel-cage and wound rotor induction machines.

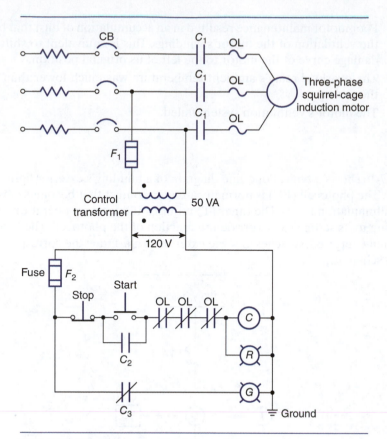

FIG. 7-17 Elementary control schematic for a three-phase, squirrel-cage induction motor, 480 V or less.

Low-Voltage, Squirrel-Cage Rotor

An elementary control schematic used for the starting and protection of 3-ϕ, LV induction machines is shown in Fig 7-17. The sequence of operation and the essential features of this control schematic are as follows:

Sequence of Operation

1. When the start pushbutton is depressed, its momentary contact closes.
2. The coil (C) of the magnetic contactor is then energized, and as a result, the main contacts (C_1) in the motor circuit plus the "seal-in" contact (C_2) are closed.
3. The motor starts to rotate.
4. The red (R) light is switched ON, indicating that the motor is operating.
5. When the momentary start pushbutton is released, its contact opens through the action of its own spring.

6. The "seal-in" contact maintains the circuit around the start pushbutton, and the motor continues to rotate.

7. When the stop pushbutton is depressed, its contact opens.

8. The circuit to the coil (C) of the magnetic contactor is broken, the coil is deenergized, and thus the main contacts in the motor circuit and the auxiliary seal-in contact are opened. The contact C_3 to the green light (G) is closed.

9. The motor stops.

Features

1. When the supply voltage is removed, the motor will stop and will not automatically restart after the control voltage is reapplied.

2. If the overload contacts are opened, the motor will stop. To restart the motor, the overload relay must be manually or automatically reset (depending on its type) and the start pushbutton must be depressed.

Remarks

1. The volt-amperes capacity of the control transformer should equal or exceed the volt-amperes (VA) requirements of the devices connected to its secondary terminals.

2. In some applications, control schematics include additional control devices for intermittent motor operation (jogging). In such cases, the motor starts and continues operation only when an instantaneous start pushbutton is kept depressed.

Low-Voltage, Wound-Rotor Motor

The control schematic of a three-phase, low-voltage, wound-rotor induction motor is shown in Fig. 7-18. The rotor resistors reduce the motor's starting current and increase its starting torque. Once the motor starts, the resistors are automatically shorted out in steps by the operation of the magnetic contactors (A, B, and C) and the time-delay relays. The time-delay relays are all of the "time-delay-on-energization" type. Briefly, the sequence of operation is as follows:

1. When the start pushbutton is depressed, the coils of the contactor M and of the time-delay relay X are energized. The motor then starts to accelerate. At starting, the per-phase rotor resistance is the sum of R_1, R_2, and R_3, and thus the starting current of the motor is accordingly reduced.

2. After t_1 s, the contact of the time-delay relay X closes, energizing relay Y and contactor A. This removes the resistors R_3 from the rotor circuit.

3. After t_2 s, the contact of the time-delay relay Y closes, energizing the relay Z and the contactor B. This removes the resistors R_2 from the rotor circuit. Similarly, after t_3 s, the contact of the time-delay relay Z closes. The contactor C then removes the remaining resistors R_1 from the external rotor circuit. At about this time, the motor operates at a steady-state condition.

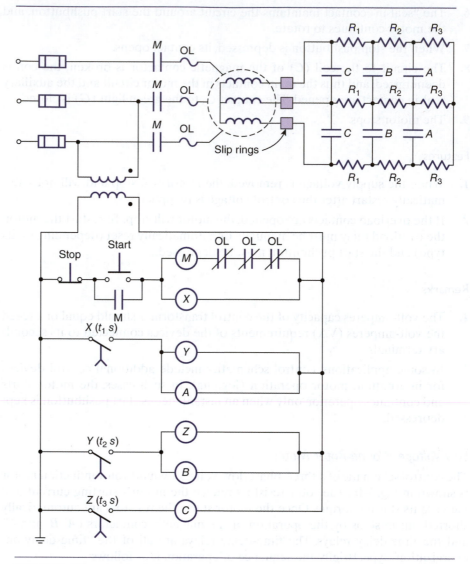

FIG. 7-18 Wound-rotor induction motor.

EXAMPLE **7-4**   A 550 kW, 480 V, 3-ϕ industrial plant is equipped with a standby 480 V, 500 kW, 3-ϕ diesel generator system. The plant's load includes two pumps of 50 kW each, which operate at different times.

a.   Draw the one-line distribution diagram.

b.   Design the control schematic, including the required interlocks, so that the diesel generator will deliver rated power and will supply only one of the two pumps that happened to be operating at the instant of power failure.

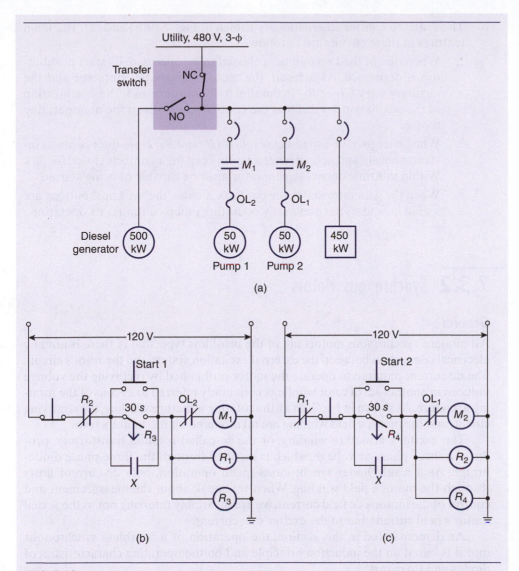

FIG. 7-19

SOLUTION

a. The one-line diagram of the industrial plant is shown in Fig. 7-19(a). Under normal operating conditions, the 500 kW diesel generator is OFF, and the entire electrical load is supplied from the utility's power lines. When the utility's power is interrupted, the contacts of the transfer switch change status and the diesel generator starts to provide rated power. The voltage-sensing devices and auxiliary relays that are required to complete the transfer of the power sources are not shown.

b. The required control schematics are shown in Figs 7-19(b) and (c). The main features of these circuits are as follows:

1. When one of the two pumps is chosen to be operated, its start pushbutton is depressed. As a result, the motor's magnetic contactor and the auxiliary relay (R_1 or R_2) in parallel to it are energized. The energization of the auxiliary relay disables the control schematic of the nonoperating motor.

2. When energized, the time-delay relays (R_3 and R_4) close their contacts instantaneously, and upon deenergization keep their contacts closed for 30 s. Within this time delay, the generator must be capable of being started.

3. When the generator starts, the contacts X close, the start pushbuttons are sealed in, and so the previously operating pump continues its operation.

7.3.2 Synchronous Motors

General

All modern synchronous motors are of the brushless type; that is, there is not any electrical connection between the external excitation system and the rotor's circuit. The dc current required to operate the motor is obtained by rectifying the voltage induced in another set of rotor windings, commonly referred to as coils of the rotating armature. As shown in Fig. 7-20(a), the rotating armature winding, the rectifying circuit, and the motor's field winding are all mounted on the motor's rotor.

The exciter's armature winding, or the so-called rotating transformer, produces a three-phase ac voltage, which is rectified through the three-phase diode-bridge. As a result, under synchronous-speed operation, only dc current flows through the motor's field winding. When we speak about the measurement and control of the motor's dc field current, we are invariably referring not to the actual motor's field current, but to the exciter's dc current.

As demonstrated in this section, the operation of a brushless synchronous motor is based on the induction principle and on the operating characteristics of diodes and thyristors.

Principle of Operation

When ac voltage is applied to the motor's stator windings (Fig. 7-20(b)), a synchronously rotating field is generated that induces a voltage in the field windings (coil F_1, F_2) of the rotor. The field produced by the rotor current attempts to align its magnetic axis to that of the stator field. As a result, the motor starts to accelerate. The frequency and magnitude of the alternating voltage induced in the rotor's windings depend on the effective turns ratio between the stator and rotor windings, and on the rotor's speed relative to the speed of the synchronously rotating stator field.

The frequency and the magnitude of the induced voltage are at a maximum when the rotor is stationary, and at zero when the rotor rotates with the same

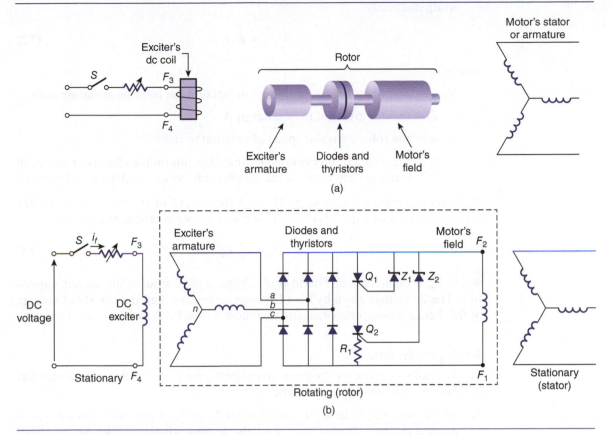

FIG. 7-20 Brushless synchronous motor: **(a)** physical representation, **(b)** circuit diagram. *Based on General Electric Canada, Inc.*

speed as the stator field. (For a detailed description of the relationships between a rotor's induced voltage and its frequency, see Chapter 3.)

When terminal F_1 is positive with respect to F_2, the motor's field winding is shorted through the rectifier's diodes. When terminal F_2 is positive with respect to F_1 and is of proper magnitude, the motor's field winding current is circulated through thyristors Q_1, Q_2, and resistor R_1. The gating pulse for the triggering or conduction of the thyristors is provided through the reverse-biased Zener diodes Z_1 and Z_2.

Up to a level of about 95% of the synchronous speed the voltage induced in the field windings *must be larger* than the threshold voltage (knee point) of the Zener diodes. Otherwise the Zener diodes will not contact.

When the rotor reaches about 95% of the synchronous speed, the control devices of the excitation circuit switch on the dc voltage supply. As a result, a voltage is generated in the rotating armature of the exciter. The voltage generated is proportional to the shaft's speed of rotation and to the effective field of the dc exciting coil.

Mathematically,

$$V_{L\text{-}n} = K_1 I_f \omega \qquad (7.2)$$

where

$V_{L\text{-}n}$ = the phase-to-neutral voltage induced in the rotating armature coils

I_f = the dc current of the exciter in A

ω = the rotor's angular speed of rotation in rad/s

K_1 = a constant of proportionality that depends on the effective number of winding turns between the rotating transformer and the excitation coil.

The average value of the voltage (V_{av}) at the output of the three-phase bridge rectifier is given by Eq. (3.71), which for convenience is rewritten here as:

$$V_{av} = \frac{3}{\pi} V_m \qquad (7.3)$$

where V_m is the maximum line-to-line voltage at the output of the exciter's armature. The dc voltage given by Eq. (7.3) *must be smaller* than the threshold voltage of the Zener diodes; otherwise, the thyristors would be driven into conduction.

Starting and Protection

The control schematic for the starting and protection of a synchronous motor has to fulfill the following requirements:

1. The motor must be brought up to about 95% of its synchronous speed as an induction machine. During this period, the dc rotor winding is disconnected from its nominal supply and is either shorted or connected across a discharge resistor.

2. At the end of this period, the dc rotor source is connected to the motor's field winding.

3. The protective devices must be inoperative during this stabilizing or synchronizing period.

4. The circuit must incorporate protective devices that provide adequate motor protection and must include provisions for field-current adjustments.

Control Schematics of Brushless Synchronous Motors

A basic control schematic for the starting, operation, and control of a brushless synchronous motor is shown in Fig. 7-21. The sequence of operation is as follows:

1. By depressing the start pushbutton, the coil of the auxiliary relay X is energized. The "seal-in" contact maintains the voltage across its coil, and the other contact energizes the motor's starter M. Thus, the voltage source is connected to the motor's terminals. The motor starts to accelerate as a 3-ϕ induction motor.

2. Following the energization of the motor's contactor, the main field application relay (56F) is energized. This relay has an adjustable time-delay contact

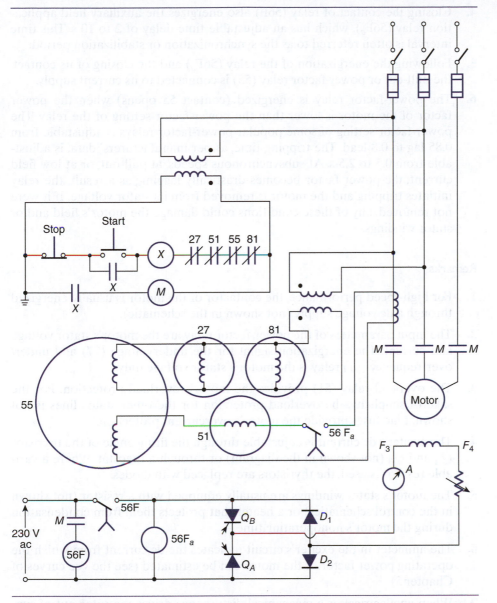

FIG. 7-21 Elementary control schematic for a 3-ϕ 4160 V, brushless synchronous motor.

(5 to 20 s) on energization. The selected time is equal to the time that it takes for the motor to reach about 95% of its nominal speed.

3.   At the end of the acceleration period—often determined in the field by trial and error—the contact of the main field application relay (56F) closes, the exciter's armature is energized, and consequently a dc voltage is applied to the motor's field winding. The rotor tries to pull into synchronism with the stator field (a natural tendency of fields to align their magnetic axes).

4. Closing the contact of relay (56F) also energizes the auxiliary field application relay (56F$_a$), which has an adjustable time delay of 2 to 10 s. This time interval is often referred to as the synchronization or stabilization period.

5. Following the energization of the relay (56F$_a$) and the closing of its contact, the pull-out or power-factor relay (55) is connected to its current supply.

6. The power-factor relay is energized (contact 55 opens) when the power factor of the motor is lower than the power-factor setting of the relay. The power-factor setting of some popular power-factor relays is adjustable from 0.85 lag to 0.8 lead. The tripping time, as per manufacturers' data, is adjustable from 0.5 to 2.5 s. At subsynchronous speeds, at pull-out, or at low field current, the power factor becomes drastically lagging; as a result, the relay initiates tripping and the motor is removed from its stator voltage. If it were not removed, any of these conditions could damage the motor's field and/or stator windings.

Remarks

1. For high-speed performance, the contactor of the motor is usually energized through a dc voltage supply (not shown in the schematic).

2. The input parameters of the power-factor relay are the motor's stator voltage and current. The energization signal for the undervoltage (27) and under/overfrequency (81) relay is the motor's stator voltage only.

3. The overload relay (51) provides the motor's overload protection. For the sake of simplicity, the overload protection for the other stator lines is not shown. Line fuses provide the motor's overcurrent protection.

4. The exciter's dc current is adjustable through the firing angle of the thyristor Q_A and Q_B (not shown in the diagram) or through a rheostat. Where a variable resistor is used, the thyristors are replaced with diodes.

5. The motor's stator windings are usually equipped with a resistor (not shown in the control schematics) or a heater that protects them from condensation during the motor's nonoperating time.

6. The ammeter in the exciter's circuit indicates the dc current from which the operating power factor of the motor can be estimated (see the vee curves of Chapter 5).

7. When an electrical or a mechanical disturbance drives the rotor out of synchronism, the armature current and its power factor change according to the level of disturbance. Protective equipment (armature overcurrent relay and power-factor relay) will sense the change in these parameters and may, according to the settings, remove the electrical supply from the stator, bringing the motor to a complete stop. If the protective equipment does not remove the electrical supply from the stator, the motor's field winding will be shorted, as in the case of the accelerating period. This implies, of course, that the external disturbance forced the rotor to run at a speed that was outside the range of 95% to 100% of synchronous speed.

A 514 r/min, 0.8 Pf leading, 4160 V, 60 Hz, 500 kW brushless synchronous motor has an effective stator-to-rotor per-phase turns ratio equal to unity. Determine:

EXAMPLE **7-5**

a. The effective value of the voltage induced in the motor's field just prior to the switching ON of the dc supply. Assume this to take place at 95% of synchronous speed.

b. The dc voltage across the motor's field winding at nominal operating conditions, given that the voltage constant between the dc exciter and its armature is 2 V/rad/s/phase.

c. How the setting of the power-factor relay controls the minimum breakdown torque developed by the motor.

SOLUTION

a. The voltage induced in the rotor's winding (V_r) is proportional to the stator voltage (V_s), the slip, and the effective turns ratio between the stator and the rotor. Mathematically,

$$V_r = sV_s \frac{N_r}{N_s} \tag{I}$$

where N_r/N_s is the ratio of rotor to stator number of turns per phase. The slip s is given by

$$s = \frac{\omega_s - \omega_a}{\omega_s} = \frac{1 - 0.95}{1} = 0.05 \text{ pu}$$

Substituting into Eq. (I), we obtain

$$V_r = 0.05 \left(\frac{4160}{\sqrt{3}} \right)(1) = 120.09 \text{ V}$$

b. The line-to-line voltage across the motor's rotating armature is

$$V_{\text{L-L}} = K\omega = 2(514)\frac{2\pi}{60}\sqrt{3} = 186.46 \text{ V, L-L}$$

The dc voltage across the motor's field windings is determined by Eq. (7.3):

$$V_{\text{av}} = \frac{3}{\pi}(186.46)\sqrt{2} = \underline{251.81 \text{ V}}$$

c. The breakdown torque of the motor, as can be seen from Eq. (5.29), depends on the motor's field current. In turn, the field current controls the power factor of the motor. Therefore, the setting of the power-factor relay controls the minimum permissible value of the field current and the minimum value of the breakdown torque.

Exercise

7-4

a. Referring to Fig. 7-20(b), what are the functions of:
    1. The diodes
    2. The stationary dc coil
    3. The rotating exciter's armature.

b. Referring to Fig. 7-21, what are the functions of:
    1. The power-factor relay (55)
    2. The undervoltage relay (27)
    3. The frequency relay (81).

7.4 Conclusion

The plant's control schematics constitute the interface between the engineer and the plant's electrical equipment. Using conventional symbols, they reveal the mode of operation and protection of each electrical apparatus. Understanding and interpreting control schematics is simple; their overall design, however, requires in-depth knowledge of the equipment's operation and protection, which is a specialty in itself.

Conventionally, control schematics show start and stop pushbuttons, magnetic contactors, relays, fuses, circuit breakers, voltage and current transformers, and many other auxiliary control devices whose complexity and worth depend on the relative importance and cost of the controlled equipment.

By far, the most widely used control devices are magnetic contactors.

Thermal overload relays are included in almost all motors rated below 200 kW. Fuses or circuit breakers generally provide short-circuit protection. The more expensive circuit breakers are tripped through protective relays, and thus they can provide additional protection from ground faults, unbalanced voltages, and so on.

Each type of motor, depending on its application, has a particular control schematic. In most cases, the control schematics are standard and can easily be found in manufacturers' catalogues. However, since each process has its own peculiarities, the design engineer must use the most economical and effective control alternative.

The controls for a squirrel-cage 3-ϕ induction motor are quite simple. Those of the wound motor are rather more complex, because their design requires the selection of the motor's external resistors, which depend on the moment of inertia of the load and motor.

The control schematics of three-phase synchronous motors are unique in that they require knowledge of the motor's operation and special types of protective devices. The control schematics of three-phase induction motors are quite simple and have been in use for a long time, while the control schematics of synchronous machines are relatively new. These machines are started and controlled through solid state devices which, since the 1980s, have completely replaced their electro-mechanical counterparts.

Control schematics have evolved considerably. Electromagnetic devices (*electrical*) were replaced by solid-state devices (*electronic*), which in turn have been supplanted by *microprocessor* controls. The basic difference between electronic and microprocessor controls is that the microprocessor controls can easily be programmed to provide not only switching functions but also continuous control of all parameters of a given process.

The symbols of the electromechanical devices are presented in Table 7-5.

TABLE 7-5   Summary of symbols for control schematics

	Symbol	Description	Notes
a.		Voltage coil of a relay, electromagnetic starter, timer, or the like	
b.	NO NC	Instantaneous contacts: normally open (NO) and normally closed (NC)	
c.	Start Stop	Momentary start and stop pushbuttons	
d.	Off On	Selector switch	
e.		Contact of a temperature sensor	
f.		Contact of a pressure sensor	
g.		Limit switch	

(Continued)

	Symbol	Description	Notes
h.	NO NC NO NC	Time contacts. Contact action is retarded when the coil is: Energized Deenergized	When a dot is placed adjacent to a time-delay contact, it implies a cyclic variation of its status while the coil is maintained in an energized state.
i.		Float switch	
j.		Contact of a relative-humidity sensor	
k.		Disconnect	
l.		Fuse	
m.	Coil Contact	Molded-case circuit breaker	
n.		Circuit breaker, operated through relays	

TABLE 7-5   *(Continued)*

	Symbol	Description	Notes
o.		Solenoid	
p.		Pump	
q.		Heat exchanger	

7.5 Review Questions

1. What is the difference between a control schematic of a motor and its wiring diagram?

2. Draw a single-pole single-throw and a single-pole, double-throw switch.

3. Explain what a thermal overload relay is.

4. Explain the differential range of a sensor.

5. Differentiate between a thermal overload relay and a fuse or low-voltage circuit breaker.

6. Draw a typical thermostat and explain the significance of the terminals X, R, Y, and W.

7. What is a "seal-in" contact?

8. What is the difference between an overload and an overcurrent condition?

9. What do the protection numbers 50 and 51 signify?

10. Draw the starting and damage characteristics of a motor and of the motor's protective devices on the same time-current base.

11. Describe the starting of a brushless synchronous motor.

7.6 Problems

7-1 Draw a 24 V control schematic that includes two pushbuttons and a NO pressure switch that upon activation, illuminates a light and provides an audible alarm. One of the pushbuttons is used to silence the audible alarm, and the other determines if the indicating light is burned out.

7-2 Draw a 120 V control schematic that incorporates a manual reset and a NO temperature contact that upon activation, energizes a fail-safe solenoid valve and an annunciator light.

7-3 Figures P7-3(a) and (b) show, respectively, the control schematic and the one-line diagram of a 3-ϕ, 480 V motor.

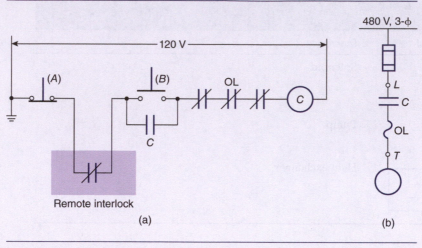

FIG. P7-3

a. Modify the control schematic so that the motor can be started and stopped from a remote control station.

b. Explain why it is improper to wire the contacts of the overload relay as shown. (*Hint*: Consider what would happen to the motor if the overload contacts opened while the coil of the magnetic contactor was accidentally grounded through its enclosure.)

c. If the pickup value of the starter's coil is 80% of its nominal voltage, what is the maximum permissible length of the wire (size #14, $R = 0.0085 \ \Omega/m$) that connects the remote interlock to the control circuit? The coil's starting current is $2 \ \underline{/-70°}$ A.

7-4 Referring to the control schematic of Fig. P7-3:

a. If the supply voltage is removed and then reapplied, will the motor start automatically? Explain.

b. When the motor's current increases above its rated value, the contacts of

the thermal overload relay will be opened and thus the motor will stop. Will depressing the start pushbutton start the motor? Explain.

c. The number one danger to a motor is its frequent starts. Modify the control schematic so that the plant's operators or electricians will not be able to restart the motor within 10 minutes after it has been stopped.

7-5 Referring to the basic circuit diagram of a brushless synchronous motor (Fig. 7-20(b)):

a. Identify the path of the current through which the motor's field current flows during the accelerating period, and explain the reasons for the thyristor's conduction.

b. Estimate the maximum permissible line-to-line output voltage of the rotating armature if the Zener's nominal voltage is 16 V.

c. Explain why the resistor R_1 increases the motor's starting torque.

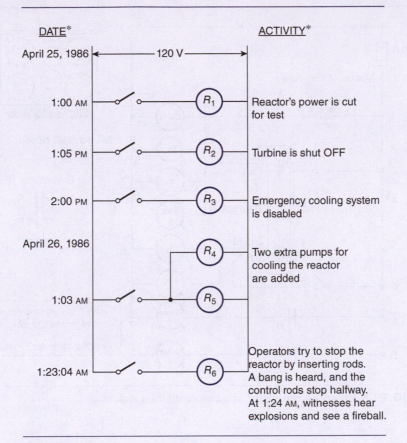

DATE*

April 25, 1986 ← — 120 V — →

ACTIVITY*

1:00 AM — R_1 — Reactor's power is cut for test

1:05 PM — R_2 — Turbine is shut OFF

2:00 PM — R_3 — Emergency cooling system is disabled

April 26, 1986

R_4 — Two extra pumps for cooling the reactor are added

1:03 AM — R_5

1:23:04 AM — R_6 — Operators try to stop the reactor by inserting rods. A bang is heard, and the control rods stop halfway. At 1:24 AM, witnesses hear explosions and see a fireball.

FIG. P7-7
Based on The New York Times

7-6 Draw a control schematic for a brushless synchronous motor that includes a dc contactor for motor control, a winding heater, a ventilating fan, three overload relays, an overtemperature-bearing protection, and a contact that is derived from a rotor-vibration detector.

7-7 Figure P7-7 shows, in a simplified version, the chronological events and the operators' actions that led to the Chernobyl nuclear accident. The switches are manually operated, and each relay's coil, when energized, implements through a circuit (not shown in the diagram) the function as described adjacent to the coil. The accident, of course, took place because the plant was producing power while the operators performed a series of steps that were permissible when and only when the reactor's output power was negligible.

a. Modify the control schematic to incorporate an interlock that would not permit the operators to do what they did while the reactor was apparently still producing power. The interlock must be a contact whose status is controlled by the temperature of the reactor.

b. Draw the logic diagram equivalent to (a).

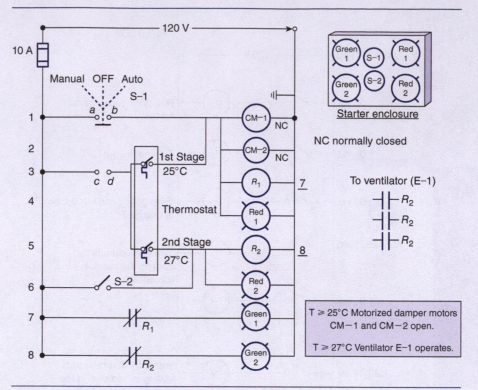

FIG. P7-8 Control schematic of a room's ventilation system.

7-8 The devices shown in Fig. P7-8 annunciate and control the operation of a room's ventilation system. The manual switches (S–1 and S–2) and the 2-stage thermostat control the operation of the dampers (CM–1 and CM–2) and the exhaust ventilator (E–1). Briefly explain the operation of the system.

(*Hint*: Switching S–1 on "Manual" setting makes contact across terminals a–b and on "Auto" setting makes contact across terminals c–d. The starter's enclosure houses all the system's relays and indicating lights).

8

Electrical Safety and Reduction in Energy Consumption

8.1 Electrical Safety

8.2 Reduction in Energy Consumption

What You Will Learn in This Chapter

A **Electrical Safety**
A description of the events and circumstances associated with electrical safety.

1 Grounding system

2 Open neutral

3 Arc flash

4 Energy stored in inductive circuits

5 Energy stored in capacitive circuits

6 Stray voltages

7 Galvanic corrosion

8 Lightning strokes

9 Mishandling medical equipment

Additional Students' Aid on the Web

•   Detailed examples demonstrating the quantitative and qualitative aspects of the above events and pertinent end of chapter problems

B **Reduction in Energy Usage**
A description of about 15 areas associated with reduction in energy usage.

1 Unbalanced voltages

2 Unbalanced line currents

3 Lighting

4 Synchronous motors

5 Reduction in peak power demand

6 Power factor

7 High-efficiency motors

8 Variable-speed drives

9 Harmonics

10 Optimization of the voltage to dc motors

11 Discharged condensers' heat

12 Heat losses

13 Heat pumps

14 New technologies

Additional Students' Aid on the Web

•   Detailed examples demonstrating the quantitative aspects of the above areas and pertinent end of chapter problems

8.1 Electrical Safety

8.1.1 Introduction

This section describes a home's partial distribution system. This discussion, in conjunction with ohms and KVL, will lead to understanding the concept of system ground, equipment ground, electrical safety, and measures to prevent or mitigate electrical injuries. In North America, about 10,000 accidents take place in hospitals every year and about the same number of personnel are injured—some fatally— outside hospitals. More specifically, the following areas of personnel accidents from electrical mishaps are summarized in Fig. 8-1. The corresponding Web Section (8W-1) includes detailed discussions, examples, exercises, and problems.

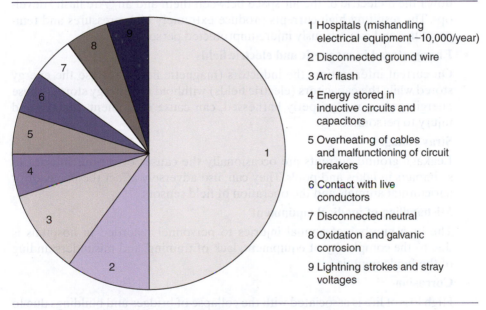

1 Hospitals (mishandling electrical equipment −10,000/year)

2 Disconnected ground wire

3 Arc flash

4 Energy stored in inductive circuits and capacitors

5 Overheating of cables and malfunctioning of circuit breakers

6 Contact with live conductors

7 Disconnected neutral

8 Oxidation and galvanic corrosion

9 Lightning strokes and stray voltages

FIG. 8-1 Approximate electrical injuries per year in North America

- **Protective devices**

  The function of protective devices (fuses or circuit breakers) is of course to protect equipment and personnel from the destructive effects of high currents. In such circumstances, these devices simply interrupt the flow of current.

- **Utilities and a home's grounding system**

  The grounding systems (utility's ground, electrical entrance ground, and the grounding of the electrical equipment enclosures) in conjunction with the protective devices help isolate the faulted (shorted) electrical devices and contribute also to the security of personnel.

- **Touching a live conductor**

  Faulted electrical equipment constitute a danger to personnel when they come in contact with such devices. The contributing factor to all these dangers is the aging of the equipment's insulation, missing ground wire, and/or improper installation. Copper thieves are seriously injured when they try, for commercial reasons, to remove live copper wires.

- **Open neutral**

  When the neutral of a 240/120 V, single-phase, 3-wire power distribution system is accidentally open (at the utility's transformer or at a home's electrical entrance), undervoltages and overvoltages develop that can damage property and injure personnel.

- **Arc flash**

  When a solid short circuit takes place between electrical conductors, it breaks down the dielectric of the air space between them and an "Arc flash" develops. The resulting high currents produce extremely high pressures and temperatures and can seriously injure unprotected personnel.

- **Energy stored in magnetic and electric fields**

  On current interruption, the inductors (magnetic fields) release the energy stored while the capacitors (electric fields) withhold the energy stored. These energies, when not properly harnessed, can cause equipment damage and injury to personnel.

- **Stray voltages**

  Leakage ground currents are occasionally the causes of serious injuries to swimmers in lakes and pools. They can also adversely affect the behavior of agricultural animals and the operation of field sensors.

- **Mishandling of medical equipment**

  The high number of annual injuries to personnel recorded in hospitals is due to the complexity of equipment, lack of training, and misunderstanding of the fundamentals.

- **Corrosion**

  High loss of life is associated with the collapse of bridges and buildings due to the corrosion of steel columns and/or reinforced steel bars.

- **Lightning strikes**

  Anyone who is outdoors in the midst of thunder and lightning has no protection. Taking shelter under a tree can perhaps mitigate the catastrophic effects if simple precautions are taken.

8.1.2 Basic Protective Devices

A protective device (a fuse or breaker), when properly selected, protects the circuits and equipment from high currents. In other words, when the current through a protective device reaches an undesirable level, commonly referred to as overload,* it interrupts the current flow. If the device is a fuse, it blows up and must be replaced, whereas if it is a breaker, it too opens up the circuit but it does not have to be replaced. It can be reused again and again. The symbols for a fuse and a breaker are shown in Fig. 8-2(a).

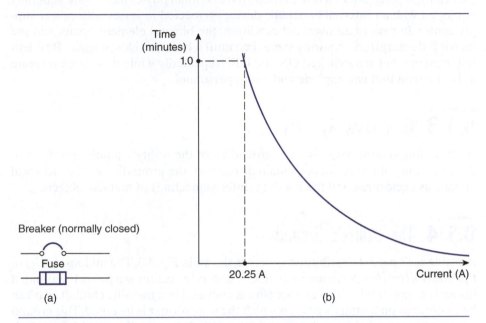

FIG. 8-2 Basic protective devices: **(a)** Symbols. **(b)** Time-current characteristics of a 15 A fuse or breaker.

How long it takes for a protective device to interrupt a given current is indicated by the device's time-current characteristic, as furnished by the manufacturer (Fig. 8-2(b)). For example, when the current increases 1.35 times higher than nominal, the device will open the current in 60 seconds. That is, a 15A fuse will blow up when the through current is [15(1.35)] = 20.25 A for a duration of 60 seconds. In general, the higher the current, the shorter the time that the device will take to interrupt the current.

These standard protective devices will be gradually replaced by so-called smart breakers. These breakers are electronic with many additional characteristics such as the ability to identify the severity and the type of short (line-to-neutral, line-to-line, etc.) and the time of its inception and duration.

---

* An overcurrent or short-circuit current is one that is more than six times the machine's nominal current. Overload is the current that is from one to six times the machine's rated current. Etymologically speaking, there is no clear-cut division between an overload and a short-circuit current. Generally, an overload current flows along the same path as the machine's full-load current, while an overcurrent flows partially along a different path.

Prolonged overloads deteriorate the winding's insulation, while uncontrolled overcurrents can damage all current-carrying and noncurrent-carrying parts of the machine.

Because of their short duration, the present electromagnetic breakers cannot interrupt the intermittent faults (repetitive restrikes) that take place between the interwindings or between the enclosures and the device's windings. Intermittent faults give rise to high transient voltages that can be higher than nominal voltages, especially in nongrounded distribution systems. Small power motors are generally equipped with an internal bimetallic device connected in series with power supply cables. In case of an overload condition, the bimetal element opens, and the motor is deenergized, requiring some time until it becomes closed again. Residential breakers that are switched ON and OFF repeatedly while downstream create a short circuit that can explode and injure personnel.

8.1.3 Grounding Systems

A grounding system comprises the grounding of the utility's supply transformer, the grounding of the residence's main disconnect, the grounding of the electrical apparatus enclosures, and the bonding (interconnecting) of metallic objects.

8.1.4 The Utility's Ground

A residence's partial distribution system is shown in Fig. 8-3. The midpoint (G_1) of the utility's transformer, commonly referred to as the center tap point, is grounded through a copper wire. This copper wire is enclosed in a metallic conduit that can be easily seen on a utility's pole on which the transformer is installed. This ground wire is connected below surface to a 3.00 m (10′-00) grounding rod whose standard diameter is 19.05 mm ($\frac{3''}{4}$).

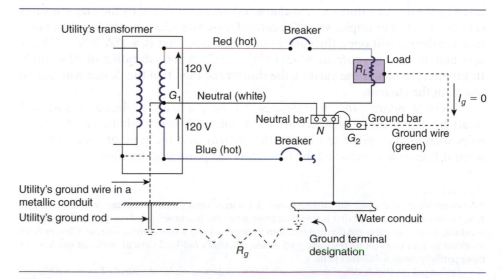

FIG. 8-3

The grounding of the distribution system, besides providing security to personnel, also mitigates and absorbs the high current impulses that are generated by the utility's switching network and/or from lightning strikes.

8.1.5 Residence's Ground

The neutral wire (white color) in the house's utility meter (N) or main disconnect switch is also grounded to an outdoor grounding rod, or most likely, to the metallic conduit through which the residence receives its water supply.

The purpose of a residence's ground is to complement that of the utility, to provide a path for the discharge of static electricity, to provide a reference point (zero volts) for measuring voltage, and in case of an open neutral, to mitigate the effects of the resulting overvoltages.

8.1.6 Equipment's Enclosure Ground

In the interior of the panel is a ground bar (G_2) to which the ground wires from the enclosures of the various appliances are connected. The ground of the building's main disconnect is connected to this grounding bar. According to the Electrical Code, the enclosures of all current-carrying equipment must be grounded. The purpose of an enclosure's grounding is to provide, in case of the apparatus's internal short circuit, a low-resistance path and thus a high current through the protective device, which in turn will interrupt the current flow.

The neutral and the ground wires must be connected to the same point *only* in the main disconnect of the premises. This point is shown in Fig. 8-3 by interconnecting the neutral (N) and the ground bar.

The equipment's enclosure ground is identified by a *green* color and is often referred to as Equipment Ground. It is one of the most critical parts of the distribution system, and as such, it should be verified annually. About 15% of electrical accidents are due to the nonexistent or improperly terminating enclosure's ground. One can easily verify it by using an ohmmeter. Just measure the resistance between the equipment's enclosure and the nearest metallic conduit. If this resistance is very high, then the ground is missing. This wire helps prevent many accidents. That is why it is often called the *Guardian Angel*. Under normal operating conditions, there is no current through the ground wires.

There is a danger of fatal injury when troubleshooting live adjacent equipment (such as dryers and washing machines) if one of the two has a disconnected ground. If there is an internal fault, if someone touches both appliances at the same time, a high current could flow through the heart of the person.

8.1.7 Bonding

Bonding, in general, refers to the interconnections of all noncurrent metallic objects (gas pipes, air conditioning ducts, exposed steel columns, etc.) with a copper wire that in turn is connected to the premise's main ground. Its main purpose is to equalize the potential between the bonded objects and provide a path for the discharge of static electricity to ground.

8.1.8 Ground Resistance

R_g represents the ground resistance. Its value depends on the number of grounding rods installed, on the type of underground structure and its chemical composition, and the season of the year. In the rainy season, of course, this resistance is quite small.

For a residence, satisfactory value is about 25 ohms. In general, the smaller the value of R_g, the better it is. (One method of measuring the ground resistance is described in Problem P8W-2 in the Web Section.)

Effects of Grounding Rods

The more grounding rods are installed in a building, the more parallel current paths are created between the building's grounding system and the grounding rod of the utility's transformer. As a result, the corresponding grounding resistance R_g is reduced. In case of an open neutral, the ground voltage is increased.

In summary, then, a grounding system consists of the following:

- The utility's ground.
- The ground of a home's or building's electrical entrance.
- The grounding of the metallic enclosures of all current-carrying equipment.
- The bonding of all noncurrent-carrying metallic objects.

8.1.9 Short Circuits

Most of the time at the inception of a fault, the short circuit is intermittent. That is, the short circuit between the load's winding and the enclosure or between the winding turns is switched ON and OFF at a very fast rate. The resulting circuit current, although it can be high, does not trip the protective device because of its short duration. In this case, the enclosure's voltage could be substantial due to inductance of the ground wire and to its possible terminating contact resistance. The analysis of intermittent faults is covered in specialized publications. In this book, only resistive loads at steady-state conditions are analyzed.

8.1.10 Floating Neutral

Refer to Fig. 8-3. Floating neutral refers to the variable voltage drop across the resistance of the neutral wire. This resistance depends on the size, length, and temperature of the wire, and the current through it depends on the loads. The current could be high depending on how well the phases are balanced and on short circuits within the connected loads. The resulting voltage across the neutral, depending on its magnitude, may overheat the wire and adversely affect the operation of voltage-sensitive equipment.

8.1.11 Touching a Live Conductor

The injuries that result from touching a live conductor depend on the magnitude of the supply voltage, the resistance of the path through which the current passes, and thus on the magnitude of the resulting current through the human body. Furthermore, the frequency of the voltage source and the duration of the current flow affect the degree of injuries. The effects of the voltage's frequency and the duration of the flow of current are discussed in specialized publications.

It is not known why it is that an individual who receives an electrical shock externally that does not look serious can die 24 hours later. For this reason, patients with electrical injuries are kept in the hospital under strict observation for 24 hours after admission. As a base of discussion, Table 8-1 lists the effects of current through the human body.

Table 8-1 Effects of Current on Humans	
Magnitude of Current in m A through the Human Heart	Effects
4	Threshold of sensation
20	Disables the nervous system (The individual cannot remove his or her hand from the touched live conductor.)
200	Could be fatal

Intentional Contact with Live Conductors

The utilities use a copper wire to ground the secondary of their transformers and also their lightning arrestors. The lightning arrestors are installed on transmission line power poles in about 5 km intervals. They protect the distribution system from high-voltage transients and lightning strikes. The lightning arrestors act as switches (open to nominal operating voltage and shorted to ground) through a copper wire for incoming high voltages.

Under normal operating conditions, the ground wires carry no current and are at ground voltage (zero voltage). In such cases, the ground wire can be removed without any electrical danger, and it can be sold to scrap dealers. Encouraged by this fact, copper thieves try to remove live conductors and in the process, they often incur injuries. In North America, about 50 such injuries take place each year, some of them ending in fatalities.

8.1.12 Open Neutral

When the neutral wire of a home's 240/120 V supply voltage is disconnected, it results in undervoltages and overvoltages in the home's appliances and lights and raises the potential of the equipment's enclosure. Although such events very seldom occur, it is important to be aware of its effects on the security of personnel and the appliances theselves. They can cause fires and injure people. In Australia, for example, a plumber working on a house water distribution system was electrocuted when the home's neutral wire was disconnected. The effects of the open neutral also depend on the ground resistance R_g.

8.1.13 Arc Flash

When a short circuit is developed between two or three live electrical conductors such as between the bus bars of a panel or switchboard, the resulting current, depending on the voltage and upstream circuit impedance, is very high. It breaks the dielectric* of the air space between the conductors, producing an arc flash. The temperature developed by an arc flash can be as high as 10,000°C, the pressure as much as 15,000 psi (103 MPa), and the accompanying explosive power as much as 10 MW.

An arc flash may result from accumulated dust between bus bars, aging of conductor insulation, loose fastening screws, underrated voltmeter voltage insulation, the accidental fall of an electrician's tool between the bare conductors, and the interruption of the current in a highly inductive load. In North America, arc flashes reportedly cause between 1500 and 3000 injuries per year, some of them fatal. The injuries from arc flashes can be prevented or at least mitigated when the safety recommendations of the Electrical Code and other authorities are followed. The best way, of course, is to work on electric equipment only when it is disconnected from the voltage supply. Nevertheless, there are instances when work has to be done on energized circuits in order to complete a process.

8.1.14 Energy Stored

Inductive and capacitive loads store energy in their respective magnetic and electric fields. This is characteristic of the basic elements; when not properly harnessed, it leads to personnel injury and electronic damage.

The governing equations and their derivations are developed in Chapter 1, Section 1.1 and constitute the basis for our discussion of follow-up electrical safety issues. For emphasis, some of the associated concepts are repeated here.

---

* The air space between two points becomes a conductor when the electric field intensity is 30,000 V/cm. When sulfur hexafluoride (SF_6) is used, its insulating property is 2.5 times higher than that of the air.

8.1.15 Energy Stored in Inductive Circuits

Refer to Fig. 8-4. When the switch S is closed, as per Ohm's law, a current (I) flows, which is given by

$$I = \frac{V}{Z}$$

Upon current interruption, the energy stored $\left(W = \frac{1}{2} LI^2\right)$ in the circuit's inductance is dissipated instantaneously across the switch. When the inductance or the current is high, this energy can be destructive. Inductance is analogous to the mass of a moving object and the electrical momentum (LI) is analogous to mechanical momentum (MU). That is,

$$LI \equiv MU$$

Just as you cannot stop a moving train without brakes, you cannot interrupt the current in a highly inductive load (large motors and transformers). Accordingly, the current interruption to a large motor takes place within the breaker's chamber which at medium voltage (MV) is often at vacuum. Similarly, the current interruption to a large MV transformer is through an insulated stick (several meters long) that opens up the transformer's switch from a reasonable safety distance, as recommended. The interruption of the current of an inductive circuit is accompanied by a flashover (arc flash).

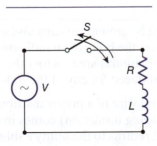

FIG. 8-4 An inductive load.

8.1.16 Energy Stored in Capacitive Circuits

The fact that the capacitors retain their energy stored $\left(W = \frac{1}{2} CV^2\right)$ on current interruption constitutes a potential safety hazard to anyone who comes in contact with their enclosure. Industrial capacitors are equipped with an internal in-parallel resistor that is supposed to discharge the energy stored, following the current interruption, within a short time (usually 2 minutes). Furthermore, the Electrical Code dictates the protective measures (grounding the enclosure of the capacitor bank) that must be taken on troubleshooting capacitor banks.

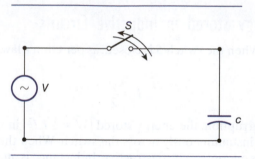

FIG. 8-5 A capacitive load.

The electrical safety concerns, the stored energy in the capacitor, could damage electrical circuits and cause fires. (The ideal capacitor circuit is presented in Fig. 8-5.)

The voltage across the switch, depending on what part of the voltage supply is closed, could be twice that of the source. This will lead to carbonization of the contacts and possible damage of equipment. Thus, a capacitor in electronic circuits should be connected in parallel to a resistor through which the energy stored will be discharged.

Liquid-containing capacitors explode when their polarity markings do not match those of the applied voltages.

8.1.17 Stray Voltages

Stray voltages are generated by ground currents and are usually less than 30 volts. They can be developed across the body of an individual in a swimming pool or in a lake, but more often in agricultural areas across the body of an animal.

Ground currents are produced for any of the following reasons:

1. The underground neutral wire of a power distribution system, which is worn out (defective or with aging insulation) comes in contact with the earth and thus some of its current returns to the utility's substation through the ground.

2. The neutral wire of a 240/120 V (3-wire, 1-phase) distribution system becomes disconnected before it reaches the premises. In such cases, the neutral current will return to the utility through the ground because that wire (white), as per the Electric Code, is grounded at the electrical entrance of the downstream network.

3. A premise's power distribution system is incorrectly designed or the wires' insulation has deteriorated.

4. One or more wires of a power line, due to adverse weather conditions, come in contact with the earth.

5. High-voltage outdoor substation transformers may develop a short to ground. This gives rise to high ground voltages and that of the enclosing metallic fence.

6. The induction currents due to utility power distribution lines induce ground voltages.

7. The sun's solar flares strike the earth. The resistance of an individual in water is about 300 ohms, and when subjected to 10 V of stray voltage, the resulting current through the human body is about 30 mA. Current of this magnitude disables the nervous system, as a result of which the swimmer cannot swim and dies from drowning.

Injuries from stray voltages are very rare but always tragic. The stray voltages and the accompaning leakage currents, depending on their magnitude, can injure individuals and adversely affect the health and productivity of milk-producing animals. In a recent incident in Quebec, Canada, ground currents increased the aggressiveness of the pigs in a farm, and at least half of them (about 200) were seriously injured from the ensuing fights. Presumably, the stray voltages were between the metallic food trough and ground.

8.1.18 Mishandling Medical Equipment

In the United States alone, there are 10,000 device-related patient injuries every year. This statistic is attributed to the complexity of equipment, misunderstanding of the fundamentals regarding their use, personnel's lack of training and experience, improper computer programming, and finally, Murphy's Law, which famously states, "Anything that can go wrong, will go wrong."

8.1.19 Cathodic Corrosion

There are many types of corrosion, but only cathodic corrosion will be discussed here and oxidation will only be mentioned. Cathodic corrosion or rusting is the deterioration of a metal due to its higher corrosion potential relative to another metal. Table 8-2 gives the approximate relative potential of some metals. The higher a metal's corrosion potential, the more of its electrons will be transferred to another metal of lower potential, provided a path of low resistance (moisture, earth, water, etc.) exists between the two.

The material damage from corrosion is estimated as being about $300 billion per year in North America; occasionally, we learn about the collapse of a bridge or a building with substantial loss of life. The metal that loses electrons is often referred to as the anode (higher potential) or a loser, while the metal that gains electrons is called the cathode (lower potential) or the winner.

The reference potential is that of gold. The corrosion potential, being the relative potential between two metals, can be negative instead of positive as tabulated here. It depends on the reference chosen.

For example, consider steel (Fe) and gold (G).

$$V_G - V_{Fe} = -0.85 \text{ V}$$

TABLE 8-2	Approximate corrosion potentials
Metal	Relative Corrosion Potential (Volts)
Gold	0.00
Nickel	0.35
Copper	0.40
Steel	0.85
Aluminum	0.90
Zinc	1.8
Magnesium	1.90

but

$$V_{Fe} - V_G = 0.85 \text{ V}$$

As Table 8-2 shows, aluminum will corrode to copper.

To prevent a metal's corrosion, electrically insulate the metal through a proper coating and/or provide a path of high resistance between it and the neighboring metals. Another technique that prevents corrosion is to use a dc voltage source (Fig. 8-6) in conjunction with the metal that is to be protected (cathode) and another that is to be sacrificed (anode). Over long periods of time, the anode should be replaced.

The rusting of steel in structures could be also due to the oxidation process whereby the iron changes to iron oxide.

$$4 \text{ Fe} + 3 \text{ O}_2 \longrightarrow 2 \text{ Fe}_2 \text{ O}_3$$

The rust formed decreases the effective area of the iron, and for a given force, its stress capability decreases.

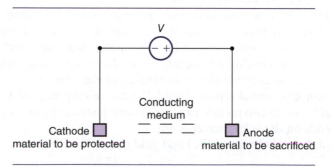

FIG. 8-6 Galvanic protection.

8.1.20 Lightning Strikes*

When lightning occurs, you must avoid seeking protection under a tree. If you do find yourself under a tree, try to step on a stone or a piece of wood (R_g very high), and use a glove or your shirt to hold your umbrella, whose rod should be made of wood.

If you are in an open field during a lightning storm and there is no shelter whatsoever (neither tree nor house), you should lie on the ground forming a loop. However, there is no absolute protection from lightning strikes. High-rise buildings are protected from lightning through a copper cable that connects the building's main grounding station to the grounding rod(s) on the roof of the building.

8.1.21 Conclusion

Electrical safety starts with understanding the grounding system. It consists of the utility's ground, the electrical entrance's ground, and, most importantly, the equipment's ground. The utility's ground (center tap at the secondary of a residence's supply transformer) is there to protect the downstream distribution system from overvoltages from upstream lightning strikes and from the utility's switching operations.

The grounding of a residence's main electrical entrance (switch or meter) serves as the utility's ground, provides a path for the discharge of static electricity as well as a return path for the ground current, and establishes a reference (zero volts) for the measurement of voltage.

The equipment's ground provides a path for the flow of static electricity and, in case of an internal fault, a path of zero resistance to the flow of current that often contributes to the breaker's current interruption. But above all, it also provides ground potential to an individual who is in contact with the shorted equipment. For this reason, it is named the Guardian Angel.

A circuit's protective device (fuse or breaker) interrupts excessive currents that result from downstream short circuits, which otherwise could cause fires and injure people. Intermittent or repetitive restrikes of shorts between the equipment's enclosure and/or between the load's interwindings are accompanied by high transient voltages, and the protective devices usually cannot isolate the malfunctioning circuit because of their very short duration.

Motors with a small power rating are protected through a properly selected bimetallic element that opens the circuit on undesirable currents, following which it will take some time until its circuit is reclosed. You should avoid contact with energized malfunctioning equipment and make sure that the affected equipment's

---

* Lightning originates from thunderclouds that usually contain positive electric charges at the top and negative charges at the bottom. As a result, electric fields are built up within the cloud, between the clouds, and between the clouds and earth. (Duration: 30 μs–30 ms, current up to 150 kA.)

ground is intact. The equipment's ground must be periodically verified. This is easily done by measuring the resistance between the equipment's enclosure and the nearest neutral wire or metallic conduit. If this resistance is small, the equipment's ground is okay. Alternatively, you can measure the voltage between the enclosure and the neutral. It should be negligible.

When you observe a flush over the utility's supply transformer or the lights become brighter or dimmer for long periods of time, a dangerous situation exists. The neutral is open, and an electrician should be called immediately.

Nature's laws reveal that magnetic fields (inductors, coils, windings relays, etc.) release the energy stored $\left(\frac{1}{2}LI^2\right)$ on current interruption while the electric fields (capacitors) retain the energy stored $\left(\frac{1}{2}CV^2\right)$.

The energy stored can be hazardous when precautions are not taken. In other words, avoid interrupting the current through highly inductive loads, just as you must not try to manually stop a large moving object.

The energy stored in capacitors can also be quite dangerous when precautions are not taken. The energy stored in a capacitor is analogous to the potential energy of an elevated object, which, on falling to a lower level, experiences reduced potential energy, which is then transformed into kinetic energy.

Similarly, a charged capacitor releases the retained energy when a path is provided for the current to flow. That is, the potential energy is changed to kinetic energy (time $\times I^2R$). This energy can be harnessed through a parallel connected resistor. If, however, an individual touches a nondischarged capacitor bank, he or she risks being electrocuted. Therefore, when you are working on a capacitor bank, you have to trip its breaker, ground the enclosure, and wait several minutes before starting to troubleshoot it.

The energy stored in inductive and capacitive loads, when not properly harnessed, not only poses risk to people, but can also damage electronic circuits and cause fires. The shorting of energized conductors creates the so-called arc flash that produces destructive high pressures and temperatures. You must not troubleshoot energized equipment, nor must you try to trace the wiring in the interior of the panels.

Sensitive medical measurements may be misinterpreted owing to displacement current. This current flows from one capacitor plate (the ceiling of the operating room) to the source, the ground, and the other capacitor plate (patient).

With regard to lightning strikes, the only way to protect yourself is to avoid being outdoors when lightning and thunder take place. Shelter under a tree is not protection. But if an individual happens to be there with an open umbrella, then injury may be mitigated by stepping on a piece of wood or a stone while holding the umbrella with the right hand through a glove or a shirt.

Copper thieves must be discouraged from attempting to remove copper because of the high danger such an activity presents. This danger has been minimized owing to stricter regulations surrounding scrap metal purchases and the utility's secure fastening of the grounding wires.

Loss of life may occur from fires that are due to overheated power-distribution cables. The overheating may be the result of undersized cables, the effects of induction currents, and placement of extension cords under carpets.

For ease of reference, Table 8-3 summarizes the main causes of electrical injuries.

TABLE 8-3	Summary of electrical safety		
No.	Cause of Injury	Description	Section
1	Breakers	Repetitive switching ON and OFF 120 V, 208 V, or 240 V circuit breakers while there is a short circuit downstream could lead to explosion and injuries.	8.1.2
2	Missing ground wire	When a short is developed within equipment, the ground wire or the so-called Guardian Angel provides a path of low resistance from the enclosure of the device to the source. This contributes to the faster interruption of the current to the faulted equipment. If this does not take place, then an individual who comes in contact with the enclosure of the faulted equipment is in danger. A home's ground wires should be verified at least annually.	8.1.6
3	Accidental contact with live conductors	This could lead to electrocution, depending on the resistance of the path from the live conductor through the human body, earth, and the source. At more than a 208 V distribution system, such an occurrence is quite destructive.	8.1.11
4	Troubleshooting adjacent equipment	When troubleshooting an energized equipment that is adjacent to another one, one of which is ungrounded, the possibility of electrocution exists when the operator happens to touch with one hand one part of the equipment and with the other hand, the adjacent one. (A path of current is created through his or her hands). Accidents of this nature are reported when homeowners try to repair a clothes dryer that is next to a clothes washer, and vice versa.	8.1.6
5	Open neutral	An open neutral in a 240 V–120 V distribution system produces in downstream circuits under- and overvoltage conditions (cause of fires). The metallic plumbing conduits and fixtures are at raised potential with respect to ground because the current that was returned to the source through the neutral is now reaching the source through the earth. This constitutes a danger to people and the plumbers who happen to work with the plumbing system.	8.1.12

6	Arc flash	Troubleshooting energized equipment must be avoided because, in case of an accidental short, destructive pressures and temperatures develop. Those who have to work with energized equipment need to wear special clothing and strictly obey the safety rules.	8.1.13
7	Interrupting the current in a highly inductive load	When the current in a highly inductive load (motor, transformer, etc.) is interrupted, the energy stored in the windings (magnetic field) is momentarily released. If precautions, such as proper breakers and safety switches, are not taken, serious injuries will result. Interrupting the current to a highly inductive load without using proper means is analogous to trying to stop a moving car without its brakes.	8.1.15
8	Capacitor	Capacitors on current interruption retain the energy stored in their electric fields, then discharge resistors are not provided or are disconnected; they can cause injury to anybody who comes in contact with them.	8.1.16
9	Stray Voltages	When somehow leakage current flows through the earth (through damaged underground neutral wire, induction phenomena, disconnected neutral, etc.), swimmers in lakes and swimming pools may be injured.	8.1.17
10	Lightning strikes	When outdoors while it is raining and thundering, there is no protection from lightning strikes. In such cases and under the shelter of a tree, the individual should hold the wooden rod of the umbrella with his or her right hand through a cloth or shirt, step on a stone or rock—and pray.	8.1.20
11	Galvanizing destruction	Metals have what is called corrosion potential. When two different metals are near each other, the metal with higher potential loses electrons to the other with the lower potential. This transfer of electrons is referred to as galvanic action, and the material that loses electrons is said to be corroded. This leads to the destruction of bridges, structures, and so on, with huge loss of life. The galvanic action is a natural phenomenon that originates from the laws of nature that try to keep everything at a dynamic equilibrium. The destructive effects of the galvanic action are aided by the oxidation of metals.	8.1.19

8.2 Reduction in Energy Consumption

This text has directly and indirectly outlined several concepts of how to minimize power demand and energy consumption. In this section, the various techniques to minimize energy consumption are described. The various methods and options that can be used to reduce energy usage and power demand are summarized in Table 8-4 and shown graphically in a sketch identified as the "bleeding heart of energy" (Fig. 8-7).

No.	Items to be Investigated	Description and Initial Cost of Required Equipment Installation	Reduction Power (kW)	Energy (kWh)	Notes*
	TABLE 8-4 Summary of options for reducing a premises' energy consumption, power demandand goals achieved.				
1	Unbalanced voltages				1
2	Unbalanced line currents				2
3	Lighting				3
4	Synchronous motors				4
5	Reduction in peak power demand				5
6	Power factor				6
7	High-efficiency motors				7
8	Variable-speed drives				8
9	Harmonics				9
10	Optimization of the voltage to dc motors				10
11	Discharged condenser's heat				11
12	Heat losses				12
13	Heat pumps				13
14	New technologies				14
15	Remarks				15

* Under the notes an individual should identify supplier of equipment, calculations, etc.

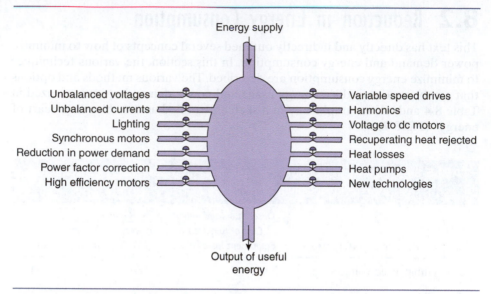

FIG. 8-7 The bleeding heart of energy.

Furthermore, Tables 1-5 and 1-6 in Chapter 1 (which present some highlights about energy and power) not only supplement the concepts of reducing the power demand and energy consumption but also give the necessary background for understanding power distribution. They constitute the nucleus of the human endeavor in understanding the concepts of power and energy. It is intended for students who may undertake it as a design project or as part of summer employment work and of course for anyone involved in reducing energy consumption per unit output or per square foot of office space. It would also be helpful to those trying to obtain the so-called green status designation for their building(s).

Implementation of some of these activities requires no initial investment, but all of them necessitate a thorough understanding of the fundamentals. The structured representations of the different areas are tabulated for easy understanding and widespread class circulation.

The corresponding Web Chapter, Section 8W.2 includes quantitative examples, exercises, and problems that illustrate each concept of energy conservation.

A prerequisite to such an undertaking is to obtain from the utility the hourly variation of the premises' power demand (kW) and energy consumption (kWh). This should be complemented by also obtaining the hourly variation of these parameters for each department or section of the plant.

8.2.1 Unbalanced Voltages

Every plant's maintenance department should periodically measure the line-to-line voltages and post them on a proper bulletin board. From this information, it is possible to calculate the unbalanced voltage in percent (X).

The increase in the temperature (ΔT) of the motor's windings above nominal is

$$\Delta T = (X)^2(2) \text{ in degrees C}$$

and the increase in the motor's rate of heat loss ΔP is

$$\Delta P = X \,(1 \text{ to } 3)$$

The cause of unbalanced voltages may be either internal or external to a plant. The internal voltage imbalance can be minimized or eliminated by reconnecting the various loads to a different circuit in such a way, if possible, that equal loads are connected to each of the phases $A, B,$ and $C.$ (See Appendix A, Section A.1.1.)

8.2.2 Unbalanced Line Currents

When 3-ϕ equipment are used instead of 1-ϕ, the line currents are reduced, become more balanced, and their negative sequence components are minimized.

When the current unbalance is minimized, and the negative sequence currents are reduced or eliminated, the results are a reduction of power losses, an increase of the motors' torque, and mitigation of resonant conditions.

8.2.3 Lighting

There are savings for using higher efficiency luminaires. This will reduce the cooling load, but, depending on the premise's geographical location, may increase usage of gas or oil heating. The power to a lighting source is transformed to heat and light, that is, energy. Usually, the utilities provide all the information required for the most efficient lighting system.

In office spaces where the power to the lighting system is more than, say, 10% of the total power demand, the possible effects of harmonics should be evaluated and proper countermeasures should be implemented.

8.2.4 Synchronous Motors

Synchronous motors, where possible, should not be used for Pf correction. Instead, substantial savings can be realized when static capacitors are used to improve the Pf. When the field current of a synchronous motor is adjusted so that it operates at unity power factor, instead of 80% leading, then the drawn line current is reduced by 25%. This will lead to important reductions in its heat loss $(3I_a^2\,R_a)$ and those of the upstream network such as cables transformers. The analysis should consider the load's torque requirements, and, as usual, the manufacturer should be consulted.

8.2.5 Reducing the Power Demand

A plant's power demand charges, depending on the utility, are about 10% to 40% of total electrical costs. The remaining costs correspond to energy usage. Therefore, a reduction of the power demand could mean important savings.

The first step in reducing power demand is to find out which load one can temporarily switch ON and OFF at selected time intervals. This step can be approached in many ways, including the following:

- Use a PLC to periodically switch ON and OFF preselected loads. This is referred to as load shedding.

- In small installations, have the maintenance department use a current transformer, a current relay, a timer, and a contactor for load shedding.

- When the process permits, manually switch OFF equipment at time intervals of peak demand and then switch them ON during intervals of low peak power demand.

- Use a so-called smart thermostat that can be programmed remotely or locally to lower the premise's heating or cooling demands at times of peak power demand.

- Use the premises' standby generator to supply power at predetermined time intervals. This would require evaluating the cost of the generator's output power ($/kW) and that of the utility in conjunction with the associated energy costs.

8.2.6 Power-Factor Improvement

Improvement in the power factor will increase the motor's output torque and reduce the rate of heat losses in the generator, transformers, and cables. It will also reduce the loading of generators, transformers, and cables. The quantitative advantages of high-power-factor operation should be tabulated. (See Chapter 5, Table 5-2.)

8.2.7 High-Efficiency Motors

In selecting new equipment, you should opt for higher efficiency equipment. In general, the cost of an equipment's losses over its lifetime is higher than the original purchasing costs. (See Chapter 3, Table 3-3.)

8.2.8 Variable-Speed Drives

In an air-flow or liquid-flow control system, the air flow is proportional to the motor's speed (N), and the power requirement is proportional to the third power of the motor's speed $(N)^3$. This is always accompanied with great savings. Consequently, all modern air-flow control systems incorporate a variable-speed motor.

8.2.9 Reduce Harmonics

The harmonics contribute to a plan's inefficiency, and when they are reduced or filtered out, they will result in some savings. Harmonics generated by the variable-speed drives can be reduced by using a 12-pulse rectifier instead of a six-pulse and ensuring that motors and transformers operate at nominal voltage. The operation of this equipment at higher than rated voltage will be accompanied by the development of higher losses and harmonics.

8.2.10 Voltage to DC Motors

No investment is required. Increase the supply voltage (V_m) by 5% through the standard off-load voltage tap settings of the supply transformer:

$$V_m \cos \alpha \, P = V_{av} \, I$$

Since the voltage (V_{av}) is increased and the motor power remains constant, the current will be reduced by 5%. At the same time, there will be a reduction in the rate of heat loss. A similar result can be obtained by adjusting the firing angle "α" of the control circuit accordingly.

8.2.11 Recuperating Rejected Heat

In some systems, the power used by the compressor of a cooling system is discharged outdoors as heat through the system's condenser. When this discharged heat can be recuperated, the result will be important savings. In premises where the cooling load is large, a so-called cooling tower is used.

Older installations such as a home's furnace have an efficiency of about 75%. Modern furnaces that incorporate a water-to-air heat exchanger have an efficiency of about 96%.

8.2.12 Heat Losses

Working and living environments are maintained at comfortable temperatures. Such temperatures will drive interior warm air to cooler outdoor environments. Minimizing this escape of heat will be accompanied by large savings. In this regard, review the basic concepts of heat transfer, namely, conduction, convection, and radiation.

For example, the steel knob of a house's exterior door will conduct 200 times more heat outdoors relative to a part of the wooden door of equal cross-sectional area:

$$(k \text{ of steel} \approx 200k \text{ of wood})$$

where k is the thermal conductivity of the material.

8.2.13 Heat Pumps

Heat pumps (ground source or air-to-air) can reduce a premise's heating and cooling energy usage. The corresponding savings are in the 25% to 65% range.

8.2.14 New Technologies

The use of photovoltaics, solar hot water heating, and wind power are areas that should be studied carefully. When subsidies are possible, the payback period is reasonable.

8.2.15 Remarks

1. **CO$_2$ Footprint**

   The reduction in energy will be accompanied by a reduction in the so-called CO$_2$ footprint. It can be shown that for each kW/h saved, a reduction of about 0.20 kg of CO$_2$ will be realized when the source of energy is natural gas.

2. **Increase in the Equipment's Life Expectancy**

   Reduction in power demand and energy consumption will also increase the life expectancy of cables, transformers, generators, and motors. This saving should be evaluated quantitatively in conjunction with the equipment's manufacturers.

3. **Subsidies**

   Usually, utilities and governments provide subsidies to anyone who invests money in projects that lead to reduced power demand and energy consumption.

Appendixes

A Three-Phase Systems   563

B Per-Unit System   587

C Laplace Transforms   597

D Solid-State Devices   601

E Basic Economic Considerations   603

F Photovoltaics   607

G Tables   611

H Bibliography   615

Answers to Problems   617

What You Will Learn in These Appendixes

A Theoretical Highlights

1 All aspects of balanced three-phase systems

2 The per-unit system

3 Laplace transforms

4 Basic economic considerations

5 Photovoltaics

6 Tables

B Outstanding Practical Highlights

1 Instantaneous, average, apparent reactive, and complex power in a three-phase system

2 Some applications of the per-unit system

3 Physical significance of Laplace transforms

4 Governing equations for economic evaluation of engineering problems

Additional Students' Aid on the Web

1 Additional problems in balanced 3-ϕ systems

2 The basics of diodes, transistors, thyristors, and insulated gate bipolar transistors

3 The basics of photovoltaics

The Appendixes include theory, examples, exercises, problems, and references on the following:

- Three-phase systems.
- Per-unit system, its definition, applications, and advantages.
- Elements of Laplace transforms.
- Solid-state devices: diodes, bipolar transistors, and thyristors.
- Review of the basic equations of economics for comparing the actual and operating costs of equipment.
- Photovoltaics.
- Protective device identification that will help an individual understand the basics of the power-distribution drawing.
- Tables about units.
- Exercises and problems.

Introduction

Power generation, transmission, and distribution occur mainly through three-phase systems. The advantage of it relative to 1-ϕ systems is economics, constant total instantaneous power, and generation of a rotating magnetic field on which the design and operation of all three-phase machines is based. This section of the Appendixes deals with many aspects of three-phase systems, including phase sequence, phasors, real, reactive, apparent and complex powers, star-delta connected loads and many examples and problems that will help in the understanding of the fundamentals.

A.1 Balanced Three-Phase Systems

A three-phase generator and its connected load are shown in Fig.A-1(a).

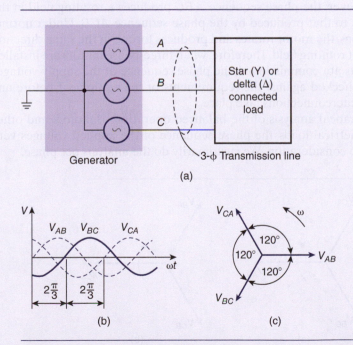

FIG. A-1 Three-phase systems: **(a)** typical generator-load distribution diagram, **(b)** generated voltages as a function of time, **(c)** voltage phasors of (b).

A.1.1 Phase Sequence

The three line-to-line voltages (V_{AB}, V_{BC}, and V_{CA}) produced at the terminals of a generator are normally balanced; each voltage phasor is equal in magnitude to the others but displaced 120° from the others in phase. The three balanced line-to-line voltages can be represented either by their instantaneous values or by their rotating phasors, as shown in Figs. A-1(b) and (c), respectively.

The angular speed of rotation of the voltage phasors for a 60 Hz system is

$$\omega = 2\pi f = 2\pi(60) = 377 \text{ rad/s} \qquad \text{(A.1a)}$$

In electrical engineering practice, the direction of rotation of phasors in the complex plane is counterclockwise.

The phase sequence of a given system of voltages can be determined by considering its instantaneous waveforms or the equivalent phasor diagrams. As shown in Fig. A-1(b), the positive maximum value of the voltages occurs first in phase A and then, successively, in phases B and C. In such cases, it is said that the order of rotation of the three-phase voltages is ABC. If the positive maximum value occurs first in phase A and then in phases C and B, the order of rotation of the three-phase voltages is said to be ACB. The order of rotation is often referred to as the *phase sequence*. Two possible phase sequences, and their directions of rotation, are shown in Fig. A-2. For phase sequence ABC or ACB, the phase voltages rotate in the same counterclockwise direction; when applied to a three-phase motor, however, the phase sequence ABC produces a rotating field in the direction opposite to that produced by the phase sequence ACB. Under normal operating conditions, the motor rotates and produces torque in the same direction as the predominant rotating field. Therefore, when three-phase motors are installed or when new plants are commissioned, the phase sequence of the supply voltages is measured and checked against the requirements of the equipment before any three-phase field interconnections take place.

For a mathematical analysis of the balanced operation of motors and other three-phase symmetrical loads, the phase sequence of the applied voltages very seldom comes into consideration. We customarily do the analysis per phase.

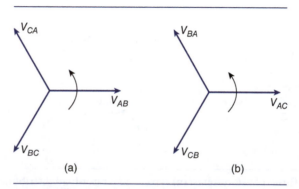

FIG. A-2 Phase sequences: **(a)** ABC, **(b)** ACB.

Voltage Unbalance*

The voltage unbalance is calculated as follows:

$$\text{voltage unbalance} = \frac{V_{max} - V_{av}}{V_{av}}(100) \qquad \textbf{(A.1b)}$$

where V_{max} and V_{av} are the maximum and average rms values of the line-to-line voltages, respectively. For example, if the rms values of the line-to-line voltages in a 3-ϕ, 208 V power-distribution system are 199 V, 209 V, and 213 V, then

$$V_{max} = 213\text{ V}, \qquad V_{av} = \frac{199 + 209 + 213}{3} = 207\text{ V}$$

and

$$\text{voltage unbalance} = \frac{213 - 207}{207}(100) = \underline{\underline{2.9\%}}$$

Some disadvantages of unbalanced voltages are overheating of motors, generation of harmonics, vibrations, and increased copper losses. A general rule is that the percentage *increase* in a motor's copper losses is about:

$$2X$$

where X is the voltage unbalance (see Chapter 8, Section 8.2.1).

A.1.2 Star–Delta Connected Systems

Three-phase generators and their loads can be connected in either a star (Y) or a delta (Δ) configuration. The essential characteristics of star- and delta-connected systems are discussed in the following sections.

Star-Connected System†

Three-phase, star-connected systems are very popular because of the advantages of grounding the neutral point and of supplying power to line-to-neutral loads. Three-phase, 4-wire distribution systems—common to most plants—require a star-connected supply.

A star-connected load is shown in Fig. A-3(a). By inspection, the line current (I_L) is equal to the phase current (I_P). That is,

$$I_L = I_P \qquad \textbf{(A.2)}$$

---

*According to published surveys of 83 utilities, 43% allowed voltage unbalance in excess of 3%, while 30% allowed voltage unbalance of 5% or higher.
†Star-connected systems are sometimes referred to as *wye* (Y) or *tee* (T) connected systems.

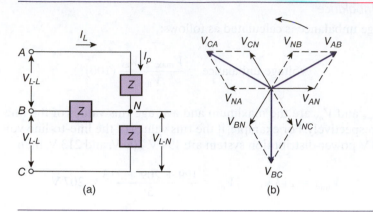

FIG. A-3 **(a)** A star-connected load, **(b)** a phasor diagram of line-to-line and line-to-neutral voltages for phase sequence ABC.

The line-to-line voltage is larger than its corresponding line-to-neutral voltage by a factor equal to $\sqrt{3}$ and leads it by 30°. Mathematically,

$$V_{\text{L-L}} = \sqrt{3}\, V_{\text{L-N}}\, \underline{/30°} \tag{A.3}$$

Equation (A.3) can easily be verified by first drawing the three balanced line-to-neutral voltages and then using KVL ($V_{AB} = V_{AN} + V_{NB}$).

Each line-to-line voltage, as seen in Fig. A-4(b), has two line-to-neutral components: one that leads the line-to-line voltage by 30°, the other that lags it by 30°. Then the question that arises is which phase voltage is represented by Eq. (A.3).

The phasor diagram of phase sequence ABC makes it clear that V_{AB} leads V_{AN}, V_{BC} leads V_{BN}, and V_{CA} leads V_{CN}. If the phase sequence is ACB, then V_{AC} will lead V_{AN} by 30°, and so on.

Delta-Connected Systems

A delta-connected load and its phasor diagram are shown in Figs. A-4(a) and (b), respectively. The line-to-line voltages are equal to the phase voltages, while the line currents are larger than the phase currents by a factor equal to $\sqrt{3}$. Mathematically,

$$V_{\text{L-L}} = V_{\text{phase}} \tag{A.4}$$

$$I_L = \sqrt{3} I_P\, \underline{/-30°} \tag{A.5}$$

The main advantage of delta-connected loads (such as can be found in some metal-processing plants) is that a single line-to-ground fault does not necessarily result in interruption of power, as is the case in a star-connected load. A delta connection is sometimes referred to as a *pi* (π) connection.

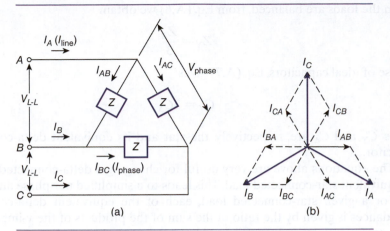

FIG. A-4 (a) A delta-connected load, (b) a phasor diagram of line and phase currents (phase sequence ABC).

Delta-Star Transformation

A three-phase delta-connected impedance (see Fig. A-5) can be changed to an equivalent star-connected impedance by using the following equation:

$$Z_{Y_1} = \frac{Z_{\Delta_1} Z_{\Delta_2}}{Z_{\Delta_1} + Z_{\Delta_2} + Z_{\Delta_3}} \qquad \text{(A.6)}$$

where Z_{Δ_1} and Z_{Y_1} are, in general, the delta- and star-connected impedances, respectively. This equation is derived from basic principles. It is easily remembered by noticing that each star-connected load is equal to the product of the two adjacent delta-connected impedances, divided by the sum of the three impedances.

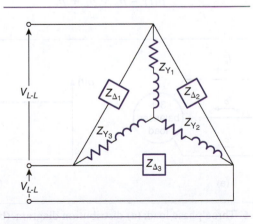

FIG. A-5 Delta-star loads.

When the loads are balanced, from Eq. (A.6) we obtain

$$Z_Y = \frac{Z_\Delta}{3} \tag{A.7}$$

In case of ideal capacitors, Eq. (A.7) gives

$$C_Y = 3C_\Delta \tag{A.8}$$

where C_Y and C_Δ are, respectively, the star and its equivalent delta-connected capacitor.

The equations above are very useful for changing a delta-connected load to an equivalent star-connected load. This leads to a simplified per-phase analysis.

For a given star-connected load, each of the equivalent delta-connected impedances is given by the ratio of the sum of the products of the y-impedances taken two at a time, and then divided by the opposite y-impedance. Considering Z_{Δ_1}, we have

$$Z_{\Delta_1} = \frac{Z_{Y_1} Z_{Y_2} + Z_{Y_1} Z_{Y_3} + Z_{Y_2} Z_{Y_3}}{Z_{Y_2}} \tag{A.9}$$

Z_{Δ_2} and Z_{Δ_3} are calculated in a similar manner.

A.1.3 Instantaneous Power

In a balanced three-phase system, the total instantaneous power is constant, while in a single-phase system—as explained in Chapter 1, Section 1.1.7—the instantaneous power changes or is said to be pulsating.

The general expression for the instantaneous three-phase power is derived as follows. Consider the balanced three-phase system shown in Fig. A-6(a). The instantaneous power ($p(t)$) delivered to the load is equal to the sum of the instantaneous powers (p_a, p_b, and p_c) drawn by the load. That is,

$$p(t) = p_a + p_b + p_c$$

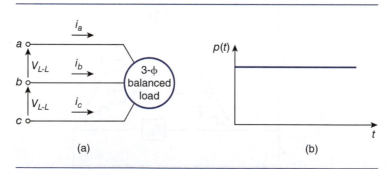

(a)   (b)

FIG. A-6 Three-phase load: **(a)** distribution diagram, **(b)** total instantaneous power.

From basic definitions, this expression can be written as follows:

$$p(t) = v_a i_a + v_b i_b + v_c i_c \tag{A.10}$$

where

v_a, v_b, v_c = the instantaneous values of the phase-to-neutral voltages

i_a, i_b, i_c = the instantaneous values of the phase-currents

The phase voltages for a phase sequence ABC are

$$v_a = V_m \cos \omega t$$
$$v_b = V_m \cos (\omega t - 120°) \tag{A.11}$$
$$v_c = V_m \cos (\omega t + 120°)$$

where V_m is the maximum value of the phase-to-neutral voltage. The corresponding phase currents for a load impedance of $|Z|\underline{/\theta}$ ohms per phase are

$$i_a = \frac{V_m}{|Z|} \cos (\omega t - \theta)$$
$$i_b = \frac{V_m}{|Z|} \cos (\omega t - 120° - \theta) \tag{A.12}$$
$$i_c = \frac{V_m}{|Z|} \cos (\omega t + 120° - \theta)$$

Substituting Eqs. (A.11) and (A.12) into Eq. (A.10), we obtain

$$p(t) = \frac{V_m^2}{|Z|} [\cos \omega t \cos (\omega t - \theta) + \cos (\omega t - 120°) \cos (\omega t - 120° - \theta)$$
$$+ \cos (\omega t + 120°) \cos (\omega t + 120° - \theta)] \tag{A.13}$$

Using the identity $\cos x \cos y = \frac{1}{2} (\cos (x + y) + \cos (x - y))$, we get

$$p(t) = \frac{V_m^2}{2|Z|} [\cos (\omega t + \omega t - \theta) + \cos (\omega t - (\omega t - \theta))$$
$$+ \cos ((\omega t - 120°) + (\omega t - 120° - \theta)) + \cos ((\omega t - 120° - (\omega t - 120° - \theta))$$
$$+ \cos (\omega t + 120° + (\omega t + 120° - \theta)) + \cos ((\omega t + 120° - (\omega t + 120° - \theta))] \tag{A.14}$$

Summing up the angles within the parentheses, we obtain

$$p(t) = \frac{V_m^2}{2|Z|} [\cos (2\omega t - \theta) + \cos \theta + \cos (2\omega t - 240° - \theta)$$
$$+ \cos \theta + \cos (2\omega t + 240° - \theta) + \cos \theta] \tag{A.15}$$

The three underlined terms are equal in magnitude and are displaced from each other by 120°. Thus, their sum is equal to zero. Therefore,

$$p(t) = \frac{V_m^2}{2|Z|} 3 \cos \theta \tag{A.16}$$

or

$$p(t) = \frac{3V_m I_m}{2} \cos \theta = 3 \frac{V_m}{\sqrt{2}} \frac{I_m}{\sqrt{2}} \cos \theta \tag{A.17}$$

Using equivalent rms values, we obtain

$$p(t) = 3V_{\text{L-N}} I_P \cos \theta \tag{A.18}$$

where $V_{\text{L-N}}$ and I_P are the rms values of the line-to-neutral voltage and phase current, respectively.

For a star-connected load, Eq. (8.18) becomes

$$p(t) = 3 \frac{V_{\text{L-L}}}{\sqrt{3}} I_L \cos \theta \tag{A.19}$$

from which

$$p(t) = \sqrt{3} V_{\text{L-L}} I_L \cos \theta \tag{A.20}$$

For a delta-connected load, Eq. (8.18) becomes

$$p(t) = 3 V_{\text{L-L}} \frac{I_L}{\sqrt{3}} \cos \theta \tag{A.21}$$

from which

$$p(t) = \sqrt{3} V_{\text{L-L}} I_L \cos \theta \tag{A.22}$$

Thus, the total instantaneous power drawn by a three-phase balanced load is constant, regardless of whether a star or a delta type of load connection is used. Instantaneous power as a function of time is shown in Fig. A-6(b).

A.1.4 Real, Reactive and Complex Power

Real Power

The real or average power (P) drawn by a three-phase balanced load is

$$P = \frac{1}{T} \int_0^T p(t) \, dt \tag{A.23}$$

where $p(t)$ is the instantaneous power and T is its period of oscillation. Substituting Eq. (A.22) for $p(t)$ after integration, we obtain

$$P = \left(\sqrt{3}V_{\text{L-L}}\, I_L \cos\theta\right)\frac{1}{T}t\,\Big|_0^T$$

from which

$$P = \sqrt{3}V_{\text{L-L}}\, I_L \cos\theta \qquad \text{(A.24)}$$

Equation (A.24) gives the average power consumed by a three-phase load, which is often referred to as the *power consumed* or the *active power*. Comparing Eqs. (A.22) and (A.24), we see that the instantaneous power drawn by a three-phase balanced load is equal to the average power consumed by the load.

The power consumed by a three-phase balanced load can also be determined by

$$P = 3I_p^2 R \qquad \text{(A.25)}$$

where R is the per-phase resistance of the three-phase load. The power reading that appears in a utility's invoice is called power demand. Power demand is the maximum of the average power recorded over any 15-minute interval of the billing period. (The time interval depends on the utility and could be 30 minutes or even 60 minutes.) Furthermore, the actual power demand is adjusted to include a power-factor penalty if applicable. For example, if a utility's minimum power-factor requirement is 0.90, while the maximum average consumption over a 15-minute interval is 150 kW at a power factor of 0.75, then the charged power demand is

$$150\left(\frac{0.9}{0.75}\right) = 180\ \text{kW}$$

Usually, the cost of power demand is in the range of 10% to 40% of the total electricity cost.

Reactive Power

By definition, the reactive power drawn by a three-phase load is

$$Q = 3V_{\text{L-N}}\, I_p \sin\theta \qquad \text{(A.26)}$$

For a star-connected load, Eq. (A.26) becomes

$$Q = 3\frac{V_{\text{L-L}}}{\sqrt{3}}\, I_L \sin\theta \qquad \text{(A.27)}$$

from which

$$Q = \sqrt{3}\ V_{\text{L-L}}\, I_L \sin\theta \qquad \text{(A.28)}$$

Similarly, for a delta-connected load, Eq. (A.26) becomes

$$Q = 3V_{\text{L-L}}\frac{I_L}{\sqrt{3}}\sin\theta$$

from which

$$Q = \sqrt{3}V_{\text{L-L}} I_L \sin \theta \tag{A.29}$$

Complex Power

By definition, the complex power drawn by a three-phase load is

$$S = 3V_{\text{L-N}} I_p^* \tag{A.30}$$

where I_p^* is the conjugate of I_p. For a star-connected system, Eq. (A.30) becomes

$$S = 3 \frac{V_{\text{L-L}}}{\sqrt{3}} I_L^* \tag{A.31}$$

from which

$$S = \sqrt{3}V_{\text{L-L}} I_L^* \tag{A.32}$$

Similarly, for a delta-connected system, Eq. (A.30) becomes

$$S = 3V_{\text{L-L}} \frac{I_L^*}{\sqrt{3}} \tag{A.33}$$

from which

$$S = \sqrt{3}V_{\text{L-L}} I_L^* \tag{A.34}$$

Thus, the real, reactive, and complex powers drawn by a balanced three-phase load are independent of the type of load connection (delta or star). For the steady-state analysis of a balanced three-phase system, it is advantageous to draw the so-called power triangles of the individual loads and then to obtain the system's power triangle by adding algebraically the real and reactive power components of the individual loads. A typical voltage–current phasor and the power triangle of an inductive load are shown in Figs. A-7(a) and (b), respectively.

A lagging power-factor load will result in a positive complex power.

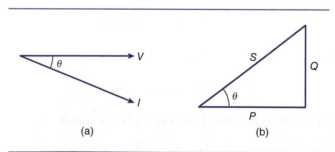

(a)　　　　　(b)

FIG. A-7 (a) Voltage and current phasors, **(b)** power triangle.

A.1.5 Measurement of Power and Power Factor

Power

The average power in a balanced three-phase system can be measured with a poly-phase meter or with two wattmeters, as shown in Fig. A-8. For a three-phase system,

$$P = \sqrt{3} V_{\text{L-L}} I_L \cos \theta \tag{A.35}$$

or

$$P = W_1 + W_2 \tag{A.36}$$

where the wattmeter readings W_1 and W_2 are given by

$$W_1 = V_{\text{L-L}} I_A \cos (\theta + 30°) \tag{A.37}$$

and

$$W_2 = V_{\text{L-L}} I_C \cos (\theta - 30°) \tag{A.38}$$

Equations (A.37) and (A.38) can easily be derived from the phasor diagram of a three-phase system supplying power to a load whose per-phase impedance has a phase angle of $\theta°$. By adding up Eqs. (A.37) and (A.38), we see that the sum of the wattmeter readings gives the average power consumed by a balanced three-phase load. This is demonstrated in Example A-1.

Power Factor

Using Eqs. (A.37) and (A.38), we can show that the per-phase power factor of the load is given by

$$\cos \theta = \frac{1}{\sqrt{1 + 3\left(\dfrac{1 - a}{1 + a}\right)^2}} \tag{A.39}$$

where $a = W_1/W_2$.

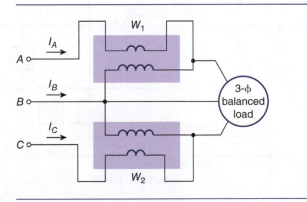

FIG. A-8 Measurement of 3-ϕ power with two wattmeters.

In many installations, a two-element kWh meter is used. On its front panel, this meter has two visible rotating disks, which derive their rotation from the power indication of their internal wattmeters (W_1 and W_2). By measuring the time (t_1 and t_2) that is required for the disks to complete one revolution, and by knowing the corresponding constants of the meter (so many kWh per revolution), the actual kW indication of each element can be calculated. If the wattmeter readings are known, then from Eq. (A.39), the instantaneous power factor of the load can be calculated. This technique—compared to the reading of standard switchboard Pf meters—is by far the most accurate method of measuring the power factor of a three-phase load. Digital meters, however, display the instantaneous power factor.

A.1.6 Three-phase, 5-Wire Distribution System

Figure A-9 shows the industrially popular three-phase, 4-wire distribution system. Between a line and the neutral, or between two lines, the single-phase loads are connected. Under normal operating conditions, the neutral conductor carries the resultant current of all loads connected between lines a, b, and c. In drawing the phasor diagrams of single- and three-phase loads of a 4-wire distribution system, the line-to-neutral voltage is usually taken as reference. The ground wire that is connected from the equipment's enclosure to the premise's main grounding terminal is not usually shown in the diagrams. After all, under normal operating conditions, it carries no current.

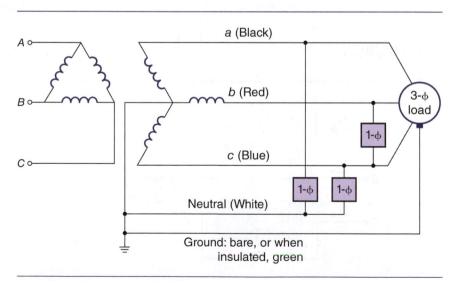

FIG. A-9 Three-phase, 4-wire distribution system.

A.1.7 Examples

Figure A-10(a) shows a 480 V generator supplying power to a delta-connected load whose impedance is $10\ \underline{/30°}\ \Omega$ per phase. For a phase sequence ABC, draw the phasor diagram of the currents and determine:

a. The phase and line currents.

b. The wattmeter readings and total power consumed by the load.

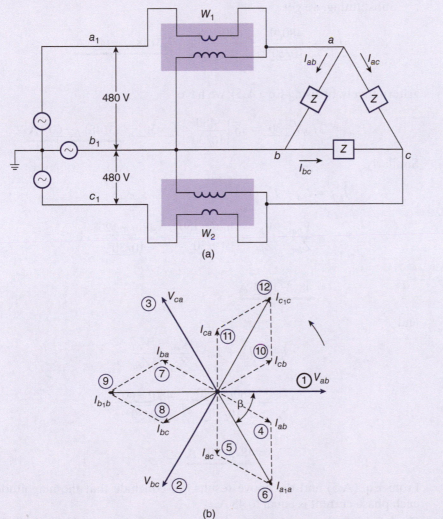

(a)

(b)

FIG. A-10 (a), (b)

SOLUTION

a. For an assumed phase sequence ABC, the generated voltages are drawn as shown in Fig. A-10(b). The encircled numbers indicate the order or the sequence of steps that are used in drawing the individual phasors. Disregarding the current through the voltage coil of the wattmeters and applying KCL, we obtain

$$I_{a_1a} = I_{ab} + I_{ac} = \frac{V_{ab}}{Z_{ab}} + \frac{V_{ac}}{Z_{ac}}$$

The phase angle of the voltages V_{ab} and V_{ac} are obtained from the phasor diagram.

Substituting, we get

$$I_{a_1a} = \frac{480\underline{/0°}}{10\underline{/30°}} + \frac{480\underline{/-60°}}{10\underline{/30°}} = \sqrt{3}(48)\underline{/-60°}\text{ A}$$

Alternatively, by using Eq. (A.5), we have

$$I_{a_1a} = \sqrt{3}I_{ab}\underline{/-30°} = \sqrt{3}\left(\frac{480}{10\underline{/30°}}\right)\underline{/-30°} = \sqrt{3}(48)\underline{/-60°}\text{ A}$$

Similarly,

$$I_{b_1b} = I_{ba} + I_{bc}$$

$$= \frac{V_{ba}}{Z_{ba}} + \frac{V_{bc}}{Z_{bc}} = \frac{480\underline{/-180°}}{10\underline{/30°}} + \frac{480\underline{/-120°}}{10\underline{/30°}}$$

$$= 48\sqrt{3}\underline{/180°}\text{ A}$$

and

$$I_{c_1c} = I_{ca} + I_{cb} = \frac{V_{ca}}{Z_{ca}} + \frac{V_{cb}}{Z_{cb}}$$

$$= \frac{480\underline{/-120°}}{10\underline{/30°}} + \frac{480\underline{/-60°}}{10\underline{/30°}}$$

$$= 48\sqrt{3}\underline{/60°}\text{ A}$$

From Eq. (A.5) and the above results, we conclude that the magnitude of each phase current is equal to 48 A.

b. By definition, the wattmeter reading is equal to the product of the voltage across its voltage coil, times the current through its current coil, times the cosine of the angle (β) between the voltage and the current. Considering W_1, we have

$$W_1 = V_{ab}I_{a_1a} \cos \beta_1 \qquad \text{(I)}$$

As seen from Fig. A-10(b), the magnitude of the phase angle between V_{ab} and I_{a_1a} is 60°. Thus,

$$\beta_1 = -60°$$

Substituting into Eq. (I), we obtain

$$W_1 = 480\,(48\sqrt{3})\cos 60° = 19.95 \text{ kW}$$

Similarly,

$$W_2 = V_{cb}I_{c_1c} \cos \beta_2$$

The phase angle between V_{ab} and I_{c_1c} is zero degrees. Thus,

$$W_2 = 480\,(48\sqrt{3})\cos 0° = 39.91 \text{ kW}$$

1. The total power consumed by the load is

$$P = W_1 + W_2 = 19.95 + 39.91 = \underline{59.86 \text{ kW}}$$

2. Alternatively, for a balanced load,

$$P = \sqrt{3}V_{\text{L-L}}I_L \cos \theta = \sqrt{3}(480)(48\sqrt{3}) \cos 30° = \underline{59.86 \text{ kW}}$$

3. Or, substituting into Eq. (A.25), we obtain

$$P = 3(48)^2(10 \cos 30°)$$

$$= \underline{59.86 \text{ kW}}$$

A balanced, star-connected load of $10\underline{/26°}$ Ω per-phase impedance is connected to a 480 V supply, as shown in Fig. A-11.

EXAMPLE **A-2**

Determine:

a. The line-to-neutral voltage V_{an}.

b. The phase current I_{an}.

c. The line current I_a.

d. The power consumed by the load.

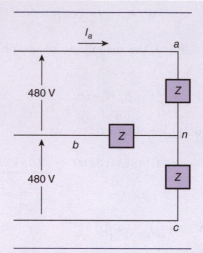

FIG. A-11

SOLUTION

a. For a phase sequence ABC, the magnitude of the line-to-neutral voltage (V_{an}) is

$$V_{an} = \frac{|V_{ab}|}{\sqrt{3}} = \frac{480}{\sqrt{3}} = \underline{277.1 \text{ V/phase}}$$

b. The phase current I_{an} is

$$I_{an} = \frac{V_{an}}{Z} = \frac{277.1}{10 \, \underline{/-26°}} = \underline{\underline{27.71 \, \underline{/-26°} \text{ A}}}$$

c. As seen in Fig. A-11, the line and phase currents are equal. Therefore,

$$I_a = \underline{\underline{27.71 \, \underline{/-26°} \text{ A}}}$$

d. The power consumed is determined by substituting the known data into Eq. (A.24). Thus,

$$P = \sqrt{3}(480)(27.71) \cos 26°$$

$$= 20.71 \text{ kW}$$

Alternatively, using Eq. (A.25), we obtain

$$P = 3(27.71)^2(10 \cos 26°) = \underline{20.71 \text{ kW}}$$

A 480 V, star-connected, three-phase ac generator supplies power to a delta-connected load through a transmission line. The per-phase impedances of the load and the transmission line are $(16 + j20)$ Ω and $(1 + j2)$ Ω, respectively. Determine:

a. The line current.

b. The magnitude of the line-to-line voltage across the load.

c. The efficiency of transmission.

SOLUTION

The given network is sketched as shown in Fig. A-12(a). Using Δ-Y transformation, the circuit is redrawn as shown in Fig. A-12(b). Because the per-phase impedances of the transmission line are equal, and because the load is balanced, we can do the analysis per phase.

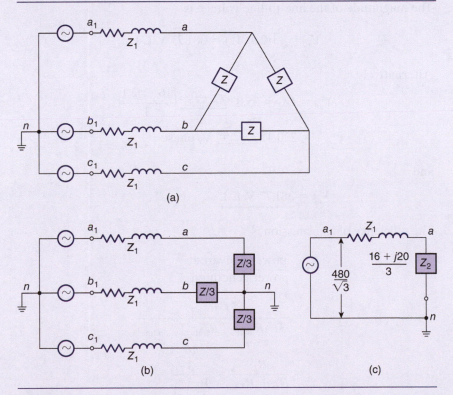

FIG. A-12

a. The per-phase equivalent circuit is shown in Fig. A-12(c). Using Ohm's law, we obtain

$$I_{a_1 a} = \frac{V_{a_1 n}}{Z_1 + Z_2} = \frac{\dfrac{480 \, \underline{/0^\circ}}{\sqrt{3}}}{1 + j2 + \dfrac{16 + j20}{3}} = 25.82 \, \underline{/-53.8^\circ} \text{ A}$$

Here, the line-to-neutral voltage is taken as reference.

b. The per-phase voltage across the load is

$$V_{an} = \frac{480 \, \underline{/0^\circ}}{\sqrt{3}} - IZ_1$$

$$= \frac{480 \, \underline{/0^\circ}}{\sqrt{3}} - 25.82 \, \underline{/-53.80^\circ}(1 + j2)$$

$$= 220.41 \, \underline{/-2.5^\circ} \text{ V/phase}$$

The magnitude of the line-to-line voltage is

$$V_{ab} = \sqrt{3}(220.41) = 381.77 \text{ V, L-L}$$

Alternatively,

$$V_{an} = IZ_2 = 25.82 \, \underline{/-53.8^\circ} \left(\frac{16 + j20}{3} \right)$$

$$= 220.41 \, \underline{/-2.5^\circ} \text{ V/phase}$$

and

$$V_{ab} = \underline{381.77 \text{ V, L-L}}$$

c. The efficiency of transmission is

$$\eta = \frac{\text{power delivered}}{\text{power generated}}$$

$$= \frac{3I^2 R_{\text{load}}}{3I^2 (R_{\text{load}} + R_{\text{transmission line}})}$$

$$= \frac{R_L}{R_L + R_{\text{T.L.}}} = \frac{\dfrac{16}{3}}{\dfrac{16}{3} + 1}$$

$$= \underline{84\%}$$

A 3-ϕ, 60 Hz, 480 V source delivers power to a balanced, star-connected load of 100 kW and 0.8 power factor lagging. Determine:

a. The magnitude of the line current and the power triangle of the load.

b. The size of the Δ-connected capacitor that, when connected across the load, will improve the power factor to 0.95 lagging.

SOLUTION

a. Referring to Fig. A-13(a), and using Eq. (A.24), we obtain

$$I_L = \frac{100 \times 10^3}{\sqrt{3}(480)(0.8)}$$

$$= 150.35 \text{ A}$$

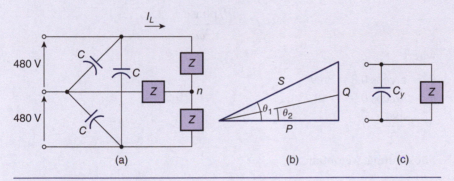

FIG. A-13

The reactive power drawn by the load is

$$Q = P \tan \theta$$

$$Q = 100 \times 10^3 \tan 36.87°$$

$$= 75 \text{ kVAR}$$

The complex power is

$$S = \frac{P}{\cos \theta} = \frac{100}{0.8}$$

$$= 125\underline{/36.87°} \text{ kVA}$$

The power triangle, with and without the capacitor bank in the circuit, is shown in Fig. A-13(b). The per-phase equivalent circuit is shown in Fig. A-13(c).

b. The per-phase star-connected capacitance that is required to improve the power factor from $\cos \theta_1$ to $\cos \theta_2$ (see Fig. A-13(b)) is found as follows. The reactive power of a capacitor is

$$Q = VI \sin \theta_c$$

$$= V \left| \frac{V}{Z} \right| \sin 90°$$

$$= V^2 \omega C$$

Also, from the geometry of Fig. A-13(b),

$$Q = (P/\text{phase})(\tan \theta_1 - \tan \theta_2)$$

From the above,

$$C = \frac{P/\text{phase}}{V^2 \omega}$$

where

$\theta_1 = \arccos 0.8 = 36.9°$
$\theta_2 = \arccos 0.95 = 18.2°$
$V = \dfrac{480}{\sqrt{3}}$

Substituting, we obtain

$$C = \frac{100 \times 10^3}{3 \left(\dfrac{480}{\sqrt{3}} \right)^2 (377)} (\tan 36.9° - \tan 18.2°)$$

$$= 485.1 \ \mu\text{F/phase}$$

Since the commercially available capacitor bank is delta-connected, we use Eq. (A.8) to obtain

$$C_\Delta = \frac{C_Y}{3} = \frac{485.1}{3} = \underline{161.7 \ \mu\text{F/phase}}$$

Usually, a capacitor bank is specified in terms of its voltage and kVAR rating. In this case, $V = 480$ V, L-L and

$$Q = 100(\tan 36.9° - \tan 18.2°) = 42.13 \ \text{kVAR}.$$

The two-element energy meter shown in Fig. A-14 is used to measure the energy consumed by a 3-ϕ, 3W load. The constant of the meter for each rotating wheel is 5 kWh/revolution. At a given instant, the rotating wheels require 30 and 50 s, respectively, to complete one revolution. Calculate the power factor of the load at that instant.

EXAMPLE **A-5**

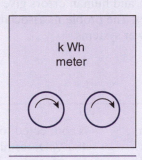

FIG. A-14

SOLUTION

Designating the time that each wheel takes to complete one revolution as t_1 and t_2, the ratio of the wattmeter's reading is

$$\alpha = \frac{W_1}{W_2} = \frac{\frac{(\text{energy})_1}{t_1}}{\frac{(\text{energy})_2}{t_2}} = \frac{\frac{5}{30}}{\frac{5}{50}} = 1.67$$

Substituting into Eq. (A.39), we obtain

$$\cos \theta = \frac{1}{\sqrt{1 + 3 \left(\frac{-0.67}{2.67} \right)^2}} = 0.92$$

A.1.8 Conclusion

Electric power is generated, transmitted, and distributed through three-phase systems. The three-power wires are color coded (Red – Blue – Black) to represent phases *a-b-c*.

The relative advantages of three-phase systems are as follows:

Economics, constant instantaneous power generates a rotating magnetic field upon which the design and operation of three-phase machines are based.

The loads could be delta or Y-connected. The line currents and the power to either of them is the same.

The neutral of the Y-connected load carries the sum of the 3-line currents, the sum of the odd harmonics' components of the line currents, and the current that results from L-N connected loads.

In contrast, the line currents in a delta-connected system carry no harmonics, but in the case of Δ-Y transformers and depending on its secondary connection, they allow the harmonic current to circulate within its primary winding.

In general, the voltages and the currents in a three-phase system are balanced. That is, they are equal in magnitude and 120 degrees to each other. In addition, the per-phase impedances of a load are equal. However, adverse weather conditions, aging equipment, and human errors give rise to unbalanced voltages, current, and load impedances. The problems associated with these conditions are analyzed in books about power systems.

A.1.9 Problems

A-1 In a 3-ϕ, 4-wire distribution system, show that:

a. The sum of the fundamental frequency components of the line currents does not appear in the neutral wire.

b. The sum of the triple harmonic currents of the line currents appears in the neutral as three times that of one line.

c. The sum of the nontriple harmonics does not appear in the neutral.

A-2 A 480/277 V, 60 Hz power-distribution system supplies power to the following loads:

- 5 kW heater
- 20 kW motor whose efficiency and Pf are, respectively, 0.92 and 0.88 lagging
- 10 kVAR capacitor bank

Determine:

a. The total real, reactive and complex power delivered.

b. The total supply current.

A-3 A three-phase 208 V, 60 Hz power line has two loads connected to it. The first is delta-connected and draws 25 kW at 0.70 power factor lagging. The second is star-connected and draws 6.25 kVA at 0.80 power factor leading. What are the total line current and the combined power factor?

A-4 A balanced star-connected load having a phase impedance of $(8 + j6)$ Ω is connected to a three-phase supply of 480 V. Determine:

a. The line-to-neutral voltage.

b. The phase current.

c. The line current.

d. The power factor of the load.

e. The power consumed by the load.

f. The equivalent delta-connected load.

A-5 The per-phase impedance of a 480 V delta-connected load is $4 + j3$ Ω. The load is supplied through a transmission line whose per-phase impedance is $0.1 + j0.3$ Ω. Determine the efficiency of transmission and the source voltage.

A-6 For the 480 V, 3-ϕ, 60 Hz system shown in Fig. P-A6:

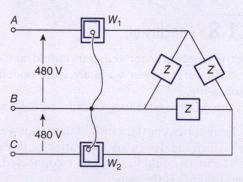

Phase sequence: ABC
$Z = (17.32 + j10)$ Ω

FIG. P-A6

a. Calculate the phase and the line currents, and draw a phasor diagram showing all phase currents, line currents, and line voltages.

b. Calculate the reading of each wattmeter and the total power.

c. Repeat (b) if $Z = 20 + j60 \ \Omega$.

A-7 The phase sequence of the voltages applied to the three-phase network of Fig. P-A7 is ABC. Determine the magnitude of the line currents by using any three of the following methods:

a. The concept of the power triangle.

b. The per-phase analysis using ohmic values.

c. The per-unit values.

d. KCL, without changing the delta load to its equivalent star.

A-8 The circuit shown in Fig. P-A8 is used to identify the phase sequence of the line voltages. Prove that for a phase sequence ABC, the voltmeter's indication is less than the line-to-line voltage, while for a phase sequence ACB, the voltmeter's indication is larger than the line-to-line voltage. (*Hint:* The voltmeter's internal impedance is relatively very high.)

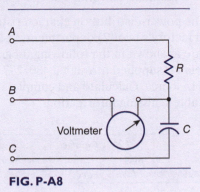

FIG. P-A8

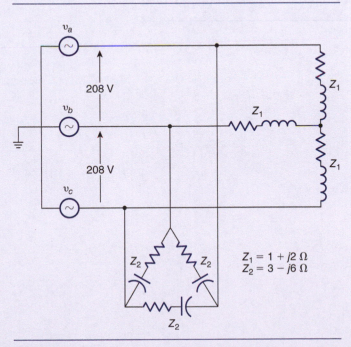

FIG. P-A7

A-9 A 3-ϕ, 480 V, 500 kW, 0.9 Pf lagging, 0.95 efficient load receives rated current through a transmission line whose per-phase impedance is $0.01 + j0.04\ \Omega$. Determine:

a. The voltage across the load.

b. The capacitor bank that when connected across the load will maintain its voltage at 480 V.

A-10 The power-distribution characteristics of a factory's two (2) departments A and B are as shown in the following table. The plan is supplied through a 480/277 V, 60 Hz source. Calculate and complete the table's missing information.

Department	Complex Power (kVA)	Average Power (kW)	Inductance (mH)	Current (A)
A				$49.1\underline{/-19.2°}$
B	$10.65\underline{/28.5°}$			
Total				

A-11 Briefly explain the following:

1. Compare and contrast the following parameters for 1-ϕ and 3-ϕ:

Instantaneous power.

Magnitude of line currents.

Cost of distribution.

Neutral wire disconnection at the utility's supply transformer.

Grounding of loads.

2. Briefly discuss the following:

1. Phase sequence and rotor rotation.

2. Source and effects of harmonics on the torque and motor losses.

3. Harmonic currents cannot flow through the supply lines of a delta-connected load.

4. For a given power-distribution system, the power supplied to the load is independent of delta or star connections of the load.

The per-unit system is a set of mathematical relationships that transforms the standard parameters of voltage, current, power, impedance, and the like into their equivalent per-unit values. Use of equivalent per-unit values greatly simplifies the understanding and solution of many engineering problems.

B.1 Calculation of Per-Unit Values

By definition, the per-unit value of a parameter is given by

$$\text{per-unit value of a parameter} = \frac{\text{actual value of the parameter}}{\text{base value for this parameter}} \qquad \textbf{(B.1)}$$

For example, if the actual value of an impedance is 2 Ω and its base value is 5 Ω, then the per-unit value of the impedance is

$$Z_{pu} = \frac{Z_{actual}}{Z_{base}} = \frac{2}{5} = 0.4 \text{ pu}$$

The per-unit value of a quantity (X) is equal to the quantity expressed in percent, divided by 100. That is,

$$X_{pu} = \frac{X\%}{100}$$

The base values of the various parameters are selected as follows.

B.1.1 Base Power (S_b)

For a single machine, in performing the per-unit computations of real, reactive, and complex power, the base power is always taken as the magnitude of the machine's complex power. The magnitude of the complex power is referred to as the *apparent power*. Mathematically,

$$S_b = (VA)_{base} \qquad \textbf{(B.2)}$$

For example, if a load draws $1000\underline{/30}$ VA, the per-unit values of the complex, real, and reactive powers are calculated as follows. By definition,

$$S_b = 1000 \text{ VA}$$

The per-unit value of complex power is

$$S = \frac{\text{actual value}}{\text{base value}}$$

$$= \frac{1000\underline{/30°}}{1000} = 1\underline{/30°} \text{ pu}$$

The per-unit value of real power is

$$P = \frac{\text{actual value}}{\text{base value}} = \frac{P_a}{S_b}$$

$$= \frac{1000 \cos 30°}{1000} = 0.87 \text{ pu}$$

The per-unit value of the reactive power is

$$Q = \frac{\text{actual value}}{\text{base value}} = \frac{Q_a}{S_b}$$

$$= j\frac{1000 \sin 30°}{1000} = j0.5 \text{ pu}$$

Thus, the complex power drawn by the load in per unit is

$$S = 0.87 + j0.5 = 10\underline{/30°} \text{ pu}$$

The base power of a transformer or generator is its rated output VA. The base power of a motor is its input apparent power. In the case of transformers and generators, their nameplate data furnishes their output apparent powers, whereas a motor's input volt-amperes can be calculated from its output power, overall efficiency, and input power factor as follows:

$$S_b = \frac{\text{output in watts}}{(\text{efficiency})(\text{power factor})} \tag{B.3}$$

For a three-phase system, the apparent power is

$$S_b = \sqrt{3}V_{\text{L-L}}I_L = \frac{P_{\text{out}}}{\eta \cos \theta} \tag{B.4}$$

Very often an educated guess or assumption has to be made regarding the overall efficiency and input power factor of a motor. In the case of a system that comprises transformers, cables, and machines, the base power is arbitrarily chosen and remains *constant* for each component of the system. In other words, once the base power is selected, it does not change throughout the system.

B.1.2 Base Voltage (V_b)

The base voltage for a motor or generator is its nameplate voltage. For a transformer, the base voltage is different on either side of the transformer and hence will be designated, depending on the winding under consideration, as

$$V_{b_H} \quad \text{or} \quad V_{b_L}$$

where V_{b_H} is the base voltage of the high-voltage winding and V_{b_L} is the base voltage of the low-voltage winding. Thus, the base voltage in a multicomponent system will change each time you advance from one side of the transformer to the other.

B.1.3 Base Current (I_b)

The base current is normally the rated or full-load current of the machine. For single-phase loads,

$$I_b = \frac{S_b}{V_b} \tag{B.5}$$

For a three-phase system,

$$I_b = \frac{S_b}{\sqrt{3}V_{b_{L\text{-}L}}} \tag{B.6}$$

The base current on either side of the transformer is different.

B.1.4 Base Impedance (Z_b)

For per-unit computations of impedance, resistance, and reactance, the base impedance is used and is given by the ratio of base voltage to base current. That is,

$$Z_b = \frac{V_{b_{L\text{-}L}}/\sqrt{3}}{I_b} \tag{B.7}$$

or

$$Z_b = \frac{V_{b_{L\text{-}L}}/\sqrt{3}}{S_b/\sqrt{3}V_{b_{L\text{-}L}}} \tag{B.8}$$

From the above,

$$Z_b = \frac{V_{b\ L\text{-}L}^2}{S_b} \ \Omega/\text{phase} \tag{B.9}$$

If the base voltage of a three-phase machine or transformer is taken as the line-to-neutral voltage, then the base power should be the per-phase apparent power. If, however, the line-to-line voltage is used, then the base power must be the total or three-phase apparent power of the machine under consideration.

As seen from the high- and low-voltage windings, the base impedance for a transformer is different. However, the per-unit values of the impedances are the same, regardless of which side is chosen as the reference.

B.1.5 Torque and Speed

The base values of a machine's torque and speed are, respectively, its full–load torque and speed.

B.1.6 Multicomponent Systems

In a multicomponent system comprising transformers, cables, and machines, a convenient base power and voltage are chosen. As already mentioned, the base power remains the same for each component of the system, while the *base voltage* and the *base current* change for either side of the transformer. Very often, in the one-line diagram of a power-distribution system, the per-unit impedances of the individual machines and transformers are given as calculated from their own base parameters. The calculations are simplified if all per-unit impedances are expressed on the same base.

To change the per-unit value of an impedance from one base (old) to a new base, the following relationship is used:

$$Z_{pu_{new}} = Z_{pu_{old}} \left(\frac{V_{b\text{-}old}}{V_{b\text{-}new}} \right)^2 \left(\frac{S_{b\text{-}new}}{S_{b\text{-}old}} \right) \qquad \textbf{(B.10)}$$

For example, if on a base of 480 V and 1000 kVA, an impedance is 0.2 per unit, then at a new base of 1200 V and 1500 kVA, the per-unit value of the impedance is

$$Z_{new} = 0.2 \left(\frac{480}{1200} \right)^2 \left(\frac{1500}{1000} \right)$$

$$= 0.048 \text{ pu}$$

B.2 Advantages of Per-Unit Values

The per-unit system is very popular in the power field of electrical engineering because of the following advantages:

1. The per-unit values of *impedance, voltage,* and *current* of a transformer are the same, regardless of whether they are referred to the low- or to the high-voltage side. This is a tremendous advantage because it helps to bypass all the confusion and misunderstanding that result when the transformer's parameters are referred to either the high- or low-voltage windings.

2. The per-unit system makes engineering calculations easier because the rated voltage and current for a single machine or a transformer, when expressed in per unit on its own machine rating, are each equal to 1.0.

3. The per-unit values of the parameters of a machine vary within a small range for all types and ratings of machines produced by different manufacturers.

This gives the designer and the application engineer a better understanding of the machine, and a tool with which similar units may easily be compared. For example, the ohmic values of the reactance of alternators could vary from 0.1 to 20 ohms; the same reactance, expressed in per unit, would be in the range of 0.10 to 1.1 per unit.

4. The per-unit value of the resistance of a machine furnishes, almost at a glance, its electrical losses in percent of its rated power. For example, a transformer operating under rated conditions at unity power factor with a winding resistance of 0.01 per unit has a copper loss of 1%:

$$I^2R = (1.0)^2(0.01) = 0.01 \text{ pu}$$

This information is very useful to a power systems engineer because he or she can estimate and locate the quantity of the various copper losses simply by looking at the one-line impedance diagram.

5. The per-unit values of the impedance of a single- or simple multicomponent power system readily provide the short-circuit current at different points on the network. For example, a transformer with a per-unit impedance of 5% would result (without taking into consideration the motor's short-circuit current contribution) in a short-circuit current (I_{sc}) of

$$I_{sc} = \frac{1}{Z_{pu}} = \frac{1}{0.05} = 20 \text{ pu}$$

Knowing the short-circuit current is very important in selecting and coordinating protective devices.

6. The per-unit system simplifies the analysis of problems that include star-delta (Y-Δ) types of winding connections. The factor of $\sqrt{3}$ is not used for the per-unit analysis. For example, consider the expression for the power

$$P = VI \cos \theta$$

When the voltage and the current are expressed in per unit, this relationship gives the total power in per unit, regardless of a delta- or star-winding connection.

7. Per-unit parameters simplify the simulation of simple or complex power-system problems on computers. Such simulations are important for transient and steady-state analyses.

EXAMPLE **B-1**

For the three-phase transformer shown in the one-line diagram of Fig. B-1, determine the per-unit values of voltages, currents, and impedances, as seen from the high- and low-voltage windings, when the transformer operates at a rated condition.

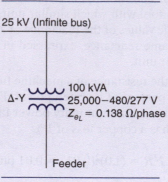

25 kV (Infinite bus)

100 kVA
Δ-Y   25,000−480/277 V
$Z_{e_L} = 0.138$ Ω/phase

Feeder

FIG. B-I

SOLUTION

Low-Voltage Winding

The base parameters are

$$V_{b_L} = 480 \text{ V}, \qquad S_{b_L} = 100 \text{ kVA}$$

and

$$I_{b_L} = \frac{S_{b_L}}{\sqrt{3} V_{b_{L\text{-}L}}} = \frac{100 \times 10^3}{\sqrt{3}\,480} = 120.28 \text{ A}$$

From Eq. (B.9),

$$Z_b = \frac{(V_{b_{L\text{-}L}})^2}{S_b} = \frac{(480)^2}{100 \times 10^3} = 2.3 \text{ Ω/phase}$$

Thus, the per-unit value of the impedance is

$$Z_{\text{pu}} = \frac{Z_{\text{actual}}}{Z_{\text{base}}}$$

$$= \frac{0.138}{2.3} = \underline{0.06 \text{ pu}}$$

The per-unit value of the current is

$$I_{\text{pu}} = \frac{I_{a_L}}{I_{b_L}} = \frac{120.28}{120.28} = \underline{1.0 \text{ pu}}$$

The per-unit value of the voltage is

$$V_{\text{pu}} = \frac{V_{a_L}}{V_{b_L}} = \frac{480}{480} = \underline{1.0 \text{ pu}}$$

High-Voltage Winding

The base parameters are

$$V_{b_H} = 25 \text{ kV}, \qquad S_b = 100 \text{ kVA}$$

Thus,

$$Z_{bH} = \frac{(V_{bL\text{-}L})^2}{S_b} = \frac{(25)^2}{0.10} = 6250.0 \ \Omega/\text{phase}$$

$$I_{bH} = \frac{100}{\sqrt{3}\,25} = 2.31 \ \text{A}$$

The ohmic value of the winding impedance referred to the high-voltage delta-connected winding is

$$Z_H = Z_L \left(\frac{N_1}{N_2}\right)^2$$

$$= 0.138 \left(\frac{25{,}000}{480/\sqrt{3}}\right)^2 = 1123.1 \ \Omega/\text{phase}$$

Changing this impedance to an equivalent star, we obtain

$$Z_y = \frac{Z_\Delta}{3} = \frac{1123.1}{3} = 374.35 \ \Omega/\text{phase}$$

Thus, the per-unit values are

$$Z_{pu} = \frac{374.35}{6250} = \underline{0.06 \ \text{pu}}$$

The per-unit value of the voltage is

$$V_{pu} = \frac{\text{actual value}}{\text{base value}} = \frac{25 \times 10^3}{25 \times 10^3} = \underline{1.0 \ \text{pu}}$$

Similarly, the per-unit value of the current is

$$I_{pu} = \frac{2.31}{2.31} = \underline{1.0 \ \text{pu}}$$

Thus, the per-unit values of the impedance, voltage, and current are the same for the high- and low-voltage sides of the transformer.

For the power system shown in the one-line diagram of Fig. B-2(a) determine: EXAMPLE **B-2**

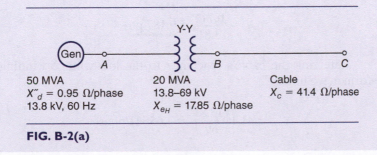

FIG. B-2(a)

a. The per-unit value of all reactances on the base of the generator rating.

b. The magnitude of the short-circuit current at points A and B. (Neglect the motor's short-circuit current contributions.)

SOLUTION

a. The base values are given by the generator rating.

$$S_b = 50 \text{ MVA} \qquad V_{b\text{L-L}} = 13.8 \text{ kV}$$

$$I_b = \frac{50 \times 10^3}{\sqrt{3}(13.8)} = 2091.85 \text{ A}$$

The base impedance at 13.8 kV is

$$Z_b = \left(\frac{13.8}{50}\right)^2 = 3.81 \text{ } \Omega/\text{phase}$$

Generator.   The per-unit value of the generator's reactance is

$$X_d = \frac{0.95}{3.81} = \underline{0.25 \text{ pu}}$$

Transformer.   Referring the transformer's reactance to the low-voltage winding, we have

$$X_{e_L} = \left(\frac{N_1}{N_2}\right)^2$$

$$= 17.85 \left(\frac{13.8}{69}\right)^2$$

$$= \underline{0.714 \text{ } \Omega/\text{phase}}$$

In per unit,

$$X_e = \frac{0.714}{3.81} = \underline{0.187 \text{ pu}}$$

Cable.   Referring the cable's reactance to the low-voltage winding of the transformer, we have

$$X_{c_L} = 41.4 \left(\frac{13.8}{69}\right)^2 = 1.656 \text{ } \Omega/\text{phase}$$

In per unit,

$$X_c = 1.656/3.81 = 0.435 \text{ pu}$$

The per-phase equivalent circuit, using per-unit values, is shown in Fig. B-2(b).

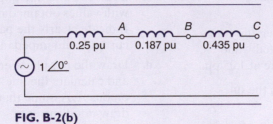

FIG. B-2(b)

b. The magnitude of the short-circuit current at A is:

Using ohmic values,

$$I = \left|\frac{V_{\text{L-N}}}{Z}\right| = \frac{13,800/\sqrt{3}}{0.95} = \underline{8.39 \text{ kA}}$$

Using per-unit values,

$$I = \left|\frac{V}{Z}\right| = \frac{1.0}{0.25} = 4.0 \text{ pu}$$

$$= 4(2091.85) = \underline{8.39 \text{ kA}}$$

The magnitude of the short-circuit current at B is:

Using ohmic values,

$$I = \frac{V_{\text{L-N}}}{Z} = \frac{13,800\sqrt{3}}{(0.95) + 0.714} = \underline{4.79 \text{ kA}}$$

Using per-unit values,

$$I = \left|\frac{V}{Z}\right| = \frac{1.0}{0.25 + 0.187}$$

$$= 2.29 \text{ pu}$$

$$= 2.29(2091.85)$$

$$= \underline{4.79 \text{ kA}}$$

Please note that when we speak of a short circuit, we usually mean a solid, three-phase short circuit.

B.3 Problems

B-1. A 3-ϕ, 4160–480/277 V, 1500/2000 kVA transformer has a leakage impedance of 0.01 + j0.0575 pu. Determine:

  a. The ohmic values of the leakage impedance.

  b. The short-circuit MVA when the transformer's secondary is shorted while the off-load voltage taps are at 1.05 pu.

B-2. For the power system shown in the one-line diagram of Fig. P-B2:

  a. Determine the per-unit value of each reactance on the base of its own machine rating.

  b. Determine the per-unit value of each impedance on the base of the generator rating.

  c. Draw the one-line impedance diagram with values obtained in (b) above. Indicate clearly the bases of voltages, currents, and impedances.

  d. Draw the per-phase equivalent circuit and calculate the line-to-line voltage on bus A. Assume that the motors draw rated currents.

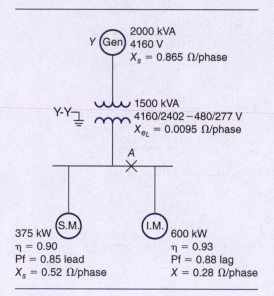

Y (Gen) 2000 kVA
4160 V
X_s = 0.865 Ω/phase

Y-Y 1500 kVA
4160/2402 – 480/277 V
X_{e_L} = 0.0095 Ω/phase

A

375 kW (S.M.)
η = 0.90
Pf = 0.85 lead
X_s = 0.52 Ω/phase

(I.M.) 600 kW
η = 0.93
Pf = 0.88 lag
X = 0.28 Ω/phase

FIG. P-B2

The transient response of a system, whether that of a simple circuit or an electric machine, can be obtained by solving the system's differential equations. The solution of a differential equation can be determined by using either the so-called classical method or the Laplace transform technique. The Laplace method uses the tabulated Laplace transforms.

The Laplace transform $\mathbf{F}(s)$ of any function $f(t)$ is found by the following equation:

$$\mathbf{F}(s) = \mathcal{L}(f(t)) = \int_0^\infty f(t)e^{-st}\,dt \qquad \textbf{(C.1)}$$

where s is a complex frequency variable defined as

$$s = \sigma + j\omega \qquad \textbf{(C.2)}$$

where ω is the angular frequency and σ is a real number approaching zero.

Find the Laplace transform of e^{-st}.

EXAMPLE **C-1**

SOLUTION

From Eq. (C.1), we obtain

$$\mathcal{L}(e^{-st}) = \int_0^\infty e^{-at}\,e^{-st}\,dt$$

$$= -\frac{e^{-t(s+a)}}{s+a}\,\Big|_0^\infty$$

$$= \frac{1}{s+a}$$

C.1 Laplace Tables

Transformation tables give the Laplace transforms of many time functions. These tables are derived from Eq. (C.1) and the properties of the frequency-domain functions. The tables can be used to find not only the Laplace transforms of the tabulated time functions, but also the time functions (inverse transformation) from their corresponding Laplace transforms.

Table C-1 includes the Laplace transforms of some functions used for the transient analysis of simple machine problems. By using Laplace transforms, a machine's transient response can easily be determined by the following procedure:

1. Write the machine's equations in their differential form.
2. Find the Laplace transform of the differential equation and bring it to the standard form that is available in the tables.
3. Take the inverse transform of (2).

The solution of differential equations using Laplace transforms is demonstrated by the following example.

TABLE C-I Laplace Tables		
No.	$f(t)u(t)*$	$F(s)$
1	k	$\dfrac{k}{s}$
2	e^{-at}	$\dfrac{1}{s+a}$
3	$\dfrac{1}{a}(1-e^{-at})$	$\dfrac{1}{s(s+a)}$
4	$L\dfrac{di}{dt}$	$LsI_{(s)} \pm LI_0^{\dagger}$
5	$\dfrac{1}{C}\int i\,dt$	$\dfrac{I_{(s)}}{sC} \pm \dfrac{V_{Co}}{s}^{\dagger}$
6	t	$\dfrac{1}{s^2}$

* The unit-step function $\omega(t)$ ensures that the time function does not exist at negative time (t)

† Where I_0 and V_{Co} are, respectively, the initial inductor current and the initial capacitor voltage.

EXAMPLE **C-2** Refer to the circuit in Fig. C-1(a). After the switch is closed, find the current $i(t)$.

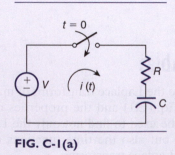

FIG. C-1(a)

SOLUTION

1. From KVL, we have

$$V = Ri(t) + \frac{1}{C}\int i(t)dt$$

2. Taking the Laplace transform of the above, we obtain

$$\frac{V}{s} = I(s)R + \frac{I(s)}{sC}$$

Solving for $I(s)$, and after simplification, we obtain

$$I(s) = \frac{V}{R\left(\dfrac{1}{CR} + s\right)}$$

3. Taking the inverse Laplace transform (No. 2 of Table C-1) of the last expression, we find the current as a function of time:

$$i(t) = \frac{V}{R}e^{-t/RC}$$

The current waveform is sketched in Fig. C-1(b).

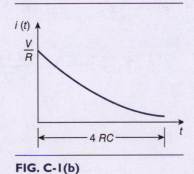

FIG. C-1(b)

C.1.1 Initial- and Final-Value Theorems

When the Laplace transform $\mathbf{F}(s)$ of a function $f(t)$ is known, then the value of the function at $t = 0$ and at $t = \infty$ can be found as follows:

$$f(t)\bigg|_{t=0} = \lim_{s \to \infty} s\mathbf{F}(s) \tag{C.3}$$

$$f(t)\bigg|_{t=\infty} = \lim_{s \to 0} s\mathbf{F}(s) \tag{C.4}$$

Equations (C.3) and (C.4) are known, respectively, as the initial- and the final-value theorems. One can easily remember these equations by noting that when $t \rightarrow 0, s \rightarrow \infty$ and vice versa. From Eqs. (C.3) and (C.4), it can be said that the time domain begins where the frequency domain ends and vice versa. These equations are often used to evaluate a machine's variable at starting ($t = 0$) and at steady state ($t = \infty$) from its equivalent Laplace transform. The Laplace transform can be easily established from basic principles. Furthermore, from Eq. (C.3) and the definition of a coil's reactance, it can be seen why a coil presents an infinite imped-ance to the propagation of upstream electrical impulses or spikes. That is, electri-cal disturbances (switching transient and lightning strikes) cannot pass through a transformer.

EXAMPLE **C-2**   The Laplace transform of a motor's speed is given by

$$\omega(s) = \frac{125}{s(s + 2)} \text{ rad/s}$$

Determine the speed of the motor at $t = 0$ and at $t = \infty$.

SOLUTION

$$\omega(0) = \lim_{s \to \infty} s \frac{125}{s(s + 2)} = 0$$

$$\omega(\infty) = \lim_{s \to 0} s \frac{125}{s(s + 2)} = \underline{62.5 \text{ rad/s}}$$

The solid-state devices (diodes, transistors, thyristors, and insulated gate bipolar transistors) constitute the basic blocks of motor controls. They are extensively discussed in Additional Students' Aid on the Web, (Section DW).

The following formulas are very useful in the economic analysis of various subjects and are based on simple interest compounded once per year.

E.1 "Future Worth" of a One-Time Investment

The future worth of a one-time investment is given by

$$F = P(1 + r)^n \qquad\qquad \text{(E.1)}$$

where

F = the future worth

P = the present worth compounded once per year

r = the rate of interest compounded once per year

n = the number of years.

E.2 "Future Worth" of Equal Annual Installments

The future worth of equal annual installments is given by

$$F = A\left[\frac{(1 + r)^n - 1}{r}\right] \qquad\qquad \text{(E.2)}$$

where

A represents the amount of equal annual installments

$F, r,$ and n have the same meaning as in Eq. (E.1).

E.3 "Present Worth" of Equal Annual Installments

The present worth of equal annual installments is given by

$$P = A\left[\frac{(1 + r)^n - 1}{r(1 + r)^n}\right] \qquad\qquad \text{(E.3)}$$

where

P is the present worth of the equal annual installments

A, r, and n have the same meaning as in Eq. (E.2).

E.4 "Present Worth" of Uniform Incremental Increases in the Annual Deposits

The present worth that results from uniform annual deposit increases is given by

$$P = A \left[\frac{(1 + r_x)^n - 1}{r_x(1 + r_x)^n} \right] \qquad \text{(E.4)}$$

where

$$r_x = \frac{1 + r}{1 + r_e} - 1 \qquad \text{(E.5)}$$

where r_e is the uniform annual percentage increase of the parameter under consideration. The other parameters are the same as in the previous expressions. Equations (E.4) and (E.5) are very useful in evaluating the present worth of equipment energy consumption, while the cost of energy consumption increases at a constant annual rate.

EXAMPLE **E-1**

One thousand dollars is deposited in a bank at a simple interest rate of 10% for five years. Find its worth at the end of the fifth year.

SOLUTION

Substituting the given data into Eq. (E.1), we obtain

$$F = 1000(1 + 0.1)^5 = \$1610.50$$

EXAMPLE **E-2**

At the beginning of each year for 10 consecutive years, a businessperson makes a bank deposit of $4000. The annual interest rate is 4%. Find:

a. The worth of these deposits at the end of 10 years.
b. The equivalent present worth of these deposits.

SOLUTION

a. Substituting the given data into Eq. (E.2), we obtain

$$F = 4000 \left(\frac{1}{0.04} \right) [(1 + 0.04)^{10} - 1]$$

$$= \$48,024.40$$

b. The present worth is found by substituting the given data into Eq. (E.3).

$$P = 4000 \left| \frac{(1 + 0.04)^{10} - 1}{0.04(1 + 0.04)^{10}} \right| = \$32,445$$

A 25-year mortgage of $50,000 is obtained at a 4% annual interest rate. Find: **EXAMPLE E-3**

a. The monthly installments.
b. The actual money paid by the borrower over the 25-year period.
c. The effective worth of the mortgage at the end of 25 years.

SOLUTION

a. From Eq. (E.3), the equivalent annual payments are

$$A = \frac{50,000(0.04)(1 + 0.04)^{25}}{(1 + 0.04)^{25} - 1} = \$3,201/\text{year}$$

Thus, the monthly payment is

$$\frac{\$3201}{12} = \$266.72/\text{month}$$

b. The actual money paid by the borrower over the 25-year period is

$$12 \times 25 \times 266.72 = \$80,015$$

c. The effective worth of the mortgage at the end of 25 years is found from Eq. (E.1).

$$F = 50,000(1 + 0.04)^{25} = \$133,292$$

That is, you will receive $133,292 if you deposited $50,000 in the bank and left it there for 25 years at an annual interest rate of 4%.

EXAMPLE **E-4**

Find the present-worth cost of 1 kW of power that is continuously consumed over a five-year period, assuming that the energy cost is 12¢/kWh and increases annually at 3%. The nominal interest rate is 4% per year.

SOLUTION

From Eq. (E.5),

$$r_x = \frac{1 + 0.04}{1 + 0.03} - 1 = 0.0097$$

and

$$A = (0.04)(365 \times 24) = \$350.40/\text{kW}$$

Substituting into Eq. (E.4), we obtain

$$P = 350.4 \left| \frac{(1.0097)^5 - 1}{0.0097 (1.0097)^5} \right| = \underline{\underline{\$1702.15/\text{kW}}}$$

The following is a brief description of the external characteristics of photovoltaic (PV) cells and how they can be used to evaluate the energy production and corresponding costs. The underlying concepts are discussed in Additional Students' Aid on the Web (Appendix, Section FW).

F.1 Equivalent Circuit of a Photocell

A typical *I-V* characteristic of a photocell is shown in Fig. F-1(a). The density of light is measured in foot-candles $\left(\dfrac{\text{Lumens}}{\text{ft}^2}\right)$ or $\dfrac{\text{Lux}}{\text{cm}^2}$, 1 Lux/cm² = 0.0929 foot-candle. It should be noted that at a particular light density, its intersection with the *x* and *y* axis designate, respectively, the cell's open-circuit voltage (V_{oc}) and short-circuit current (I_{sc}). Then from Thévenin's theorem, the equivalent circuit of a photovoltaic cell is as shown in Fig. F-1(b).

$$R_{th} = \frac{V_{oc}}{I_{sc}} \qquad \text{and} \qquad V_{th} = V_{oc}$$

At 10 $\dfrac{\text{Lux}}{\text{cm}^2}$, we obtain

$$V_{oc} = 0.3\ V \qquad \text{and} \qquad I_{sc} = 0.15\ mA$$

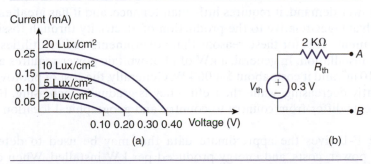

FIG. F-1 Photovoltaic cell: **(a)** *I-V* characteristics. **(b)** Thévenin's equivalent at 10 Lux/cm².

Thus,

$$R_{th} = \frac{0.3}{0.15 \times 10^{-3}}$$

$$= 2 \text{ k}\Omega$$

The load resistance (R_L) for maximum power transfer is equal to R_{th}. In this case,

$$R_L = 2 \text{ k}\Omega$$

In practice, many cells are connected in series and in parallel in order to increase their output current and voltage. Such a setup constitutes a module, and several modules form an array. In large photovoltaic systems, a tracking device is often used to tilt the PV arrays toward the sun. This increases their output energy by a factor in the range of 30% to 100%.

The efficiency of the PV cells is about 20%. When, however, dirt or part of the module is in the shade, the power produced is substantially reduced.

F.2 Economic Considerations

The use of photovoltaics for producing electricity is not economical, but it is very popular for many reasons: It is renewable, it generates electricity at periods of peak power demand, it requires little maintenance, and it has great environmental advantages relative to the production of electricity through fossil fuels and/or uranium. It is for these reasons that governments and/or utilities subsidize their installation. In general, a kW of photovoltaic power requires a space of about 10 m², and it costs about $8000/kW. Generally, the cost of photovoltaics is constantly decreasing, while their efficiency is gradually increasing. Both of these aspects differ from country to country and geographical location where installed.

Table F-1 gives the approximate data that may be used to determine potential power, costs, and energy produced per kW installed. When clouds, leaves, or dirt cover part of a photovoltaic module, the unit's output power is reduced.

TABLE F-I Highlights of photovoltaic modules

No.	Description	Approximate Data	Notes
1	Power available	1 kW/10 m² of surface area	
2	Cost per kW installed	$8000	
3	Payback period	4–40 years, depending on the amount of subsidy	
4	**City**	**Potential kWh/kW**	
	Montreal (Quebec)	1185	
	Moscow (Russia)	803	
	Berlin (Germany)	848	
	Rome (Italy)	1283	
	Los Angeles (United States)	1485	
	Cairo (Egypt)	1635	

Based on data from Natural Sources Canada, 2007. Photovoltaic Potential and Solar Resources Map of Canada

EXAMPLE F-I

A home in Los Angeles has a roof area of 100 m², and the total cost of installing a photovoltaic system is $8000/kW. The cost of energy is $0.15/kWh. Estimate:

a. The total power produced.

b. The payback period.

c. The payback period if the utility buys the energy produced at $0.60/kWh.

d. Repeat (c) above when the installation is equipped with a tracking system whose cost is $6000 and provides an increase in output energy of 50%.

SOLUTION

a. Since 1 kW of PV power requires 10 m² of roof space, then the total power produced is

$$P = \frac{1(100)}{10} = 10 \text{ kW}$$

b. The installation cost is 10(8000) = $80,000.
The potential output energy from Table 1 is

$$10(1485) = 14,850 \text{ kWh/year}$$

and the corresponding savings (C_1) are

$$C_1 = 14,850 \,(0.15) = \$2227.50/\text{year}$$

The payback period (T_1) is

$$T_1 = \frac{80,000}{2227.5} = 35.9 \text{ years}$$

c. The savings (C_2) at $0.60/kWh are

$$14,850(0.60) = \$8910/\text{year}$$

and the payback period (Y_2) is

$$Y_2 = \frac{80,000}{8910} = 8.98 \text{ years}$$

d. When a tracking system is installed, the output energy is

$$1.5(1485)(10) = 22,275.00 \text{ kWh/year}$$

The corresponding savings (C_2) are

$$C_2 = 22,275.00(0.60) = \$13,365.00/\text{year}$$

and the payback period (T_2) is

$$T_2 = \frac{86,000}{13,365.0} = 6.44 \text{ years}$$

EXAMPLE **F-2** Refer to Fig. F-1(a). For a light intensity of 5 Lux/cm^2, calculate
a. The load impedance for maximum power to the load.
b. The power and efficiency of the cell in part (a).
c. The output voltage and current when the circuit efficiency is 80%.

SOLUTION

a. $V_{oc} = 0.2 \text{ V}, I_{sc} = 0.10 \text{ mA}$

$$R_{th} = \frac{0.2}{0.10 \times 10^{-3}} = 2 \text{ k}\Omega$$

From the maximum power transfer theorem,

$R_L = R_{th}$

$R_L = 2 \text{ k}\Omega$

b.

$$P_L = I^2 R_L = \left(\frac{V}{R_L + R_{th}}\right)^2 R_L$$

$$= \left(\frac{0.2}{2(2 \times 10^3)}\right)^2 (2 \times 10^3) = 5 \text{ }\mu\text{W}$$

$$\eta = \frac{R_L}{R_{th} + R_L} = \frac{2 \times 10^3}{(2 + 2)10^3} = 50\%$$

c. For a circuit efficiency of 0.80,

$$\frac{R_{L_1}}{2 \times 10^3 + R_{L_1}} = 0.8$$

From which $R_L = 8 \text{ k}\Omega$,

$$I = \frac{0.2}{(2 + 8)10^3} = 20 \text{ }\mu\text{A}$$

$$V_o = 8(20)10^3 \times 10^{-6} = 0.16 \text{ V}$$

G.I Constants and Conversion Factors

Constants

Permeability of free space:	$\mu_0 = 4\pi \times 10^{-7}$ H/m
Permittivity of free space:	$\varepsilon_0 = 8.854 \times 10^{-12}$ F/m
Velocity of light:	$C = 2.998 \times 10^8$ m/s

Conversion Factors

Energy	1 Joule (J) = 1 watt second
	= 0.7376 foot-pound (ft-lb)
	= 10^7 ergs
	= 9.48×10^{-4} British Thermal Units (BTU)
Length:	1 m = 3.281 ft
	= 39.37 in.
Mass:	1 kg = 2.205 lb_m
	= 0.0685 slug
Force:	1 newton (N) = 0.225 lb_f
	= 7.23 poundals
	= 10,000 dynes
	= 102 grams
Torque:	1 newton · meter (N · m) = 0.738 ft · lb
	= 141.7 oz · in.
Power:	1 watt (W) = 1.341×10^{-3} hp
Moment of inertia:	1 kg · m^2 = 23.7 lb · ft^2
	= 0.737 slug · ft^2
Magnetic flux:	1 weber (Wb) = 10^8 maxwells (lines)
Magnetic flux density:	1 tesla (T) = 1 Wb/m^2
	= 10,000 gauss
	= 64.5 kilolines/m^2
Magnetizing force:	$\dfrac{1 \text{ ampere-turn}}{\text{meter}} = \dfrac{0.0254 \text{ ampere-turn}}{\text{in.}}$

G.2 ANSI Standard Device

Function Numbers

Table G-1 presents a partial list of *American National Standards Institute* (ANSI) standard device function numbers.

Table G-1	ANSI standard device function numbers (partial list)
Device Number	**Definition and Function**
26	Apparatus thermal device functions when the temperature of the shunt field or the amortisseur winding of a machine, or that of a load-limiting or load-shifting resistor, or of a liquid or other medium, exceeds a predetermined value; or if the temperature of the protected apparatus, such as a power rectifier, or of any medium, decreases below a predetermined value.
27	Undervoltage relay is a device that functions on a given value of undervoltage.
32	Directional power relay is one that functions on a desired value of power flow in a given direction, or upon reverse power resulting from arc-back in the anode or cathode circuits of a power rectifier.
38	Machine's bearing temperature protection.
49	Machine, or transformer, thermal relay is a device that functions when the temperature of an ac machine armature, or of the armature or other load-carrying winding or element of a dc machine, or converter or power rectifier or power transformer (including a power rectifier transformer), exceeds a predetermined value.
50	Instantaneous overcurrent or rate-of-rise relay is a device that functions instantaneously on an excessive value of current, or on an excessive rate of current rise, thus indicating a fault in the apparatus or circuit being protected.
51	AC time overcurrent relay is a device with either a definite or an inverse time characteristic that functions when the current in an ac circuit exceeds a predetermined value.
52	AC circuit breaker is a device used to close and interrupt an ac power circuit under normal conditions or to interrupt this circuit under fault or emergency conditions.
55	Power-factor relay is a device that operates when the power factor in an ac circuit is above or below a predetermined value.
56	Field application relay is a device that automatically controls the application of the field excitation to an ac motor at some predetermined point in the slip cycle.

Table G-1	(Continued)
Device Number	Definition and Function
59	Overvoltage relay is a device that functions on a given value of overvoltage.
63	Liquid or gas pressure, level, or flow relay is a device that operates on given values of liquid or gas pressure, flow, or level, or on a given rate of change of these values.
81	Frequency relay is a device that functions on a predetermined value of frequency—either under or over or on normal system frequency—or rate of change of frequency.
86	Locking-out relay is an electrically operated hand or electrically reset device that functions to shut down and hold an equipment out of service on the occurrence of abnormal conditions.
87	Differential protective relay is a protective device that functions on a percentage or phase angle or other quantitative difference of two currents or of some other electrical quantities.

Circuits

For any chapter, refer to the corresponding section of the Internet.

1 Boylstad, R. *Electronic Devices and Circuit Theory*. Columbus, OH: Charles Merrill, 1982.

Electric Machines

2 Chapman, S. *Electrical Machinery Fundamentals*. New York: McGraw-Hill, 2012.

3 Fitzgerald, A. E., Kingsley, C., and Umans, S. *Electric Machinery*, 6th Ed. New York: McGraw-Hill, 2003.

4 General Electric Company. *Permanent Magnet Manual*. Edmore, MI: Magnetic Materials Business Section.

5 Gross, C. *Electric Machines*. Boca Raton, FL: CRC Press, 2006.

6 Sarma, M. S. *Electric Machine, 2e.* Stamford, CT: Cengage Learning, 1997.

7 Slemon, G. R., and Straughen, A. *Electric Machines*. Reading, MA: Addison-Wesley, 1980.

8 Wild, T. *Electric Machines, Drives and Power Systems.* Englewood Cliffs, NJ: Prentice Hall, 2006.

Controls

9 Harwood, P. B., *Control of Electric Motors*. New York: John Wiley, 1952.

10 Kosow, I. L., *Control of Electric Machines*. Englewood Cliffs, NJ: Prentice Hall, 1973.

Power Systems

11 Stevenson, W. *Elements of Power System Analysis*. New York: McGraw-Hill, 1982.

Electrical Safety

12 John Webster Medical Instrumentation Applications and Design. Hoboken, NJ: John Wiley, 2010.

Step Motors

13 *Bodine Electric Company: Handbook*. Chicago, IL.

Control Motors

14 Baldor Motors (AC, DC and Servo Motors). International division, member of the ABB group.

Chapter 1

1-1 a. 133%; b. 7.5 s.

1-2 a. 5.77 V; b. 0 V; c. 8.33 W.

1-3 73.52 V.

1-4 $I_1 = 10.29$ A, $I_2 = 0.37$ A, $I_3 = 1.1$ A.

1-5 13.6 V.

1-6 1.29 A.

1-7 a. 0.349 J; b. 1.0 ms, 0.28J.

1-8 2 ms, ∞.

1-9 $S_T = 19.48 \underline{/-13.1}$ kVA, $P_T = 18.97$ kW,
$Z = 2.22 \underline{/-13.1}$ Ω, $I = 93.67 \underline{/-13.1}$ A.

1-10 $S_T = 30.34 \underline{/23.2}$, $P_T = 27.89$ kW,
$L_T = 1.49$ mH, $I_T = 145.86 \underline{/-23.2}$ A.

1-11 a. 37.1 kVAR; b. $2293/year

1-12 a. $705.88; b. $930.

1-13 36.04 kVAR.

1-16 a. 736.12 W; b. 802.37 W; c. 8.26%;
d. 887.43 VA.

1-17 $I_T = 26.84$ A; b. 5.05 kW.

1-19 b. 1.45 mV/m, rms.

1-20 $-(UB)(L_1 + L_3(\sin\theta))\alpha_y$

1-21 a. 3023.9; b. 63.7 μWb(max);
c. 0.57 T (rms); d. 150.9 A/m; e. 90.1 A.

1-22 0.75 mWb.

1-23 $\dfrac{-\mu_0 b^2 i_1 i_2}{\pi a(a+b)}\alpha_y$

1-25 b. (1) 0.34 H; (2) 0.67 J; (3) −59.57 N;
(4) 0.22 T.

1-26 a. 1.68 H; b. 83.96 J; c. 339.5 J.

1-27 a. $-2L_{12}I_m^2 \sin 2\beta \sin^2(\omega t - \theta)$ N · m

1-29 a. 0.5 V/m, 1.33 μ-A/m. b. 226;
c. 358.1 A/m

Chapter 2

2-1 b. 222.93 V.

2-2 b. 2509.8 V; c. 4.57%, 0.949;
d. 0.05 $\underline{/50.2°}$ pu.

2-3 a. 300.96 A; b. 0.98/0.97; c. 0.90 leading.

2-4 a. 1551.6 W, 387 W; b. 322.5 W.

2-5 a. 240 V; b. 2.04%, −0.7%, 4.2%.

2-6 a. 0.013 pu, 4.07%; b. 1.05 pu.

2-7 a. $I_{AB} = 4$ A; b. $I_{AB} = I_{CB} = 3.46$ A;
c. $I_a = 167.1$ A, $I_{AB} = I_{BC} = I_{CA} = 11.1$ A.

2-8 a. $0.016 + j0.026$ pu, $0.043 + j0.087$ pu;
b. 55.2 kW; c. 27.4 kV.

2-9 a. 90.7 kVA; b. $I_a = 70.77$ A, $I_A = 4.68$ A,

$I_b = 95.2$ A, $I_B = 6.3$ A,

$I_c = 68$ A, $I_C = 4.5$ A.

2-10 a. (1) 277.1 A, 554.3 $\underline{/120°}$ A,
−277.1 $\underline{/60°}$ A; (2) 0 V; (3) fuse in
line c will blow; b. (1) 277.1 A,
277.1 $\underline{/-120°}$ A, 277.1 $\underline{/120°}$ A; (2) 277.1 V;
c. (1) $i_c = \infty$ A; (2) 0 V; (3) fuse in line
c will flow; d. (1) 277.1 A; (2) 277.1 V.

2-11 a. 25 kVA; b. 20 kVA.

2-12 a. 680 kVA, 70.6%; b. 0.99;
c. 10.75 kA.

2-13 a. 4350.8 V; b. 1.365 A, 3123 A;
c. 655 kVA, 360 kVA; d. 312 A;
e. 3.3 kA.

2-14 (1) 441.9 V, 4 A, 1202.8 A;
(2) 0 V, 35.6 A, 10.7 kA;
(3) 489.2 V.

2-15 a. $54,259, $54,516; b. $44,928, $47,129;
c. $48,660, $49,928.

Chapter 3

3-1 a. 3.33%; b. 4; c. $I_a = 11.8$ A.

3-3 a. (1) 1200 r/min; (2) 0.05 pu;
(3) 309.84 A; (4) 64.82 A, 0.98 Pf,
200.6 N · m, 0.90; (5) $0.05 + j0.07$ pu,
$0.05 + j0.12$ pu; (6) 0.2425 pu, 213.2 A;
(7) 447.3 N · m; b. 312.58 A.

3-4 a. 0.05 pu; b. 0.25 pu; c. 1.224 pu;
d. 2.6 pu.

3-5 a. $R_1 = 0.08$ Ω, $R_2 = 0.13$ Ω, b. 230.8 N · m.
$X_1 = 0.17$ Ω, $X_2 = 0.25$ Ω
$R_m = 2.47$ Ω, $X_m = 20.67$ Ω;

3-6 a. 0.382 N · m/rad/s; b. 0.155 Ω/phase;
c. 1665 r/min.

3-8 a. $I_A = 199.1$ A, $I_B = 166.1$ A,
$I_C = 129.2$ A, $I_n = 75.9$ A;
b. $I_A = 102.8$ A, $I_B = 94.3$ A,
$I_C = 76.5$ A, $I_n = 75.93$ A.

3-9 a. 230 kVAR; b. 0.99 lag.

3-10 a. 0.993 pu; b. 1793 kVAR; c. 1.9 pu.

3-11 a. 16.8 A; b. 112.1 N · m.

3-12 a. 0.81; b. 1.71 pu; c. 266.5 kVAR.

3-13 a. 1140 r/min; b. 2.2 s; c. 1.34 s.

3-14 a. 949.7 N · m.

3-17 84.85 V, 125 Hz.

3-19 $11,414, $6,304.

3-20 $10.39 \, \underline{/-147°}$ A, −7.24 kW

Chapter 4

4-1 a. 29.84 A, 0.99 lagging; b. 8.86 N · m;
c. 0.18 N · m; d. 8.70 N · m.

Chapter 5

5-1 a. 520.8 V, L-L; b. 5.7°; c. 5.13 kW,
3.57 kW.

5-2 a. $49.23 \, \underline{/25.8°}$; b. 4289.8 V, L-L, −5.6°;
c. $0.02 + j0.10$ pu.

5-3 a. 7.5 A; b. 99.5 A.

5-4 a. −86.2°; b. −3.8°.

5-5 a. (1) 2441.97 V, L-L;

(2) −25.7°;

(3) 9.6 A;

(4) 203.96 A;

(5) 777.2 kW, 777.2 kVA, 0 kVAR;

b. $1.25 I_f$: (1) 3052.46 V, L-L;

(2) −20.3°;

(3) 12 A;

(4) 240.5 A;

(5) 777.2 kW, 916.6 kVA,
485.9 kVAR;

$0.751 I_f$: (1) 1831.5 V, L-L;

(2) −35.4°;

(3) 7.2 A;

(4) 245.1 A;

(5) 777.2 kW, 934 kVA,
518 kVAR.

5-6 a. 1203.7 V, L-L; b. 21.3 $\underline{/-103°}$ A;
c. 7.8 A; d. 24.3 $\underline{/-121.2°}$ A, −18.2°;
e. 1367.2 V, L-L.

5-7 a. 4 poles; b. 0.03 + j0.35 pu;
c. 14.45 kW; d. 2257.5 N · m.

5-8

Questions	Parameter	Operating at Rated Conditions, Unity Pf	Constant Field Excitation	
			Increase in the Input Power of the Prime Mover by 25%	Decrease in the Input Power of the Prime Mover by 25%
1	Excitation voltage in pu	$\sqrt{2}$	$\sqrt{2}$	$\sqrt{2}$
2	Torque angle	−45°	−62.1°	−32°
3	Armature current in pu	1.0 $\underline{/0°}$	1.30	0.78
4	Field current in A	142	142	142
5	Real power in pu	1.0	1.25	0.75
	Reactive power in pu	0	0.34	0.20

5-9 a. 1100 V, 7 A; b. 15; c. 18.5 A.

5-10 a. 1.0; b. −27.5°; c. 1.26 pu;
d. 298.4 kW, 27.5%.

5-11 a. 39.8°; b. 62%; c. 0.18 pu.

Chapter 6

6-1 120.96 N · m.

6-2 a. 48.41 A, 132.7 N · m, 781.10 r/min;
b. 126.10 N · m, 822.1 r/min.

6-3 a. 1335.33 A; b. 1.74 Ω; c. 82.62 Ω

6-4 a. 11.7%; b. 7.4%.

6-5 a. 1946.3 r/min; b. 71.1 N · m; c. 14.1 kW;
d. $281.13\left(\dfrac{i_a}{25 + i_a}\right)$.

6-6 2000 r/min.

6-12 a. 7.58; b. 278.4 kVA; c. 0.74 lag.

6-13 a. 53.67°; b. 516.5 A; c. 0.57 lag.

Appendixes

A-2 a. 26.74 kW, 1.73 kVAR, 26.80 $\underline{/3.7}$ kVA;
b. 32.23 $\underline{/-3.7}$ A.

A-3 0.81 lagging, 102.86 A.

A-4 a. 277.13 V/phase; b. 27.71 $\underline{/-36.9°}$ A;
c. 27.71 A; d. 0.8 lagging; e. 18.43 kW;
f. 30 $\underline{/36.9°}$ Ω.

A-5 0.93, 557.3 V, L-L.

A-6 a. 24 A, 41.57 A; b. 9.98 kW, 19.95 kW,
29.93 kW; c. −1.27 kW, 4.72 kW,
3.45 kW.

A-7 48 A.

A-9 a. 446.9 V, L-L;   b. 415 kVAR.

A-10

Department	Complex Power (kVA)	Average Power (kW)	Inductance (mH)	Current (A)
A	40.82 $\underline{/19.2}$	38.55	4.92	49.1 $\underline{/-19.2}$
B	10.65 $\underline{/28.5}$	9.36	27.38	12.81 $\underline{/-28.5}$
Total	51.36 $\underline{/21.1}$	47.91	4.3	61.78 $\underline{/-21.1}$

B-1 a. $0.0015 + j0.009 \, \Omega/\phi$;   b. 27 MVA.

B-2 a. Generator 0.1 pu,
Transformer $0.15 + j0.062$ pu,
Induction Machine 0.89 pu,
Synchronous Machine 1.11 pu.

A

Accelerating torque, 279
Alternators
  open-circuit characteristic
    (OCC), 393–394
  short-circuit characteristic
    (SCC), 394–396
  synchronous machines, 392–397,
    399–402
  zero-power-factor (ZPF) lagging
    characteristic, 396–397, 399–402
American National Standards
  Institute (ANSI) standard
  device function numbers,
  612–613
Ampere's law, 255–256
Apparent power, 45, 177–178
Arc flash, 546
Armature current
  synchronous machines, 365,
    380–381, 383–384
  DC machines, 440–442
  demagnetization, 383–384
  field current versus, 380–381,
    383–384
  magnetization effects on,
    383–384, 440–442
Armature reaction (AR),
  440–442, 462
Armature resistance, 444–461
Armature rheostat, 447
Armature windings, 251–252,
  431–432
Asynchronous generators, 298–301
  equivalent circuit, 299–300
  excitation of, 298–299
  induction machines, 298–302
  shaft rotation of, 298
Autotransformers, 153, 211–219, 302
Auxiliary winding
  compound DC machines, 459–462
  permanently connected, 337–339

removed after starting, 340–341
single-phase (1-ϕ) motors, 337–341
starting and running capacitor
  with, 340–341
starting capacitor with, 339–340
Average (real) power, 19, 43–44,
  570–571

B

B-H curves
  electromagnetic circuit analysis,
    100–103
  magnetic domains and, 75–78
Base parameters, 178–179
Base values, 587–590
Bonding, 543–544
Breakdown (maximum) torque,
  271–273
Brushless synchronous motor
  schematics, 524–530

C

Capacitance, 23–29
Capacitors
  auxiliary winding with, 340–341
  types of, 28–31
Cathode corrosion, 549–550
Charge-capacitor relationship, 24
Circuit breakers, 516–517
Coefficient of coupling, 113–114
Commutation, 442–444
Complex power, 45–46, 572
Compound DC generators, 472–473
Conductors, contact with live, 545
Constant of proportionality, 80
Constant torque, 279
Contacts (electrical), 504–506
Control configurations
  single-phase (1-ϕ) motors, 348
  three-phase (3-ϕ) motors,
    302–319

Control schematics, 497–536
  design of, 498–499
  electrical contacts, 504–506
  electromagnetic relays, 499–503
  indicating lights, 506
  industrial timers, 507–509
  protection and, 512–519
  stop-start pushbuttons, 506–507
  synchronous motors, 524–530
  temperature sensors, 509–512
  thermal overload relays, 503–504
  three-phase induction motors,
    519–524
  wiring diagrams, 498
Controllers, 311
Conversion factors, 611
Curie point (temperature), 76
Current
  armature, 365, 380–381, 383–384
  base (per-unit), 589
  displacement, 27–28
  excitation, 169–170, 210–211
  field, 366–367, 378–389
  induction machines, 304,
    302–307
  inrush, 173–174
  Kirchhoff's law (KCL), 39–41, 97
  primary, 171–173
  three-phase (3-ϕ) systems, 190
  transformers, 169–174, 190,
    210–211
  starting, 19, 27, 302–307
  unbalanced, 557
Current transformers (CT),
  228–230
Cylindrical rotor generators,
  389–404
  alternators, 392–397
  equivalent circuits, 389–391
  parameter measurement,
    397–404
  phasor diagrams, 389–391
  regulation, 392–393

Cylindrical rotor machines,
    see Synchronous machines
Cylindrical rotor motors, 355–389
    equivalent circuits, 363–366
    field current, 366–367, 378–389
    phasor diagrams, 367–373
    power and torque development,
        374–378
    principles of operation, 361
    speed control, 378
    starting, 362–363

D

DC excitation, 128–129
DC machines, 425–496
    armature current reaction,
        440–442
    commutation, 442–444
    equivalent circuits, 444–462
    flux distribution, 438–440
    generators, 432–445, 464–476
    magnetic system, 438–439
    motors, 427–432, 433–464
    mutual inductance and,
        436–438
    open-circuit characteristics,
        464–476
    power considerations, 433–435
    principles of operation, 429–433
    rectifiers, 477–489
    speed control, 477–489
    starting, 462–464
    steady-state analysis, 427–435
    voltage and torque relationships,
        436–438
Delta- (Δ-) connected load
    configurations, 190, 566–567
Delta–delta (Δ–Δ) transformers,
    204
Delta–star (Δ–Y) load
    transformations, 567–568
Delta–star (Δ–Y) transformers,
    194–203
Demagnetization (degaussing)
    armature current, 383–384
    British fleet example, 83–84
Differential range, 511–512

E

Economical analysis, equations
    for, 603–606
Eddy-current loss, 103–104

Efficiency
    induction machines, 284–285
    transformers, 174–176, 237
Electric field flux density, 11
Electric field intensity, 10–11
Electric potential, 3–7
    average value of waveforms, 6–7
    effective value of waveforms, 5–6
    instantaneous value of
        waveforms, 4–5
    root-mean-square (rms) value,
        5–7
Electrical circuits
    capacitance, 23–29
    electric field parameters, 10–11
    electric potential of, 3–7
    harmonics, 57–61
    impedance, 32–34
    inductance, 16–23
    Kirchhoff's laws, 37–41
    Ohm's law, 35–37
    phasors (complex numbers)
        and, 7–10
    power and, 41–46
    power factor, 47–57
    resistance, 12–16
    supercapacitors, 29–31
    theorems for, 61–63
    voltage, 3–4
Electrical injuries, 539–540
Electrical pressure, 3
Electrolyte capacitors, 28–29
Electromagnetic circuits, graphical
    analysis of, 100–102
Electromagnetic relays, 499–503
Electromagnetism, 134–135
Electromotive force (emf), 2
Electrostatic capacitors, 28
Energy
    content of magnetic materials,
        93–95
    electric circuits, 19
    exerted on current-carrying
        conductor, 123–124
    force and, 123–127
    magnetic circuits, 123–127,
        131–133
    stored in a coil, 19, 131–133
    torque and, 125–127, 133–134
Energy consumption, 555–560
    DC motor voltage, 559
    harmonics reduction, 559
    heat losses, 559–560

heat pump efficiency, 560
high-efficiency motors, 358
lightning, 557
power demand reduction, 558
power-factor improvement, 558
recuperating heat rejection, 559
synchronous motors, 557
unbalanced line currents, 557
unbalanced voltages, 556–557
variable-speed drives, 558–559
Energy storage, safety of, 546–548
Equivalent circuits
    asynchronous generators, 299–300
    DC machines, 444–462
    induction machines, 264–268,
        290–300
    magnetic coils, 109–110
    magnetically coupled coils,
        119–123
    parameter measurement, 290–298
    photovoltaic (PV) cells, 607–608
    polarity and, 119–123
    power angle, 365
    rotor parameters, 366
    running rotor conditions, 266–268
    single-phase (1-φ) motors, 334–335
    stationary rotor conditions,
        264–266
    stator parameters, 364–365
    synchronous machines, 363–366,
        389–391
    transformers, 160–169
Excitation current
    harmonics of, 210–211
    transformers, 169–170, 210–211
    waveforms of, 169–170
Excitation voltage, 364

F

Ferroresonance, 233–234
Field current
    armature current magnitude
        versus, 380–381, 383–384
    cylindrical rotor motors, 378–389
    DC generators, 466–470
    effective, 466–470
    power factor versus, 378–380
    reactive power versus, 381–382
    real power and, 382
Field resistance, 444–456
Field rheostat, 447–448
Final-value theorem, 599–600
Floating neutral, 544–545

Flux
   armature reaction, 440–442
   DC machines, 438–442
   density, 11, 69, 439–440
   distribution, 438–440
   electric, 11
   magnetic system effects on,
      438–439
   magnetic, 64–69
   primary current in transformers
      and, 171–173
   voltage–flux relationship, 91–92
Force
   energy and, 123–127
   exerted by an electromagnet,
      127–131
   torque and, 125–127
Fringing, 67, 97
Function numbers of ANSI
   standard devices, 612–613
Fuses, 514–515

G

Gauss's law, 23
Generators
   asynchronous, 298–301
   DC, 432–445, 464–476
   synchronous, 389–404
Ground resistance, 544
Grounding systems, 542–544

H

Harmonics
   excitation current, 210–211
   reduction of, 559
   steady-state, 58–61
   three-phase (3-ϕ) systems, 210–211
   transient, 58
Heat losses, 559–560
Heat pump efficiency, 560
Heat rejection, recuperation of, 559
Hysteresis loop, 77–78
Hysteresis loss, 104–106

I

Ideal transformers, 157–160
Impedance
   base (per-unit), 589
   capacitance and, 25
   current reactance and, 32–34
   inductance and, 17–18

leakage, 220, 236–237
   transformers, 220, 236–237
In-line resistors or inductors, 302
Indicating lights, 506
Inductance
   coefficient of coupling, 113–114
   electrical circuits and, 16–23
   magnetic circuits and, 111–119
   mutual, 111–113
   self-, 16–23, 111–112
Induction machines, 249–330
   asynchronous generators, 298–301
   control schematics, 519–524
   controls, 302–319
   efficiency, 284–285
   equivalent circuits, 264–268,
      290–298
   industrial considerations, 277–290
   mechanical loads, 279–282
   motor characteristics, 286–290
   NEMA motor classification,
      277–279
   parameters, 281–282, 290–298
   plugging, 319
   power factor, 285–286
   principles of operation, 255–256
   reduction of starting current and
      torque, 302–307
   rotating magnetic field, 255–261
   rotor windings, 251–255
   slip, 261–264
   soft start, 317–318
   stator (armature) windings,
      251–252
   three-phase (3-ϕ) motors, 251–290
   torque and power relationships,
      269–277
   variable frequency drives
      (VFD), 307–317
   voltage, 282–284
   winding configurations, 302–307
Induction principle, 85–86
Initial-value theorem, 599–600
Inrush current, 173–174
Instantaneous power, 41–43, 568–570
Instrument transformers, 153,
   225–230
Inverters, 311–314

K

k-factor, transformers, 234–236
Kirchhoff's laws
   current (KCL), 39–41, 97

electrical circuits, 37–41
   magnetic circuits, 97–99
   voltage (KVL), 37–39, 97
kVA rating, 177–178, 236

L

Laplace transforms, 8, 597–600
Lap-type armature winding, 431
Leakage flux, 67–68, 97–98
Leakage impedance, 220
Lightning, 551, 557
Linear differential equation
   (LDE), 8
Locked-rotor test, 291–292, 296–298
Losses
   eddy-current, 103–104
   heat, 559–560
   hysteresis, 104–106
   magnetic, 103–109
   three-phase, single-conductor
      power distribution, 106–108
   transformer, 108–109

M

Magnetic circuits
   B-H curves and magnetic
      domains, 75–78
   coils, 109–110, 119–123, 131–133
   electromagnetism, 134–135
   electromagnets, 100–102
   energy and, 123–127, 131–133
   energy content of materials, 93–95
   equivalent circuits, 109–110,
      119–123
   force and, 123–130
   inductance, 111–119
   Kirchhoff's laws, 97–99
   magnetic fields, 64–69, 71–75
   magnetic losses, 103–109
   magnetomotive force (mmf),
      70–71
   measurement of magnetic
      properties, 83
   Ohm's law, 95–96
   permeability, 78–81
   reluctance, 82–85
   torque and, 125–127, 133–134
   transducers, 133
   voltage and, 85–92
Magnetic fields
   electric and magnetic intensity
      relationships, 72–73

Magnetic fields (*Continued*)
induction machines, 255–261
intensity, 11, 71–75
magnetic flux and, 64–69
rotating, 255–261, 359–361
single-phase (1-ϕ) motors, 341–345
synchronous machines, 359–361
three-phase (3-ϕ) motors, 255–261
Magnetic flux, 64–69, 438–442
Magnetic flux density, 69, 439–440
Magnetic levitation, 134–135
Magnetization effects of armature
current, 383–384, 440–442
Magnetizing reactance, 364
Magnetizing voltage, 364
Magnetomotive force (mmf), 70–71
Maxwell's theorem, 27–28
Mechanical loads
effects of changes on motor
parameters, 281–282
induction machines, 279–282
torque proportional to speed,
279–282
Mechanical momentum, 16
Motors
DC, 427–432, 433–464
high-efficiency, 358
three-phase (3-ϕ) induction,
251–290
single-phase (1-ϕ), 331–352
synchronous, 355–389
Mutual flux, 67
Mutual inductance
DC machines, 436–438
magnetic circuits, 111–113

N

Nameplate data, 236–238
National Electrical Manufacturers
Association (NEMA) motor
classification, 277–279
No-load test, 292–296
Nodes, 39

O

OFF-load voltage tap changers, 237
Ohm's law
electrical circuits, 35–37
magnetic circuits, 95–96
One-line diagrams, 180–181, 193–194
Open-circuit characteristics (OCC)
alternators, 393–394

DC generators, 464–476
Open-circuit test, 164–167
Open neutral wires, 546
Operating range, 511–512
Optical transformers: current (CT)
and potential (PT), 229–230

P

Parallel combinations, 15, 22
Parallel operation of transformers,
153, 219–224
Permeability, 17, 78–81
electric circuit inductance, 17
magnetic circuits, 78–81
Per-unit values, 153, 178–187, 587–596
advantages of, 590–591
base parameters, 178–179
base values, 587–590
multicomponent systems, 590
one-line diagrams for, 180–181
transformers, 153, 178–187
Phase rotation, transformers,
222–224
Phasor diagrams
single-phase transformers,
172–173
synchronous machines, 367–373,
389–391, 407–409
Phasors, 7–10
complex, 8–10
polar and rectangular forms, 7–8
Photovoltaic (PV) cells, 607–610
economic considerations,
608–610
equivalent circuits, 607–608
Plugging, induction machines, 319
Polarity of magnetically coupled
coils, 119–123
Potential transformers (PT),
226–228, 230
Power
average (real), 19, 43–44, 570–571
base (per-unit), 587–588
complex, 45–46, 572
cylindrical rotor motors, 374–378
DC machines, 433–435
electric circuits, 41–46
inductance and, 19
induction machines, 269–277
instantaneous, 41–43, 568–570
kVA rating, 177–178
measurement of, 573
reactive, 44–45, 571–572

salient-pole synchronous
machines, 409–412
single-phase source, 41–46
three-phase (3-ϕ) systems,
189–190, 568–574
torque angle for, 411–412
torque relationships, 269–277
Power angle, 365
Power demand reduction, 558
Power factor
effects of, 49–50
electric circuits, 47–57
improvement using, 51–57, 558
induction machines, 285–286
measurement of, 573–574
penalty calculation, 50–51
Protection, control schematics and,
512–519
Protective devices, 540–542
Pull-in torque, 278
Pushbuttons (stop-start), 506–507

R

Reactance
current, 32–34
magnetizing, 364
synchronous, 365, 397–398
Reactive power, 44–45, 571–572
Rectifiers, 308–311
DC machine speed control
using, 477–489
single-phase, full-wave, 478–481
three-phase, full-wave, 481–489
Regulation of transformers, 176–177
Relays
electromagnetic, 499–503
protective, 515–516
thermal overload, 503–504
Reluctance, magnetic circuits, 82–85
Resistance
armature, 444–461
DC machines, 444–461
electrical circuits, 12–16
external effects on speed, 447, 452
field, 444–456
Resistance temperature detector
(RTD), 16, 509–510
Resonance
capacitance and, 28
ferroresonance, 233–234
human, 233–234
inductance and, 21–22

linear, 234
nonlinear, 234
transformers, 233–234
Revolving fields, single-phase
    motors, 333–334, 346–347
Right-hand rule, 66
Robot design and
    electromagnetism, 135
Root-mean-square (rms) value, 5–7
Rotating magnetic fields
    induction machines, 255–261
    synchronous machines, 359–361
Rotors
    equivalent circuits and,
        264–268
    induction machines, 251–255,
        264–268
    revolving fields and, 334
    running, 266–268, 335
    single-phase motors, 334
    stationary, 264–266, 334
    synchronous machines, 359
    windings, 251–255

S

Safety, 539–554
    arc flash, 546
    bonding, 543–544
    cathode corrosion, 549–550
    electrical injuries, 539–540
    energy storage, 546–548
    floating neutral wires, 544–545
    ground resistance, 544
    grounding systems, 542–544
    lightning strikes, 551
    live conductor contact, 545
    open neutral wires, 546
    protective devices, 540–542
    short circuits, 544
    stray voltages, 548–549
Salient-pole synchronous
        machines, 405–412
    direct-axis resistance, 405–406
    phasor diagrams, 407–409
    power development, 409–411
    quadrature-axis resistance,
        405–406
    steady-state analysis, 406–407
    torque angle for maximum
        power, 411–412
Scott-connection transformers,
    207–209

Self-inductance, 16–23, 111–112
    electric circuits, 16–23
    magnetic circuits, 111–112
Series AC/DC motors, 349
Series combinations, 15, 22
Series DC motors, 450–451
Shaded-pole motors, 346
Short circuits, 544
Short-circuit characteristic (SCC),
    alternators, 394–396
Short-circuit test, 162–164
Shunt DC generators, 471–476
Shunt DC motors, 445–447
Single-phase (1-ϕ) motors, 331–352
    auxiliary winding, 337–341
    control, 348
    equivalent circuits, 334–335
    full-wave rectifiers, 478–481
    magnetic fields, 341–345
    revolving fields, 333–334,
        346–347
    series AC/DC, 349
    shaded-pole, 346
    split-phase, 345
    starting methods, 337–341
    stepper, 347–348
    torque, 335–337
    use and types of, 332–333
Single-phase (1-ϕ) systems
    one-line diagrams for, 180–181
    phasor diagrams, 172–173
    transformers, 153–155, 209–210,
        238
    transformation to three-phase
        (3-ϕ), 209–210
Slip, induction motors, 261–264
Soft start, induction machines,
    317–318
Speed
    control, 378, 477–489
    cylindrical rotor motors, 378
    DC machines, 446–447, 451–452,
        477–489
    external field resistance and,
        447, 452
    rectifiers, 477–489
    torque proportional to, 279–282,
        446, 451–452
Speed voltage, 89–91
Split-phase motors, 345
Squirrel-cage rotor windings,
    252–253, 520–521
Stacking factor, 97

Star- (Y-) connected load
    configurations, 190–191,
        565–566
Star-delta (Y–Δ) starters, 302–303
Star–delta (Y–Δ) transformers, 205
Star–star (Y–Y) transformers,
    205–207
Starting current, 19, 27, 302–307
Starting methods, *see also*
        Auxiliary winding
    DC motors, 462–464
    induction machines, 317–318
    single-phase (1-ϕ) motors, 337–341
    synchronous motors, 362–363
Starting torque, 277–278, 307–307
Stator resistance, 291, 365
Stator (armature) windings, 251–252
Steady-state analysis
    DC machines, 427–435
    salient-pole synchronous
        machines, 406–407
Steady-state harmonics, 58–61
Stepper motors, 347–348
Stiffness of synchronous machines,
    412–416
Supercapacitors, 28–31
Superconductivity, 13
Superposition theorems, 61
Synchronous machines, 353–424
    control schematics, 524–530
    cylindrical rotor generators,
        389–404
    cylindrical rotor motors, 355–389
    cylindrical rotor theory, 412–415
    energy consumption, 557
    rotating fields, 359–361
    rotors, 359
    salient-pole machines, 405–412
    stiffness of, 412–416
Synchronous reactance, 365,
    397–398

T

Temperature coefficient, 13
Temperature sensors, 509–512
Terminal voltage, 364
Thermal overload relays, 503–504
Thermistors, 509–510
Thermostats, 510–511
Thevenin's theorem, 61–63
Three-phase (3-ϕ) systems
    balanced, 563–586

Three-phase (3-ϕ) systems
(*Continued*)
current and voltage
considerations, 190
delta- (Δ-) connected load
configurations, 190, 566–567
delta–star (Δ–Y) load
transformations, 567–568
four-wire distribution system, 574
full-wave rectifiers, 481–489
harmonics of exciting current,
210–211
induction machines, 249–330
one-line diagrams for, 193–194
phase sequence, 564–565
phase shifts (angular
displacement), 191–192
power considerations, 189–190,
568–574
star- (Y-) connected load
configurations, 190–191, 565–566
two-winding transformers, 153,
189–211
types of, 194–210
Time constant, 20–21, 27
Timers (industrial), 507–509
Torque
accelerating, 279
breakdown (maximum), 271–273
constant, 279
cylindrical rotor motor
development, 374–378
DC machines, 436–438, 446,
451–452
energy and force and, 125–127
induction machines, 269–282,
302–307
maximum, 278–279
mechanical loads and, 279–282
motor classification and, 277–279
mutual inductance and, 436–438
power relationships, 269–277
pull-in, 278
rotating transducer
development of, 133–134
single-phase (1-ϕ) motors, 335–337
speed, proportional to, 279–282,
446, 451–452
starting, 277–278, 302–307
three-phase (3-ϕ) motors, 125–127
voltage relationships, 436–438
Torque angle, 411–412

Transformers
autotransformers, 153, 211–219
current (CT), 228–230
delta–delta, 204
delta–star, 194–203
efficiency, 174–176, 237
equivalent circuits, 160–169
excitation current, 169–170,
210–211
harmonics, 210–211
ideal, 157–160
inrush current, 173–174
instrument, 153, 225–230
k-factor, 234–236
kVA rating, 177–178, 236
magnetic losses, 108–109
nameplate data, 236–238
one-line diagrams, 180–181,
193–194
open-circuit test, 164–167
optical, 229–230
parallel operation, 153, 219–224
parameters of, 214
per-unit values, 153, 178–187
phasor diagram, 172–173
potential (PT), 226–228, 230
primary current and flux,
171–173
principle of operation, 155–157
regulation, 176–177
Scott-connection, 207–209
short-circuit test, 162–164
single-phase (1-ϕ), 153–155,
209–210, 238
star–delta, 205
star–star, 205–207
three-phase (3-ϕ) systems, 153,
189–193, 209–210
two-winding, three-phase, 153,
188–211
voltage, 85–88, 176–177, 190, 221,
237–238
wiring diagrams, 230–236, 239
Transient harmonics, 58
Turns ratios, transformers, 220–221
Two-winding, three-phase
transformers, *see* Three-phase
systems

V

Variable frequency drives (VFD),
307–317

Variable-speed drives, 558–559
Voltage
base (per-unit), 588–589
capacitor relationship, 24–25
DC motors, 559
designation, 4
electric potential and, 3–4
equal phase shifts between, 221
flux relationship to, 91–92
induction machines, 282–284
Kirchhoff's law (KVL), 37–39, 97
magnetic circuits, 85–92
mutual inductance and, 436–438
OFF-load tap changers, 237
regulation, 176–177
safety and, 548–549
speed, 89–91
three-phase (3-ϕ) systems, 190
torque relationships, 436–438
transformer, 85–88, 176–177, 190,
221, 237–238
unbalanced, 556–557
winding connections and,
237–238

W

Waveforms
excitation current, 169–170
values, 4–7
Wave-type armature winding, 431
Windings
armature, 431–432
auxiliary, 337–341
configurations, 302–307
connections, 237–238
induction motors, 237–238,
302–307
rotor, 251–255
single-phase (1-ϕ) motors,
337–341
squirrel-cage rotor, 252–253
stator (armature), 251–252
wound-rotor, 255
Wires, safety and, 544, 546
Wiring diagrams, 230–236, 239, 498
Wound-rotor windings, 255, 521–522

Z

Zero-power-factor (ZPF) lagging
characteristic, alternators,
396–397, 399–402